HISTORICAL TOPICS

FOR THE

MATHEMATICS CLASSROOM

National Council of Teachers of Mathematics

Library of Congress Cataloging-in-Publication Data

Historical topics for the mathematics classroom / [editorial panel:
John K. Baumgart ... et al.]. — 2nd ed.
     p.   cm.
   Includes bibliographies and index.
   ISBN 0-87353-281-3
   1. Mathematics—History. I. Baumgart, John K.   II. National
Council of Teachers of Mathematics.
QA21.H559   1989
510'.9—dc20

89-12208
CIP

Printed in the United States of America

Contents

PREFACE   vii

PREFACE TO FIRST EDITION   ix

I   THE HISTORY OF MATHEMATICS AS A TEACHING TOOL
    Phillip S. Jones, University of Michigan, Ann Arbor,
    Michigan   1

II   THE HISTORY OF NUMBERS AND NUMERALS
    Bernard H. Gundlach, Colorado College, Colorado Springs,
    Colorado   18

    CAPSULES 1–27   36

    Babylonian Numeration System, 36• Egyptian Numeration System, 38•
    Roman Numerals, 40 • Greek Numeration System, 42 • Chinese-
    Japanese Numeration System, 43 • Mayan Numeration Systems, 45 •
    Hindu-Arabic Numeration System, 46 • Origin of Zero, 49 • Number
    Beliefs of the Pythagoreans, 51 • Figurate Numbers, 53 • Amicable
    Numbers, 58 • Perfect, Deficient, and Abundant Numbers, 59 • Mer-
    senne Numbers, 61• The Infinitude of Primes, 62• Prime and Composite
    Numbers, 64 • Pythagorean Triples, 66• The Euclidean Algorithm, 69•
    Incommensurables and Irrational Numbers, 70 • Eratosthenes, 72 •
    Numerology and Gematria, 74 • Al-Khowarizmi, 76 • Fibonacci Num-
    bers, 77 • Fermat's Last Theorem, 79 • Magic Squares, 80 • Large
    Numbers, 82• Algebraic and Transcendental Numbers, 83• Transfinite
    Numbers, 85

III   THE HISTORY OF COMPUTATION
    Harold T. Davis, Trinity University, San Antonio, Texas   87

    CAPSULES 28–49   117

    The Abacus, 117• Finger Reckoning, 120• The Rhind Papyrus, 123• The
    Moscow Papyrus, 127• Korean Number Rods, 128• The Peruvian Quipu,
    129 • Multiplication and Division Procedures, 130 • Fractions, 135 •
    Decimal Fractions, 137• Origins of Symbols for Operations, 139• Casting
    Out Nines, 140 • Napier's Rods, 141 • Logarithms, 142 • Nomography,
    145 • Percent, 146 • Radical Symbol, 147 • The Number π, 148 • The
    Number e, 154 • Pascal's Triangle, 156 • Probability, 158 • Calculating
    Prodigies, 159 • The Modern Digital Computer, 161

IV   THE HISTORY OF GEOMETRY

    Howard Eves, University of Maine, Orono, Maine   165

    CAPSULES 50–65   192

    Compass and Straightedge Constructions, 192 • Duplication of the Cube, 197 • The Trisection Problem, 199 • The Quadrature of the Circle, 201 • The Golden Section, 204 • Non-Euclidean Geometry, 207 • The Witch of Agnesi, 210 • Four-dimensional Geometry, 211 • The Pythagorean Theorem, 215 • *Pons Asinorum*, 219 • Regular Polyhedra, 220 • History of the Terms "Ellipse," "Hyperbola," and "Parabola," 222 • Geometry in China, 225 • The Cycloid, 227 • Polar Coordinates, 229 • The Nine-Point Circle, 230

V   THE HISTORY OF ALGEBRA

    John K. Baumgart, North Park College, Chicago, Illinois   232

    CAPSULES 66–90   260

    Equations and the Ways They Were Written, 260 • The Binomial Theorem, 264 • Continued Fractions, 267 • Oughtred and the Slide Rule, 271 • Horner's Method, 274 • Solution of Polynomial Equations of Third and Higher Degrees, 276 • Vectors, 279 • Determinants and Matrices, 281 • Boolean Algebra, 284 • Congruence (Mod m), 288 • Complex Numbers (The Story of $\sqrt{-1}$), 290 • Quaternions, 295 • Early Greek Algebra, 298 • Hindu Algebra, 301 • Arabic Algebra (820–1250), 305 • Algebra in Europe (1200–1850), 309 • Function, 312 • Mathematical Induction, 313 • Fundamental Theorem of Algebra, 316 • Descartes's Rule of Signs, 318 • Symmetric Functions, 321 • Discriminant, 323 • Interest and Annuities, 325 • Exponential Notation, 327 • Rule of False Position, 332

VI   THE HISTORY OF TRIGONOMETRY

    Edward S. Kennedy, American University of Beirut, Beirut, Lebanon   333

    CAPSULES 91–97   359

    The *Almagest* of Ptolemy, 359 • Angle, 362 • Right Angles, 363 • Angular Measure, 364 • Sine and Cosine, 368 • Tangent and Cotangent, 371 • Trigonometric Identities, 374

VII   THE HISTORY OF THE CALCULUS

    Carl B. Boyer, Brooklyn College, Brooklyn, New York   376

    CAPSULES 98–120   403

    Archimedes and His Anticipations of Calculus, 403 • Simon Stevin, 405 • Johann Kepler, 407 • Bonaventura Cavalieri, 409 • Pierre de Fermat, 410 • John Wallis, 413 • Isaac Barrow, 415 • Leibniz, 418 • Newton's *Dot*-age versus Leibniz' *D*-ism, 421 • Gauging—Volumes of Solids, 424 • Finite Differences, 427 • Archimedes and the "Method of Exhaustion," 430 • Convergence, 432 • The Origin of L'Hospital's Rule, 435 • Calculus of Variations, 439 • A Jungle of Differential Equations, 441 • Maclaurin and Taylor and Their Series, 443 • From $n!$ to the Gamma Function, 446 • Partial Derivatives, 448 • Multiple Integrals and Jacobians, 450 • Calculus in Japan, 452 • Infinitesimals in India, 454 • The Definite Integral, 457

VIII   DEVELOPMENT OF MODERN MATHEMATICS
        R. L. Wilder, University of Michigan, Ann Arbor,
        Michigan   406

 IX   THE SCIENCE OF PATTERNS
        Lynn Arthur Steen, Saint Olaf College,
        Northfield, Minnesota   477

CONTRIBUTORS OF CAPSULE ARTICLES   489

APPENDIX: Resources beyond This Yearbook
        Charles V. Jones, Ball State University,
        Muncie, Indiana   491

BIBLIOGRAPHY A   499

BIBLIOGRAPHY B   516

INDEX   521

Preface

In response to a continuing demand for historical materials for mathematics instruction, the National Council of Teachers of Mathematics is publishing this updated edition of *Historical Topics for the Mathematics Classroom*. Originally published in 1969 as the Thirty-first Yearbook of the NCTM, it was a pioneering effort to assist in the teaching of mathematics from a historical perspective. The original "capsule" format remains unchanged, but a new chapter has been added that discusses contemporary mathematics, and additional bibliographic listings cover the proliferation of historical material in the twenty years between 1969 and 1989.

Preface to First Edition

In response to long-felt need, a yearbook on "the use of the history of mathematics in the teaching of mathematics" was first proposed during the NCTM presidency of Phillip S. Jones; and a special planning committee was appointed, composed of John K. Baumgart, Duane E. Deal, E. H. C. Hildebrandt, and Arthur E. Hallerberg.

The committee's proposal was approved by the Board of Directors, and early in 1963 the editorial panel was appointed, consisting of the original planning committee with the addition of Irvin H. Brune and Bruce R. Vogeli. Because of the pressure of other responsibilities, Professor Brune and Professor Hildebrandt later withdrew from the editorial panel and Professor Vogeli's contributions were intermittent. The advice and counsel of Professor Hildebrandt during the early stages of work was particularly helpful.

The following principles have guided us in the preparation of this yearbook:

1. A previous course in the history of mathematics is not assumed as a prerequisite for the reader of this book.

2. The book should include not only usable historical materials but also some indication of how they may be used by the teacher or student.

3. The general topics should include something of significant mathematical value for all different grade levels whenever possible.

4. The yearbook should provide basic materials in a form that is specific and complete enough to make historical topics immediately available for classroom use. At the same time it should be open-ended in encouraging the teacher or student to do further reading or study in the same or related topics.

5. While the book should be directed to the teacher and student in the classroom, its contents should also make it usable, as a reference work or for collateral reading, in college courses both in the history of mathematics and in the teaching of mathematics.

6. The difficulties inherent in the oversimplification of historical development should be avoided as much as possible; however,

it is not the role of this book to be the place where subtle distinctions or obscure priorities are to be debated. The primary objective of this book is to make available to mathematics classes important material from the history and development of mathematics, with the hope that this will increase the interest of students in mathematics and their appreciation for the cultural aspects of the subject.

The major portion of this book was envisioned quite naturally as being of mathematical and historical content; nevertheless it was felt essential that the form of presentation should differ from that of the customary "history" of mathematics. The format of this book, therefore, is one of *overviews* and *capsules*. The first chapter is an overview of both the "why" and the "how" of using the history of mathematics in the classroom. The remaining overviews are devoted to those topics that are significant for elementary and secondary school mathematics (including also the junior college years). Hence there are overviews on numbers and numerals, computation, geometry, algebra, trigonometry, calculus, and modern mathematics. The purpose of each overview is to orient the reader by giving him a rather general picture of the historical development of the subject at hand. Many of the details and interesting sidelights must necessarily be omitted, but perhaps something else is gained in seeing more clearly the path that leads from first beginnings to the mature development of a particular subject area.

Details, however, are important. So the overview is followed by capsule treatments of a number of the important topics within that area. These capsules make easily accessible the pertinent facts related to important theorems, concepts, and developments in mathematics. While they augment the overview, they may be read independently. In most cases they include references for further reading.

Several areas have purposely been given less emphasis than might be expected:

One has to do with the second principle mentioned above. How detailed should suggestions be as to the manner in which particular items could actually be used in the classroom? Careful consideration will show that many of the topics have aspects relating to a number of different grade levels. To include only some of these might tend to limit the applications that might be made; to try to include them all would overtax both our resourcefulness and the size limitations of this book. In such cases, therefore, our intent has been to emphasize the mathematical content of the material and to leave the method of bringing it into the individual classroom in the hands of the person most qualified to make this decision—the teacher in charge of the classroom.

Another area receiving less emphasis than might be expected is that of biography. We are well aware that students are interested in the "men of mathematics" as people. Anecdotes and incidents in the lives of these men are often of great value in the classroom. But it was our feeling that this sort of material is reasonably available already and that to have included it here would have forced the exclusion of other materials less likely to come before the teacher in accessible form.

Originally we had hoped that the Bibliography would be annotated. Its final size has precluded this. Some indication of books that would be helpful for various purposes is given in the Appendix ("Resources Beyond This Yearbook"), beginning on page 479.

Also, with considerable regret, we have reduced the number of diagrams to a minimum and, with a few exceptions, have left out all pictures, illustrations, and reproductions. Just as we have indicated that the topics presented here are open-ended, leaving more to be done by the student, so we should like to think of this yearbook as an encouragement for additional work by others. There are many historical topics left untouched. For example, there are no capsules for the modern mathematics section—these had to be omitted both for lack of space and for want of contributors. Also, there is opportunity for the development of additional teaching and learning aids for the classroom. Posters and wall displays; illustrations and pictures in the media of slides, overhead transparencies, and printed materials; enrichment units and self-study items for the mathematics laboratory—all of these present new challenges for the imaginative teacher and author!

It is in order to say something about the mathematical symbolism used in this book. The development of a meaningful, adequate, and consistent system of notation in the various branches of mathematics is part of the history of mathematics, and we have endeavored to include this. But concern for unambiguous terminology and symbolism is relatively new. The "lines" of Euclid were sometimes lines and sometimes segments—and T. L. Heath has given us the geometry of the Greeks without a single instance of using any modern notation for segment, ray, line, arc, or congruence! It therefore has seemed appropriate to abstain from such symbolism when discussing these concepts. Sometimes we have purposely written "four" instead of "4" so that the reader would not unknowingly think with symbolism that the ancients did not have. Often (but not always!) we have acquiesced to the demands of modern printing and permitted fractions to be written as "2/3" when we might have preferred "$\frac{2}{3}$."

In choosing name forms, spellings, and dates we have forced some sort of internal consistency on our contributors, well realizing that sometimes our choice has been quite arbitrary. (We have endeavored, of course, to use the form or dates that seem at present to have received the most authoritative acceptance.)

The structure of this book, with its overviews and overlapping capsule presentations, has brought with it a certain amount of repetition. We feel this is in order, to permit the contributors to say what is necessarily related to a given topic. Again, while striving for consistency in the main, we have tried to allow the contributors to reflect the different points of view, versions, and conflicting accounts that are a part of the history of mathematics.

Finally, we have tried to refrain from including those items of pseudohistory that the best current scholarship has refuted or invalidated. We are also quite certain that contributors, referees, and editors have not been 100 percent effective in such elimination! We strongly urge each reader to become acquainted with Carl Boyer's "Myth, the Muse, and Mathesis" (*Mathematics Teacher*, April 1964); for this excellent essay makes abundantly clear that the history of mathematics is not a static subject.

A great many persons have been directly involved in the writing of this yearbook. Their very number has been the cause of some delays, but we feel that the project has benefited thereby. The time and work contributed by these people, moreover, is a significant indication of the unselfish support they are willing to give to the NCTM.

We are particularly indebted to those who have contributed the overviews for this book. Working under the explicit restrictions of limited space, the denial of documentary footnotes, and the exhortation to cover all the essential material, these authors have made contributions that give the reader an unusual insight into the development of the major areas of mathematics.

Likewise, the eighty-four contributors of the capsules have given unusual cooperation in bringing together their materials to amplify the topics discussed in the overviews. We regret that space limitations did not permit us to publish all the items that were submitted. In some cases contributions from two persons were combined into one capsule when the nature of the topics suggested this. Capsules with no author listed are the work of the editorial committee.

Early in the project Francis T. Wilson, Jr., checked the content of a large number of periodical articles and evaluated them for possible use by members of the committee. Particularly helpful in the choice of topics and materials for the capsule sections were the following, who served on various subcommittees with members of the panel: Rodney Hood, Wendell Johnson, Lyman Peck, Cecil B. Read, Ralph Shively, and Harold C. Trimble.

The names of the writers of the overviews and capsules are noted elsewhere in the yearbook. Here we should like to recognize those persons who have contributed in various other ways to this project. Some served as readers and referees; others suggested items for the yearbook or persons to help in its work; still others submitted materials for possible use. While their specific contributions cannot be recognized individually, we do indicate our appreciation for their assistance by listing the following names: Barbara Abbring, Thomas Abbring, M. H. Ahrendt, Charles E. Allen, Richard V. Andree, S. Louise Beasley, Richard F. Bellman, Clifford Bernstein, Annis M. Bleeke, Benjamin Bold, Lyle D. Bonnell, Wray G. Brady, Br. T. Brendan, J. D. Bristol, G. Ross Buckman, Jr., William E. Bugg, Paul C. Burns, William O. Buschman, Given Carey, Mary K. Carrothers, Laura E. Christman, Burton H. Colvin, Ruth M. Cordell, Thaddeus Dillon, Richard R. Eakin, P. D. Edwards, Earl Eidson, Charles R. Eilber, Van Phillip Estes, Harley Flanders, Marilyn B. Geiger, Richard W. Geist, E. Glenadine Gibb, Arnold Gleisser, Samuel Goldberg, Emil Grosswald, Louise B. Haack, Sidney G. Hacker, Theodore Hailperin, Doyle Hayes, Leon Henkin, Joyce Higgins, Peter Hilton, Franz Hohn, Fr. Barnabas Hughes, Sara L. Hughes, Mary Ingraham, Lee Irwin, Randall E. Johnson, Margaret Joseph, M. L. Juncosa, Karl S. Kalman, Jerome D. Kaplan, William R. Kinslow, Gary B. Kline, Morris Kline, Esther Drill Krabill, Daryl Kreiling, Mary Laycock, Daniel B. Lloyd, Richard Marzi, Edna McNutt, Virginia T. Merrill, Vera C. Moomaw, Marian A. Moore, Marston Morse, R. V. Murphy, Stephen A. Nemeth, Carroll V. Newsom, Oystein Ore, Jim Pearson, Clarence R. Perisho, Robert E. Pingry, Henry O. Pollak, John Riordan, Hartley Rogers, Jr., David Rosen, Tom Sampson, Arthur F. Scheid, Robert C. Seber, Carl Shuster, Sr. M. Francetta Coughlin, J. Laurie Snell, D. J. Struik, Theresa Taffee, Larry W. Thaxton, G. L. Thompson, Andries van Dam, John C. Van Druff, Cecilia T. Welna, A. B. Willcox, Robert Wilson.

In addition, some four hundred participants in National Science Foundation mathematics institutes in the summer of 1963 sent in suggestions, ideas, and offers of assistance for this project. While we cannot include their names, we can assure them that their comments were read by the major writers for the yearbook and so in an indirect way they have made a contribution to this work.

It seems appropriate here to express appreciation for the work of three persons who over a long span of years have played an important role in keeping the history of mathematics before the members of the National Council. Vera Sanford, Phillip S. Jones, and Howard Eves have served as editors of the history section of the *Mathematics Teacher*, and we are all indebted to them for the resources they have given us month by month. We have been continually impressed by the wealth of historical material that has appeared in this journal over the years. A compilation of articles from this section would make a worthwhile yearbook all by itself.

We should like to express our thanks to those presidents of the NCTM whose administrations have encompassed the production of this yearbook. Phillip S. Jones, Frank Allen, Bruce E. Meserve, Donovan A. Johnson, and Julius H. Hlavaty were helpful with their suggestions, generous with their support, and patient with our delays. James D. Gates, executive secretary of the NCTM, has been most helpful in the administrative details involved in this project.

Finally, our deepest appreciation goes to those persons who have made this book a reality: Charles R. Hucka, who directs the Council's publishing program; Julia A. Lacy, who supervised the editorial process; Dorothy C. Hardy, who did the major copy editing; and Gladys F. Majette, who gave substantial help at every stage of production.

The Editorial Panel

John K. Baumgart
North Park College
Duane E. Deal
Ball State University
Bruce R. Vogeli
Teachers College, Columbia University
Arthur E. Hallerberg, *Chairman*
Valparaiso University

REFERENCE SYMBOLS

All bibliographical references are given in one listing at the end of the yearbook. To eliminate the necessity of footnotes, the reader is directed to any entry in this listing by a text citation given between slashes. For example, /D. E. SMITH (a) : II, 156–92/ refers to pages 156–92 of Volume II of the first entry under the author's name in the Bibliography, where the full bibliographical information appears. If some of this information has already been given in the text, the citation is more brief.

Another method has been adopted for referring the reader to any of the 120 "capsules" interspersed throughout the book. These capsule discussions of limited topics are numbered consecutively, and reference is made by simply giving the capsule's number, in boldface type, within square brackets—for example, [12].

Note to the Reader

For an explanation of the style here adopted for referring to the Bibliography and to individual "capsules," see "Reference Symbols" on the preceding page.

I

The History of MATHEMATICS as a Teaching Tool

an overview by

PHILLIP S. JONES

I have more than an impression—it amounts to a certainty—that algebra is made repellent by the unwillingness or inability of teachers to explain why. . . . There is no sense of history behind the teaching, so the feeling is given that the whole system dropped down ready-made from the skies, to be used only by born jugglers.—Jacques Barzun, *Teacher in America.*

Teaching so that students understand the "whys," teaching for meaning and understanding, teaching so that children see and appreciate the nature, role, and fascination of mathematics, teaching so that students know that men are still creating mathematics and that they too may have the thrill of discovery and invention—these are objectives eternally challenging, ever elusive.

The "sense of history" cited by Barzun will not, of itself, secure these objectives. However, properly used, a sense of the history of mathematics, coupled with an up-to-date knowledge of mathematics and its uses, is a significant tool in the hands of a teacher who teaches "why." There are three categories of whys in the teaching of mathematics. There are the *chronological,* the *logical,* and the *pedagogical.*

Teaching for Meaning

Chronological whys.—The first of these categories includes such items as why there are sixty minutes in a degree or an hour, and why sixty seconds in a minute; it looks for the sources of such words as "zero" and "sine" as well as "minute" and "second." These and many other historical facts, large and small, not only give the why of a particular question but also, in the hands of a skillful and informed teacher, trigger discussions about the necessity and arbitrariness of definitions and of undefined terms, about the psychological bases of mathematical systems, and about the development of extended definitions, such as those of "number" or "sine," which change with the development of new systems, being generalized and rephrased to include the new ideas as well as the old.

Logical whys.—The second category mentioned—the logical "whys" —may seem to be independent of the history of mathematics, but actually history may contribute greatly to developing the logical insights of students. Logical whys include an understanding of the nature of an axiomatic system as well as the logical reasoning and proofs that clothe the axiomatic skeleton with theorems. It is important that our students grow to understand this structure, but for many topics the direct and minimal statement of axioms and proofs is neither the way these ideas developed historically nor the way perceptions grow in the minds of many of our students. One of this century's leading mathematicians, Henri Poincaré, is quoted on this subject by Jacques Hadamard /104/.

> To understand the demonstration of a theorem, is that to examine successively each syllogism composing it and ascertain its correctness, its conformity to the rules of the game? . . . For some, yes; when they have done this, they will say, I understand.
> For the majority, no. Almost all are much more exacting; they wish to know not merely whether all the syllogisms of a demonstration are correct, but why they link together in this order rather than another.

It is not true that a historical approach is the *only* way to communicate this sense of why theorems hang together as they do—this "mathematical motivation" for a proof. The historical approach is not even always the *best* way to communicate these insights. But it can often help tremendously! The authors of the article "On the Mathematics Curriculum of the High School" /see under title: 192/ were on the right track, even though they may have slightly exaggerated the utility of what they called "the genetic method," when they made the following statement:

> "It is of great advantage to the student of any subject to read the original memoirs on that subject, for science is always most completely assimilated when it is in the nascent state," wrote James Clerk Maxwell. There were some inspired teachers, such as Ernst Mach, who in order to explain an idea referred to its genesis and retraced the historical formation of the idea. This may suggest a general principle: The best way to guide the mental development of the individual is to let him retrace the mental development of the race—retrace its great lines, of course, and not the thousand errors of detail.
>
> This genetic principle may safeguard us from a common confusion: If *A* is logically prior to *B* in a certain system, *B* may still justifiably precede *A* in teaching, especially if *B* has preceded *A* in history. On the whole, we may expect greater success by following suggestions from the genetic principle than from the purely formal approach to mathematics.

More recently John Kemeny /76/ has pointed out that

> in the historical development of mathematics, it is usually, though by no means always, the case that a certain body of mathematical facts is first discovered, and then one or more people perform the very important task of systematizing this information by specifying a minimal number of axioms and deriving the other facts from these. It is, therefore, clear both that some acquaintance with axiomatic mathematical systems is an important part of mathematical education, and that mathematics is something over and above mere development of axioms.

A portion of the history of mathematics can often display Kemeny's "something over and above." Recapitulating all the errors of the past would be confusing and uneconomical. However, it is more than merely consoling to find that Descartes called negative numbers "false" and avoided their use; that Gauss had a "horror of the infinite"; that Newton wrote four approaches to his fluxions, probably because none of them quite satisfied him conceptually; that Hamilton really thought he was doing something quite different logically when he invented the abstract, modern, "ordered-pair" approach to complex numbers. Not only do these stories offer the consolation that great men once had difficulties with what today are fairly-well-clarified concepts. They

3

also show how mathematics grows and develops through generalizations and abstractions; and they indicate that the mathematics of a few years from now will no doubt be different from ours while still including today's ideas, perhaps in altered form.

Pedagogical whys.—Pedagogical "whys"—our third category—are the processes and devices that are not dictated by well-established arbitrary definitions and do not have a logical uniqueness. They include such devices as "working from the inside out" in removing nested parentheses. This is not logically necessary, but experience and a little thought suggest that it is less likely to lead to errors. Similarly, there may be many logically sound algorithms for a particular process, say, finding rational approximations to $\sqrt{26}$. Pedagogically, I would choose the "divide, average, and repeat" process, sometimes called merely the "division" process, for the first one to teach. There are several reasons for this. It is closer to the basic definition of square root (in fact, thinking the process through calls for continually reviewing this definition); it is easier to understand; and it is easier to recall. Later the time-honored square-root algorithm may be taught in several contexts—for its connection with the square of a binomial or a multinomial, or for its connection with the expansion of a binomial with a fractional exponent into an infinite series, or for its connection with iterative processes.

Historical ideas may help in both the selection and the explication of such pedagogically motivated processes. For example, there is good evidence that the division process for finding square roots was used by the Babylonians. It is clear that the approximation to $\sqrt{a^2 + b}$ as $a + (b/2a)$ was known to the early Greeks and can be motivated by reference to the square diagram associated with $(a + b)^2$ in Euclid's *Elements*, Book II, 4. The repeated use of this to get successively better rational approximations is to be found in the work of Heron (or Hero) of Alexandria. Heron, unlike earlier Greek philosopher-mathematicians such as Pythagoras, Euclid, and Apollonius, was also interested in mechanisms (he invented the first jet engine), surveying, and the related calculations. (The formula for the area of a triangle which bears his name, "Heron's formula," involves a square root and was no doubt used by him, though it is probably actually attributable to Archimedes.) The actual algorithm as we teach it involves an approximation and an iteration of that approximation. It also involves a process (marking off periods of two, etc.) that depends on the use of a place-value numeration system which, though partially achieved by the Babylonians, did not develop to the extent of being ready for this algorithm until the

4

rise of Hindu mathematics. Here is a prime example of a situation where a genetic approach does more than help the teacher to select a good pedagogical sequence and good teaching devices. Exploitation of this story, with its diagrams and processes, may really contribute to an understanding of such important concepts as irrational numbers, rational approximations to them, iterative processes, and base-ten numeration.

RECOMMENDATIONS FOR USE OF THE HISTORY OF MATHEMATICS

It is easy to collect testimonials about the importance and value, both for teachers and for students, of some study of the history of mathematics.[1] Recommendations for the inclusion of some study of history in teacher-training programs are to be found in many studies and committee reports in many countries.[2] A survey by John A. Schumaker /416, 418–19/ of trends in the education of secondary school mathematics teachers shows that the percent of teacher-education institutions offering such a course increased from 44 to 52 over the period from 1920/21 to 1957/58. In the latter year such a course was required of prospective teachers majoring in mathematics in 12 percent of the institutions. K. A. Rybnikov, chairman of the Department of the History of Mathematics and Mechanics at the University of Moscow, states that such a course is required of all mathematics majors there.

In spite of all these claims and recommendations, however, the history of mathematics will not function as a teaching tool unless the users (1) see significant purposes to be achieved by its introduction and (2) plan thoughtfully for its use to achieve these purposes. The remainder of this chapter will enumerate such purposes tnd illustrate methods of achieving them. We shall use historical anecdotes and vignettes or sketchily outlined sequences to do this. The aim is not to present the history of mathematics itself but to suggest in a concrete way how and for what purpose historical materials may be used. We believe that this concreteness is desirable even though some

---

[1] Several of these are to be found in an excellent article, "The History of Mathematics and Its Bearing on Teaching," which is the ninth chapter of *Teaching Mathematics in Secondary Schools* /see under title/. This article includes, also, suggestions for using the history of mathematics as a teaching tool, as do articles by Jones /(d)/ and the Freitags /(b/.

[2] See particularly /AAAS: 1028; BUNT: 664; DODD: 313; KEMENY: 76; MAA (a): 276; and MAA (b): 635, 637, 643/.

5

clarity may be lost because of the fact that the sketches generally illustrate more than one important objective.

THE DEVELOPMENT OF NUMERATION SYSTEMS

The story of the development of different systems of numeration and of the binary system in particular is an excellent one to illustrate the reasons why persons do mathematics, as well as to illustrate the connections between mathematics and the physical world. The basic elements of a modern positional or place-value numeration system are these: an arbitrary integer greater than 1 selected as a base, b; a set of b distinct digits including 0; and multiplicative and additive concepts. (The multiplicative concept is that each digit is to be multiplied by the power of the base corresponding to the position in which the digit is written, and the additive concept is that the number represented by these symbols will be the sum of these products.) A final necessity is a decimal point or other scheme for marking the position of the "units digit" or the position associated with b^0.

Base-and-place-value systems.—The various elements composing the complete concept of a positional numeration system first appeared in history at different times and in different countries. Egyptians, Greeks, and Romans grouped their symbols by tens and in a sense were using the idea of a base. The ancient Babylonians made the first use of a positional system; but their system, using base sixty, was ambiguous because they used repetition in forming symbols and they lacked a zero and a "sexagesimal point." The symbol ▼▼ could represent 2, 2/60, $1 \cdot 60 + 1$, or even $1 \cdot 60^2 + 0 \cdot 60 + 1$. The Hindu-Arabic system's separate symbol for each of the whole numbers less than ten, the base, eliminated the first two sources of ambiguity. The Babylonians, however, anticipated by over three thousand years the development of place-value notation for numbers less than 1. It was not until the time of Simon Stevin of Bruges (1585) that the Hindu-Arabic numeration system was extended to decimal fractions with a scheme for marking the point separating representations of integers from representations of fractions. Such a historical analysis of the components of our numeration system, along with an investigation of its advantages and a realization of the dependence of our computational algorithms upon the system, can add substantial understanding as well as interest to the teaching of arithmetic.

The binary system.—Moreover, several very important concepts and insights can be simply demonstrated by detailing the develop-

6

ment of the binary system. An anthropological study of the western tribes of the Torres Straits, published in 1889, tells of one tribe which had only two number words, *urapun* and *okosa*, for 1 and 2 respectively. Out of these two words, by grouping, they manufactured names for 3, 4, 5, and 6: *okosa urapun, okosa okosa, okosa okosa urapun,* and *okosa okosa okosa.* This was the totality of their number names, with the exception of *ras,* which was used for any number greater than 6. Although not so ancient in years, this primitive society resembles prehistoric ones; and its story suggests two things. It suggests that number names and systems of numeration undoubtedly existed in prehistoric times (prehistoric notched bones and bronze counters confirm this) and that no doubt this numeration developed out of the needs of daily life, commerce, taxation, and trade.[3]

In spite of the antiquity of some of the components of a base-and-place-value numeration system, a general and abstract perception of its structures is relatively recent. The Torres Straits number words, grouping by twos, may suggest a primitive binary system to us, looking at it from the vantage point of today, but the first published description of a binary system appeared in 1703. This was written by Gottfried Wilhelm von Leibniz, the famous German philosopher and developer of the calculus, who was partially motivated by an attempt to explain semimystical symbols found in an ancient Chinese work. Leibniz, however, was so intrigued by this notion that he wrote a generalized and extended discussion of the manner in which one could use any arbitrarily chosen number as the base for the development of a numeration system. His interest in religion and philosophy led him to feel a particular interest in the binary system, where any integers as large as you wish may be written using only two symbols, 0 and 1. Leibniz saw a parallel to the story of the creation of the universe in Genesis, in which we learn that God, whom he associated with 1, created the universe out of nothing, or a void, which he associated with 0. It is even reported that Leibniz caused a medal to be struck commemorating this idea.

Concepts illustrated by introducing the binary system.—At any rate, for our students these stories further illustrate several points that are now commonly accepted in mathematics, viz., that (1) connections are frequent between mathematics and philosophy and

---

[3] For details of the Torres Straits numeration system, see /CONANT/. For an account of genuine prehistoric mathematics, see /VOGEL: I, 7–20/. For a theory of a ritual (religious) origin for both counting and geometry, see /SEIDENBERG (a, b)/.

even between mathematics and religion; that (2) practical, social, economic, and physical needs often serve as stimuli to the development of mathematical ideas; but that (3) intellectual curiosity—the curiosity of the man who wonders, "What would happen if . . . ?"—leads to the generalization and extension and even abstraction of mathematical ideas; and that (4) perception of the development of general, abstract ideas, along with perception of the patterns or structure in mathematics, may be one of the most practical goals of mathematical instruction.

This fourth point is very well illustrated by the fact that, although Leibniz proposed no use whatsoever for the binary system, it has in recent times played a critical role in the development of high-speed electronic digital computers and in the development of information theory and data processing of all sorts. In this case the insights and perceptions associated with mathematical theory as expounded by Leibniz, and not the narrow, immediate application to counting as employed by the natives of the Torres Straits, led to recognition of the utility of the binary system in a context that had never been anticipated by its inventor.[4]

The concept of "model" in mathematics.—Another point illustrated by this historical sequence is basic in modern mathematics, namely, that (5) the concept of a "model" is important in both pure and applied mathematics. The model in this case is very simple both to describe and to understand. Electrical circuits are capable of two states: they are either open or closed. Electron tubes or electric lights are either illuminated or not illuminated: they are on or off. The binary numeration system uses only two symbols. In every position there is either a 1 or a 0. The parallel, when one thinks of it from the vantage point of history and of the perception of a general structure, is clear and obvious. If each position in the numeration system were made to correspond to an electrical circuit, and if the digits 0 and 1 were thought of as corresponding to an open or a closed circuit or switch, one would have a correspondence between a set of physical elements and a set of mathematical concepts and symbols. Each can be thought of as a model for the other. If the models are well chosen, any operation or state in the one system has a corresponding operation or state in the other. Hence one may draw conclusions by operating in one system and interpreting the conclusions in the other.

[4] For further details and references to the binary system, see /JONES (b)/.

8

An understanding of the role of matched "models" (mathematical and physical or social or economic) in the applications of mathematics not only supports the importance of pure mathematics and the role of intellectual curiosity; it also helps students to develop a better understanding of the nature of mathematics itself. Mathematicians may be stimulated by physical problems and assisted by geometric diagrams, but the mathematics they create is an abstraction. This "model" concept itself is quite modern, but it has an extensive historical background, a study of which contributes to its understanding. For example, mathematicians for years thought of Euclidean geometry not as merely one possible model for physical space around us but as the true and ideal system abstracted from a system of physical points and lines. Its axioms were viewed as necessary and self-evident, probably on the basis of observations of the physical world and geometric diagrams. The source and role of axioms are viewed differently today, however.

MATHEMATICS AS AN ART

Today mathematicians frequently liken mathematics and its creation to music and art rather than to science. J. W. N. Sullivan /7/ wrote:

> It is convenient to keep the old classification of mathematics as one of the sciences, but it is more just to call it an art or a game. . . . Unlike the sciences, but like the art of music or a game of chess, mathematics is a free creation of the mind, unconditioned by anything but the nature of the mind itself. . . . There is nothing necessary in any of the fundamental postulates or definitions of any branch of mathematics. They are arbitrary: which does not mean that their choice was *psychologically* arbitrary, but that it was *logically* arbitrary.[5]

This view of mathematics certainly was not held by the ancient Egyptians or Babylonians, nor even by the Pythagoreans, who thought the basis of music, of philosophy, and indeed of the universe, lay in number. Not even Euclid, who organized the first successful axiomatic structure and did such a fine job that it persisted with little change for over two thousand years, had this view of his accomplishment.

If mathematics is an art, some appreciation of this fact, and of the

---

[5] See also /WHITEHEAD/. The second chapter begins: "The science of Pure Mathematics, in its modern developments, may claim to be the most original creation of the human spirit. Another claimant for this position is music." This interesting chapter is reprinted in /NEWMAN: I, 402–16/.

relation of mathematics to the world of physical reality, can be as much a part of the liberal education of a doctor, lawyer, or average intelligent citizen as is some appreciation of the humanities.

Non-Euclidean Geometry

For persons whose exploration of modern advanced mathematics must be limited, a survey of the history of non-Euclidean geometry or of the relations between mathematics and man's view of his universe can be a vehicle for communicating a valuable, though partial, insight.

Non-Euclidean geometry was born of intellectual curiosity, without concern for practical applications; these appeared many years after its birth. In this respect it resembles an art. Its story, like others we are reciting, illustrates more than one of the major concepts and insights that discussions of historical background serve to convey. Dissatisfaction with Euclid's parallel postulate existed almost as soon as it was enunciated. Attempts to find better or simpler statements equivalent to it (as a postulate) accompanied attempts to prove it as a theorem dependent on other postulates. Although there was some earlier work by Arabic geometers, the first person to develop an extensive list of theorems in non-Euclidean geometry was Girolamo Saccheri, who never knew what he had actually done. He thought that his book *Euclides ab omni naevo vindicatus* ("Euclid Freed of All Flaws"), published in 1733, had proved the parallel postulate by assuming a different postulate and showing that it would lead to a contradiction. The "contradiction" he finally found was not a real contradiction but, rather, a theorem of non-Euclidean geometry which was different from the analogous theorem of Euclidean geometry. Without the generalized perception of the importance and arbitrary nature of postulates and the possibility of the existence of different geometries based on different sets of consistent, but arbitrary, postulates, Saccheri could not have had any real understanding of what he had accomplished.

Nikolai Ivanovich Lobachevsky and Janos Bolyai, who in the nineteenth century were the real inventors of non-Euclidean geometry, did not themselves comprehend the revolution in mathematics that was to grow out of their invention of another geometry. Out of their work developed (1) an understanding of the nature and importance of axioms, (2) the concept of the possibility of many different mathematical systems, and (3) attempts to axiomatize all branches of

10

mathematics—attempts resulting in different algebras, different geometries, the need for proofs of consistency and independence, and finally in recent years the whole body of philosophical, logical studies comprehended under the new name "metamathematics."

Not only did the creators not comprehend these connections with philosophy and logic and the foundations of mathematics, but they certainly never anticipated the fact that one hundred years later the physicists, when formulating their theory of relativity, would find a non-Euclidean geometry just the tool they needed for rephrasing and simplifying the original statements of Einstein.

This fascinating story, with its range of connections all the way back to ancient Greek philosophy—notably, to Plato's understanding of the nature and importance of definitions and to Aristotle's work in logic—typifies again the connections of mathematics with other aspects of our culture, especially philosophy and the philosophical science of logic. It also illustrates two facts: (1) that often the applications of a mathematical concept or system are unforeseen by the inventors and may follow years later in unpredicted ways and (2) that over the years mathematicians' perceptions of their own subject change and develop.

The relations between mathematics and the physical world are an interesting and sometimes complicated two-way street. Physical needs and intuitions based upon ideas of such physical things as points and lines may serve well to stimulate the thinking and intuition of a mathematician; on the other hand, mathematics developed as pure theory or extended and generalized into abstract theory may at the hands of other persons serve as a vehicle for the discovery and extension of unforeseen physical or even social and economic theories.

SUCCESSIVE EXTENSIONS AND GENERALIZATIONS

The process of making successive extensions and generalizations has gone on over the centuries, as we have seen. Let us now look at this process as a pedagogical necessity and a pedagogical problem in the design and teaching of any modern mathematics curriculum. Although people are today urging that we begin at a much more advanced level than we have in the past, no one has suggested that we should begin in the kindergarten with discussions of the entire real and complex number systems. If we assume that it will be pedagogically desirable to study and teach positive integers before dealing with negatives, fractions, irrationals, etc., we automatically set up a situation in which we are going to be responsible for helping children to understand how number systems were extended repeatedly in order to overcome difficulties such as the impossibility of subtracting a larger number from a smaller one, of dividing 3 by 2 (in the domain of integers), of finding the square root of -4, etc. The pedagogical difficulty is that what has previously been deemed impossible suddenly is made possible and the same word that we have used in the past, "number," now has a new and extended meaning as well as new properties and uses.

A helpful example of this process of extension and generalization can be traced in the development of modern trigonometric functions. These grew out of Greek astronomy, where a conception of planets moving in circular orbits developed a need for determining the lengths of chords of circles if one knew the length of the corresponding arcs. Significant segments of early Greek geometry, especially those dealing with the construction of regular polygons, were useful (at least in part) in the computation of tables of chords by Ptolemy, who in this computation was extending work begun by Hipparchus. Hindu mathematicians of about A.D. 500 saw the greater utility of tables of half-chords and not only computed them but gave us, indirectly, the word "sine." This has come to us as the English version of the Latin word used to translate an Arabic word which, through an error, was used for the Hindu *jya*, "half-chord." In later stages the sine has been variously defined as "one side of a right triangle with a given acute angle and a hypotenuse of one," "the ratio of the side opposite an acute angle to the hypotenuse of a right triangle," "the ratio of the ordinate to the radius vector determined by an angle at the center of a circle," "the ordinate of a point on a unit circle determined by a line segment 'wrapped' around the circle," "the limit of an infinite series corresponding to a given real number substituted in the series," "the limit of an infinite series corresponding to a complex number substituted in the series," and "the inverse function of a relation determined by a definite integral." All the tables computed under these varying definitions of the sine are in some way equivalent to or derivable from one another. All these definitions and all these tables could be used to solve problems involving chords of circles and sides of triangles both right and obtuse. But the more abstract and general definitions, which do not depend on the notions of angles, triangles, or even necessarily of circles or arcs, are the definitions that particularly stress the special properties of these trigonometric functions. Amongst these special properties the most important is, perhaps, their periodicity.

12

The study of periodic phenomena such as vibrating strings, electrical impulses, radio waves, sound waves, and the development of physical theories such as the wave theory of light have made it important to have periodic mathematical models, functions, and relations. As the trigonometric functions were progressively through history separated from their dependence on circles and triangles and ratios, and as they were generalized and abstracted to be sets of ordered pairs of numbers matched with one another by such devices as summing infinite series, they became more useful. Their use was no longer restricted to situations with angles and triangles, but they became functions of real (or complex) numbers, which, in turn, could represent time intervals or distances or whatever was useful. These abstract functions were then studied for their intrinsic properties, such as their periodicity, rates of change, maxima and minima, to such an extent that they were well understood as separate entities and were ready for use in constructing a mathematical model of periodic phenomena. Alfred North Whitehead, in commenting on this story, wrote, "Thus trigonometry became completely abstract; and in thus becoming abstract, it became useful" /48/.

It may be possible to exaggerate the importance of this idea that the generalizations and abstractions of mathematics are really the useful parts because it is they that make the mathematics applicable to physical situations as yet unknown and perhaps even unforeseen. However, this concept of mathematics is far too little appreciated, and the history of mathematics serves a valuable purpose if it makes this aspect of the subject understood by future citizens. It should be clear that an understanding of this role of abstraction and generalization is a goal to be sought, not a guide for initial pedagogical procedures. It is not our contention that the introduction to new mathematical topics should be general and abstract, especially at the more elementary levels; it is our contention, however, that as the subject grows and develops we should lead students not only to understand generalizations and abstractions but also to appreciate their importance and their role. A sufficiently concrete and detailed tracing of the history of the development of a generalized idea is one of the best ways to teach an appreciation of the nature and role of generalization and abstraction.

Intuition, Induction, and Analogy

However, balance demands that we also stress the importance of intuition, induction, and analogy in the conception and perception of mathematical ideas. The fascinating story of Srinivasa Ramanujan may help to illustrate this idea (as well as the internationalism of mathematics). Ramanujan was a poor, largely self-taught Hindu mathematical genius who lived from 1887 to 1920. His story, along with stories of Gauss, Galois, Euler, Archimedes, and many others, can further serve to stimulate interest when used as a subject for reading and class or mathematics-club reports. One of the remarkable things about Ramanujan is the fact that in almost complete isolation from other mathematicians he developed for himself substantial portions of mathematics. Some of his developments were awkwardly and incompletely done. Some of them were even incorrect. But, in spite of this, an almost unparalleled imagination and intuition led him to many remarkable discoveries. When he was being brought to England, his sponsor, the famous mathematician G. H. Hardy, wrote of a question that was bothering him:

> What was to be done in the way of teaching him modern mathematics? The limitations of his knowledge were as startling as its profundity. . . . All his results, new or old, right or wrong, had been arrived at by a process of mingled argument, intuition, and induction, of which he was entirely unable to give any coherent account.
>
> It was impossible to ask such a man to submit to systematic instruction, to try to learn mathematics from the beginning once more. I was afraid too that, if I had insisted unduly on matters which Ramanujan found irksome, I might destroy his confidence or break the spell of his inspiration.[6]

The caution here expressed illustrates the fact that all mathematicians recognize the role of imagination and intuition in leading to the conjectures they must later test for truth, then prove if found to be probably true. A story told by Hardy of a conversation with Ramanujan when visiting him in a hospital also illustrates the mathematician's natural inclination to ask, "What would happen if . . . ?", to generalize, and after each new idea to ask himself, "What more can I make of that? To what new idea does it lead me?" The story is that as Hardy was talking to Ramanujan, he remarked that he had ridden to the hospital in taxicab number 1729 and that this number seemed a rather dull one. "No," Ramanujan replied, "it is a very interesting number; it is the smallest number expressible as a sum of two cubes

[6] For this anecdote and more about the life of Ramanujan, see /NEWMAN: I, 366–76/.

14

in two different ways." Hardy goes on to say, "I asked him, naturally, whether he knew the answer to the corresponding problem for fourth powers."

Interdependence in Mathematics

Another theme in the history of mathematics is illustrated by our brief reference to the development, simultaneously, of non-Euclidean geometry by a Hungarian and a Russian who had no connections with or knowledge of each other. The frequent occurrence of simultaneous discoveries in mathematics illustrates the growing and maturing nature of mathematical knowledge and the fact that frequently new discoveries are generated by earlier ones. Not only are new discoveries generated by earlier ones, but often earlier ones are so necessary as a preparation for the next stages that when the preparatory stages have been completed a number of persons will see the next step. The discovery of logarithms by the Scotsman John Napier and the Swiss Jobst Bürgi, as well as the development of a geometric representation of complex numbers by the Norwegian Caspar Wessel, the Swiss Jean Robert Argand, and the German Carl Friedrich Gauss, all at much the same time, furnishes additional illustrations of simultaneity and internationalism.

Mathematics advances by the interplay of many devices and approaches. In spite of what has been quoted previously about the bareness of mere statements of axioms and proof, the modern axiomatic method can be a tool and a motive for extending the bounds of mathematics itself. For example, Whitehead and Russell /(b): v/, in the Introduction to their famous *Principia mathematica,* say:

> From the combination of these two studies (the work of mathematicians and of symbolic logicians) two results emerge, namely (1) that what were formerly taken, tacitly or explicitly, as axioms, are either unnecessary or demonstrable; (2) that the same methods by which supposed axioms are demonstrated will give valuable results in regions, such as infinite number, which had formerly been regarded as inaccessible to human knowledge. Hence the scope of mathematics is enlarged both by the addition of new subjects and by a backward extension into provinces hitherto abandoned to philosophy.

The variety of ways in which today's students can participate, both vicariously and directly, in the formulation of the axioms needed to advance a proof and in the fumbling for conjectures that follow from a set of assumptions, has been discussed in many articles; they do not necessarily have a strong historical connection. However, the examination of Euclid's axioms and their probable sources, and of some of his theorems and the gaps in their proofs occurring from the use of diagrams, is an instructive way to point out the nature and need of an axiomatic system.

TEACHER AND STUDENT USE OF THE
HISTORY OF MATHEMATICS

This then leads to the question, How do you use the history of mathematics to accomplish all the fine things suggested above—to increase the understanding of mathematics itself; of its relation to the physical world; of how and why it is created, grows, develops, changes, and is generalized; of the internationalism of its appeal; of the practicality of generalizations, extensions, and abstractions; and of the nature of structure, axiomatization, and proof?

There is no neat, simple, general answer to this. The age and background of the student and the ingenuity of the teacher join in determining the approach that may be used. It may be that a brief historical talk or anecdote is indicated. Or the class may be given—with some discussion and buildup—a problem such as the famous one propounded by Girolamo Cardano (also known as Cardan), "to find two numbers whose sum is 10 and whose product is 40." He said of the problem, "This is obviously impossible." (Why?) But then he went on to say, "Nevertheless, let us operate." Completing the square in the same manner as he had with earlier "possible" problems, he arrived at $5 + \sqrt{-15}$ and $5 - \sqrt{-15}$ as the new "sophisticated" numbers that would satisfy the conditions.

Some historical stories and topics are useful for outside reports, independent study, and club programs. Other situations provide the setting for encouraging student "discovery." Certainly a student who has seen the sequence of squares of integers associated with Pythagorean diagrams for figurate numbers (see [10]) is on the verge of discovering for himself the formula $n^2 + 2n + 1 = (n + 1)^2$ and the fact that the sum of the n successive odd integers beginning with 1 is n^2, as well as the fact that the second differences of a quadratic function tabulated for equal increments of the dependent variable are constant. History can do more than present stimulating problems and ideas; it can also help students to perceive relationships and structure in what appears to be a tangled web of geometry, algebra, number theory, functions, finite differences, and empirical formulas.

16

For the teacher, a historical view helps to determine what "modern mathematics" should really be. History shows that contemporary mathematics is a mixture of much that is very old—counting, for example, and the Pythagorean theorem, which is still important—with newer concepts, such as sets, axiomatics, structure. These newer aspects may be more significant for their clarifying and unifying value than for their explicit content. If a perception of structure in an old system makes a segment of mathematics easier to comprehend and extend, if new symbolism and terminology help to integrate and systematize a body of knowledge that was growing unwieldy, then these new perceptions of structure and symbolism make the teaching, learning, and using of mathematics easier. Often (not always) "modern mathematics" may be merely a modern perception of several old topics. The important thing is neither to throw out all that is old nor to add whatever is new but to develop and pass on to our students new syntheses of old ideas and systems as well as to introduce new concepts and systems that are appropriate. Insight into the development (history) of ideas can serve to improve both the curriculum maker's choices and the teacher's power to communicate insights and stimulate interest.

II

The History of
NUMBERS and NUMERALS

an overview by

BERNARD H. GUNDLACH

O f all known forms of life on earth, man is the only one to have developed a systematic procedure for storing up useful information and passing it on from one generation to the next. A considerable part of this information is related to form and quantity. A language to express form and quantity and their various relationships is a necessity. It is as a language that we wish to treat mathematics, at least in its earlier developmental phases.

One approach might be to follow the rather generally accepted procedure of fixing a historical starting point coincident with the earliest written documents that have been found, deciphered, and dated. In the discipline of mathematics most such documents show clearly that symbolisms and procedures, as well as the problems contained in them, are much older than the documents themselves. In most cases the latter constitute an already highly polished product, revealing few if any of the first faltering steps that must have been taken in the infancy of mathematics. Our approach, therefore, is different. To reconstruct the early beginnings of this language, we turn for help to a special branch of anthropology—namely, ethnography.

As the scientific description of individual cultures, ethnography includes the study of existing civilizations that happen to have developed in relative isolation, little influenced by the cultural mainstreams that emerged from China, India, Mesopotamia, and Egypt.

Such isolated civilizations, some of them still at the level of the Early Stone Age when they were first studied some hundred years ago, have been found in Africa, Australia, South America, Indonesia, and various other widely scattered locations. We know that different civilizations have developed at vastly different rates in different places. It is plausible, therefore, to assume that ethnographic studies of primitive civilizations—now rapidly becoming either extinct or modernized—can furnish valuable clues to an understanding of the earlier, and perhaps the earliest, stages of our own civilization, where traces of these stages have been almost erased by the heavy footsteps of later achievements.

It appears quite certain that on the path to more advanced levels of civilization enumeration preceded numeration, and numeration, in turn, preceded number.

ENUMERATION

By "enumeration" we mean here simply keeping track of the objects in a collection or set by a one-to-one matching of the objects with other objects used as checks or counters. If an early herdsman wished to determine whether or not his flock was still intact at the end of a day, a check could be made by matching each animal with one object from a well-known and ordered collection that was readily available. In most primitive civilizations studied, this well-known and ordered collection consisted of a sequence of parts from the human body. No language was needed to effect such a one-to-one matching. For example, the Bugilai of British New Guinea used the following sequence, named simply by ticking off the following, one by one, with the right-hand index finger:

> Left-hand little finger
> Left-hand ring finger
> Left-hand middle finger
> Left-hand index finger
> Left-hand thumb
> Left wrist
> Left elbow
> Left shoulder
> Left breast
> Right breast

All that was needed in order to check whether a herd of suitable size was complete was to remember which was the last body part touched. If that part was, for example, the left elbow, then the last animal checked should correspond to "left elbow." The order of the sequence was fixed and natural. Every man carried it with him.

Observe that not only is a concept of number not implied by such a procedure; it is not even necessary to have spoken words for the various parts of the body. Many primitive peoples have used similar procedures /BERGAMINI: 18–19/. Some could go as far as 31 using all fingers, all toes, and eleven other parts of the body. One very advanced tribe was able to reach 100.

NUMERATION

With the creation of a language that contained words for the various body parts, it was natural that these words, rather than the actual parts of the body, should be used in the enumeration process; and this change marks the transition to "numeration." In the Bugilai dialect, words we would call "number words" for the beginning numbers had as original meanings the physical features already listed:

1	*Tarangesa*	Left-hand little finger
2	*Meta kina*	Left-hand ring finger
3	*Guigimeta*	Left-hand middle finger
4	*Topea*	Left-hand index finger
5	*Manda*	Left-hand thumb
6	*Gaben*	Left wrist
7	*Trankgimbe*	Left elbow
8	*Podei*	Left shoulder
9	*Ngama*	Left breast
10	*Dala*	Right breast

Thus the gradual use of spoken languages marked a great step forward. Referring again to the Bugilai, we find that, once words were available for the various body parts, it was no longer necessary to move through the time-consuming sequence of physical actions. It was enough merely to say the corresponding word names in order. If the last object in a set to be matched was, say, *podei* ("left shoulder"), then every time this set had to be checked one had only to listen and note whether the last object checked did or did not correspond to *podei*. This does not mean, however, that *podei* actually had become the name for the cardinal number 8.

This type of matching procedure might be thought of as qualitative rather than quantitative. Indeed, in this primitive form, numeration seems to reside entirely in the things enumerated rather than in the human mind. All that is required is an ordered sequence of signs that can be reproduced at will. With the invention of language, words took the place of objects in the ordered sequence. As time went by, these words were learned by heart. The use of number words, however, does not in itself imply the concept of cardinal number, though it undoubtedly did lead to it. Ethnographic experiments with primitive people have shown that the mastery of an ordered sequence of number words does not lead necessarily to the concept of a cardinal number.

NUMBER

We do not have enough evidence to fix the period in early history when the epochal discovery of cardinal number was made. The earliest written documents in our possession show that the concept was equally present in China, India, Mesopotamia, and Egypt. All these documents contain the question "How many . . . ?" This question can be answered best in terms of a cardinal number. Therefore, when these early documents were being written, and probably much before that time, the concept of cardinal number already had been formed.

If we attempt to reconstruct the type of situation from which the cardinal-number concept must have sprung, we are forced to realize at the very outset that the fundamental concept we express today by using the word "set" must have been one of the earliest abstractions made by man. When the Bugilai said *manda* ("thumb of the left hand"), they did not as yet mean "hand" or "complete set of five objects." They merely knew that at a certain point in the natural sequence the sound *manda* had to be made. But when the more sophisticated island dwellers of Nicobar, who used both fingers and toes as their natural counting sequence, said *hean umdjome* (which means "one man") for twenty, the set concept stood out in full relief. Use of the set concept becomes clearer still when we look at the same tribe's expression for one hundred—*tanein umdjome*—where *tanein* is the word for five. A multiplicative idea is foreshadowed here; at least there seems to be a rather useful understanding of the idea of equivalent subsets.

In due course it must have been observed by some of the more alert peoples that the order of the objects in the sets to be matched was unimportant. The next step after this appears to have been the difficult one, namely, to realize that the last ordinal number name enounced not only assigned a certain name to the last object in the set to be matched but also told how many objects there were in that set altogether. Today, of course, it is well known that the cardinal

21

number of a set is independent of the nature of the objects it comprises as well as of the order in which these objects may be arranged.

EARLY NUMERATION SYSTEMS

A more formal outgrowth of numeration is found in the formation of numeration systems. In cultures where the fingers of one hand had been used in earlier phases of numeration, the basic grouping number for the numeration system became five. Where the fingers of both hands were used, the basic grouping number became ten. Where both fingers and toes or the parts of both hands and arms were used, it became twenty. The need for a numeration system arose with the following question (which would not, of course, have been phrased in these terms): *What is to be done when the finite ordered sequence of counters* (fingers or other body parts) *is exhausted, yet more objects remain to be matched?*

It has been assumed that objects other than fingers or body parts—objects such as pebbles, dried seeds, or notches cut with a flint knife in a stick—also were used as counters during the early stages of enumeration. If, for example, a herd consisting of more than ten or twenty sheep was kept track of by means of a one-to-one matching with a pile of pebbles, the number of pebbles in the pile could easily be adjusted to the size of the herd. Pebbles were everywhere; when the herd grew, it was easy to place more pebbles in the matching pile. Such objects, however, do not possess the two essential features needed by early man—features provided by a set of body parts—namely, a definite order and "absolute finiteness."

So we return to the fruitful question that led to numeration systems: What is to be done when these counters are exhausted and more objects remain to be matched?

In principle three different methods have been developed to solve this problem, although only one of them has survived in modern society.

One way to continue is simply to extend the ordered sequence of counters. Certainly, in procedures in which the original counting sequence had been the fingers of one hand, extensions to the fingers of the other hand, or to the toes, or to still other body parts, were easy. But the real advantage of numeration over enumeration consisted in using the words in place of the actions, and the need for an increased vocabulary would have presented difficulties to many if not to most of the early peoples.

Actually a simpler way, and one that lent itself especially well to

22

FIGURE II-1

written representation, was extension by repetition. For a simple example, consider the case of "counting men" found in cave drawings dating back to the Middle Stone Age. These men used the fingers of both hands. When a counting man had raised all his fingers, a second counting man was brought into the picture, then a third, and so on, to continue the count on the same one-to-one matching basis. A count of thirty-five was recorded as shown in Figure II-1. Although the base number obviously is ten, such a system is nonpositional. It may be considered an additive system. Since addition is both commutative and associative, it makes no difference where in the lineup a certain counting man stands.

Consideration of the symbols used in many of the oldest known systems of numeration (e.g., Babylonian [1], Egyptian [2], early Greek [4], and Roman [3]) indicates the use of a single stroke for each item counted. This corresponds to a raised finger for each item. New symbols, in turn, were used for ten (all fingers raised on both hands). Repetition of symbols would permit representation of any number.

Again, a great number of repeated symbols would be necessary for large numbers. An obvious extension of such an approach was the use of a third new symbol for ten of the second new symbols. The Roman numerals I, X, C, M immediately come to mind as typical of such a system. An earlier illustration is the Egyptian hieroglyphic system, where the symbols shown in Figure II-2 represented successive powers of ten. A nonpositional additive numeration system based on ten is certainly a natural, if not an inevitable, development.

Though considerably superior to previous methods of extending numeration, such systems had obvious drawbacks when used for the expression of large numbers or for certain computational procedures.

The third extension procedure is that which has resulted in our present numeration systems. These systems are positional; that is,

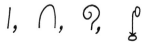

FIGURE II-2

they are based on place value, and they originate with the choice of a certain number as base [**5, 6, 7**]. Let the choice be ten, and let us see how the positional pattern might have developed from counting men who used the fingers of both hands. (A similar development might be given in terms of "counting" on an abacus.)

Development of Positional Numeration

The positional pattern involves two distinct ideas. First, when a counting man has raised *all* his fingers, he will have to return to the original position (both hands closed into fists) before he is able to continue his counting procedure. By itself this action is not sufficient to sustain a positional pattern, although in a way it is the basic idea involved in base-ten positional numeration. Because his attention is concentrated on carefully carrying out the one-to-one matching between the fingers of his hands (in a certain order) and the objects to be matched, he will be in a very poor position to remember how many times he has raised all his fingers and started again. This is especially true when the set to be matched is large. What he needs, therefore, is a second idea, consisting of a sort of "memory" or "memory record" for the "all" counts—one that is readily available and easy to read.

There is evidence that various types of "memory" procedures have been used at one time or another. The one we wish to bring out here is that of calling on a second man to keep track with his fingers of the "all" counts of his mate. This second man will record by one raised finger every "ten" count that the first counting man has made. This will enable the team of two men to record counts through ninety-nine. At the count of ninety-nine there arises a certain difficulty. It can be overcome, and obviously has been overcome, simply by extending the memory principle. This principle can be stated as follows: Each man in the counting lineup must be ready at any moment to record another count, regardless of whether such a count appears to be imminent. When this principle is applied to a count of ninety-nine, the sequence shown in Figure II-3 will result. From the first position, with each

One more count: then

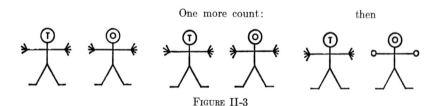

Figure II-3

24

counting man extending nine fingers, an additional count of one results in the second position and then in the third.

At this moment the "memory man," or "tens man," as he may be called, is in the same situation the original counting man was in when he had raised all ten of his fingers and was ready to return to the closed-fist position. The tens man is now in a situation in which he could not record another "all" count if that should become necessary. He, too, is now in need of some sort of recording memory, and this means that a third man will be needed. This third man will appear as shown in Figure II-4. The numeral 100, placed underneath the counting men with each digit naming the number of fingers shown on the man above, makes clear without need for many words that a zero numeral is needed in positional numeration.

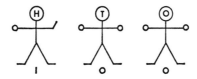

FIGURE II-4

While "0" is a placeholder in a positional numeral, it is much more than this. It is the symbol for the cardinal number of the empty set—that is, a numeral for the cardinal number zero. The counting-man sequence makes clear that zero is both. Figure II-4 relates two of the fundamental meanings of zero in one comprehensive and comprehensible picture.

HISTORICAL SEQUENCE AND MODERN SYNTHESIS

It has been our thesis that enumerations evolved into numeration through the use of spoken and, later on, written language. Languages have played a leading role also in bringing about the change in mathematical emphasis from numeration to number. In retrospect, it may appear strange that there could have existed a serious and rather consistent concern with numeration before the concept of number actually had been formed. The strangeness, of course, is merely the fault of the purely retrospective view. From a modern viewpoint, we tend to think of numeration as being concerned with ways of expressing numbers—that is, with creating symbols for certain ideas. It seems natural that the concept of number should precede concern with symbols that express the concept. However, we hope to have shown that this was not the way in which numeration actually developed. Num-

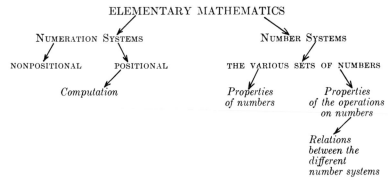

FIGURE II-5

eration evolved from nonverbal enumeration simply because language appeared as a simplifying procedure.

This does not mean, however, that we should try to reenact the historical sequence of events in teaching young children. Today's children come, even to kindergarten, with considerable facility in the use of language and a fairly complete number-word vocabulary. All we really need to do for these modern children—but we must do it straightforwardly and consistently—is to let them discover as soon as possible that the last number word given to the last object of a set to be counted not only names that object as the last one of an ordered sequence but also names the cardinal number of the whole set.

In the modern approach to the teaching of elementary mathematics, we distinguish from the outset between numerals and numbers. We also separate the study of number into two different aspects or branches. One aspect is concerned with what is called "properties of numbers." The other deals with "number systems"—that is, with the properties of operations on numbers and the relationships these operations have with the sets of numbers for which they are defined. If we omit from the following account any concern with geometry, we can draw the clarifying scheme shown in Figure II-5.

Historically, concern with computation preceded by many centuries concern with properties of numbers. In turn, the concern with properties of numbers preceded concern with number *systems* by almost two thousand years. It is from this latter concern, however, that the powerful unifying and simplifying patterns of modern mathematics are generated. Most historians ascribe the chief concern for computation to the Oriental peoples and the chief concern with number systems to those who more or less followed the predominantly Greek tradition. It appears, however, that the essence of the modern approach is due to a synthesis of both points of view—a synthesis that has been achieved rather recently.

THE MEANINGS OF ZERO

Consider the role of zero with respect to the scheme portrayed in Figure II-5. "Zero as a placeholder" would be a concept pertaining to numeration, while "zero as the cardinal number of the empty set" would come under number. We can only conclude that zero is both, depending on whether one is thinking primarily about numeration or about number. It may be stated here that beyond the elementary realm there may be many more and different meanings for zero; but we shall concentrate only on these two and look at each in turn from a historical vantage point.

Nonpositional numeration, which has been shown to have preceded positional numeration by considerable time in most civilized regions of the ancient world, is purely additive. When various individual symbols have been combined to form a numeral, they name the number to be expressed as the sum of the numbers expressed by the individual symbols. Since addition is both commutative and associative, the order in which the individual symbols appear in the numeral cannot possibly affect the sum; hence any order can be used.

For example, in the nonpositional numeration system of the Classical Greek period (*c.* 600–300 B.C.) an ordinal alphabet system was used. The principle of this system was most probably transmitted to Greece by the Phoenicians, a people of navigators and tradesmen. In such a system each letter of the alphabet plays a double role. It serves as a numeral as well as a letter. In some numeration systems of this type, like the Hebrew system, no special sign was used to indicate whether a composition of letters denoted a name or a number. In the Greek

α	β	γ	δ	ε	ς	ζ	η	θ
1	2	3	4	5	6	7	8	9

ι	κ	λ	μ	ν	ξ	ο	π	ϙ
10	20	30	40	50	60	70	80	90

ρ	σ	τ	υ	φ	χ	ψ	ω	ϡ
100	200	300	400	500	600	700	800	900

FIGURE II-6

27

system shown in Figure II-6 (also see [4]), apostrophes were often used to show that the letter was to be interpreted as a numeral.

Obviously this is a base-ten system, yet it is nonpositional. If the number 345 is represented by Greek letter-numerals from the table, the representation could be τμє (300 + 40 + 5 = 345). It could be represented just as easily by єμτ (since 5 + 40 + 300 = 345). We quickly realize that there are six different ways to represent the number.

On the other hand, if we were to make corresponding variations in a positional numeration system with a base of six or greater, the six numerals 345, 354, 435, 453, 534, and 543 would represent six different numbers.

Let us now represent the number 305, again using the Greek letter-numerals. Obviously, there are two ways: єτ (for 5 + 300) or τє (for 300 + 5). Is there need for a zero symbol in nonpositional numeration?

Zero as a placeholder.—On the other hand, positional or place-value numeration could not possibly function adequately without a symbol for an empty place or position. In pre-Babylonian Sumer, where a positional numeration system with base sixty was in use (by a very few experts) as early as 3500 B.C., an empty place was sometimes indicated by actually leaving an empty place in the numeral [1]. Thus the Sumerian numeral shown in Figure II-7 might stand for

FIGURE II-7

$(4 \times 60) + (5 \times 1) = 245$. It might stand equally well, however, for $(4 \times 60^2) + (5 \times 1) = 14{,}405$. Only a careful study of the context might reveal the intended meaning of the numeral. Since in most positional and nonpositional numeration systems some form of the abacus was used for computation, the idea of leaving an empty place in a numeral probably was suggested by an empty groove in a pebble abacus or by an empty wire in the more sophisticated form.

It is impossible to trace precisely the development of our present numeration system, the so-called Hindu-Arabic numeration system [7]. This system is named after the Hindus, who probably invented it, and after the Arabs, who later transmitted it. In A.D. 825 the Persian mathematician al-Khowarizmi [21] described a Hindu system complete with positional value and a zero. From about 900 we have a

١٢ ٣٤ ٥٦٧٨٩٠

FIGURE II-8

well-established set of numerals of Arabic origin. These are shown in Figure II-8. The dot stands for zero in the sense of the empty place, the placeholder.

More recently, sets of Hindu numerals have been found that date from before 800 and not only have a zero symbol almost exactly like the one we use today but also assign a definite name to it. (These earlier symbols are the model of the set in Fig. II-8.) The name for the zero symbol was *sunya*, which meant then, as it does today in Sanskrit, "empty" or "blank." It has absolutely no implication of a meaning like "void" or "nothing," however. Available evidence seems to indicate that the placeholder concept of zero preceded the number concept [8].

Zero as a number.—Zero as a number had a rather different history. Here again, it will be best to keep the developmental diagram in mind and to consider separately the two branches of the development of the number concept—that of the "properties of numbers" and that of the "properties of operations on numbers." Although the Greeks of the sixth, fifth, and fourth centuries B.C. were the chief developers of the study of properties of numbers, they never recognized zero as a number. For them the set of whole numbers began with one, to which their number lore assigned such attributes as "male," "reason," "the essence of number," "the origin of all things," "the divine principle." Even where mystical aspects were less pronounced than in the original Pythagorean tradition, with its concepts of "polygonal numbers," "perfect numbers," "abundant and deficient numbers," and "amicable numbers" [10, 11, 12], to mention just a few, zero had no place in Greek deliberations. Likewise, when Greek mathematicians took a few halting steps toward development of a number system, as in Eudoxus' rather admirable treatment of irrationals, zero played no part.

As nearly as we can tell, tracing back in history, zero was first recognized as a number by Hindu and Arab mathematicians of the period from about A.D. 500 to 1100, with the initiative probably belonging to the Hindus rather than to the Arabs. This came about in two ways: (1) through attempts to solve certain quadratic equations of a type that might be described as being of the form $AX^2 - BX = 0$, where one root is zero while the other is some rational number different from zero, and (2) through a more and more systematic study of the properties of operations on numbers.

With regard to the first point, most early mathematicians accepted zero as a possible solution, although most did not assign much importance to such a solution, since zero was not very meaningful as a solution to practical problems.

With regard to the second, several very important and well-documented statements exist, of which two will be quoted here. The Hindu mathematician Mahavira (about A.D. 850 in Mysore, India) wrote a great book, *Ganita-Sara-Sangraha* ("The Compendium of Calculation"), containing this statement: "A number multiplied by zero is zero, and that number remains unchanged which is divided by, added to, or diminished by zero." While this statement already seems to contain the core of the concept of zero as "the identity number of addition," it is interesting to observe also that Mahavira considers division by zero as having the same effect as addition and subtraction of zero—namely, as having no effect on the number on which it operates as a divisor. This rather gross misconception was changed in the works of the great Hindu mathematician Bhaskara (teaching in Ujjain, India), who states some 300 years after Mahavira, whose works he knew, that "a definite number divided by cipher [zero] is a submultiple of nought." He then goes on to illustrate his meaning by writing (we use modern notation here), "$10 \div 0 = 10/0$," and, "$3 \div 0 = 3/0$," and says, "These fractions of which the denominator is cipher are termed *infinite quantities*." Almost all the minor writers of this period recognize zero as the identity number of addition and—under certain restrictions—of subtraction, but most of them either entirely avoid the problem of dividing by zero or declare the result of such a division to be meaningless.

In summary, we can say that zero became fully recognized as a number only from the time of Bhaskara on. We can safely reassert that in modern elementary mathematics zero is both a placeholder and a cardinal number, the role it takes depending on whether one talks and thinks about numeration or about number systems.

"THE GREEK GENIUS"

"The Greek genius" did not happen spontaneously. Once the Greeks were settled in the Peloponnesus and on the western shores of Asia Minor, they began to travel. Probably they learned about ships and navigation from the Phoenicians, the traders of the ancient world. Soon they were off to faraway places. On these travels they made contact with many more ancient cultures—in India, in Mesopotamia, and in Egypt. They learned and partially absorbed ways of life that had taken thousands of years to develop. Knowledge, wisdom, and religion often were indistinguishable in these ancient cultures. What the early Greek travelers brought home from their trips abroad was a curious and intricate mixture of various religious cults and philosophies of life grown under conditions very different from those familiar to the Greeks. They accumulated also a tremendous wealth of knowledge pertaining to practically all aspects of life. Deeply woven into it all was knowledge of numeration and number, astronomy and (as we would call it now) astrology, and an abundance of geometric patterns and designs. Here we are concerned only with numeration and number.

It may be conjectured that the Greeks of the sixth and fifth centuries B.C. were not very much interested in numeration—if, indeed, they were interested in it at all. This was true in spite of something that appears quite certain: the Greek travelers to the East did come into intimate contact with positional numeration systems, like those of the Babylonians, which were vastly superior in design and manageability to their own nonpositional numeration system. Their minds apparently were not inclined toward the mechanical and rote aspects of elementary mathematics but rather were fascinated by suspected underlying reasons and possible justifications.

One school of historians of mathematics holds that the obvious failure of the Greeks to recognize and adopt an efficient numeration system when they saw one constitutes "a dark blotch on the otherwise shining shield of Greek mathematics." However, in the absence of firsthand information, a different interpretation is also possible. The original Pythagorean brotherhood (c. 550–300 B.C.) was a secret aristocratic society whose members preferred to operate from behind the scenes and, from there, to rule the social and intellectual affairs of the marketplace with an iron hand. Their nobly born initiates were taught entirely by word of mouth. Written documentation was not permitted, since anything written might give away the secrets largely responsible for their power. Among these early Pythagoreans were men who knew more about mathematics (as it was then available) than most other people of their time. It is difficult to accept the idea that these men were outright stupid with respect to problems of numeration. Instead of assuming that there was a "blind spot" with regard to numeration, it seems at least as plausible to suppose that they had recognized clearly that a base-ten positional numeration system would quickly make computational skills available to people in all walks of life. This would rapidly democratize mathematics, so to speak, and thus diminish at least one of the powerful holds they had over the masses, who for computation had to consult experts or use complicated tables—and both of these sources of help could be controlled by the brotherhood. There are historical precedents for this explanation. In both the Babylonian and the Egyptian civilizations computations were handled by a small and exclusive group of experts, frequently the priests. Their special and carefully guarded skills and knowledge gave them influence and power. The Pythagoreans may simply have followed their example.

However this may have been, the Pythagoreans certainly did not refine and propagandize numeration but concentrated—aside from their magnificent work in geometry—on studying the properties of numbers, in particular the positive integers. They thereby missed or knowingly passed by the much more significant study of the properties of operations on numbers, which might have led them to create a structure of number systems similar to that which they created for geometry.

The Greek language of that early day had two words of special significance for our purpose: *logistike* and *arithmetike,* which might be rendered by "logistic" and "arithmetic."

Greek "logistic."—Logistic was concerned with numeration and computation, the latter being various means and ways by which an unwieldy numeral could be changed to a simpler one, as when we change a multiplication numeral like 8×27 into the more convenient place-value numeral 216. In keeping with the aristocratic and antidemocratic philosophy of the brotherhood, logistic was considered an occupation unworthy of a gentleman; thus it was left largely to people of the lower classes who made a special trade of computation, using abaci and tables which the Pythagoreans might have designed for them. The Pythagoreans would tell these tradesmen how such tables and devices were to be *used* but never how to *make* them or what the hidden patterns were which made them possible.

Greek "arithmetic."—The nobly born youth who wished to be initiated into the secrets of the brotherhood emphasized the study of *arithmetike. Arithmetike,* as you will have already guessed, was not at all the "arithmetic" of our own day but, rather, what we would describe as "theory of numbers" or possibly as "higher arithmetic."

To appreciate the preoccupation of the Pythagoreans with properties of numbers, we must keep two things in mind: (1) The Greeks had inherited from the earlier Eastern cultures an almost inextricable mixture of genuine number knowledge, myths, and religious beliefs. (2) The prevailing numeration system of this period made use of the standard Greek alphabet supplemented by special symbols so as to make a set of twenty-seven characters [4]. Although there was no difficulty in determining when the symbols represented a number instead of a word, it was possible to use the numerical value of each letter to assign a unique number to any given word. Since as numerals λ stood for 30, o for 70, γ for 3, and σ for 200, the word λογοσ (*logos*), meaning "reason," "idea," or "thought," has the numerical value of 373 in our notation (30 + 70 + 3 + 70 + 200).

As far as we know, the Pythagorean brotherhood had only male members, despite a doubtful anecdote telling us that Pythagoras, the grand master of the movement, had his beautiful girl friend admitted as a member. It is not surprising to learn that, according to Pythagorean number lore, odd numbers were considered male! Men, at least men of correct birth, could have *logos*. Women, associated with even numbers, evidently could not. Some of the first few odd and even numbers were associated with such human attributes as "opinion" (two); "justice" (four, since it was the first perfect square); "marriage" (five, since it represented the "union" of the first odd and even numbers). "One" was not considered a number itself but was taken to be the (divine) generator of all numbers. The list of such number-lore associations is extensive [9].

Regardless of what mystical reasons may have motivated the early Pythagorean investigators, they discovered many curious and fascinating number properties. The odd and even numbers have been mentioned already. Since the general Greek outlook toward mathematics was more geometric than arithmetical, and since in their earlier work, at least, the Greeks considered only whole numbers, it is no wonder that they attempted to represent numbers as geometric patterns [10].

Next to the circle, which many considered the most nearly perfect of all plane figures, the square was most important. Square arrays of dots, probably formed with pebbles in earlier versions, led the Greeks

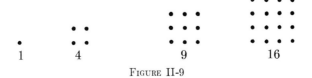

1   4   9   16

FIGURE II-9

to numbers that were perfect squares—that is, to numbers which, when expressed in various ways as the products of two numbers, would have two equal factors. See Figure II-9.

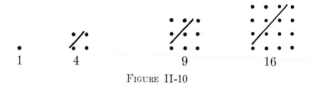

1   4   9   16

FIGURE II-10

By removing the pebbles or erasing the dots above the diagonals shown in Figure II-10, they obtained configurations for triangular

1   3   6   10

FIGURE II-11

numbers, as shown in Figure II-11. The triangular sequence of numbers can be generated by the formula

$$\frac{n(n+1)}{2},$$

which for $n = 1, 2, 3, \cdots$ yields the numbers 1, 3, 6, 10, 15 $\cdots$.

Other number properties—partly discovered, partly inherited by the Pythagoreans from Eastern lore—pertained to perfect, deficient, and abundant numbers [12].

The use by various numeration systems of letters of the alphabet in the double role of letters and numerals gave rise to a sort of occult number magic (it really should be "numeral" magic) known as "numerology" or "gematria" [20]. To the numerologist, two words were equivalent if they added up to the same number when the individual letters were interpreted as numerals. Unfortunately, such number mysticism is not confined to Greek mythology, nor to the ancient world. For example, in practically every period of history some Christian writers have been able to show, by equating names and numerals, that archenemies of their own thinking are signified by "the number of the beast" in the Book of Revelation, 666! (Rev. 13:18.)

The Greeks' concern with prime numbers [15] was considerably

deeper and more serious. It was known that, with the exception of
one and two, any whole number that is not prime can be expressed as
a product of primes. The Greeks not only formalized these findings
but established what later became known as "the fundamental theorem
of arithmetic" (where the term "arithmetic" is taken in its original
Greek sense)—namely, that a composite number can be expressed as a
product of primes in one and only one way (not counting, of course,
the possible permutations of the factors). This theorem is known also
as the "unique factorization theorem." Indeed, whether 144 is factored
as

$$12 \times 12$$
$$(4 \times 3) \times (4 \times 3)$$
$$(2 \times 2 \times 3) \times (2 \times 2 \times 3)$$

or as

$$9 \times 16$$
$$9 \times (4 \times 4)$$
$$(3 \times 3) \times (2 \times 2 \times 2 \times 2)$$

or as

$$8 \times 18$$
$$(2 \times 4) \times (2 \times 9)$$
$$(2 \times 2 \times 2) \times (2 \times 3 \times 3),$$

in every case the final result gives the same set of factors—four 2's
and two 3's. We point out in this connection that the unique factorization theorem does not apply to complex numbers, of the form $a + bi$.

Euclid presented a proof in his *Elements* to show that the set of
prime numbers is infinite—that is, that there is no greatest prime [14].
In spite of many attempts, some of them by the greatest mathematicians of their time, so far no one has been able to devise a practical
test for checking the primality of large numbers, nor has a truly
general prime generator been discovered [19].

With due respect to a very few isolated Greek arithmeticians, it
must be pointed out that the only numbers accepted by the vast majority of Greek mathematicians were the natural numbers; they
interpreted what are now known as rational numbers as ratios of
natural numbers. A few of the truly great Pythagoreans did conceive
of numbers that could not be expressed as ratios of natural numbers.
As usual, they arrived at this concept by way of a geometric situation.
The foremost of these mathematicians was Eudoxus (408–355 B.C.).

Eudoxus showed that the measure of the diagonal of the unit square
could not be expressed as the ratio of two natural numbers—as we
would say today, that the symbol $\sqrt{2}$ does not represent a rational
number. He developed an ingenious theory of "equal ratios," which

with just a few minor refinements could have become the basis for the real number system. Probably Eudoxus was not understood by more than a very few contemporaries; it is doubtful whether any of them (and this may well include Eudoxus himself) could have foreseen the tremendous implications of this discovery.

To most of the Greek mathematicians the very idea of incommensurable quantities [18] was disagreeable and fearful. Eudoxus' theory of equal ratios was soon discarded and forgotten. More than two thousand years elapsed before the German mathematicians Dedekind and Cantor [26, 27] took up the work where Eudoxus had left off and brought it to completion, creating the real number system and, thereby, a legitimate "place" for imaginary and complex numbers. Thus the Greek genius was no more concerned with number systems than with numerational systems. While the mathematical contributions of many ancient cultures were numeration, a principal Greek contribution was *arithmetike*, knowledge of the properties of numbers.

Today, however, neither *logistike* nor *arithmetike* comprises the core of the approach to the teaching of elementary mathematics. The modern approach is definitely oriented toward the structural properties of number systems (not of numeration systems)—that is, toward the patterns and properties of operations on numbers which provide unity, simplicity, and continuity from the system of the whole numbers through the system of the complex numbers.

Capsule 1   Barry D. Vogeli

BABYLONIAN NUMERATION SYSTEM

SOMETIME before 2000 B.C. the Babylonians developed a base-sixty or sexagesimal system of numeration which employed the positional principle. Actually, this system was a mixture of base ten and base sixty in which numbers less than 60 were represented by using a simple base-ten grouping system and numbers 60 and greater were designated by the principle of position with the base of sixty.

The Babylonians wrote on clay tablets, using a stylus. In early times a cylindrical stylus was used to impress number signs into the wet clay, a circular form representing 10 and a half-moon representing a unit. This system was then completely superseded by use of a stylus with a triangular end, which produced the characteristic wedge-shaped, or cuneiform, impressions.

The symbol ▼ represented the unit and was repeated for numbers up to 9. The symbol ◀ stood for 10 and was repeated and used with the unit symbol, as needed, to represent numbers from 11 to 59. Numbers 60 and above were represented in terms of the symbols for numbers from 1 to 59, using the principle of position to indicate multiples of powers of 60. In the Old Babylonian texts (1800 to 1600 B.C.) no symbol for zero was used, but a blank space was left for any missing power of 60. Some of the texts from the Seleucid period (first three centuries B.C.) contain a separation symbol, ⩍, used to indicate such an empty space between digits. Such symbols, however, were never used (at least not in extant documents) at the end of a numeral. This created the greatest disadvantage of the Babylonian system, for except from the context there is no way to determine, for example, whether the combination ◀▼ meant 11 or 11 · 60 or 11 · 60^2. Still, this system of numeration was superior to other ancient systems, since it was the first to employ the concept of place value.

The same place-value principle was also used advantageously to represent fractions. The value of $1 + 30/60$, or $1\frac{1}{2}$, would be represented by ▼⩩ . Since no separatrix ("sexagesimal point") was used, however, this combination could also be interpreted to mean $1 \cdot 60 + 30$—i.e., 90—or $1/60 + 30/(60^2)$—i.e., $90/3{,}600$ or $1/40$. The disadvantages of this ambiguity were considerably balanced by the flexibility permitted in the use of tables for computational purposes [34].

In the Old Babylonian period the subtractive principle was sometimes used in representing numbers. Thus 19 would be represented as 20 minus 1, with the sign for *lal* ("subtract"), ◀▲, written between the 20 and the 1.

The origin of the base of sixty cannot be determined with certainty. A plausible theory associates this with the values found in certain early systems of weights and measures in which a larger unit was 60 times the smaller. Although the division of the circumference of the circle into 360 parts originated in Babylonian astronomy in the last centuries B.C., the sexagesimal numeration system was developed many centuries before this and thus probably had nothing to do with astronomical concepts. Actually, the sexagesimal system is applied consistently only in mathematical or astronomical contexts. In tables referring to various economic matters (dates, measures of weights, areas, etc.), mixtures of base sixty, base ten, and other bases are found.

In using modern notation to convey the form of the sexagesimal system it is customary to separate the "digits" (from 0 to 59) by commas and to use a semicolon for the "sexagesimal point." Thus 2,34;15 represents $2 \cdot 60 + 34 + 15/60$. The position of the separatrix must always be inferred from the context in translating from the original tablets; as already stated, no symbol was employed by the Babylonians for this purpose.

For Further Reading

AABOE (b): 5–20
MIDONICK: 45–55
NEUGEBAUER (a): 5, 14–20

RUDNICK
VAN DER WAERDEN: 37–41

Capsule 2   Diana La Mar

EGYPTIAN NUMERATION SYSTEM

THE earliest known Egyptian numerals are inscribed in hieroglyphic form on a royal mace dating back to about 3400 B.C., when Menes united the Lower and Upper kingdoms of Egypt. These symbols were used to denote large numbers associated with the spoils of war: the capture of 120,000 human prisoners, 400,000 cattle, and 1,422,000 goats.

The early Egyptian system used a base of ten but had no symbol for zero. The appropriate number of vertical strokes represented the numbers from 1 to 9. Individual symbols were used for successive powers of ten through and sometimes beyond 1,000,000. These symbols were used in combination and repeated as required to express any

	Hiero-glyphic	Hier-atic		Hiero-glyphic	Hier-atic		Hiero-glyphic	Hier-atic		Hiero-glyphic	Hier-atic		Hiero-glyphic	Hier-atic
1...	**I**	I	10...	∩	∧	100...	℮	⌐	1,000...	⚘	∬
2...	**II**	ԡ	20...	∩∩	⋋	200...	℮℮	⌐	2,000...	⚘⚘	⁝
3...	**III**	Ⴑ	30...	ⁿ∩	ʼ∧	300...	℮℮℮	⌐	3,000...	⚘⚘⚘	ꮗ
4...	**IIII**	—	40...	∩∩∩∩	⋅	400...	℮℮℮℮	⌐	4,000...	⚘⚘⚘⚘	ꮗ
5...	III II	Ⴑ	50...	∩∩∩ ∩∩	⌐	500...	℮℮℮ ℮℮	⌐	5,000...	⚘⚘⚘ ⚘⚘	ꮗ ꮗ
6...	III III	ꭡ	60...	∩∩∩ ∩∩∩	ᚎ	600...	℮℮℮ ℮℮℮	⌐	6,000...	⚘⚘⚘ ⚘⚘⚘	ꮗ
7...	IIII III	Ⴣ	70...	∩∩∩∩ ∩∩∩	⅃	700...	℮℮℮℮ ℮℮℮	⌐	7,000...	⚘⚘⚘⚘ ⚘⚘⚘	ꮗ
8...	IIII IIII	=	80...	∩∩∩∩ ∩∩∩∩	ꮗ	800...	℮℮℮℮ ℮℮℮℮	⌐	8,000...	⚘⚘⚘⚘ ⚘⚘⚘⚘	ꮗ
9...	III III III	ꮶ	90...	∩∩∩ ∩∩∩ ∩∩∩	ꮃ	900...	℮℮℮ ℮℮℮ ℮℮℮	ꮑ	9,000...	⚘⚘⚘ ⚘⚘⚘ ⚘⚘⚘	ꮗ

Fɪɢ. [2]-1.—Eɢʏᴘᴛɪᴀɴ Nᴜᴍᴇʀᴀʟꜱ

number. The representations for the different symbols have been variously interpreted: 1, vertical staff; 10, heel mark or arch; 100, scroll, coil, or chain; 1,000, lotus flower; 10,000, pointed finger; 100,000, burbot (fish) or tadpole; 1,000,000, an astonished man with arms outstretched or a picture of the cosmic deity Hh. The hieroglyphic and hieratic forms of the symbols are pictured in Figure [2]-1.

Since the simple additive principle was used, the symbols could appear in any order. In later years a multiplicative principle was used as an aid in writing large numbers; for example, 120,000 was represented by the symbols for 120 placed before the lotus flower.

The hieroglyphic form was used mainly on monuments of stone, wood, or metal. A more cursive writing script was developed when other media, particularly papyrus, came into use for writing. As a result of the more rapid use of the reed pen, this representation was more rounded. This form, known as the hieratic (religious or priestly), evolved into a somewhat more abbreviated style, known as the demotic (popular), by about the eighth century B.C.

The hieratic form and the later demotic form of the numerals represented a different system from the hieroglyphic in principle as well as in appearance. A collection of similar symbols (seven vertical strokes, for example) was now represented by a single symbol. Thus a distinctive mark was used for each of the numbers 1 through 9, for each of the first nine multiples of 10 and of 100, etc. This sys-

39

tem, which is sometimes referred to as a cipherization or coding system, thus required no more than three symbols to express any number less than 1,000—one for units, one for tens, and one for hundreds. Again, no symbol for zero was necessary.

There were, of course, some variations in the symbols from one scribe to another. The additive and repetitive principles were sometimes invoked in special ways: in one of the forms the symbol for 4 was one horizontal stroke and the symbol for 8 was two horizontal strokes. The demotic system represented no great change over the hieratic except for some greater simplicity in the form of the symbols.

The important mathematical documents of ancient Egypt which are available to us now were written on papyrus and made use of the hieratic system of numerals [30].

In Egyptian mathematics the use of fractions was restricted almost entirely to the so-called unit fractions, those with a numerator of 1. In hieroglyphics such unit fractions were represented by placing the symbol $\bigcirc$ over the symbol for the denominator. Special symbols were used for ½ and ⅔. In the hieratic form a dot was usually placed over the denominator to indicate a unit fraction, although again each of certain simple fractions, such as ½, ⅓, and ⅔, had its own symbol.

For Further Reading

BOYER (j) KARPINSKI: 5
CAJORI (d): I, 11–18 VAN DER WAERDEN: 17–20

Capsule 3   William Heck

ROMAN NUMERALS

VERY little is definitely known concerning the origin of the Roman notation for numbers. The Romans never used the successive letters of their alphabet for numeration purposes as was the practice in some other early civilizations.

Before the ascendancy of Rome (that is, until about 500 B.C.), the Etruscans ruled the city. The Etruscans used numerals that resemble the letters of their own alphabet and also the numerals used by the Romans. In the earliest inscriptions on stone monuments the "one" was a vertical stroke. The "five" was generally V, perhaps representing a hand. This naturally suggests X (two V's) for "ten." But it is also possible that the use of the X came from crossing off the ones, as is done in tallying today. If this was the origin of X for "ten," then V (and the sometimes-used inverted V) for "five" would follow naturally as half of X. There is no definite information on the origin of L for "fifty." The Roman word for "one hundred" was *centum*, and that for "one thousand" was *mille*. This may account for the use of C for "one hundred" and M for "one thousand." An early symbol used for that was C|Ɔ, and the D for "five hundred" may be taken from the right-hand portion of the symbol.

The dominating feature in Roman notation is the principle of addition. (The systematic use of the subtractive principle, illustrated in writing IV for 4 and IX for 9, was seldom applied by the old Romans or even during medieval times; its use appears to have become common only after the invention of printing with movable type.) There are sporadic occurrences in Roman notation of the principle of multiplication, according to which VM does not stand for 5 less than 1,000, but for 5 · 1,000, or 5,000. The thousandfold value of a number was indicated in some instances by a horizontal line placed above it. Strokes placed on the tops and sides indicated hundred thousands.

Roman numerals were commonly used in bookkeeping in some European countries long after the modern Hindu-Arabic system became generally known. (In 1300 the use of Hindu-Arabic numerals was forbidden in the banks of certain European cities. The argument was that these numerals were more easily forged or changed than Roman numerals.) Use of Roman numerals continued in some schools until about 1600 and in bookkeeping for another century.

For Further Reading

CAJORI (d): I, 30–37   D. E. SMITH (a): II, 54–64

GREEK NUMERATION SYSTEM

Of the several systems of numeration used by the Greeks, two are worthy of mention.

The earlier system is known as the Attic (since the symbols occur frequently in Athenian inscriptions) or Herodianic (after the name of the writer who described them in the second century A.D.); it was used as early as 600 B.C. In this system I was used for 1, Γ for 5, Δ for 10, H for 100, X for 1,000, and M for 10,000. The last five symbols are simply the initial letters of the corresponding Greek number words, forms of which are preserved in English in the prefixes "penta-," "deca-," "hecto-," and "kilo-" and in the word "myriad." Some consolidation was accomplished by combining symbols. Figure [4]-1 shows, for example, how the symbol for 5 was used in combination with the symbols for 10 and 100 to indicate 50 and 500. This system used the additive principle, with any number represented by the minimum group of symbols whose values add up to the number. The representation of 10,517 is also shown in Figure [4]-1.

Γ̣	Γ̣	MᖴΔΓΠ
50	500	10,517

Figure [4]-1

The other noteworthy system of numeration employed by the Greeks is often called the Ionic system. This came into general use beginning about 200 B.C., although Athens retained the earlier system for another century.

The Ionic system is an additive base-ten system that employs twenty-seven symbols, as shown in the overview in Figure II-6. These symbols are the twenty-four letters of the Greek alphabet augmented by three additional forms from the Phoenician or obsolete Greek: *digamma* for 6, *koppa* for 90, and *sampi* for 900. Figure [4]-2

				β	γ
͵α	͵β	͵γ	M	M	M
1,000	2,000	3,000	10,000	20,000	30,000

Figure [4]-2

shows the extension of this system to indicate numerals in the thousands.

As can be seen in the figure, special devices were used to denote large numbers. Multiples of 1,000 through 9,000 were indicated by preceding each of the first nine letters with an accent or stroke attached to the lower or upper left. Tens of thousands were indicated in a multiplicative form, using the symbol M (myriad) for 10,000 below the numeral to be multiplied, as shown; and 120,000, for example, would be indicated by writing the symbol for 12, ιβ, above a M.

Various methods were used to distinguish numerals from words, an accent at the end of a number sign or a stroke over it being the two most commonly used.

Notation for fractions also varied considerably, and some fractions were subject to misconstruction. Unit fractions (numerator of 1) were ordinarily represented by the letter symbol for the denominator with an accent above it. Other fractions were represented sometimes by following the numerator with an accented denominator, sometimes by writing the accented denominator twice. Later the fractions were written in a fashion somewhat similar to modern notation—one numeral above the other—but here the denominator appeared on top, and usually there was no line between denominator and numerator.

For Further Reading

CAJORI (d): I, 21–29   HEATH (c): 14–24

Capsule 5   Barry D. Vogeli

CHINESE-JAPANESE NUMERATION SYSTEM

THE traditional Chinese numeration system is a base-ten system employing nine numerals and additional symbols for the place-value components of powers of ten. Some of the symbols used in the system, which dates back to the third century B.C. and was later adopted by the Japanese, are shown in Figure [5]-1. Numerals were written from the top downward or from left to right. For example, the combination

七百二十六

represents 7 · 100 + 2 · 10 + 6, or 726. Strictly speaking, the term "multiplicative principle" should not be applied here, since no place-value component stands by itself as a numeral /NEEDHAM: 13–15/.

1	2	3	4	5	6	7	8	9	10	10^2	10^3
一	二	三	四	五	六	七	八	九	十	百	千

FIG. [5]-1.—CHINESE-JAPANESE NUMERALS

At least as early as 200 B.C. the Chinese calculated by means of sticks laid on a table. The arrangements of the actual calculating rods led to forms for a second written system of numeration. This again was a base-ten system, but it required no place-value components. Referring to Figure [5]-2, it will be seen that

$$\pm \| \equiv \pi$$

represents 8,237 and

$$|T|| \equiv 0 -$$

represents 103,261.

Units Hundreds Ten thousands	0	1	2	3	4	5	6	7	8	9
	O	I	II	III	IIII	IIIII	T	⊤⊤	⊤⊤⊤	⊤⊤⊤⊤
Tens Thousands Hundred thousands	0	—	=	≡	≣	≣	⊥	⊥	⊥	⊥

FIG. [5]-2. SECOND NUMERATION SYSTEM

Before the eighth century A.D. the place where a zero would be required was always left vacant. A circular symbol for zero is first found in a document dating from 1247, but it may have been in use a hundred years earlier.

The modern commercial form of numeration, which has been in use since the sixteenth century, uses symbols similar to the counting-rod forms but with place-value components.

For Further Reading

NEEDHAM: 5–17

MAYAN NUMERATION SYSTEMS

ANTHROPOLOGISTS have shown that the Mayan culture of Yucatan and Central America was highly advanced in such areas as astronomy, the calendar, architecture, and commerce. This culture began its great advance around the fourth century A.D., and it had mathematics as an important cornerstone.

Notable among the accomplishments of the Mayas was the development of a vigesimal (base-twenty) system of numeration with positional notation and a special symbol for zero.

Two basic types of numeration systems were used, each with variations. One employed distinctive head-variant symbols or hieroglyphs for the numbers 0 through 13, with the numbers 14 through 19 (occasionally also 13) formed by affixing the lower jaw of a death's-head symbol for 10 to one of the symbols for 4 (or 3) through 9. /MIDONICK: 467–69./

More commonly used was the second system, which employed a dot (pebble) for 1, a bar (stick or rod) for 5, and a special symbol for 0. This symbol, seen in Figure [6]-1, somewhat resembles a shell design; but it is thought to be, more probably, a conventionalized front view of a closed fist. Numbers from 1 through 19 were represented additively by using appropriate combinations of the dots and bars symbolizing 1 and 5, 19 being represented by four dots (1's) and three bars (5's). At 20 positional notation began, with numerals read vertically from top to bottom; 20 was represented by a dot over the zero symbol.

The straight vigesimal system was used for civil and business purposes; but it was modified to simplify calendrical calculations, with the third positional value being considered as 360 rather than 400.

FIG. [6]-1.—MAYAN NUMERALS. The first column represents a value of 13 and the second (seven 20's plus 11) a value of 151. Using calendric count, with a value of 360 rather than 400 for the third position, the third and fourth columns represent a value of 360 and 7,202, respectively. For business purposes, they would represent 400 and 8,002.

(Successive places after the third had, again, a multiplicative value of 20.) This modification made a closer approach to the calendar year of 365 days.

The calendar year consisted of eighteen months of 20 days each and an additional "month" of 5 days. The sacred year of the Mayas, on the other hand, consisted of thirteen months of 20 days each. The interplay of these two systems must have led to a considerable amount of arithmetic computation. A Mayan stele (stone pillar bearing inscriptions) would often give an elaborate representation of a date, using both the simple dot-and-bar and the more complex head-variant forms; the latter were associated with the deities and patron gods of the various months.

For carrying out computations, the dot-and-bar system actually became a form of abacus, with number symbols physically being joined or taken away; multiplication and the inverse, division, were subject to such simple rules that they can be thought of as purely mechanical processes /SANCHEZ: 21–42/.

The development in the Mayan numeration system of the properties that have been described here forms an interesting sidelight in the history of mathematics, even though it did not affect the mainstream of Western civilization.

For Further Reading

BIDWELL

KINSELLA and BRADLEY

MIDONICK: 465–95

RICHESON

SALYERS

SANCHEZ

Capsule 7   Carl V. Benner

HINDU-ARABIC NUMERATION SYSTEM

IT HAS been commonly recognized that the numerals and the numeration system used in our daily life are of comparatively recent origin. The number of systems of notation employed before the Christian era was about the same as the number of written languages. These systems were primarily restricted to a form for recording the number of objects involved in all sorts of daily living and for recording the intermediate or final results of arithmetical computation performed with the assistance of tables or some mechanical aid. Not only does our present system permit a simple and concise way of recording numbers; even more importantly, the notation itself assists with tremendous power in performing computations.

The essential components of our system are the basic symbols, which include a symbol for the zero concept, and their incorporation into a positional or place-value system. One wishes that the history of the origins of this notation, now most commonly designated as the Hindu-Arabic system, contained details of its development. Unfortunately, the details of its exact formation are missing, and one can only conjecture about the manner in which the transition from older notations took place. It is possible, however, to perceive a number of factors that may have assisted in this development.

There are examples of symbols, a few of which are at least somewhat similar to those of today, which have been found in India; some of the earliest are found on stone columns of a temple built during the time of King Asoka, a Buddhist, about 250 B.C. However, no symbol for zero appears, and the place-value principle was not used at that time. Different symbols were apparently used for the various multiples of 10, 100, and so on.

Later the mathematics of the Hindus was often presented in versified style. Word and letter forms were used for representing numbers, and these were combined with a series of different words for the various powers of ten (compare with 327 as "three hundreds two tens seven"). No zero was necessary at this stage, since the corresponding positional value would simply be omitted if it were not needed. If the names of the position-places were omitted throughout, however, obvious difficulties arose; so a zero symbol became necessary.

The exact influence of the Babylonians, Greeks, and possibly even the Chinese on the Hindus has not been determined. The earlier Babylonian sexagesimal system was essentially a positional system, although the zero concept was by no means fully developed. When the Greeks continued the development of astronomical tables, they explicitly chose the Babylonian sexagesimal system to express their fractions, rather than adopting the unit-fraction system of the Egyptians. The repeated subdivision of a part into 60 smaller parts necessitated that sometimes "no parts" of a given unit were involved, and so a symbol had to be used for such a designation [8]. The Chinese, as early as the Shang period (before 1000 B.C.), used only nine symbols, with base-ten place-value components, and it has been suggested that the empty blanks of the Chinese counting boards could have been the stimulus for the invention of a zero symbol. /NEEDHAM: 5–17./

Specifically, very little is known concerning the exact mode of evolution of the Hindu notation, and no one knows what suggested certain of the early numeral forms used in India. Most historians have placed the final development of this system, with full and systematic use of the zero and the principle of place value, probably sometime between the fourth century A.D. and the seventh century.

About 800 the system had been brought to Baghdad and adopted by the Arabs, who were to play an important role in familiarizing other parts of the world with it. The Arabs never laid claim to the invention, always recognizing their indebtedness to the Hindus both for the numeral forms and for the distinguishing feature of place value. The Persian al-Khowarizmi, writing about 825, gave an account of the system in which he specifically ascribed it to the Hindus.

The movement of the Arabs across the northern shores of Africa and then up to Spain carried with it these Arabic works as well as Arabic translations of many of the important Greek works. Although the original Arabic work by al-Khowarizmi is lost, a twelfth-century Latin translation, probably by Adelard of Bath, an English monk, is extant. This work, entitled *Liber algoritmi de numero Indorum* and usually referred to as *Liber algorismi,* brought the system and its computational forms to the Western world. The opening words of the translation, *Algoritmi dixit* ("Al-Khowarizmi says"), gave rise to the term "algorithm" (or "algorism") for various sorts of computational processes.

Among the thirteenth-century works of Europeans which were to aid in describing the new system were the *Liber abaci* of Fibonacci (Leonardo of Pisa) and the *Algorismus vulgaris* of Sacrobosco (also known as John of Halifax and John of Holywood). Manuscript copies of these works written by students of mathematics of the thirteenth to fifteenth centuries are found in many European libraries. Standardization of the forms of the digit symbols was a result of the invention of printing with movable type in the middle of the fifteenth century, after which time many commercial arithmetics were printed. (Theories that have attempted to associate the number of strokes in the form of a numeral with the number represented by it have no foundation in historical fact.)

Interestingly enough, the forms of the modern Arabic numerals are not the same as the Hindu-Arabic forms of the Western world. For example, their numeral representation for five is 0, and their zero is represented by a dot.

The use of decimal fractions was not a part of the original Hindu development. The first systematic treatment of these was given in 1585 by Simon Stevin [36].

For Further Reading

DATTA and SINGH: 9–121   SMITH and KARPINSKI
KARPINSKI: 38–60   VAN DER WAERDEN: 51–61
NEEDHAM: 5–17

Capsule 8   Lloyd C. Merick, Jr.

ORIGIN OF ZERO

ALTHOUGH the great practical invention of zero has often been attributed to the Hindus, partial or limited developments of the zero concept are clearly evident in a variety of other numeration systems that are at least as early as the Hindu system, if not earlier. The actual effect of any one of these earlier steps in the full development of the zero concept—or, indeed, whether there was any actual effect —is by no means clear, however.

The Babylonian sexagesimal system [1] used in the mathematical and astronomical texts was essentially a positional system, even though the zero concept was not fully developed. Many of the Babylonian tablets indicate only a space between groups of symbols if a particular power of sixty was not needed, so the exact powers of sixty that were involved must be determined partly by context. In the later Babylonian tablets (those of the last three centuries B.C.) a symbol was used to indicate a missing power, but this was

49

used only inside a numerical grouping and not at the end. When the Greeks continued the development of astronomical tables, they explicitly chose the Babylonian sexagesimal system to express their fractions, rather than the unit-fraction system of the Egyptians. The repeated subdivision of a part into 60 smaller parts necessitated that sometimes "no parts" of a given unit were involved, so Ptolemy's tables in the *Almagest* (c. A.D. 150) include the symbol $\bar{o}$ or $\bar{o}'$ for such a designation. Considerably later, approximately 500, Greek texts used the omicron, o, first letter of the Greek word *ouden* ("nothing"). Earlier usage would have restricted the omicron to symbolizing 70, its value in the regular alphabetic arrangement [4] /Aaboe (b): 104/.

Perhaps the earliest systematic use of a symbol for zero in a place-value system is found in the mathematics of the Mayas of Central and South America [6]. The Mayan zero symbol was used to indicate the absence of any units of the various orders of the modified base-twenty system. This system was probably used much more for recording calendar times than for computational purposes.

Perhaps the earliest Hindu symbol for zero was the heavy dot that appears in the Bakhshali manuscript, whose contents may date back to the third or fourth century A.D., although some historians place it as late as the twelfth. Any association of the more common small circle of the Hindus with the symbol used by the Greeks would be only a matter of conjecture.

Since the earliest form of the Hindu symbol was commonly used in inscriptions and manuscripts in order to mark a blank, it was called *sunya*, meaning "void" or "empty." This word passed over into the Arabic as *sifr*, meaning "vacant." This was transliterated in about 1200 into Latin with the sound but not the sense being kept, resulting in *zephirum* or *zephyrum*. Various progressive changes of these forms, including *zeuero*, *zepiro*, *zéro*, *cifra*, and *cifre*, led to the development of our words "zero" and "cipher." The double meaning of the word "cipher" today—referring either to the zero symbol or to any of the digits—was not in the original Hindu. In early English and American schools the term "ciphering" referred to doing sums or other computations in arithmetic [7].

For Further Reading

Boyer (1)

D. E. Smith (a): II, 69–72

Van der Waerden: 56–57

NUMBER BELIEFS
OF THE PYTHAGOREANS

For Pythagoras and his followers the fundamental studies were geometry, arithmetic, music, and astronomy. The basic element of all of these studies was number—not in its practical, computational aspects, but as the very essence of their being.

As Aristotle put the matter (*Metaphysics* i. 5. 985^b):

> Since, then, all other things seemed in their whole nature to be modeled on numbers, and numbers seemed to be the first things in the whole of nature, they [the Pythagoreans] supposed the elements of number to be the elements of all things, and the whole heaven to be a musical scale and a number. And all the properties of numbers and scales which they could show to agree with the attributes and parts and the whole arrangement of the heavens, they collected and fitted into their scheme; and if there was a gap anywhere, they readily made additions so as to make their whole theory coherent.

The beginnings seem to have been in Pythagoras' study of the tones produced by a lyre string. The concordant intervals of the musical scale can be expressed in terms of the ratios of the corresponding lengths of the string, which are 1:2 for the octave, 2:3 for the fifth, and 3:4 for the fourth. The smallest whole numbers that fit this pattern are 6, 8, 9, and 12; 9 is the arithmetic mean of 6 and 12; 8 is the harmonic mean of 6 and 12. The sum of the numbers 1, 2, 3, and 4 forms the triangular number 10, the "perfect number," which in turn generates all the other combinations of number and figure that make up the cosmos. For the Pythagoreans a point was one, a line two, a surface three, and a solid four, since these are the minimum number of points necessary to define each of these dimensions. And so ten, the sum of these numbers, was a sacred and omnipotent power, the holy *tetraktys* ("fourness": 1 + 2 + 3 + 4), by which they took their oath. All of this led to the "harmony of the spheres" and, more generally, the belief that the things of the world are arranged and held together by principles of harmony and number.

51

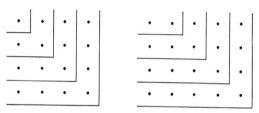

FIG. [9]-1.—GNOMON PATTERNS

Thus the roots were in the scientific observation of sense data. But from these roots grew the Pythagorean mysteries. "One" (the monad) is the beginning of number. Numbers that can be divided into two equal parts are even; those that can be divided only into two unequal parts are odd. Odd numbers are associated with the limited, even numbers with the unlimited. And the pairings bring harmony and balance: the odd-limited-good versus the even-unlimited-evil, the male versus the female, the one versus the many. It was clear to the Pythagoreans that the odd numbers were good because odd numbers added to 1 gave perfect squares in the gnomon pattern shown in Figure [9]-1: $1 + 3 = 2^2$, $1 + 3 + 5 = 3^2$, and so forth. In contrast, as the second pattern shows, even numbers added to 2 give rectangles: $2 + 4 = 3 \times 2$; $2 + 4 + 6 = 4 \times 3$, and so forth. (See also [10].)

Perceptive accounts of the convictions of the Pythagoreans have been given /DE SANTILLANA: 51–80; VAN DER WAERDEN: 92–102/; it is hard for us to recapture these convictions. Yet as one places himself among the Pythagoreans, he finds himself among contemporaries. Not again, at least until the time of Galileo, do people begin to look for the essence of things in number. Galileo, like the modern scientist, finds a model for his observations in a set of numbers,

$$\left\{ (s,\ t) \mid s = \frac{gt^2}{2} \right\},$$

the mathematical expression for the free fall. We have progressed since Galileo. As one thinks of energy as a probability distribution he finds himself almost, if not quite, back among the Pythagoreans.

For Further Reading

DE SANTILLANA: 51–80
SHANKS: 121–37

D. E. SMITH (a): II, 69–77
VAN DER WAERDEN: 92–102

FIGURATE NUMBERS

THE geometrical or physical representation of numbers by points (or pebbles) in a plane and the investigation of their resulting properties was a natural study for the early Pythagoreans. Special cases of these figurate numbers were the polygonal numbers, the numbers whose geometric representation took on the form of the various polygons.

The simplest polygonal numbers are the triangular and square numbers. The triangular numbers represent the successive sums of the consecutive counting numbers: $1; 1 + 2; 1 + 2 + 3; \ldots ; 1 + 2 + 3 + \ldots + n$. The square numbers, of course, are $1, 4, 9, \ldots , n^2$. Triangular and square numbers are shown, with dots, in Figure [10]-1; in the early manuscripts α's were used.

The most complete discussion of these numbers was given by a Greek, Nicomachus of Gerasa (*c.* A.D. 100), in *Introductio arithmetica*, the earliest extant manuscript of which dates back to the tenth century. It is quite generally accepted that there is little in this work which is original with Nicomachus himself; but his work brought together the results of previous generations in a reasonably clear and concise manner, judged by the standards of that day, and his "art of arithmetic" was to remain the standard work of its class for many centuries.

Nicomachus presents his material in the form of definitions and statements of general principles, with explanations and many illustrations. He does not include proofs or demonstrations of his results in the sense of Euclid; reliance on the physical form and the illustrated cases was apparently deemed sufficient proof. His generalizations in some matters (e.g., perfect numbers [12]) are not correct.

I	3	6	10	I	4	9	16
T_1	T_2	T_3	T_4	S_1	S_2	S_3	S_4

FIG. [10]-1.—TRIANGULAR AND SQUARE NUMBERS

53

Today we find it convenient to introduce simple algebraic symbolism to aid in expressing and proving various results. For example, the square numbers may be represented as $S_1 = 1$, $S_2 = 4$, $S_3 = 9$, $S_n = n^2$. But the basic relations were given by Nicomachus in words, not in symbols: "If you add any two consecutive triangles that you please, you will always make a square, and hence, whatever square you resolve, you will be able to make two triangles of it."

The following results are implicit in the work of Nicomachus. The student of elementary algebra today should have little difficulty in establishing them, algebraically as well as geometrically.

1.

$$1 + 2 + 3 + \cdots + n = n(n + 1)/2.$$

Hence

$$T_n = n(n + 1)/2.$$

2. The sum of consecutive odd numbers beginning with 1 is a square number:

$$1 + 3 + 5 + \cdots + (2n - 1) = n^2$$

$$= S_n.$$

(See Fig. [10]-2.)

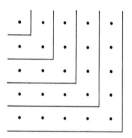

Fig. [10]-2.—Sums of Consecutive Odd Numbers

3.

$$S_n = T_n + T_{n-1}.$$

4. Eight times any triangular number plus 1 is a square number (given by Plutarch, *c.* A.D. 100).

54

5. The pentagonal numbers, $P_1 = 1$, $P_2 = 5$, $P_3 = 12$, $\cdots$, have been represented physically in two forms, as seen in Figures [10]-3 and -4. The basic pentagonal form readily illustrates the relation

$$P_n = 3 \cdot T_{n-1} + n,$$

which yields the result

$$P_n = (n(3n - 1))/2.$$

The second form does not emphasize the pentagonal form but does illustrate the relation (easily satisfied algebraically but less so geometrically)

$$P_n = S_n + T_{n-1}.$$

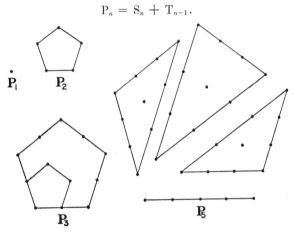

Fig. [10]-3.—Basic Pentagonal Form. $P_n = 3\,T_{n-1} + n$.

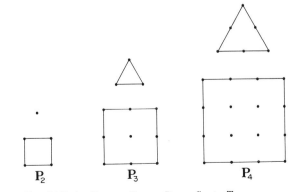

Fig. [10]-4.—Second Form. $P_n = S_n + T_{n-1}$.

55

6. Numbers of the form $n(n + 1)$, such as 2, 6, 12, 20, $\cdots$, can be defined as "oblong numbers." (Nicomachus used the term "heteromecic" for this special kind of rectangular number.) The sum of the first n positive even numbers is an oblong number: $2 + 4 + 6 + \cdots + 2n = n(n + 1)$.

7. Any oblong number is the sum of two equal triangular numbers.

8. Two successive square numbers plus twice the included oblong number produces a square number.

9. "When the successive odd numbers are set forth indefinitely, beginning with 1, observe this: The first one makes the potential cube: the next two, added together, the second; the next three, the third; the four next following, the fourth, and so on" /NICOMACHUS: Book II, 20/. That is:

$$1, \quad \underbrace{3, 5,}_{} \quad \underbrace{7, 9, 11,}_{} \quad \underbrace{13, 15, 17, 19,}_{} \cdots$$

$$1 \qquad 8 \qquad\quad 27 \qquad\qquad 64$$

$$1^3 \qquad 2^3 \qquad 3^3 \qquad\qquad 4^3$$

From this result, which has usually been credited to Nicomachus himself, the next follows easily:

$$1^3 + 2^3 + 3^3 + 4^3 + \cdots + n^3 = T_n^2$$
$$= (n(n + 1)/2)^2.$$

10. Nicomachus does not stop with pentagonal numbers but continues the pattern by writing down the following table:

Triangular	1	3	6	10	15	21	28	36	45	55
Square	1	4	9	16	25	36	49	64	81	100
Pentagonal	1	5	12	22	35	51	70	92	117	145
Hexagonal	1	6	15	28	45	66	91	120	153	190
Heptagonal	1	7	18	34	55	81	112	148	189	235
Octagonal	1	etc.								

Referring to the above table, Nicomachus /Book II, 12/ points out that

the pentagons are the sum of the squares above them in the same place in the series, plus the elementary triangles that are one place further back in the series; . . . the hexagonals are similarly the sums of the pentagons above them in the same place plus the triangles one place back. . . . Thus each polygonal number is the sum of the polygonal in the same place in the series with one less angle, plus the triangle, in the highest row, one place back in the series.

Symbolically, $Hx_n = P_n + T_{n-1}$; $Hp_n = Hx_n + T_{n-1}$; and so forth.

Although the generalization does not appear in the work of Nicomachus, Hypsicles (*c.* 180 B.C.) is credited with the result (here stated in modern form) that the nth k-gonal number is given by

$$\left(\frac{n}{2}\right)(2 + (n - 1)(k - 2)).$$

This geometrical sort of representation can be—and was—extended to space by forming pyramids (today one may think of piles of cannon balls or spheres of any kind). Successive pyramidal numbers with a triangular base are 1, 4, 10, 20, 35, and so forth; those with a square base are 1, 5, 14, 30, and so forth. Other bases are, of course, possible; but the result may be a truncated pyramid (a pyramid in which the top layer is not a layer of one).

Following the tradition of the Pythagoreans, Nicomachus does not consider unity, or 1, a number. He makes this statement: "Unity may appear to be potentially a triangle, and 3 the first actually." Similarly, unity is potentially a square, a pentagonal number, and so forth.

It should be noted that Euclid in the *Elements* does not consider such figurate numbers. For Euclid the product of two numbers (lengths or segments) represented an area; and the product of three, a volume. Thus Euclid's geometrical representation was not that of a pattern of points but of magnitudes of various sorts: segments, areas, and volumes.

Two other names warrant mention in this capsule.

The Greek Diophantus (*c.* A.D. 75 or 250), author of the classic *Arithmetica*, also wrote a treatise on polygonal numbers, only a fragment of which is known. In it he generalizes Statement 4 above. He also proposes this problem: Given a number, find in how many ways it can be a polygonal number. (Note, in the table given earlier, that 55 is both a triangular and a hepatagonal number; 81 is both square and heptagonal.) The manuscript breaks off in the middle of this problem.

Much later (*c.* 1636), the Frenchman Pierre de Fermat applied polygonal numbers to the summation of certain series.

For Further Reading

BEILER: 185–99

BELL (*b*): 161–65

BRADFIELD

HEATH (a): 124–27, 245–59

———— (c): 43–50, 66–69, 507–9

HOGBEN: *See index,* Polygonal
    Numbers
MIDONICK: 565–82

NICOMACHUS OF GERASA: 241–57
D. E. SMITH (a): II, 24–26
VAN DER WAERDEN: 98–100

Capsule 11   I. A. Barnett

AMICABLE NUMBERS

"AMICABLE numbers" are pairs of numbers such that each one is the sum of the proper divisors of the other. An equivalent definition is that two numbers are amicable if their sum is the sum of *all* the divisors of either of the numbers.

The smallest pair of amicable numbers is $220 = 2^2 \cdot 5 \cdot 11$ and $284 = 2^2 \cdot 71$. The proper divisors of 220 are 1, 2, 4, 5, 10, 11, 20, 22, 44, 55, and 110, which add up to 284; the proper divisors of 284 are 1, 2, 4, 71, and 142, which add up to 220.

A reference to amicable pairs is given by Iamblichus of Chalcis (*c.* A.D. 325), who ascribes his knowledge to the early Pythagoreans. Pythagoras reportedly said that a friend is "one who is the other I, such as are 220 and 284."

Amicable numbers were studied extensively by the Arabs, and by the ninth century Arab mathematicians had formulated rules for the discovery of amicable numbers. Results spread, through the Arabs, to western Europe.

Leonhard Euler made a systematic search for amicable numbers, and in 1747 he published a list of thirty pairs. Later he extended the list to sixty pairs. But in spite of his work and the work of many other mathematicians through the centuries one relatively small pair, 1,184 and 1,210, long escaped attention ($1,184 = 2^5 \cdot 37$, and $1,210 = 2 \cdot 5 \cdot 11^2$). A sixteen-year-old boy, Nicolo Paganini, discovered this pair in 1866. Over four hundred pairs have now been found, including 2,620 and 2,924; 5,020 and 5,564; and 6,232 and 6,368.

Recently mathematicians have extended this idea to amicable triplets—three numbers such that the sum of the proper divisors of any one of them is equal to the sum of the two remaining numbers.

A "chain" of numbers is said to be amicable if each is the sum of the proper divisors of the preceding number, the last being considered as preceding the first of the chain. The five numbers 14,288, 15,472, 14,536, 14,264, and 12,496 form an amicable chain of numbers.

For Further Reading

BARNETT (a)

BEILER: 26–30

DICKSON: I, 3–50

ORE (c): 96–100

ROLF

Capsule 12   I. A. Barnett

PERFECT, DEFICIENT, AND ABUNDANT NUMBERS

THE Greeks called a number such as 6 or 28 a "perfect" number because the sum of the proper divisors in each case is equal to the number; the proper divisors of 6 are 1, 2, and 3, and their sum is 6.

Although Euclid in his *Elements* did not actually name any perfect numbers, he gave the following theorem (Book IX, 36):

> *If as many numbers as we please beginning from a unit be set out continuously in double proportion, until the sum of all become prime, and if the sum multiplied into the last make some number, the product will be perfect.*

In essence Euclid is saying the following: Calculate the partial sums of the series $1 + 2 + 4 + 8 + 16 + \ldots$; if a sum turns out to be a prime, multiply it by its last summand and obtain a perfect number.

Thus, since $1 + 2$ is a prime, multiply it by the last summand, 2, and obtain the perfect number 6. Again, $1 + 2 + 4$ is a prime; multiply it by 4 and obtain the perfect number 28.

In modern symbolism, the above theorem states that a number N of the form $(2^n - 1) \cdot 2^{n-1}$ is a perfect number if the factor $(2^n - 1)$ is prime. If n assumes the value 2, 3, 5, or 7, the expression $(2^n - 1)$ takes on the value 3, 7, 31, or 127—all of which are prime. For these values of n we obtain the perfect numbers 6, 28, 496, and 8,128.

The Neoplatonists Nicomachus of Gerasa (c. A.D. 100) and Iamblichus of Chalcis (c. 325) listed these perfect numbers and concluded that they follow a pattern: They alternately end in a 6 or an 8, and there is one perfect number for each interval from 1 to 10, 10 to 100, 100 to 1,000, and 1,000 to 10,000. They conjectured that both parts of the pattern would continue, but in this they were wrong. The fifth perfect number (discovered in the fifth century) corresponds to $n = 13$ and is 33,550,336, with 8 digits rather than 6; and the sixth perfect number, like the fifth, ends with a 6.

In 1961, the twentieth perfect number was found. It contains 2,663 digits in its decimal representation and corresponds to the case where $n = 4,423$. Twenty-three perfect numbers are now known. The largest of these is $2^{11,212} (2^{11,213} - 1)$, which contains 6,751 digits. Whether there is an infinitude of perfect numbers is still an open question.

In 1757 Leonhard Euler, a Swiss mathematician, proved that every even perfect number must be of Euclid's form, given above.

It has been proved also that every even perfect number must end in 6 or 28; if it ends in 6, the digit preceding it must be odd /BARNETT (b): 15–16/. No one has as yet discovered an odd perfect number; but it is known that none exists below 10^{20}, and this limit may already have been extended.

In his *Arithmetica* Nicomachus also described "deficient" and "abundant" numbers. Deficient and abundant numbers are numbers that differ from perfect numbers in that the sum of the proper divisors is, respectively, less than or greater than the number. Thus 8 is a deficient number because $1 + 2 + 4$ is less than 8, while 12 is an abundant number because $1 + 2 + 3 + 4 + 6$ is greater than 12. The first six abundant numbers are 12, 18, 20, 24, 30, and 36. There are only twenty-one abundant numbers between 1 and 100, and all of them are even. The first odd abundant number is $945 = 3^3 \cdot 5 \cdot 7$.

The following are a few propositions that may be proved without much difficulty: (1) A prime or a power of a prime is a deficient number. (2) Any divisor of a perfect number or a deficient number is deficient. (3) Any multiple of a perfect number or an abundant number is abundant.

For certain abundant numbers the sum of the proper divisors may be a multiple of the number. For example, $120 = 2^3 \cdot 3 \cdot 5$, and the sum of the proper divisors is $1 + 2 + 3 + 4 + 5 + 6 + 8 + 10 + 12 + 15 + 20 + 24 + 30 + 40 + 60 = 240 = 2 \cdot 120$. A number N for which the sum of the divisors, S, is a multiple of N is called a "multiply-perfect" number of order k if $S = kN$.

The first three multiply-perfect numbers of order three are 120, 672, and 523,776. The problem of finding such multiply-perfect numbers was formulated by Marin Mersenne (1588–1648) in a letter to the philosopher and mathematician René Descartes. Some years later Descartes sent Mersenne a list of nine multiply-perfect numbers. Well over five hundred are now known.

For Further Reading

BARNETT (b): 15–16

BEILER: 11–25

DICKSON: I, 3–50

NEWMAN: I, 498–518

ORE (c): 71, 91–96, 359–59b

SIERPINSKI

VAN DER WAERDEN: 92–98

Capsule 13   I. A. Barnett

MERSENNE NUMBERS

A NUMBER of the form $2^n - 1$, $n = 1, 2, 3, \ldots$ is called a Mersenne number (denoted by M_n) after the Franciscan friar Marin Mersenne (1588–1648). If $2^n - 1$ is prime, it is called a Mersenne prime. It is easy to show that if $2^n - 1$ is prime, then n itself is prime. However, $2^n - 1$ is not necessarily prime for every prime number n; for example, for $n = 11, 2^n - 1 = 89 \cdot 23$.

Each Mersenne prime immediately leads to a perfect number [12]. Euclid's formula for a perfect number, $(2^n - 1) \cdot 2^{n-1}$, depends upon the first factor's being a prime number. This has been a motivation in the search for Mersenne primes.

In 1750 Leonhard Euler proved that M_{31} is prime. In 1876 the French mathematician Édouard Anatole Lucas showed that M_{127}, a number named by 39 digits, is prime. It has been conjectured (but not proved) that the number of Mersenne primes is infinite. Since there are certain tests that can be applied to determine whether a given M_n is prime or composite, the use of computers has permitted the checking of extremely large Mersenne numbers.

In the early 1950's $M_{2,281}$ was found to be prime, but $M_{8,191}$ (after a hundred hours of machine computation) was found to be composite. In 1961 the American mathematician Alexander Hurwitz found that $M_{4,253}$ is prime; this is the first known prime to possess more than 1,000 digits in decimal expansion. The machine time to test $M_{4,423}$ (leading to the twentieth perfect number, containing 2,663 digits) was about fifty minutes on an IBM 7090.

The largest known prime is the Mersenne prime $2^{11,213} - 1$, containing 3,376 digits.

For Further Reading

BEILER: 11–25   SHANKS: 14–29, 194–98

Capsule 14   Fred W. Lott

THE INFINITUDE OF PRIMES

IT IS natural to ask, concerning the set of primes, $\{2, 3, 5, 7, 11, \ldots\}$, whether or not the set is infinite. Is it possible that eventually one reaches a positive integer such that all those greater are composite numbers?

The answer to this question was known to the ancient Greeks, and the proof that the set of primes is infinite appears as Proposition 20 in Book IX of Euclid's *Elements*. (Many students of mathematics think of Euclid's writings as pertaining only to geometry and are surprised to find that this famous work also contains, in geometric language, information on algebra and the theory of numbers as well.)

The proof is remarkably simple and provides an excellent illustration of the indirect method of proof in mathematics. In essence, Euclid's method of proof is to suppose that the set of primes, $\{2, 3, 5, \ldots, p_n\}$, is finite. Form the number

$$N = (2 \cdot 3 \cdot 5 \cdot \ldots \cdot p_n) + 1.$$

The number N must be either prime or composite. If N is prime, then the supposition is false because we have found a prime not in the set (since it is greater than any one of them). If N is composite, let q be one of its prime factors. Now q cannot be any of the primes in the given set, since there is always a remainder of 1 when N is divided by each of these; hence the supposition that the set of primes is finite is not tenable. For example, $(2 \cdot 3 \cdot 5 \cdot 7 \cdot 11) + 1 = 2{,}311$, which is an additional prime;

$$(2 \cdot 3 \cdot 5 \cdot 7 \cdot 11 \cdot 13) + 1 = 30{,}031$$
$$= 59 \cdot 509,$$

both of which are new primes. Essentially, then, for any given set of primes, Euclid has given a method of finding one additional prime not in that set. This demonstration shows exactly what is meant when the set of primes is said to be infinite.

With this question settled, investigation turned to another: If p_n is the largest prime number that we know, how far beyond p_n will we have to look before we are certain to find another prime? Euclid's method assures us that the next prime, p_{n+1}, will not be greater than $(2 \cdot 3 \cdot 5 \cdot \ldots \cdot p_n) + 1$. Since the number $M = (2 \cdot 3 \cdot 5 \cdot \ldots \cdot p_n) - 1$ could just as well have been used in Euclid's proof, we can be sure that $p_{n+1} \leq (2 \cdot 3 \cdot 5 \cdot \ldots \cdot p_n) - 1$. In 1845 Joseph Bertrand verified that for any integer m (where $7 < m < 6{,}000{,}000$) there exists at least one prime between $m/2$ and $m - 2$. Betrand's postulate, as this came to be known, was proved by P. L. Tchebycheff in 1851. For any known prime, p_n, we can let $m = 2p_n$. Then there will be some prime, p, such that $p_n < p < 2p_n - 2$. This enables us to be certain that $p_{n+1} < 2p_n$, a result which is a substantial improvement over that of Euclid.

Not only is the set of primes infinite. In 1837 P. G. Lejeune Dirichlet showed that every set of integers of the form $an + b$, where a and b are relatively prime integers, contains an infinite number of primes. Thus the set of integers of the form $3n + 1$, that is, the set $\{1, 4, 7, 10, 13, \ldots\}$, contains infinitely many primes, as does the set of the form $6n + 5$, $\{5, 11, 17, 23, \ldots\}$, and so forth.

For Further Reading

TIETZE: 1–20, 121–55.

Capsule 15   James Fey

PRIME AND COMPOSITE NUMBERS

THE classification of positive integers greater than 1 as either prime numbers or composite numbers is of great importance in mathematics. The fundamental theorem of arithmetic is that every positive integer greater than 1 can be expressed uniquely as a product of powers of prime numbers (apart from the order in which the factors are written). Because of this theorem, many properties of integers can be reduced to properties of prime numbers, a consideration that makes study of prime numbers especially important.

The early Greeks excluded 1 (unity, the monad) from the set of primes because they did not even consider it a number; they considered it to be the *principle of number*, the beginning or generator of numbers. Euclid and Aristotle accepted 2 as a prime, but the earlier Pythagoreans did not. To them 2, the dyad, was not a number at all but only the *principle of "even."*

Today the customary exclusion of 1 from the set of prime numbers permits greater simplicity in stating theorems and formulas concerned with prime numbers.

Euclid made one of the first significant contributions to the theory of prime numbers when he proved that the set of primes is infinite [14]. Somewhat after Euclid's time, about 230 B.C., Eratosthenes developed the first systematic, although basically empirical, method of testing for primeness: the "sieve" that bears his name [19]. For almost twenty centuries thereafter few important results were obtained in the search for a general method of testing for primeness.

In 1638 Pierre de Fermat admitted knowing no method for testing the primeness of a number n except by trial of each number less than $\sqrt{n}$ as a divisor. He later expressed the belief that each number of the form $2^{2^n} + 1$, now referred to as the Fermat number F_n, is prime for integral values of n. In 1732 Leonhard Euler proved that F_5 (i.e., $2^{32} + 1$) is composite, and since then many more values of F_n have been shown to be composite. Actually, no values of F_n other than those for $n = 0, 1,$ 2, 3, or 4 have been found to be prime, and students of the problem now seem inclined to the opposite conjecture—that there are no Fermat primes other than the five already found. Surprisingly enough, the prime Fermat numbers are related to the conditions under which a regular polygon of n sides can be constructed by compass and straightedge only /ORE (c): 346–52/.

Considerable time was spent during the eighteenth and nineteenth centuries in testing numbers for primeness and in finding the prime factors of composite numbers. The result was a series of factor tables, beginning with that of Johann Heinrich Rahn, in 1659, which covered the numbers up through 24,000, and being climaxed by the still unpublished table of J. P. Kulik (1773–1863), which includes all numbers up to 100,000,000. Kulik spent twenty years of his life on the table, unassisted.

In 1772 Euler gave the formula $n^2 - n + 41$, which yields a prime number for all integral values of n from 0 to 40 but fails for $n = 41$. Similarly, in 1879 E. B. Escott gave the formula $n^2 - 79n + 1,601$, which gives primes for all $n = 0, 1, 2, \ldots, 79$ but fails in the case of $n = 80$. It has been proved that no such polynomial function can ever produce only primes.

Another question concerns the distribution of primes among the integers: What can be said about the number of primes occurring in a given interval? Jacques Hadamard and C. J. de La Vallée-Poussin independently proved, in 1896, that the sum of the natural logarithms of all primes $\leq n$ equals n asymptotically. From this follows the fundamental theorem that if $\pi (n)$ is the number of primes less than the integer n, then

$$\lim_{n \to \infty} \frac{\pi(n)}{n/\log_e n} = 1.$$

This fact had been conjectured earlier by Carl Friedrich Gauss after observation of tables of primes.

One of the fascinating features of prime-number theory is the simplicity with which some of its most difficult problems can be stated. A number of these simple relations remain conjectures, neither proved nor disproved. One example is the conjecture of Christian Goldbach, made in 1742 in a letter to Euler, that every even number ≥ 6 can be represented as the sum of two primes and every odd number ≥ 9 as the sum of three odd primes. It has also been conjectured that there are infinitely many pairs of "twin primes" of the form $p, p + 2$.

In 1961 the largest prime known was $2^{3,217} - 1$, expressed by 969 digits in decimal notation. By 1965 this had been replaced by

$$2^{11,213} - 1,$$

expressed by 3,376 digits. Although the use of computers has allowed mathematicians to investigate very large numbers for primeness and to test many conjectures about primes, the fruitful subject of prime numbers holds many problems that are answerable only in theory.

For Further Reading

Eves (f)

Ore (c): 50–85, 346–52

Capsule 16   James Fey

PYTHAGOREAN TRIPLES

A Pythagorean triple may be defined as any set of three positive whole numbers x, y, and z such that $x^2 + y^2 = z^2$. Familiar examples are (3, 4, 5) and (5, 12, 13). The search for such triples and for general formulas that generate them has had two main sources of motivation, extending back over thousands of years.

The converse of the Pythagorean theorem assures us that if we have three such numbers a triangle with corresponding numbers as the lengths of the sides has a right angle opposite the longest side. As a problem in number theory, there is the question of expressing any "square number" as the sum of two perfect squares.

One specific formula that is attributed to Pythagoras (c. 540 B.C.) by Proclus (c. A.D. 470) was probably derived from the properties of figurate numbers [10]. Any perfect square can be represented geometrically by a square array of dots, as shown in the first drawing in Figure [16]-1. The addition of a right-angle configuration to such a square produces, in turn, another square, as shown in the second drawing. Such a configuration (or the corresponding number) is called a "gnomon."

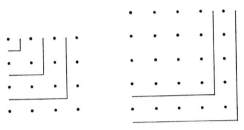

FIGURE [16]-1

This idea can be used to obtain a formula generating Pythagorean triples. We begin with any square figure representing n^2. The gnomon placed around this figure will be $2n + 1$, and the resulting figure will represent $(n + 1)^2$. Written in symbols,

$$n^2 + (2n + 1) = (n + 1)^2.$$

Since we are interested in the case where the number of units forming the gnomon $2n + 1$ is itself a square number, we use the following equation: $2n + 1 = m^2$. Then

$$n = (m^2 - 1)/2 \quad \text{and} \quad n + 1 = (m^2 + 1)/2;$$

and we obtain

$$m^2 + ((m^2 - 1)/2)^2 = ((m^2 + 1)/2)^2,$$

where m must necessarily be odd. Here, of course, the hypotenuse of the triangle is always 1 more than the longer of the two legs.

Another formula, which is attributed to Plato (c. 380 B.C.) and can be derived in a somewhat similar fashion /HEATH (c): 47–48/, is

$$(2m)^2 + (m^2 - 1)^2 = (m^2 + 1)^2.$$

This form is such that it appears it could be obtained from the equation previously given by multiplying by 4; but it is more general, since here m can be either even or odd.

A general formula for obtaining all primitive Pythagorean triples (such that the three numbers contain no common factor) was given by Euclid in his *Elements,* Book X, 28, Lemma 1. In modern notation it can be expressed as

$$(2uv)^2 + (u^2 - v^2)^2 = (u^2 + v^2)^2,$$

where u and v are relatively prime, $u > v$, and u and v are of different parity (one odd and one even).

The designation "Pythagorean triples" is misleading if one thereby places their first appearance at the time of Pythagoras. Neugebauer and Sachs /38–41/ have described the cuneiform tablet Plimpton 322, which dates from 1900 to 1600 B.C. This tablet clearly demonstrates that the Babylonians were familiar with at least fifteen different sets of triples. The size of some of these triples (e.g., 12,709, 13,500, 18,541) indicates that these were not obtained by trial and error. The method of arriving at the values is not indicated by the table, but from the particular sets of values given there is strong evidence that the Babylonians were familiar with the general parametric representation of the primitive Pythagorean triples, as given above, in the form attributed to Euclid /Eves (c): 35–37/.

On the other hand, the often-repeated statement that the Egyptian *harpenodoptai* ("rope stretchers") somehow used the 3, 4, 5 relationship to set off right angles for the construction of pyramids or temples is not based on historical proof. Thus far no documentary evidence has been found which shows that the Egyptians associated the $x^2 + y^2 = z^2$ relation with the sides of a right triangle.

The Hindu *Sulvasutras,* religious writings, include an indication that triples such as (15, 36, 39) were used in constructing temples between 800 and 500 B.C. There is no evidence that a proven general formula was used to obtain these triples.

Formulas somewhat similar to those given above were known to Diophantus of Alexandria and to Brahmagupta (*c.* 628) and Bhaskara (*c.* 1150) in India. However, their formulas were rather complicated special solutions to the problem answered in general by Euclid.

For Further Reading

Barnett (b): 48–49

Euclid: I, 356–60; II, 61–66

Eves (c): 35–37
    [5th ed. 27–30]

Hart

Jones (f)

Neugebauer (a): 36–42

Neugebauer and Sachs: 38–41

Ore (c): 165–208

Van der Waerden: 78–80

THE EUCLIDEAN ALGORITHM

THE Euclidean algorithm for finding the greatest common (integral) divisor, G.C.D., of two positive integers and testing two integers for relative primeness is found in Book VII of Euclid's *Elements*. (Euclid devoted the seventh, eighth, and ninth books of the *Elements* to arithmetic.) However, it is felt that much of the material is based on the work of the Pythagoreans and other predecessors and that the process was undoubtedly known much earlier.

The algorithm was applied to the solution of problems early in history. In the arithmetic of Aryabhata in India (c. A.D. 500) one finds indeterminate problems in several unknowns. Brahmagupta (c. 628) gave a perfected method for their solution, based on the Euclidean algorithm.

Simon Stevin (1548–1620) applied Euclid's method of finding the G.C.D. of a set of numbers to that of polynomials; for example, he found the G.C.D. of $x^3 + x^2$ and $x^2 + 7x + 6$ by the process of continual division. Johann Hudde (1633–1704) is the author of an ingenious method for finding equal roots of an equation (if they exist). His method included finding the G.C.D. of the original and a derived equation.

The French mathematician Gabriel Lamé (1795–1870) showed that the number of divisions in the algorithm, assuming decimal notation, is at most five times the number of digits in the smaller number. Leopold Kronecker (1823–1891) showed that no Euclidean algorithm can be shorter than the one obtained by least absolute remainders.

These are but a few of the many properties of Euclid's powerful algorithm, which has become a basic tool in modern algebra and number theory.

For Further Reading

CAJORI (e): 58, 148, 180   ORE (c): 41–49, 122, 142, 193
EVES (c): 119, 131 [5th ed. 107, 113]

INCOMMENSURABLES
AND IRRATIONAL NUMBERS

DETAILS concerning the discovery of the existence of incommensurable quantities are lacking, but it is apparent that the early Greeks found it as difficult to accept incommensurable quantities as to discover them.

Two segments are commensurable if there is a segment that "measures" each of them—that is, is contained exactly a whole number of times in each of the segments. The fact that there are pairs of segments for which such a measure does not exist provides the incommensurable case. This is reflected in later mathematics in the concept of irrational number, a number that cannot be expressed as the ratio of two integers. This concept was gradually clarified, as will be seen.

Although Proclus (c. A.D. 450) appears to ascribe the discovery of incommensurable quantities to Pythagoras, it was more probably made by some later Pythagorean in the period between 500 and 375 B.C. Since the philosophy of the Pythagorean school was that whole numbers, or whole numbers in ratio, are the essence of all existing things, followers of that school could only regard as a "logical scandal" the discovery and then the revelation of geometrical magnitudes whose ratio could not be represented by pairs of integers. Proclus wrote: "It is told that those who first brought out the irrationals from concealment into the open perished in shipwreck, to a man."

The earliest known proof, transmitted by Aristotle, concerns the diagonal and side of a square. It is an indirect proof, based only on the Pythagorean theorem and the fact that if the square of a number is even the number itself must be even.

Suppose that the diagonal AC and a side AB of a square are commensurable, with the ratio m to n expressed in smallest numbers. Then $AC^2/AB^2 = m^2/n^2$. By the Pythagorean theorem, $AC^2 = 2AB^2$, so $m^2 = 2n^2$. Since m^2 is even, m must be even; correspondingly, n must be odd because m and n have no common factor. Now let $m = 2h$; then $m^2 = 4h^2$, or $4h^2 = 2n^2$. Since $n^2 = 2h^2$, n^2 must be even, and therefore n is even. The conclusion that n is both odd and even is impossible, and therefore the diagonal and side cannot be commensurable.

70

It is possible that the first pair of segments found to be incommensurable was the side and diagonal of a regular pentagon, that figure being a favorite of the early Pythagoreans because its diagonals form the star pentagon, the distinctive mark of their society. The proof of this is again indirect /JONES (e): 123–27/, but it is more geometrical and involves an intuitive sort of limit argument. This proof may have been the approach of Hippasus (*c.* 400 B.C.), who was one of the Pythagoreans supposedly lost at sea.

This same geometric procedure can also be adapted to the side and diagonal of a square, discussed before. Here there is an association with the so-called side and diagonal numbers described by Plato (*c.* 400 B.C.). The ratios of associated pairs of these numbers give successively closer and closer rational approximations to $\sqrt{2}$; in fact, they are the approximations obtained by computing successive convergents of the continued fraction form of $\sqrt{2}$ /JONES (e): 187–91/.

Theodorus of Cyrene (*c.* 390 B.C.) is credited with showing that the sides of squares whose areas are the nonsquares 3, 5, 6, . . . , 17 are not commensurable with a side of 1. In modern language, we would say he proved the irrationality of the nonsquare integers from 3 through 17. There are several conjectures as to why he stopped at 17 /VAN DER WAERDEN: 141–46/.

The discovery of incommensurables resulted in a need to establish a new theory of proportions which would be independent of commensurability. This was accomplished by Eudoxus (*c.* 370 B.C.); it is his work that formed the basis for Book V of Euclid's *Elements*/HEATH (d): II, 116–24/. It is on the basis of the definition given by Eudoxus and of the so-called method of exhaustion, also attributed to Eudoxus, that Euclid could give a proof of a theorem such as that in Book XII, 2, "circles are to one another as squares on the diameter" /JONES (e): 282–84/.

The details of the transition from a theory of proportion which included incommensurable quantities to a clear realization of the concept of an irrational number would cover a wide range of mathematical topics: computational devices for finding approximations for roots of numbers, operations and manipulations of expressions involving radicals, the development of analytic geometry with the need to associate a number with every point on the line, algebraic symbolism and the theory of equations, and the development of the calculus, with the problems of limits and continuity, to list several of the possible topics /JONES (e): 469–71, 541–43/.

71

The approaches of two men, in the nineteenth century, played an important part in the development of the theory of irrationals. These men were Georg Cantor (1845–1918), who worked with fundamental sequences of rational numbers, and Richard Dedekind (1831–1916), who used the concept of a "partition" or "cut" of all rationals into two sets to define a real number. Both published first papers on these topics in 1872.

In mathematics today the studies of both algebra and analysis include detailed developments of the system of real numbers, of which the irrationals form an important subset.

For Further Reading

EUCLID: II, 116–24　　　　　　　　　NIVEN (b): 42–43, 52–61
JONES (e)

Capsule 19　Carl Benner

ERATOSTHENES

ERATOSTHENES was born about 275 B.C. in Cyrene, a Greek colony in Egypt. After studying in Athens, he returned to Alexandria and became one of its greatest scholars. A versatile genius, he contributed to geography, mathematics, poetry, literature, and history.

In arithmetic, Eratosthenes developed a sieve for finding prime numbers. One form of it follows:

> Write down the numerals for all the positive integers (to as far as you wish to go) in their natural order. Disregard 1. Starting with 2, cross out every second numeral after 2; starting with 3, every third numeral after 3. The next uncrossed numeral after 3 is 5. Retain 5, but cross out every fifth numeral after 5. Repeat this procedure for all uncrossed numerals after 5 (7, 11, 13, etc.), counting also the numerals that have

already been crossed out. Eventually, the remaining uncrossed numerals name prime numbers.

In the method actually used by Eratosthenes /MIDONICK: 572–73/ the work was simplified by writing down initially only the numerals for the odd numbers. Beginning with 3 and passing over two numerals at a time, the multiples 9, 15, 21, . . . were crossed out; beginning with 5 and passing over four numerals at a time, the multiples 15, 25, 35, . . . were crossed out; and so on. It has been conjectured that the method may have been carried out more mechanically by writing the numerals for the odd numbers vertically in sets of fifty (1 to 99) for each successive hundred. Since any multiple of 3 plus 102 is also a multiple of 3, moving one step to the right and one step down immediately locates another multiple of 3. Similarly, the drawing of various diagonal or oblique lines would automatically cancel out the respective multiples in successive or perhaps alternating columns. /FISHER./

In geometry, Eratosthenes developed a mechanical method for solving the famous Delian problem of the duplication of the cube [51]. A sliding mechanism automatically determined the two mean proportionals between two given line segments /VAN DER WAERDEN: 230–31/. The device could be adapted to find as many mean proportionals as desired.

Of even greater interest is Eratosthenes' method of investigating the size of the earth. (He was not the first to measure the earth; Aristotle and Archimedes, earlier, were aware of other measures.) Eratosthenes undertook to find the circumference of the earth by using the arc of a great circle passing through Syene (now Aswan) and Alexandria, cities that lie approximately on the same meridian. He took the length of this arc as 5,000 stadia. (How the length was ascertained is not stated in any of the ancient writings; but it is possible that the official pacers employed by Alexander and other military leaders in planning their campaigns had made reports of all such standard distances and that Eratosthenes, as librarian at Alexandria, had access to their records.)

He observed that at Syene at noon and at the summer solstice the sun cast no shadow from an upright gnomon. This was verified by noting that the sun penetrated to the bottom of a deep vertical well dug there. At Alexandria at the same time the inclination of the sun's rays to the vertical was one fiftieth of a complete circle, or

7°12′. Eratosthenes concluded that the circumference was 50 × 5,000 stadia, or 250,000 stadia. When converted by the usually accepted factors, this is equivalent to a diameter of 7,850 miles, which is only 50 miles less than the polar diameter as known today.

Eratosthenes is said to have estimated also the distances of both the moon and the sun from the earth, as well as the height of various mountains.

For Further Reading

FISHER

HEATH (c): 162–64, 343–46

MIDONICK: 572–73

NEWMAN: I, 205–7

VAN DER WAERDEN: 160–62, 228–34

Capsule 20   William B. Wetherbee

NUMEROLOGY AND GEMATRIA

THE idea of a symbolic correspondence between numbers and objects or philosophical concepts was held by almost all of the ancient cultures, and traces of it are still found today.

Such number mysticism has been promoted by many factors, including natural curiosity, superstition, search for entertainment, and even search for a philosophy of life. The forms of such number mysticism have varied all the way from the simple consideration of interesting number curiosities to a complete dependence upon number as determining one's destiny in life.

The Pythagoreans were strongly devoted to number speculation in both philosophy and nature. Pythagoras (*c.* 540 B.C.) supposedly studied Egyptian and Babylonian mathematics and mysticism and, with his followers, expanded this mysticism of the East. One form of their number study involved what might be called the geometrical aspects of number, and they studied such categories as the odd, even, triangular, square, and polygonal numbers [**10**]. Numbers also took on certain characteristics because of the nature of their divisors, and so perfect, abundant, deficient, and friendly numbers were studied [**11, 12**].

One of the forms of numerology prevailing over the years is that commonly designated as "gematria." This involves assigning number values to the letters of the alphabet in some systematic way. This scheme can be traced to the actual use of letters as numerals by Hebrew, Greek, and other Eastern civilizations.

If two names had the same numerical value when their letters were converted to numbers and then summed, the individuals designated by the names were considered to be somehow akin. Success or failure in various activities was augured from the pertinent number values.

The names of the Bible have been a favorite field for gematria. Most famous is the beast of the Book of Revelation, which was named by 666. The number appears in Rev. 13:18 and is described thus: "Here is wisdom. Let him that hath understanding count the number of the beast; for it is the number of a man, and his number is Six hundred three score and six."

For nearly two thousand years numerologists have been trying to fit men's names to 666. "Nero Caesar" was one of the first names which added up to 666 in the letter symbols of the Aramaic language. During the Middle Ages and the Reformation the idea of "beasting" one's archenemy was a favorite device; one could always resort to a slight change or omission in the spelling if this were needed to produce the desired result. Michael Stifel, one of the greatest German algebraists of the sixteenth century, predicted the end of the world on October 3, 1533, from an analysis of Biblical writings, and reasoned that Pope Leo X was the beast of the Book of Revelation. Some years later John Napier showed that 666 could stand for "the Pope of Rome," but Father Bongus, a Jesuit contemporary, declared that it stood for "Martin Luther." During World War I the numerical value of Kaiser Wilhelm's name was found to be 666.

Some old tombstones bear both a man's name and his gematria number.

When written in Greek the letters of the word "amen" add up to 99, and in certain Christian manuscripts the number 99 is written at the end of a prayer to signify "Amen."

For Further Reading

DANTZIG (b): 40–41   D. E. SMITH (a): II, 54

Capsule 21   Dorothy Schrader

AL-KHOWARIZMI

FROM an intellectual viewpoint, the ninth century A.D. was essentially a Moslem century. The activity of the scholars of Islam was far superior to that of any other group. Flourishing under Caliph al-Mamun, the Moslem mathematician, astronomer, and geographer Mohammed ibn-Musa al-Khowarizmi influenced mathematical thought more than did any other medieval writer on the subject. He is generally known for his work in algebra and astronomy and for his mathematical tables; but perhaps one of his greatest achievements was his introduction of the so-called Hindu-Arabic notational system to the Western world. His writings were the main channel by which the "new" numerals became known in the West, and through them he revolutionized the common processes of computation.

Al-Khowarizmi's best-known book, *Hisab al-jabr w'al-muqabalah* ("science of restoration [or reunion] and opposition"), was written to show "what is easiest and most useful in arithmetic such as men constantly require in cases of inheritance, legacies, partition, lawsuits, and trade, and in all their dealings with one another." This work, containing **79** pages of inheritance cases, **16** pages of measurement problems, and **70** pages of algebra, begins with a brief recapitulation of the place-value number system based on ten and then goes on to teach the use of these positional numerals in computation.

There is, for example, a systematic treatment of surds. Al-Khowarizmi noted that, to double a root, one must multiply the square by four; to triple it, by nine; and to halve it, by one-fourth. He gave examples of the multiplication and division of surds, concluding the series by saying, "Likewise with other numbers," thus indicating that the reader was to deduce a general rule from the examples given. He used this same technique in mercantile problems, where many of the examples imply a use of proportionality. "If ten items cost six [coins], what will four cost?" After seeing how several such problems are solved, one is to do "likewise with other numbers."

It should be noted that throughout the treatise numbers are expressed in words, not symbols, numerals being used only in some diagrams and a few marginal notes. The methods of computation thus taught verbally could be applied only to the then new place-value system of numeration used by the Moslems. Thus the promulgation of this work of al-Khowarizmi gave strong impetus to the scholars of the West to learn what we now call the numerals of the Hindu-Arabic system and to become skillful in manipulating them.

For Further Reading

KHOWARIZMI   SARTON (a): I
MIDONICK: 418–34   STRUIK (e): 55–60

Capsule 22   Sam E. Ganis

FIBONACCI NUMBERS

FIBONACCI, as Leonardo of Pisa is called, included the following problem in his *Liber abaci* (1202):

> What is the number of pairs of rabbits at the beginning of each month if a single pair of newly born rabbits is put into an enclosure at the beginning of January and if each pair breeds a new pair at the beginning of the second month following birth and an additional pair at the beginning of each month thereafter?

It is easily seen that the answer for the first year is 1, 1, 2, 3, 5, 8, 13, 21, 34, 55, 89, 144. These numbers are the first twelve terms of what is now called the Fibonacci sequence.

In 1611 Johann Kepler observed what certainly must have been known to Fibonacci himself, that $f_n + f_{n+1} = f_{n+2}$, where f_n is the nth Fibonacci number $(f_0 = 0, f_1 = 1)$.

Luca Pacioli published in 1509 a book called *De divina proportione*, which was illustrated by Leonardo da Vinci. In it he treated the number $(1 + \sqrt{5})/2$, which appears as a ratio in various plane and solid figures. This number is the positive root of the equation $x^2 - x - 1 = 0$, which may be obtained from the proportion $1/x = x/(x + 1)$. In its original context this is related to the problem considered by Euclid and the early Pythagoreans of dividing a segment in mean and extreme ratio [54]. The decimal expansion of $(1 + \sqrt{5})/2$ is 1.61803398 $\cdots$, which is often designated by the Greek letter ϕ (mathematicians also use the Greek letter τ). The reciprocal of ϕ is .61803398 $\cdots$, which is exactly $\phi - 1$, while ϕ^2 is 2.61803398 $\cdots$.

The striking relationship between the Fibonacci sequence and the "golden ratio" was first established by a Scottish mathematician, Robert Simson, in 1753. Simson proved that

$$\phi = \lim_{n \to \infty} \frac{f_{n+1}}{f_n}.$$

In 1843 J. P. M. Binet discovered the formula

$$f_n = \frac{\phi^n - (-\phi)^{-n}}{\sqrt{5}},$$

making the connection between Fibonacci numbers and the golden ratio explicit.

A classical geometrical paradox is to use three lines to dissect a checkerboard (8 squares on a side) into two triangles and two trapezoids, which are then reassembled into what appears to be a 5-by-13 rectangle of 65 squares. Charles Lutwidge Dodgson (pseudonym, Lewis Carroll) used the following variation of Simson's identity,

$$f_{n-1} \cdot f_{n+1} - f_n^2 = (-1)^n,$$

for the case $n = 6$, to explain the paradox.

Édouard Anatole Lucas (1842–1891) studied Fibonacci numbers in some detail and has credit for the following two identities:

$$\sum_{k=1}^{n} f_K = f_{n+2} - 1 \quad \text{and} \quad f_{n+1} = \sum_{k=0}^{n} \binom{n-k}{k}.$$

He also introduced the related Lucas numbers, g_n, defined by $g_0 = 2, g_1 = 1, g_n + g_{n+1} = g_{n+2}$. The first several terms of this sequence are 2, 1, 3, 4, 7, 11, 18, 29, 47, 76, 123. It is easy to to show that $g_n = f_{n-1} + f_{n+1}$, where the backward extension

$$f_{-n} = (-1)^{n+1} \cdot f_n$$

is made as required.

The identities derivable from Fibonacci and related numbers, many of which offer a number of nonobvious examples of proof by mathematical induction, are of great variety and appear in many branches of mathematics. Of special interest seem to be the number-theoretic properties, such as random numbers, primes, factorization properties, and so forth.

For Further Reading

ARCHIBALD (a)   SCHAAF (d): 49–50
GARDNER (a)   STRUIK (e): 2–4
————(c): 89–103   VOROBYOV

Capsule 23   James Fey

FERMAT'S LAST THEOREM

THE history of Pythagorean triples [16] goes back to 1600 B.C., but it was not until the seventeenth century A.D. that mathematicians seriously attacked, in general terms, the problem of finding positive integer solutions to the equation $x^n + y^n = z^n$.

The French mathematician Pierre de Fermat (1601–1665) conjectured that there are no positive integer solutions to this equation if n is greater than 2. Fermat's now-famous conjecture was inscribed in the margin of his copy of the Latin translation of Diophantus' *Arithmetica*. The note read:

> To divide a cube into two cubes, a fourth power, or in general any power whatever into two powers of the same denomination above the second is impossible, and I have assuredly found an admirable proof of this, but the margin is too narrow to contain it.

Despite Fermat's confident proclamation the conjecture, called "Fermat's last theorem," remains unproven. Fermat gave elsewhere a proof for the case $n = 4$. It was not until the next century that Leonhard Euler supplied a proof for the case $n = 3$ / STRUIK (e): 36–40/, and still another century passed before Adrien Marie Legendre and P. G. Lejeune Dirichlet arrived at independent proofs of the case $n = 5$. Not long after, in 1839, Gabriel Lamé established the theorem for $n = 7$.

In 1843 the German mathematician Ernst Kummer submitted a proof of Fermat's theorem to Dirichlet. Dirichlet found an error in the argument, and Kummer returned to the problem. After developing the algebraic "theory of ideals," Kummer produced a proof for "most small n." Subsequent progress in the problem has utilized Kummer's ideas, and many more special cases have been proven. It is now known that Fermat's conjecture is true for all $n < 4{,}003$ and many special values of n, but no general proof has been found.

Fermat's conjecture generated such interest among mathematicians that in 1908 the German mathematician P. Wolfskehl bequeathed DM100,000 to the Academy of Science at Göttingen as a prize for the first complete proof of the theorem. This prize induced thousands of amateurs to propose solutions, with the result that Fermat's theorem is reputed to be the mathematical problem for which the greatest number of incorrect proofs have been published. However, these faulty arguments have not tarnished the reputation of the genius who first proposed the proposition.

For Further Reading

BARNETT (b): 25–27   EVES (c): 195, 292–93, 311–12
BELL (b)   [5th ed. 175, 266–67, 281]

Capsule 24   William Heck and James Fey

MAGIC SQUARES

ALTHOUGH little is known concerning the early history of magic squares—square arrays of numerals in which the rows, columns, and diagonals have the same sums—they seem to be of Chinese origin. A common one, shown in Figure [24]-1 with modern numeral notation, is attributed to the semilegendary engineer-emperor Yu the Great (*c.* 2200 B.C.). The tradition is that once when Yu was standing by the Yellow River there appeared a divine tortoise, on whose back was the symbol now known as the *lo shu.* The symbol was first given in the *I Ching*, one of the oldest Chinese classics, written about 1100 B.C. Here the square array of numerals was depicted by means of knots in strings, black knots for even numbers and white knots for odd numbers.

4	9	2
3	5	7
8	1	6

FIGURE [24]-1

Magic squares gradually found their way to Japan, India, and the Middle East, usually retaining a connection with mysticism. They appeared in Arabia during the ninth century and in India during the eleventh century (or earlier), and they are to be found in Hebrew writings of the twelfth century. Today magic squares still serve as charms in Tibet, India, and much of southeast Asia.

The Byzantine writer Manuel Moschopoulos is credited with introducing magic squares to Europe during the fifteenth century. The mystical squares were connected with alchemy and astrology, and a magic square engraved on a silver plate was used as a charm against the plague.

One of the first magic squares to appear in print was shown in Albrecht Dürer's well-known engraving of about 1500 entitled "Melancholia." This had four numerals across and four down, and the sums were 34.

By the seventeenth century the mathematical theory of constructing magic squares was studied seriously in France. Antoine de La Loubère, envoy of Louis XIV to Siam in 1687–88, discovered a method of constructing a magic square of any odd order. In 1686 Adamas Kochansky, a Pole, succeeded in extending magic squares to three dimensions.

Interest in magic squares revived during the latter part of the nineteenth century, and the squares were applied to problems in probability and analysis. The related subject of Greco-Latin squares, pioneered by Leonhard Euler, has recently produced important applications in the design of experiments. Thus an idea with its origins deep in mysticism, frequently considered as a mere pastime, has become a useful part of contemporary mathematics.

For Further Reading

Eves (c): 48–50
[5th ed. 179–80]
Gardner (c): 130–40

D. E. Smith (a): I, 28–29; II, 591–98

Capsule 25   Leland Miller

LARGE NUMBERS

In a letter to King Gelon of Syracuse, Archimedes posed this question: How many grains of sand would it take to fill the entire universe? Under the assumption of a finite universe, Archimedes constructed a geometric proof to show the number must be finite.

But the Greek numeration system of that time (Attic or Herodianic—see [4]) was inadequate to represent the large numbers required; it went only as far as one myriad (10,000). Archimedes, in order to express the large number that he claimed exceeded the number of grains of sand in the universe, developed an extension of the traditional system. This extension enabled him to express numbers that in decimal notation would require up to 80 thousand million million digits, although in his proof he needed only 10^{51}. His method can be carried farther to name numbers as large as desired.

The Archimedean numeration system was a significant accomplishment in the attempt to express large numbers. Most earlier systems were very limited in scope. For instance, the largest uncompounded number in the Hebrew version of the Old Testament is *r'vavah*, which is equal to 10,000. Nonetheless, simple expression of arbitrarily large numbers was not possible until the introduction of place-value numeration.

Although base-ten numeration has been in use in the West for more than eight hundred years, the assigning of common names to large numbers has taken place very slowly. The word "million," meaning a thousand thousand, appeared first in Italy during the thirteenth century. The word "billion" is of French origin, and it was not adopted in English until the seventeenth century. Today these common names go up to "vigintillion" (10^{63}).

Beyond vigintillion, one is at liberty to choose his own names for large numbers. Edward Kasner has recently introduced the names "googol" and "googolplex":

$$1 \text{ googol} = 10^{100}.$$

$$1 \text{ googolplex} = 10^{10^{100}} = 10^{\text{googol}}.$$

Kasner claims that the number of grains of sand on the beach at Coney Island is much less than a googol—a far less pretentious claim than that of Archimedes twenty-two hundred years earlier.

For Further Reading

ARCHIMEDES: 221–32

P. J. DAVIS (b): 22–26

HEATH (c): 19–20

NEWMAN: I, 420–29

READ (a)

SMITH (a): II, 80–88

Capsule 26   James Fey

ALGEBRAIC AND TRANSCENDENTAL NUMBERS

AN "ALGEBRAIC number" is a number, real or complex, that satisfies a polynomial equation of the form

$$a_n x^n + a_{n-1} x^{n-1} + \cdots + a_1 x + a_0 = 0,$$

where the a_k are integers. A number that is not algebraic is called "transcendental." Many real and complex numbers are algebraic—in

83

numbers underwent a series of technical refinements under the leadership of Russell, Ernst Zermelo, and John von Neumann.

However, Cantor's conception of transfinite number remains fundamental.

For Further Reading

G. CANTOR   WILLERDING
NEWMAN: III, 1576–90, 1593–1611

III

The History of COMPUTATION

an overview by

HAROLD T. DAVIS

T he art of computation originated in the basic needs of human life and thus is found in the earliest records of the race. It was practiced by the ancient Egyptians and the Babylonians. It appears in the oldest traditions of the Orient. The art has been discovered even in such an isolated culture as that of the Mayas.

THE ORIGIN OF COMPUTATION
AND THE NEED FOR IT

The origin of computation, or calculation, is indicated by the origin of the very word "calculate," which is derived from the Latin *calculus* and is related to the Greek *chalix*, both of which mean a small stone or pebble. The use of pebbles is mentioned by Herodotus (fifth century B.C.), who said: "The Greeks write, and calculate with pebbles, by moving the hand from left to right; the Egyptians do contrariwise."

Early peoples found two primary needs for calculation. The first was a need for the basic enumerations of business transactions such as the counting of herds, the exchange of money, and the apportionment of land. Another was the need for a calendar by which man

87

could keep an account of the seasons. "At the time when the Pleiades are rising, begin your harvest," said Hesiod (eighth century B.C.), "and plow again when they are setting. The Pleiades are hidden forty nights and forty days."

Man thus turned his eyes to the sky. The movement of the moon was carefully measured. Among the stars primitive astronomers soon noticed five "wanderers." These seemed to have motions of their own as they wandered among stars whose movements seemed fixed in relation to one another. For any such body the Greeks used the word *planes* ("wanderer"), from which we derive our word "planet."

In Egypt, where the flooding of the Nile was a matter of primary importance, the natives needed to determine the time of the beginning of the inundation caused by the great river. Attention was therefore fixed upon the heliacal rising of Sirius (that is, the time when Sirius appears with the sun), since this coincided very nearly with the beginning of the flood. This computation later became very important in fixing some of the chronology of Egyptian history.

We see in these ancient records the origin of our own spectacular development of computation. The use by modern business, engineering, and science of the great calculating engines of today is a development of the primitive applications of the art of computing. We shall now try to set forth in its historical perspective the story of how all these modern wonders have come about.

DEVELOPMENT OF NUMERATION SYSTEMS

The first requirement in computation is a system of numeration—a way of writing numbers. This must be sufficiently flexible for arithmetical operations and yet comprehensive enough to include the application of the most advanced analysis. Some of the steps in the development of a satisfactory numeration system are worth noting in this context. (For details concerning specific numeration systems, the reader should refer to Chapter II and its capsules.)

Since it is obviously impossible to assign a special symbol to every integral number, a primary requirement is the choice of a convenient radix, or base, for the system. Since man was created with ten fingers, the most obvious choice is ten; and this base was widely used by primitive people—although not exclusively, as we shall see. (The word "digit" is itself derived from the Latin *digitus*, which means "finger.")

Many references in literature indicate the widespread use of finger symbolism for representing numbers and in reckoning. The Venerable

Bede (673–735), an English Benedictine monk, in a work on computing time, *De temporum ratione*, describes positions of the fingers to indicate numbers up to 10,000. By placing the hand in various positions on the body he was able to indicate numbers up to 1,000,000. This type of representation was used in many countries and was known among the early Greeks [29].

The wish has often been expressed that man had been created with twelve fingers instead of ten. Doubtless this would have resulted in a duodecimal system of numbers, which would have provided greater flexibility than the decimal system, since 12 has four factors and 10 has only two. With respect to this, however, we should note the curious fact that the binary system, based upon the radix 2, has now come into prominent use; this is the system used exclusively by the great calculating machines of today.

The choice of a particular base for a numeration system did not always depend upon a correspondence with certain parts of man's body, however. In some cases the number relations involved in other physical situations seem to have been a determining factor. One plausible explanation of the origin of the sexagesimal system among the Babylonians associates it with one of the systems of units of weights and measures, in which the larger unit was 60 times as large as the smaller [1] /AABOE (b):20; NEUGEBAUER (a):17–20/.

Among the Mayas a "modified" vigesimal system apparently resulted from the form of the Mayan calendar, which divided the Mayan year into 18 months of 20 days each, supplemented with 5 additional days. This gave successive places the values, after the units, of 20, 18×20, 18×20^2, etc. [6].

It should be pointed out that a given people did not necessarily adhere to a single base. The Babylonians, for example, used mixed systems in matters concerning dates and measures of weights and areas but a consistent sexagesimal place-value notation for the purely mathematical and astronomical texts. (Many of the Babylonian cuneiform tablets concern economic items. Some of these tablets contain modifications of the customary number symbols, including a symbol for 100 that is the same impression used for 10 except for being bigger.)

One of the many accomplishments of Archimedes (287–212 B.C.) was his development of a system to represent large numbers which made essential use of powers of ten. In his interesting work known as *The Sand-Reckoner*, Archimedes chose as his unit a "myriad," which is equal to 10,000 (in algebraic notation, 10^4). Numbers up to a "myriad of myriads" (i.e., 10^8) were called numbers of the first order. Archimedes then used this myriad of myriads as the unit of the second order and continued in this fashion to develop an elaborate scheme involving successive orders, periods, and other divisions [25].

It is interesting to note that Archimedes made an application of this system by computing the number of grains of sand in the universe. This number, he estimated, was less than 10^{63}, a quantity which in the older arithmetics of our day was called a "vigintillion." A similar computation was made a few years ago by the distinguished astrophysicist A. S. Eddington, who estimated that the number of hydrogen atoms in the universe is of the order of 10^{79}.

In astronomical work the Greeks used a mixed sexagesimal-decimal system. An excellent example of this is found in the tables of Ptolemy's *Almagest* (c. A.D. 150; see [91]), where the ordinary alphabetic notation was used not only for all integers (even those greater than 60) but also for the fractions, which were written sexagesimally. This inconsistency, of course, is still found in modern degree measure or astronomical notations when one writes 132° 15′ 23″. It has been further compounded in more recent times with the division of seconds into tenths. The notation 132° 15′ 23.4″ is a double reminder of the interplay of different cultures in one notational system and of the inconsistencies of mankind.

The well-known Roman numeral system was essentially a base-ten system, although it used additional notational symbols involving the intermediate values of 5, 50, 500, etc. Here again the division of certain monetary and weight systems into parts other than tenths complicated things. The copper *as* of the Romans weighed one pound and was divided into 12 *unciae,* or "ounces"; successive parts were divided into halves. Thus fractional units of twelfths, twenty-fourths, forty-eighths, etc., were involved. Certain versions of the Roman abacus, which we will describe shortly, made a provision for this variation in fractional parts.

The delay in developing adequate and efficient computational procedures is quite understandable when one reflects upon the slow development of the various numeration systems, the variations between them, and the inconsistencies within a given system. Rather obviously, mechanical aids or special devices were essential in earliest times if computations were to be carried out.

Several of the early aids to computation will be considered next. All of these point to the fact that a consistent place-value system with a reasonable number of basic symbols, including one for the all-important concept of zero, was necessary before reasonable computational techniques and algorithms could follow.

COMPUTATION BY THE ABACUS

The earliest device known for carrying out the ordinary operations of arithmetic is the abacus [28]. Its name is derived from the Greek word *abax*, or *abakion*, which meant a board strewn with dust or sand for reckoning or drawing figures. As we have seen from the comment of Herodotus, pebbles were often used in computation.

In time the pebbles and the sand-reckoner gave place to a table upon which there were ruled lines to guide counters that represented numbers. The Roman abacus consisted of a plate of metal with grooves along which the counters moved. The decimal system was used when computing with integers, and another system was used for fractions. The Roman abacus had two grooves for fractions. The first of these measured units of twelve, since the monetary and weight systems were in twelfths, and the second groove provided counters representing fractions as shown below.

$$\mathcal{L} \left(\tfrac{1}{24}\right) \qquad \mathcal{D} \left(\tfrac{1}{48}\right) \qquad \cup \left(\tfrac{1}{72}\right)$$

An example of the Roman abacus is shown in Figure III-1.

The abacus, in time, became the principal computing machine of Western nations. Its use also spread into the Orient. It was known in China as early as the sixth century A.D. It was in popular use in Japan in the seventeenth century, although doubtless it was known

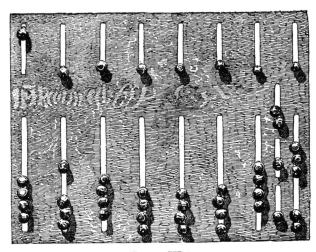

FIGURE III-1

91

Figure III-2

there much earlier. In China the abacus is called the *suan phan,* or arithmetic board; and in Japan, the *soroban.*

The modern abacus differs little in its essential features from its Roman prototype, but its form has changed; the grooves have been replaced by wires and the counters by beads that slide upon them. Figure III-2 shows the modern Japanese soroban. Its general resemblance to the Roman abacus (except for the provision for fractional parts) is evident.

In both the Roman abacus and the Japanese soroban there are four counters in the lower panel and one in the upper corresponding to each line. Those in the upper panel have five times the value of those in the lower.

The basic use of any form of the abacus is always the same, regardless of which base is involved and whether the lines or wires are vertical or horizontal. In the operation of addition, one of the given numbers is recorded on the abacus. Using the second number, and proceeding in either direction, you then "add in" the appropriate number of beads on their proper columns. The moment a full column is attained, it is cleared and replaced by a single counter in the proper adjacent column.

One weakness of calculation by the abacus is that each step erases the preceding step, so that there is no way to check the answer except by recomputing. Multiplication and division are performed by repeated addition and subtraction. Knowledge of various shortcuts and considerable practice are essential for efficient operation of the abacus.

As an instrument for doing ordinary computations, the abacus is rapid and efficient and is still widely used, especially in the Orient. An interesting example of its speed was afforded in 1946 by a contest between Kiyoshi Matsuzaki, a Japanese champion operator of the soroban, and Thomas N. Wood, an American said to be one of the most expert operators of the electrical desk calculator in Japan. In operations of addition (adding fifty numbers of from three to six digits), subtraction, and division Matsuzaki defeated his opponent. Wood won in multiplication; but multiplication with the desk calculator was faster only when both factors exceeded ten digits. In a composite problem involving all the operations the soroban won a decisive victory. (It should be noted, however, that a different account of this contest raises a question concerning the skill of the desk-calculator operator /MAISH/.)

MULTIPLICATION AND DIVISION

We turn next to a consideration of the evolution of methods of multiplication and division in practical arithmetic. Since such methods are intrinsically tied in with whatever numeration system is being used, it should not be surprising to find that early procedures were quite different from our present methods.

Several problems in the Rhind papyrus [30] indicate that by 1650 B.C. the Egyptians were using a method for multiplication that required only the doubling of successive numbers and then the addition of the proper multiples. Since "doubling" a number written in hieroglyphics could be accomplished by simply rewriting each symbol of the original number (and making the substitution of the next higher unit when this became necessary), multiplication depended only on being able to add.

A variation of this method can be traced down to the Middle Ages in the operation of "duplation and mediation" (doubling and halving) [34]. The justification for these procedures is given most easily in terms of the binary representation of the numbers, although of course this was not known to the Egyptians.

Multiplication was accomplished by the Babylonians (at least as early as 2000 B.C.) by reference to appropriate multiplication tables, which undoubtedly had first been obtained by addition. The use of tables of reciprocals (values of $1/N$ for given values of N, both expressed sexagesimally) reduced the operation of division to that of multiplication [34]. The tables of reciprocals also permitted a treatment of fractions that was, as we shall see, a considerable improvement over the handling of fractions by the Egyptians.

Examples of multiplication as handled by the Greeks are given by a mathematician of the fifth century A.D., Eutocius of Ascalon, in his commentary on Archimedes' *The Measurement of the Circle*. Since the numerals were expressed in alphabetic form, each digit of the multiplier, beginning with the highest, was applied successively to each digit of the multiplicand, also beginning with the highest. The final step was to add these values (see [**34**]). The basic form is therefore quite similar to ours today.

If it is true, as Heath has conjectured /(c): 28/, that the Greeks had little need of the abacus for calculation, expert reckoners must have mastered the necessary addition and multiplication tables. (These would have been considerably more extensive than ours today because of the increased number of symbols.) It should, of course, be remembered that the ability to carry out such operations was not possessed by the ordinary citizen nor the small tradesman.

The Hindu-Arabic system of numbers, with its principle of place value and its zero, began to invade Europe near the end of the thirteenth century. Early arithmeticians, recognizing its simplicity, set to work to devise methods for the multiplication and division of numbers. The Hindus had experimented with the problem, but their vague mysticism and obscure ways of recording their computation in verse greatly hindered general understanding and acceptance.

It was really not until the end of the fifteenth century that arithmetic began to assume a somewhat modern form. The book by Nicomachus of Gerasa entitled *Introductio arithmetica* (c. A.D. 100) provided a multiplication table going as far as 10 × 10, but it contained no rules for multiplication and division. Fibonacci's *Liber abaci* (1202) was noteworthy for its use of the Hindu-Arabic numbers, but it added little to the art of computation.

The first printed arithmetic appeared in Treviso, Italy, in 1478.

The Italians, emulating the Hindus before them, became interested in devising schemes for multiplication and division. Luca Pacioli describes some of these methods in his work entitled *Summa de arithmetica, geometrica, proportioni et proportionalita* but customarily referred to as the *Sūma*, which was published in 1494 /D. E. Smith (a): II, 107–17/. He lists eight different forms for multiplication, some of them having fanciful names such as *castellucio* ("the method of the little castle") and *graticola* or *gelosia* ("latticed multiplication"). The latter was so named because it suggested the gratings placed in Venetian windows to protect the dwellers from the gaze of a curious public.

Similarly, various schemata were advanced for division, and four of these were given by Pacioli. The system most highly regarded was called the "galley" method, both because in its completed form the work resembled such a vessel and because the method was thought to be the swiftest, as the galley was the swiftest of boats. This system of division made use of the "scratch" method of writing down the work [**34**].

Our present arrangement of figures in long division developed gradually after the introduction of Hindu-Arabic numerals. By the end of the seventeenth century the galley method was looked on mainly as a curiosity, and the modern form was fairly well established /D. E. SMITH (a): II, 128–44/.

It is curious that for many years, except in the previously mentioned case of the Babylonians, multiplication tables seem to have been little used as an aid in computation. Boethius (*c.* 480–524), in his book entitled *Institutis arithmetica*, gave currency to the small table of Nicomachus; but he did not extend it.

The early Italian mercantile arithmetics often gave tables with products of all primes to 47×47 and sometimes to 97×97. The first large multiplication table was apparently published in Munich in 1610 by Herwart von Hohenberg. But we hear nothing more about such tables until 1820, when A. L. Crelle published his *Rechentafeln* in Berlin, giving the values of the products of all pairs of integers between 1 and 999. This work went through a number of editions and was one of the most valuable aids to computation before the advent of modern computing machines. Division tables giving the values of x/y, for $x < y$, were less common, the largest not exceeding values for x and y much beyond 100. There were, of course, extensive tables of reciprocals.

FRACTIONS

Computation by fractions presented a major problem to many of the early mathematicians. This was particularly true among the Egyptians, who resorted almost entirely to the use of "unit fractions" —fractions with 1 as the numerator. The "part of 12" meant 1/12 and was indicated by writing the denominator with a special symbol above it for "part": 𓂋 . In the hieratic form a dot was placed above the numeral symbol. A special symbol, ☥ , was used for one important nonunit fraction, 2/3. In the following discussion, a single bar over a numeral will indicate the "part of" the denominator (e.g., $\bar{5}$ means "fifth part," or 1/5), and 2/3 will be represented by $\bar{\bar{3}}$.

As has been previously noted, multiplication for the Egyptian computer depended on doubling. The doubling of any unit fraction with even denominator was easily accomplished by halving this denominator. For twice $\bar{3}$ the Egyptians used their special symbol. For doubling the remaining values with odd denominators, special tables were needed. The Rhind papyrus begins by giving such a table for double $\bar{N}$, for all odd values of N up to 101. For example, $2/43 = \bar{42} + \bar{86} + \bar{129} + \bar{301}$.

If the denominator is divisible by 3—that is, if it is of the form $3k$—the double is always $\overline{2k} + \overline{6k}$.

Unit fractions were used in division problems also. Early problems in the Rhind papyrus involve the dividing of 6, 7, 8, or 9 loaves of bread equally among 10 men. The Egyptian answers are $\overline{2} + \overline{10}$, $\overline{3} + \overline{30}$, $\overline{\overline{3}} + \overline{10} + \overline{30}$, and $\overline{\overline{3}} + \overline{5} + \overline{50}$ loaves each [**30**]/VAN DER WAERDEN: 19–30; NEUGEBAUER (a): 74–78/.

The Greeks often followed the Egyptians in expressing fractions as the sum of two or more submultiples. Thus we find Archimedes writing "1/2 1/4" for 3/4 and Heron (or Hero) of Alexandria, centuries later, writing "1/2 1/17 1/34 1/51" for 31/51. The Greeks by no means limited themselves to such unit fractions, however.

For some common fractions the Greeks had special symbols, but for others the fractions were written with one accent on the numerator and two on the denominator, which was written twice. Thus we have

$$\iota\gamma'\kappa\theta''\kappa\theta'' = \frac{13}{29}.$$

Diophantus sometimes used a form similar to ours today, but he wrote the denominator above the numerator and without the horizontal line. A number of other variations were used by the Greeks.

We have already noted the superiority of the Babylonians over the Egyptians in their use of "sexagesimal fractions." Although the absence of a zero and a separatrix meant that the symbol "3" might actually mean 3/60 or even $3/60^2$, in computation this afforded little difficulty. (Recall that our multiplication today is first of all carried out independently of the position of the decimal point.) The use of tables of reciprocals with the operation of addition took care of problems in division.

In their scientific work, particularly in astronomy, the Greeks abandoned the fractional system of the Egyptians and turned instead to the more convenient sexagesimal system of the Babylonians. To illustrate, we find in Ptolemy's *Almagest* the following:

$$\text{The chord of } \kappa\delta = \kappa\delta \cdot \nu\varsigma \cdot \nu\eta,$$

which translates into the following:

$$\text{The chord of } 24° = 24 + \frac{56}{60} + \frac{58}{3600}$$

$$= 24.94944 \text{ units if expressed decimally.}$$

Since the radius of the circle used by Ptolemy consisted of 60 units, the chord of 24°, divided by 60, would equal 2 sin 12°. That is,

$$\sin 12° = \frac{24.94944}{120}$$
$$= 0.207912,$$

which is correct to the last place.

As we have said earlier, the sexagesimal system prevailed among astronomers down to modern times and, of course, is still used in spherical trigonometry.

Our own methods for handling fractions have not been attained easily [35], as one might assure himself were he able to inspect a series of arithmetics dating from the end of the sixteenth century. Bizarre forms for the addition, subtraction, multiplication, and division of fractions are found in these early works. The Treviso arithmetic includes such fractions as 3345312/4320864; such fractions were not unusual in the period that followed.

The writing of fractions with the numerator over the denominator appears to have originated with the Hindus, but a dividing line was not in general use before the sixteenth century. The use of the solidus, as in 2/3, did not attain acceptance until the nineteenth century, when it was advocated by Augustus De Morgan. The use of a colon, as in 2:3, is found in the work of Gottfried Wilhelm von Leibniz as early as 1676.

The delay in the introduction of decimal fractions until the sixteenth century perhaps seems strange. This delay may have been partially due to the time-honored sexagesimal division of the angle, which was perpetuated in the trigonometric tables. Furthermore, the mathematical tables were specifically constructed so that fractions could be avoided. In trigonometric tables, as we shall see later, a radius of 10^{15} or even 10^{22} was used as a reference value so that the appropriate value could be given to a large number of places without need for any sort of fractional part. The most influential of the commercial arithmetics published in Germany, that of Adam Riese in 1522, gave a table of square roots assuming that the numbers had been multiplied by 1,000,000, whence the square roots were multiplied by 1,000. Thus the square root of 2 is given simply as 1 414, with no decimal.

The invention of decimal fractions [36] is usually assigned to Simon Stevin. He set forth this system in 1585 in a seven-page pamphlet, published first in Dutch and then in a French edition under the title *La Disme*, by which is now known /D. E. SMITH (c):I, 20–24/. Unfortunately, Stevin's notation was clumsy, and his work did not receive the attention that it deserved.

Since Stevin did not introduce the decimal point, it is an interesting question to ask who did. Historians of mathematics are not in agreement about the answer; among the candidates for this honor one finds the names of Pellos (1492), Bürgi (1592), Pitiscus (1608), Kepler (1616), and Napier (1617).

Certainly considerable credit is due John Napier, who in his *Rabdologia* recommended the use of "a period or a comma" as the separator of units and tenths and used the comma in his division. Both the comma and the period were used in several of his later works. (See [36].)

It should be mentioned that today many European countries use the comma instead of the dot for the decimal point. In England the position of the decimal point and the position of the dot used to indicate multiplication are the reverse of their position in this country. For example, $2 \cdot 5$ (with the raised dot) is a decimal, whereas 2.5 is an expression of multiplication.

COMPUTATION OF ROOTS

It is a matter worthy of special comment that the most important analytical tool of the early computers was a method for the extraction of square roots.

The importance of the computation of square roots is illustrated by the work of Ptolemy, who achieved the construction of a table of chords by means of an ingenious use of successive square roots. That he had in his possession an excellent method for computing such roots is shown by the fact that, in evaluating the chord of 120°, which involves the square root of 3, he approximated it as follows: $\sqrt{3} = 1 + 43/60 + 55/60^2 + 23/60^3$, which is 1.7320509, expressed decimally, and is in error by only one unit in the last place. Even more remarkable than this was his ability to evaluate such a quantity as the chord of 72°, which is equal to 2 sin 36°—that is,

$$\frac{\sqrt{5 - \sqrt{5}}}{\sqrt{2}}.$$

One observes also that the extraction of square roots remained the principal and perhaps the exclusive tool of calculators until the seventeenth century. This is attested by the fact that Henry Briggs (1561–1631) made fundamental use of square roots in the computation of his table of logarithms. We shall discuss his work more fully in a later section of this overview.

How did these men achieve their approximations? Little is said by early mathematicians about methods for finding square roots. Theon (c. A.D. 390), father of the philosopher and mathematician Hypatia, describes a computation that is roughly equivalent to our method based upon the formula

$$(a + b)^2 = a^2 + 2ab + b^2.$$

Thus, in finding the square root of 4,500, he writes:

$$\sqrt{4,500} = \sqrt{67^2 + 11}$$

$$= 67 + \frac{x}{60} + \frac{y}{60^2}.$$

By an involved method of trial and error he finds that $x = 4$ and $y = 55$, which yields the approximation of 67.0819 in decimal notation, one unit too small in the last place. (We should continually remind ourselves that the Greeks expressed their numbers only in the sexagesimal system, not decimally.)

It is impossible to believe that better methods were not known to Ptolemy. One of the most efficient formulas ever devised is commonly referred to as "Heron's method," although the method had most probably been used long before by the Babylonians. The attribution refers to Heron of Alexandria, who certainly flourished before Ptolemy, although the exact dates are uncertain. Heron was a practical engineer and undoubtedly had to solve many problems involving square roots. He gave the first known proof of the formula for the area of a triangle expressed in terms of its sides, and this involves a square root.

Heron's method is one of successive approximations based on the formula

$$x_{n+1} = \frac{1}{2}\left(x_n + \frac{N}{x_n}\right),$$

where N is the number whose square root is desired and x_1, x_2, etc., are successive values. The speed of convergence is phenomenal. Thus, if $N = 3$ and $x_1 = 2$, we have

$$x_2 = \frac{1}{2}\left(2 + \frac{2}{3}\right) = \frac{7}{4} = 1.375,$$

$$x_3 = \frac{1}{2}\left(\frac{7}{4} + \frac{12}{7}\right) = \frac{97}{56} = 1.73214 \cdots,$$

$$x_4 = \frac{1}{2}\left(\frac{97}{56} + \frac{168}{97}\right) = 1.73205085 \cdots,$$

and this is correct to seven places. The next approximation will yield a value correct to seventeen places.

Heron's method was a special application of a much broader iterative method discovered independently many centuries later by Isaac Newton and Joseph Raphson, which is sometimes called the Newton method but should be referred to as the Newton-Raphson method. In its current form it is applied to the computation of some root, $x = c$, of the equation $f(x) = 0$, by the following iterative process:

$$x_1 = x_0 - \frac{f(x_0)}{f'(x_0)}, \qquad x_{n+1} = x_n - \frac{f(x_n)}{f'(x_n)},$$

where x_0, x_1, x_2, $\cdots$, are successive approximations of c. This form, used by Raphson, differs somewhat in method but not in principle from that actually used by Newton.

An interesting clay tablet of the Old Babylonian period gives some of the earliest results in connection with $\sqrt{2}$, but unfortunately no details of the computational methods are included. The tablet consists simply of the drawing of a square with its two diagonals. Three numbers in sexagesimal notation are given: 30 along one of the sides and the two numbers 1; 24,51,10 and 42; 25,35 along one of the diagonals. It is easily verified that 1; 24,51,10 is an excellent approximation for $\sqrt{2}$, while the value 42; 25,35 is the product of the side 30 with the given approximation. From this it may be deduced that the early Babylonians possessed arithmetical techniques sufficient for obtaining a good approximation to $\sqrt{2}$, as well as knowledge of a specific case of the Pythagorean theorem some twelve hundred years before the time of Pythagoras /AABOE (b): 25–27/.

A very interesting problem involving square roots has been suggested by a curious matter found in *The Measurement of the Circle*, written by Archimedes. This is the following inequality: $265/153 < \sqrt{3} < 1{,}351/780$. The remarkable thing about this approximation is that the two fractions are approximants in the continued fraction expansion of $\sqrt{3}$. (See [68].) No indication is given by Archimedes as to how these values were obtained /HEATH (c): 309–10/.

The problem of extracting cube roots was almost untouched by early mathematicians. One reference is made by Heron, who gave the value $4 + 9/14$ as an approximation of the cube root of 100. That cubic irrationalities were of interest, however, is shown by the existence of such devices as the conchoid of Nicomedes (*c.* 240 B.C.) and the cissoid of Diocles (*c.* 180 B.C.), which respectively provided approximate solutions of two of the classical Greek problems. (See [51, 52].)

100

The Hindus had a method for computing cube roots based on the expansion of $(a + b)^3$, and the Arabs gave a similar method in their arithmetics. Fibonacci in the thirteenth century and Tartaglia and Cardano in the sixteenth century, as well as others, became interested in the problem. However, there does not appear to have been an adequate table of cube roots available before the publication in 1814 of the *New Mathematical Tables* of Peter Barlow.

Late in the sixteenth century François Viète (also known as Francis or Franciscus Vieta) advanced a method for approximating the roots of certain equations of the third degree, the fifth degree, etc., one of which had $n = 45$ for its largest exponent.

It might be of some interest to note that Heron's formula for finding square roots, given earlier, can be generalized to include the evaluation of the pth root of a number N. The generalized formula follows:

$$x_{n+1} = \frac{p-1}{p}\left[x_n + \frac{N}{(p-1)x_n^{p-1}}\right].$$

Unfortunately, the formula is not as handy to use for values of p greater than 2, because N must be divided by a power of the approximation. For example, in computing to twelve decimal places a table of cube roots, the author of this overview actually found it more convenient to use a table of fifteen-place natural logarithms and to compute the antilogarithms from a fifteen-place table of the exponential function.

Computation of the Trigonometric Functions

The art of table making may be said to have originated with Claudius Ptolemy, whose great work, the *Almagest*, we have already mentioned several times. In this we find for the first time a table of chords (equivalent to a table of sines), although it is known on the authority of Theon that Hipparchus, who flourished around 140 B.C., was in possession of a similar table. Although Hipparchus is generally credited with laying the foundations of trigonometry, it was reserved for Ptolemy to present the subject in an essentially complete form /Aaboe (b): 101–26; Brendan/.

Ptolemy's table gives the lengths of chords subtended in a circle with a radius of 60 units by arcs varying from 0° to 180° through increments of half a degree. His method of computation was very ingenious and was achieved by an application of what is known today as Ptolemy's theorem: *The product of the diagonals of a quadrilateral inscribed in a circle is equal to the sum of the products of the opposite sides.*

By subtle reasoning Ptolemy discovered a valuable inequality. If AB and BC are chords subtended respectively by arc AB and arc BC, and if $AB < BC$, then the following inequality holds: $BC/AB < $ arc $BC/$arc AB. A present-day student of trigonometry can easily verify that, for example, $\sin 11°/\sin 10°$ ($= 1.0988$) $< 11°/10°$ ($= 1.1000$).

Ptolemy's objective was to evaluate with a high degree of accuracy the chord of $1°$. As we have said earlier, the main computing tool available to him was the extraction of square roots. By this means, however, he could find to any desired degree of accuracy the chords of such arcs as $120°$, $90°$, $72°$, $60°$, $36°$, $18°$, and of any differences or sums of them. He could also find the chords corresponding to one-half, one-fourth, one-eighth, etc., of these arcs. Hence, from the values of the chords of $72°$ and $60°$, Ptolemy found the chord of $12°$ and then, by subdivision, the chords of $(1\frac{1}{2})°$ and $(\frac{3}{4})°$. His remaining problem was to use these values to compute the chord of $1°$.

This he achieved by means of the inequality already given, for he could write

$$\frac{\text{chord } 1°}{\text{chord } (\frac{3}{4})°} < \frac{\text{arc } 1°}{\text{arc } (\frac{3}{4})°} = \frac{4}{3},$$

$$\frac{\text{chord } (1\frac{1}{2})°}{\text{chord } 1°} < \frac{\text{arc } (1\frac{1}{2})°}{\text{arc } 1°} = \frac{3}{2};$$

that is,

$$\tfrac{2}{3} \text{ chord } (1\tfrac{1}{2})° < \text{chord } 1° < \tfrac{4}{3} \text{ chord } (\tfrac{3}{4})°.$$

Since these bounding values, for a circle of unit radius, differ only in the seventh place, a six-place table of chords was assured. Ptolemy, however, had divided his radius into 60 parts, these parts into 60 other units, and these in turn into 60 parts; so he stated his approximation in sexagesimal units as follows:

$$\text{chord } 1° = 1 \cdot 2' \cdot 50''$$

$$= 1 + \frac{2}{60} + \frac{50}{3,600}$$

$$= \frac{3,770}{3,600}.$$

Since $\sin (\frac{1}{2})° = $ chord $1°/120$, we see that the approximation of Ptolemy was equivalent to $\sin (\frac{1}{2})° = 0.0087269$, which is in

102

error by 3 in the last place. It is also interesting to observe that chord 1° is approximately equal to $\pi/3$, from which we get Ptolemy's value for π, namely, $377/120 = 3.14166. \ldots$

That interest in the values of the trigonometric functions was general among early mathematicians is attested by the fact that tables of sines are found in India dating from as early as the sixth century, apparently influenced by Ptolemy's table of chords.

Similar tables appeared among the Arabs in the year 772, when a Hindu astronomer visited the court of Caliph al-Mansur, at Baghdad, with astronomical tables probably taken from the writings of the Indian mathematician Brahmagupta. This work contained the important Hindu table of sines, or half-chords.

The first step in the construction of tables among the Arabs appears to have been taken in the next century by the Arab astronomer al-Battani, who was familiar with the work of Ptolemy. He was the first to prepare a table of cotangents. A few years later Abu'l-Wefa introduced the new functions of secant and cosecant and devised a method by means of which he computed the sine of half a degree to nine decimal places. His formula was the following:

$$\sin\left(\frac{1}{2}\right)^\circ = \sin\left(\frac{15}{32}\right)^\circ + \frac{1}{6}\left[\sin\left(\frac{18}{32}\right)^\circ - \sin\left(\frac{12}{32}\right)^\circ\right],$$

which was derived from an inequality similar to that of Ptolemy. (For this derivation see /DAVIS *et al:* I, 20–21/.)

The greatest activity in the computation of tables of the trigonometric functions occurred in the period prior to the invention of logarithms and was stimulated by a revival in trigonometry. This was initiated in the fifteenth century by Georg Peurbach and his celebrated pupil Johann Müller, more generally called "Regiomontanus" from the name of his birthplace, Mons Regius (Königsberg).

Peurbach, dividing the radius of the circle into 600,000 parts, undertook the computation of sines to every minute of arc. Regiomontanus completed the table with a radius of 1,000,000 parts. Of lesser significance was the publication of a table of sines by Nicolaus Copernicus in his celebrated work *De revolutionibus orbium coelestium*, first published in Nürnberg in 1543. This table used a radius of 100,000 and gave the values, with differences, for every minute of arc.

The first table of tangents appeared in the *Canon foecundus*, published in 1553 by Erasmus Reinhold, and the first table of secants in the *Tabula beneficia*, published about the same time by Francesco Maurolico, abbot of Messina in Sicily.

Perhaps the greatest labors ever undertaken in the matter of table making, at least until modern times, were those of Rheticus (as Georg Joachim was commonly called), Valentin Otho, and Bartholomäus Pitiscus. Rheticus, a student of Copernicus, conceived the idea of computing tables of the trigonometric functions for every ten seconds of arc with a radius equal to 10^{15} and for every second in the first and last degrees of the quadrant. He completed the table of sines and commenced the construction of tables of tangents and secants, but he died in 1576 before they were finished. It is said that he kept several computers in his employ for twelve years. This gigantic work was finally completed by his pupil Otho and republished in 1613 by Pitiscus of Heidelberg, who also added a few of the first lines computed to a radius of 10^{22}.

These tables were computed essentially after the manner of Ptolemy by the continuous bisection of arcs for which sines were known.

About this time, however, a new method was introduced by Viète, who showed how to compute the roots of equations of degrees higher than two and hence how to approximate numerically the trisections, quinquisections, etc., of angles. In his collected works, published in Leyden in 1646, we find formulas for calculating the chords of multiple arcs and also for finding the chords of fractional arcs. In modern notation these two methods are equivalent, first, to the identities

$$\sin 3a = 3 \sin a - 4 \sin^3 a,$$
$$\sin 5a = 5 \sin a - 20 \sin^3 a + 16 \sin^5 a,$$
$$\text{etc.,}$$

and, second, to the solution of these regarded as equations in which $\sin 3a$, $\sin 5a$, etc., are given and $\sin a$ is to be computed.

The expansion of $\sin x$ in the series

$$\sin x = x - \left(\frac{1}{3!}\right)x^3 + \left(\frac{1}{5!}\right)x^5 - \cdots$$

came later, about the time of Newton and Leibniz.

Discovery and Computation of Logarithms

There is no exaggeration in the often-repeated statement that the invention of logarithms, by shortening the labors, doubled the life of the astronomer. The introduction of logarithms, which completely revolutionized the art of computation, occurred near the beginning of the seventeenth century. This invention is universally attributed to John Napier, laird of Merchiston, whose Merchiston Castle was near Edinburgh.

104

One advantage of logarithms is found in the fact that multiplication can be performed by addition. The idea was not entirely new, however, since other methods can be used in the same manner. Thus one observes the formula

$$ab = \tfrac{1}{4}(a + b)^2 - \tfrac{1}{4}(a - b)^2,$$

which is a very useful device for multiplication when a table of "quarter squares" is available. (Many such tables have been published to facilitate application, the most extensive being that of J. Blater, published in 1888, which gave the quarter squares of all integers from 1 to 200,000.)

The following formula is similarly useful as a device for multiplication:

$$\sin A \sin B = \tfrac{1}{2} \cos (A - B) - \tfrac{1}{2} \cos (A + B).$$

Indeed, it has been conjectured that this formula may have been the origin of Napier's ideas, since he constructed his tables as the logarithms of the sines of angles.

In 1614 Napier published his first work, entitled *Mirifici logarithmorum canonis descriptio*, which gave a description of the nature of logarithms and a table of the logarithms of natural sines. This little book was the result of twenty years of steady labor, devoted to the specified purpose of reducing the time and increasing the accuracy of carrying out the numerical computations of multiplication, division, and the extraction of roots. A second work, published posthumously in 1619 and entitled *Mirifici logarithmorum canonis constructio*, gave the details of how he computed his logarithms.

Although the concept of logarithm is often associated today with that of exponent, the original presentation by Napier was not based on this relation. Napier arrived at his logarithms by considering the velocities of two points, each on a separate line, moving in the same direction.

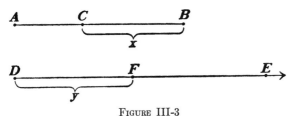

FIGURE III-3

The first line (see Fig. III-3) is a segment of fixed length, $AB = a$; and the second line, DE, is infinite in extent. One point moves along

105

DE with constant velocity equal to the initial velocity of the other point, which moves along *AB* with a velocity that diminishes proportionally to its distance from *B*. If *C* and *F* are corresponding positions of the two points at any given time and *y* and *x* are the distances as indicated, then *y* is said to be the Napierian logarithm of *x*[40].

Napier originally called the number (length) *DF* an "artificial number," but he later invented the word "logarithm," using two Greek words, *arithmos*, meaning "number," and *logos*, meaning "ratio."

Logarithms to the base ten—"common logarithms," as they are now called—were computed by Henry Briggs, professor of geometry of Gresham College, London. His interest was so stirred by the methods of Napier that he made a journey to Merchiston Castle. A consultation between the two men resulted in the suggestion that the logarithm of 1 should be 0 and that the logarithm of 10^{10} should be 10, or—what is the same thing—that the logarithm of 10 should be 1.

Briggs went to work with great zeal, and in 1624 he published tables to fourteen places of the logarithms of numbers between 1 and 20,000 and from 90,000 to 100,000 /DAVIS *et al.:* I, 35–38/. The gap between 20,000 and 90,000 was filled by Adriaen Vlacq, who, however, reduced the number of places to ten. In addition to this heroic computation Vlacq gave also the logarithms of the sines, tangents, and secants for every minute of arc. His work was published in 1628 as a second edition of the tables of Briggs.

David Eugene Smith has concluded that "it is unquestionably true that the invention of logarithms had more to do with the use of decimal fractions than any other single influence" /(a): II, 244/. Despite the fact that no explicit use of decimal fractions was made by Napier in the first tables he published, the modifications made by Briggs made the knowledge of decimals essential for the practical use of logarithms.

Briggs introduced the term *mantissa*, which is a Latin word meaning "addition" or "makeweight." The word *characteristic* was suggested by Briggs and used by Vlacq.

The discussion of logarithms would be incomplete without a mention of the work of one of Napier's contemporaries, Jobst (or Joost) Bürgi, a native of Switzerland. (The name is sometimes given in its Latin form, Justus Byrgius.) In 1620 Bürgi published *Arithmetische und geometrische Progress-Tabulen*, a work that contained a table of what are essentially antilogarithms. Since this book appeared six years after the *Descriptio*, Napier was first to discover the idea of logarithms, although the two works were done independently.

106

Computation of Mathematical Tables in Modern Times

Because of the vast scope of the subject, we can give here only a brief and inadequate account of the history of table making following the computations of Briggs and Napier. The magnitude of these computational efforts is shown in the following table, which gives the number of mathematical tables computed between 1500 and 1960.

TABLE III-1

Number of Mathematical Tables Computed Between 1500 and 1960

Period	No. of Tables	Period	No. of Tables
1500–1600.......	9	1850–1900......	762
1600–1700.......	70	1900–1925......	807
1700–1800.......	84	1925–1950......	1,999
1800–1850.......	223	1950–1960......	1,143
		Total........	5,097

We notice from these data that the twentieth century had already produced, at the end of the sixth decade, approximately 77 percent of all the tables computed since 1500—and that approximately 29 percent of the total for this century were produced in a single decade, after electronic computers became generally available. There are some who have expressed the belief that these new computing machines will make mathematical tables obsolete. But it is to be observed that the production of new tables continues, at the rate of more than a hundred each year.

In order to keep track of the tables and provide a source of information about them, a magazine called *Mathematical Tables and Aids to Computation* was founded in 1943, under the editorship of R. C. Archibald. (In 1960 the name was changed to *Mathematics of Computation*.)

In 1946 a remarkable volume entitled *An Index of Mathematical Tables* was issued in London by the Scientific Computing Service under the supervision of L. J. Comrie, a celebrated computation expert. This work was the joint production of A. Fletcher, J. C. P. Miller, and L. Rosenhead. In its pages a large number of tables are surveyed and their ranges and extent described in detail. A new edition of this valuable work appeared in 1962.

The computation of tables has sometimes been regarded as a lowly

107

occupation for a mathematician. Peter Barlow, in the Introduction to his *Tables* (discussed earlier), makes the following comment about his own labors:

> From computing tables little is to be expected of mathematical repu-
> tation; nothing being more requisite for the execution of such an under-
> taking than a moderate skill in computation and a persevering industry
> and attention; which are not precisely the qualifications a mathematician
> is most anxious to be thought to possess.

In contrast to this, however, a casual survey of *An Index of Mathe-matical Tables*, mentioned above, will show that among the two thousand or more computation experts whose work is represented appear the names of some of the most distinguished men in the history of mathematics—among others, for example: J. C. Adams, George Airy, M. H. Andoyer, F. W. Bessel, Augustus De Morgan, L. E. Dickson, Leonhard Euler, Carl F. Gauss, C. J. Jacobi, A. M. Legendre, Karl Pearson, and T. J. Stieltjes. Who would hesitate to have his name included in such a list?

Along with their theoretical contributions, these men found pleasure in working with the building blocks that then, as now, must form the foundations of both applied and theoretical science.

And if we should add to this list the names of those analysts who contributed the ingenious algorithms by which the functions were reduced to final tables, we should find included a considerable number of the most distinguished mathematicians of their time.

Computation of Special Constants

The computation of certain constants has occupied the attention of mathematicians for many centuries. (Our attention here is briefly on methods of determining approximations for these values rather than on the development and significance of the constant.) Certainly the two most important constants are π [44] and e [45].

The first recorded scientific attempt to measure π is that by Archi-medes /91–98/. His method was to inscribe and circumscribe polygons about a circle, determining their perimeters and using them as upper and lower bounds to the length of the circumference. By using polygons of ninety-six sides, he established the inequality $3\ 10/71 < \pi < 3\ 1/7$. Since the first of these numbers equals $3.140845\ \ldots$ and the second equals $3.14285\ \ldots$, we see that Archimedes scarcely achieved three-place accuracy. This he could have done if he had taken either the

108

arithmetic average or the geometric average of the values, both of which equal 3.14185.

As we saw earlier, under "Computation of the Trigonometric Functions," Ptolemy gave the approximation $377/120 = 3.1466 \cdot \cdot \cdot$. Another early approximation was the fraction $333/106$, which is actually the third approximant in the continued-fraction expansion of π. A very unusual approximation based upon these two values is attributed to Adriaen Anthoniszoon (1527–1607), who naively added the numerators and the denominators of the two fractions and thus obtained $(377 + 333)/(120 + 106) = 710/226 = 355/113 = 3.14159292$, which is in error by less than three units in the seventh place. This approximation, which is the fifth approximant in the continued-fraction expansion of π, was known to the Chinese as early as the fifth century A.D.

Many people have attempted to approximate π, including such famous mathematicians as Fibonacci and Viète. The most successful of these was Ludolph van Ceulen, who in 1610 attained a 35-place approximation. This was regarded as such a great achievement that the number was carved on his tombstone in St. Peter's churchyard at Leyden. For many years π was called the Ludolphian number, and it is still commonly so designated in Germany [44].

A new era in the computation of π was inaugurated with the discovery of infinite series near the end of the seventeenth century and, in particular, with the derivation in 1671 by James Gregory of the expansion

$$\text{arc tan } x = x - \tfrac{1}{3}x^3 + \tfrac{1}{5}x^5 - \cdots ,$$

which appears to have been discovered independently by Leibniz in 1673. Although the series converges much too slowly for $x = 1$ to be of any use, other formulas were obtained. The most famous of these was due to John Machin, who in 1706 derived the following, now called Machin's formula:

$$\tfrac{1}{4}\pi = 4 \text{ arc tan } \tfrac{1}{5} - \text{ arc tan } \tfrac{1}{239}.$$

Other (similar) formulas are the following:

$$\tfrac{1}{4}\pi = \text{ arc tan } \tfrac{1}{2} + \text{ arc tan } \tfrac{1}{5} + \text{ arc tan } \tfrac{1}{8}$$

$$= 2 \text{ arc tan } \tfrac{1}{3} + \text{ arc tan } \tfrac{1}{7}$$

$$= 3 \text{ arc tan } \tfrac{1}{4} + \text{ arc tan } \tfrac{1}{20} + \text{ arc tan } \tfrac{1}{1,985}.$$

William Shanks, using Machin's formula, published in 1873 an approximation to 707 places. This prodigy of computation remained unchallenged for nearly three quarters of a century; but in 1945, D. F. Ferguson, in England, using the last formula above, found that Shanks's value was wrong beginning with the 528th place.

Computing π with the new electronic machines now became a game; and in 1949, using the ENIAC of the Ballistic Laboratories at Aberdeen, Maryland, G. W. Reitwiesner computed π to 2,035 places. Since then, in 1961, π has been computed to 100,265 places.

It may be observed that motivation for such extended calculation of π is not to be found in any practical need for greater accuracy. Suppose we were to compute the circumference of space in terms of electron-diameters; assume the estimate of Edwin Hubble for the radius of space as 2.7×10^{10} parsecs (1 parsec equals 3.258 light-years); and assume the estimate of J. J. Thomson for the diameter of an electron, 2×10^{-13} centimeters. Under these assumptions, π is needed to only 41 places!

The value of e has also been extensively computed, although its approximation is much more readily attained than that of π, since it is defined by the rapidly converging series

$$e = 1 + 1 + \frac{1}{2!} + \frac{1}{3!} + \frac{1}{4!} + \cdots .$$

The following continued fraction, first given by Euler, has been extensively used:

$$e = 1 + \cfrac{2}{1 + \cfrac{1}{6 + \cfrac{1}{10 + \cfrac{1}{14 + \cdots}}}} .$$

In 1926 D. H. Lehmer, using an extension of this method, computed the value of e to 709 decimal places. With the advent of the great computers there has been no difficulty in extending this value to 10,000 or more places. The extraction of the roots of small numbers like 2, 3, 5, etc., to a large number of places has also been carried out, often quite beyond any usefulness as far as the actual approximation is concerned.

It should be pointed out, however, that some of these approximations to a large number of places have been used in certain theoretical studies. For example, since no structure has been observed in the decimal approximation of π, it has been a matter of some interest to see whether special sequences occur. In this connection we have some curiosity about the value of what we might call Brouwer's number. Attacking the general validity of the law of the excluded middle (i.e., that, of the two propositions "A is B" and "A is not B," one proposition must always hold irrespective of what A and B may be), L. E. J. Brouwer considered the decimel development of $\pi = 3.14159 \cdots$. He denoted by d the nth digit after the decimal point and then defined the number $N = (-\frac{1}{2})^m$, where m is the position of d when d equals 0 and the next nine digits are the sequence 1, 2, 3, 4, 5, 6, 7, 8, 9. Brouwer then argued that we cannot say that N is either positive or negative, since it may be that no such sequence exists. Only vast approximations might be able to answer this question.

Development of Computing Machines

The first calculating machine was the abacus, which we have already described. The second was undoubtedly a device for multiplication made by the discoverer of logarithms. This is known to mathematicians under the engaging title "Napier's bones," but it is perhaps more properly called "Napier's rods." This device the inventor described in the 1617 work entitled *Rabdologia* (from the Greek word *rabdos* meaning "rods"), which had a considerable circulation and attracted for a time even more attention than his logarithms. Napier's bones are really a glorified multiplication table, which can be arranged in such a way that only additions are necessary to complete the multiplication process [**39**].

Napier's bones were later placed on parallel cylinders so that they could more readily be put into place for multiplication. A set of these, said to have belonged to Napier, is preserved in the Science Museum of South Kensington, England.

The next progress in mechanical computation was made by Edmund Gunter, a colleague of Briggs. He constructed in 1620 a logarithmic scale two feet in length and multiplied numbers by means of a pair of dividers. For example, he would add the length between 1 and 2 to the length between 1 and 3 and thus find that the sum was equal to the length between 1 and 6. In this manner he obtained the product of 2 and 3.

William Oughtred extended Gunter's idea, replacing the dividers by two logarithmic scales, one of which would slide by the other. This he invented as early as 1622, but he did not describe it until ten years later. Even Newton got into the game by suggesting a runner for the slide rule (1675), but this idea was not actually put into practice until a century thereafter—by John Robertson in 1775. Many modifications

111

of the rule since 1900 have greatly increased its usefulness in more complex computations. [**69.**]

The first computing machine that might be called the prototype of those in use today was invented by the French mathematician, philosopher, and physicist Blaise Pascal in 1642. His machine was designed to do addition and subtraction. It consisted of a large rectangular box on top of which were six wheels. Since Pascal's machine was used principally for calculations involving English currency, the two right-hand wheels were numbered for pence and shillings and the others were numbered from 1 through 9 for pounds. The wheels were connected to recording drums, which in turn activated numerical wheels that could be read through holes at the top of the machine. Subtraction was accomplished by addition, the conumber of the subtrahend being added to the minuend. For example, $197{,}642 - 34{,}576 = 197{,}642 + (65{,}424 - 100{,}000) = 163{,}066$.

Leibniz, another versatile genius, one of the discoverers of the differential and integral calculus, designed a computing machine in 1671 and completed it in 1694. One important innovation introduced by Leibniz was the sliding carriage. Others were the delayed carry, rotations in different directions for addition and subtraction, latches to provide protection against overrotation, and a mechanism for "erasing." Modifications in 1820 by Charles X. Thomas (French), in 1875 by Frank S. Baldwin (American), and in 1878 by W. T. Odhner (Swedish) led to the development, by about 1900, of the essential features of the present desk calculator.

The most ambitious project undertaken during the nineteenth century in the making of computing machines was that of Charles Babbage. Babbage was the Lucasian Professor of Mathematics at Cambridge from 1828 to 1839, but he resigned his position to work on his "difference engine." The government contributed £17,000 to the project, and Babbage used much of his own resources (£6,000) in the quest. He was unsuccessful and never completed his machine. However, after a new government subsidy had been denied him in 1842, Babbage, instead of being discouraged, enlarged his ideas and began work on what he called his "analytic engine." This machine also was never finished; but his son, H. P. Babbage, completed part of the engine in 1906 and published 25 multiples of π to 29 figures as a specimen of its work. His failure was not so much in the design of his machine; rather, in the words of Professor Howard H. Aiken, who developed the theory of the Automatic Sequence Controlled Calculator (ASCC) of Harvard (1944), Babbage failed "because he lacked the

112

machine tools, electrical circuits, and metal alloys" so essential in modern machines. The engine, sometimes referred to as "Babbage's Folly," should have been called much more accurately "Babbage's Vision."

A new dimension was given to computing machines in 1888 by William S. Burroughs, who designed a machine that printed its figures. This was similar to one invented by Henry Pottin, in Paris, that was patented in England in 1883 and in the United States in 1885. It printed totals and subtotals.

Herman Hollerith, while employed by the U.S. Bureau of the Census, developed in 1880 the forerunner of tabulating and sorting machines. He invented a machine for sorting cards. The cards contained holes which permitted their distribution by electromagnetic relays activated by contacts made through the holes. The principle governing this machine was adopted by IBM (International Business Machines Corporation) in its early computers (1929).

From this time on, progress was rapid. Because machines working on the relay principle were relatively slow, they were replaced in 1944 by electronic mechanisms in the form of vacuum tubes. Unfortunately, the number of such tubes made the machines bulky; and the heat generated, when many of them were in operation, presented a serious problem. In 1948 the invention of the transistor was announced by the Bell Laboratories. This revolutionary device is a small crystal that operates in the same manner as a vacuum tube. But it is much smaller, has a longer life, uses much less current, and consequently generates almost no heat. In the most modern computers (since 1961) vacuum tubes have been replaced by transistors, and this change has resulted in much greater efficiency of operation.

The enormous calculating power of these new machines is derived very largely from their memories (storage capacity) and their great speed. Their development, both under government sponsorship and through the continuous research of private companies, has been so rapid that one machine is rendered obsolete by another almost before it has been completed.

Except for the slide rule, all the computing devices just described are what are called "digital calculators." The slide rule belongs to a second class, called "continuous variable computers." Since the numerical answers of the latter are read from graphs or scales, their accuracy is much less than that of the digital calculators, which can deliver answers to many decimal places.

Subsequent invention in this field has led to the development of what are called "analogue computers," in which the analogue between electrical circuits and the mechanism of mechanical devices is utilized to transfer the computation to machines operated electronically. These differential analyzers have completely superseded their mechanical analogues in modern computing laboratories.

Many other ingenious computers have been invented, usually designed for special purposes. One of these was Kelvin's tide-predicting machine (1872), which is essentially a harmonic synthesizer in which a curve is constructed from its harmonic components. The solution of algebraic curves and the solution of systems of linear equations have also attracted inventors, who have devised both mechanical and electronic machines with which to attack these problems.

Some Concluding Remarks on the Art of Computation

Modern analytical methods.—As the reader is now well aware, the history of computation with numbers is one of vast extent. The art began with ancient civilizations and has had a steady growth through the centuries. But until the discovery of logarithms mathematicians were greatly restricted in the problems they could solve, since they had few analytical tools with which to work. As we have observed earlier, the most powerful of these was a method for the extraction of square roots. How have modern mathematicians extended analytical methods, and what are a few of the newer techniques? We shall give a brief answer to these questions.

Soon after the discovery of the differential and integral calculus near the close of the seventeenth century, new tools were available for the computers. The first of these was the great discovery of Brook Taylor that many functions, by use of the differential calculus, could be expanded into an infinite series of powers of the variable.

Taylor's theorem had been foreshadowed by several special developments, such as the expansion of the binomial $(1 + x)^p$ for any value of p, generally credited to Newton (1676); the series for arc tan x by James Gregory in 1671; the expansion of

$$\ln (1 + x)$$

by Nicolaus Mercator in 1668; and the expansion of

$$\frac{1}{2} \ln \frac{1 + x}{1 - x}$$

by John Wallis in 1695. These were special series for which the theorem of Taylor finally supplied a general method.

114

Unfortunately, the power of the theorem was not realized until, half a century after its discovery, it was exploited by Joseph Louis Lagrange. And even then it was not used rigorously. The question of the convergence of the series was not systematically investigated until the time of Niels Henrik Abel and Augustin Louis Cauchy in the early years of the nineteenth century.

Another powerful aid to computation was the development of the calculus of finite differences. This subject was initiated by an interpolation formula developed independently by Gregory and Newton around 1670. Its more general aspects were explored in 1717 by Taylor, who thus became the founder of this new branch of mathematics. The calculus of finite differences was extensively studied by Lagrange and Pierre Simon Laplace and was further developed by the mathematicians of the nineteenth century. Most prominent of these was George Boole, who wrote the first treatise on the subject in 1860.

The importance of the calculus of finite differences in modern calculation cannot be exaggerated. Most of the techniques used today in computing laboratories require the use of finite differences. Thus the solution of ordinary differential equations, both linear and nonlinear, makes use of the Adams-Bashford method, the Runge-Kutta method, the Milne method, and the method of continuous analytic continuation —to name only a few—all of which employ finite differences. The "method of relaxation" has been employed effectively in the solution of boundary-value problems associated with partial differential equations when these equations have been expressed in terms of finite differences.

The past century has also seen the rapid development of nomography, the use of diagrams (called "nomograms") to obtain rapid approximation solutions from complex analytical expressions [41].

Many other analytical techniques have been developed. Unfortunately, these great methods are too technical to be described readily, and thus a historical account of their origins would be unprofitable. We shall content ourselves by giving the names of a few of them, as follows: analytical continuations, Laurent series, continued fractions, asymptotic series, expansions in series of sines and cosines (Fourier series) and in other systems of orthogonal functions, iterative methods, and inversion formulas such as that of Lagrange.

Social implications of modern computing machines.—We shall conclude our history of computation by considering the impact of the

115

great calculators upon modern society. This question was first presented in a systematic way by Norbert Wiener in a work entitled *Cybernetics,* published in 1948. The title was derived from the Greek word for "governor" and was designed to denote "the entire field of control and communication theory, whether in the machine or in the animal." The amazing analogy between the computing machine and the human brain suggested that in certain areas the machine could do what man could do, and in many instances do it as well or better.

Subsequent developments have led to a social revolution. About this somewhat controversial matter F. J. Murray makes the following interesting comment /I, 222/:

> The finiteness of experience would seem to indicate that we can almost always obtain a mechanism which will stimulate any mechanistic relationship between an organism and its environment, no matter how complex this relationship may be. There still remains a subtle distinction, however, between the mechanism and the organism, which is usually accepted or denied on a philosophical basis. In this connection, the term "thinking machine" is used. A robot can be set up with a number of possibilities of action to respond to various environmental circumstances. It can be constructed to keep score on the results of these possibilities by prescribed evaluation processes and to vary its response on the basis of such scores. This type of response to environment is usually considered to be a very important aspect of "intelligence" in an organism.

If we return for a moment to the first section of this overview, we see that calculation began with two problems, one concerned with the needs of practical living, the other with the requirements of astronomy and a burgeoning science. As problems became more complex, more powerful methods of computation were required. Ultimately came the period of mechanical invention, which began to flower in the early years of the twentieth century. The objective at first was to provide machines to lessen the burdens of business accounting and incidentally to help in solving scientific problems.

But with the invention of servomechanisms, vacuum tubes, memory units, transistors, and the like, man suddenly discovered he had created an instrument that far transcended many of his own powers. Science took a vast step forward. The machine could perform prodigies of deductive reasoning, in the sense that it could solve complicated problems formulated in terms of Boolean algebra. It could remember complicated chains of operations; it could form images and draw pictures of them; it could be adapted with some success to the problem of translating languages. The machine could solve the most difficult equations; in fact, it could do most of those things that man has required both in his daily living and in his highest scientific endeavor.

The impact of such a machine upon society was not long in being felt. One application after another was found for it, far beyond the dreams of the original inventors. Many jobs hitherto performed by human energy were turned over to the new machines. The prospect of automation became a reality in many lines of human endeavor. And all of this, as we have seen in our historical account, has come about as a direct consequence of man's first calculations, his struggle with the mystery of fractions, his mathematizing of measurement, his study of the planets.

Our story of computation with numbers must thus conclude without an ending, for society does not know what the end will be. Some believe that we have created a Frankenstein's monster that will ultimately destroy us. Others, however, see in the powers of the machine an Arabian genie that will create new wonders and a higher standard of living for the human race.

Capsule 28

THE ABACUS

NOWADAYS the abacus is likely to be thought of as a primitive calculating device that has found its way into the elementary classroom for instructional purposes. Actually it was—and is—a rapid means of computation in the hands of a skilled operator, and it still continues in use in such countries as India, China, Japan, and Russia.

Historically, it has formed the most basic mechanical aid to computation, particularly in times when numeration systems were still unsuited to written computation.

Three basic forms of the abacus-principle have been in use over the centuries: the ancient dust board (*abax*, in Greek—which probably gave its name to the abacus); the ruled table, with disks or counters placed on or between lines to indicate the proper number; and, finally, the portable frame with counters or beads that are moved along wires or wooden rods.

The exact origins of the various forms and their transitions from country to country are not clear. A considerable number of details are given in /D. E. SMITH (a): II, 156–92/. Even when there is little reference to a people's use of some such device or little evidence of it, as with the Babylonians, it has been customary to assume that one was employed.

The essential idea of any of the various forms of the abacus is that counters (pebbles, markers, or disks) on one line indicate units; on the next line, assuming that a base-ten system is in use, the counters indicate tens; on the next, hundreds; then thousands, and so forth. In its most rudimentary form this means that the moment ten counters appear on a line, they are replaced by a single counter on the proper adjacent line; similarly, one counter would be replaced by ten when moving in the opposite direction. This action is probably the basis of the terms "carry" and "borrow" used in arithmetic today. On the Roman abacus, discussed also in the overview, counters placed in proper position could represent intermediate values, such as V, L, D, and the like; and in this case five counters in the bottom portion of a line would be replaced by one counter in the top panel. The spaces between the lines on the counting table were similarly used for such intermediate values.

In the more primitive form, a board was strewn with dust or sand, on which lines and marks could be drawn with a stick. Erasures were easily accomplished with the finger or hand.

An interesting but little-used form of the abacus was developed by Gerbert, who later (in 999) became Pope Sylvester II. This contained twenty-seven columns; in each column just one marked counter would be placed—for example, a counter with some form of the numeral 4 on it would be used instead of four actual counters. The counters were called *apices;* there was no zero counter, however. The physical assistance of the actual number of counters was lost in this abacus, and it never became popular.

Counter reckoning, the use of the counting table or so-called line abacus, was very common in western Europe from the 1500's into the 1600's. Here the lines were usually placed in a horizontal position, and lines denoting thousands and millions were marked with a cross or star for easy identification. This practice later led to the use of commas before groups of three digits in writing with Hindu-Arabic numerals.

118

Many of the early printed arithmetic books of about 1500 included instructions for both kinds of computing, mechanical and algorithmic. In 1522 Adam Riese, a German, published *Rechnung auff der Linien vnd Federn* ("reckoning on lines and with the pen")—the latter, of course, referring to the algorithmic form of computation.

Robert Recorde in *The Ground of Artes* (*c.* 1540–42) first described calculating with the pen and then devoted forty pages to describing the same art with counters /STRUIK (e): 4–6/. The controversy between the abacists and algorists in the Middle Ages over the relative merits of the mechanical versus the algorithmic form of computation was long and heated. Those who were affected or influenced by the customs of the Italian merchants tended to abandon the abacus (in its various forms) sooner than those who were in closer contact with German counting houses.

The portable-frame type of abacus (an early form using sliding buttons in grooves which goes back at least to the time of the Romans) is represented today by the *suan-phan* of China and the *soroban* of Japan. The Chinese form developed with five counters below and two above, as did the Roman version. The Japanese soroban took the form of five below and one above after the 1868 political revolution, and since 1940 the most popular form has been four below and one above. /KOJIMA: 11–23./

A description of one of the early forms of the Chinese abacus suggests that in this case a single ball was moved up and down each vertical column, which was divided into nine horizontal divisions. By moving these balls up and down the operator could place any number into the abacus and carry out the usual operations. One can only speculate about an earlier development of a Cartesian coordinate system if this form had continued in use! /NEEDHAM: 77./

A considerable number of literary references and interesting word derivations are associated with various forms of the abacus. The Greek historian Polybius wrote, "Those in the courts of kings . . . are in truth exactly like counters on a counting board. For these, at the will of the reckoner, are now worth a copper and now worth a talent." Boethius (*c.* 510) called it the "table of Pythagoras" (*mensa Pythagorica*), and Adelard of Bath (*c.* 1120) even went so far as to ascribe its origin to Pythagoras. (The term "table of Pythagoras" is applied also to the square array of a multiplication table.)

Shakespeare spoke disparagingly of "this counter-caster" in *Othello* (Act I, scene 1, line 31); and in *The Winter's Tale* (Act IV, scene 3) we find, "Fifteen hundred shorn, what comes the wool to? . . . I cannot

119

do't without counters." Sir Thomas Hobbes wrote, "Words are wise men's counters, they do but reckon with them, but they are the money of fools" (Leviathan, Pt. I, chap. iv).

The word "exchequer," as used in "Court of the Exchequer," is derived from the checkered cloth that covered the table at which accounts were taken. Calculations were made on the lines of the cloth when necessary. This dates back to the time of the Norman kings (c. 1100).

The bench or table on which the lines were drawn and the counters placed was known in Germany as a *Rechenbanck*, or *Banck*, from which we get our words "bank" and "banker" (possibly through the French). And the word "bankrupt" means "broken board"—the instrument used by the banker or money changer who failed in business was broken, figuratively if not actually. The "counter" in the store still remains, although it is now used for displaying goods rather than for casting accounts.

For Further Reading

KOJIMA: 11–23
SANFORD (d): 87–93

D. E. SMITH (a): II, 156–92

Capsule 29   R. S. Mortlock

FINGER RECKONING

MOST early peoples developed some system for representing natural numbers by means of various positions of the fingers and hands. Such systems were used by the Greeks, Romans, Arabs, Hindus, and many others. In the fifth century B.C., for example, Herodotus indicated that the Greeks had such a system; Pliny (first century A.D.) remarked that the fingers of the statue of Janus on the forum in Rome represented the number of days in a year.

In Europe in the Middle Ages, finger numbers were in common use as an international medium of communication. These were used in bargaining at international fairs and on other occasions when language barriers occurred. In the Orient today, finger numbers are still in common use for the same reason.

Reference to the table of finger numbers seen in Figure [29]-1 will give the solution to the riddle sometimes attributed to the English

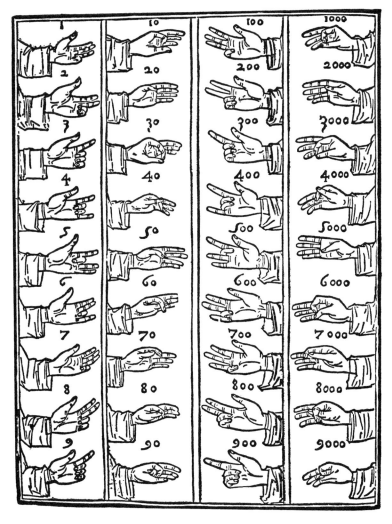

Fig. [29]-1.—Finger Numbers (From Luca Pacioli, *Summa de arithmetica, geometrica, proportioni et proportionalita,* 2d ed., Venice, 1523. Courtesy of D. E. Smith Collection, Columbia University.)

scholar Alcuin of York (735–804): "I saw a man holding eight in his hand, and from the eight he took seven, and six remained." The Roman satirist Juvenal (first century A.D.) tells us, "Happy is he indeed who has postponed the hour of his death so long and finally numbers his years upon his right hand" /D. E. SMITH (a): II, 197/. (Numbers up to a hundred were represented on the left hand alone.)

In the Middle Ages the finger representation of numbers less than a hundred was fairly uniform and in common international use, but the representation of numbers greater than that was less uniform. In some cases the symbols for hundreds and thousands were interchanged.

In the eighth century an English Benedictine monk, Bede the Venerable, gave a description of the use of fingers and hands, with various positions on the body, to represent numbers up to a million.

From finger representation there developed forms of finger computation. These extended from simple counting to special cases of multiplication. Some of these methods were in common use in the Middle Ages, and one such method was still used by some Russian and French peasants in the early twentieth century. (It is sometimes referred to as "European peasant multiplication," not to be confused with the "Russian multiplication" discussed in [34].) For example, to multiply together any two of the numbers 6, 7, 8, 9, and 10, first subtract 5 from each and raise as many fingers on each hand as given by the two resulting digits. Add the number of raised fingers for the number of tens and multiply together the number of closed fingers (including the thumbs) for the number of units.

Figure [29]-2 pictures the multiplication of 7 and 8: $(7 - 5 = 2)$ plus $(8 - 5 = 3)$ gives 5 as the number of raised fingers, for a value of

$$2 \ + \ 3 \ = \ 50$$

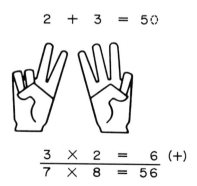

$$\frac{3 \ \times \ 2 \ = \ \ \ 6 \ (+)}{7 \ \times \ 8 \ = \ 56}$$

FIG. [29]-2.—MULTIPLICATION OF 7 AND 8

50; $2 \times 3 = 6$ is the multiplication of the closed fingers for a unit value of 6; and the total, of course, is 56.

In a somewhat similar fashion the product of two numbers from 10 to 15 can be found. Simple algebraic identities can be formed to show the validity of these rules in the general cases /D. E. SMITH (a): II, 201–2/.

For Further Reading

JONES (i)

NEWMAN: I, 434–41

ORE (c): 5–7

SANFORD (d): 76–78

D. E. SMITH (a): II, 196–202

SMITH and GINSBURG: 23–24

Capsule 30   *Harriet D. Hirschy*

THE RHIND PAPYRUS

A SCOTTISH scholar and antiquary, A. Henry Rhind, was spending the winter of 1858 in Egypt for his health. As he was browsing through a shop in Luxor, he discovered and bought an ancient Egyptian papyrus reported to have been found in some ruins in Thebes. When Rhind died a few years later, the document was transferred to the British Museum. The roll, originally 18 feet by 13 inches, was found broken into several "books," with a number of fragments missing. Many of these fragments were later discovered in the possession of the New York Historical Society, and a translation could be completed /GUGGENBUHL (b): 406–10/.

The Rhind papyrus is a collection of mathematical examples copied by the scribe Ahmes (the name is sometimes given as A'h-mosè or Ahmose) around 1650 B.C., in the reign of A-use Re, of the Hyksos dynasty. The scribe explains these writings as a copy of earlier writings of the time of Ne-ma'et-Re (Amenemhet III), which would place the work in the later half of the nineteenth century B.C. In the title inscription the scribe sets forth his purpose, to show "accurate reckoning of entering into things, knowledge of existing things all mysteries . . . se-

123

crets all." The writing is hieratic, a less formal script than hieroglyphic, using general symbols rather than the pictures of the latter [2].
The document is divided into three parts after the introduction: arithmetical problems; geometrical problems; and miscellaneous problems,
including some area and volume applications.

The Rhind Mathematical Papyrus, published in 1927 /CHACE/, includes a transcription of the text of this document into hieroglyphics
and a translation into English, as well as a great deal of helpful commentary.

The entire work emphasizes the two concepts that particularly char-
acterize the mathematics of the early Egyptians: (1) the consistent use
of additive procedures and (2) computations with fractions which relied
almost entirely on the "unit fraction." The unit fraction, a fraction with
numerator of 1, we will write as $\bar{5}$, for the "part of 5" or 1/5; the only
important nonunit fraction, which seemingly was used whenever possible, was 2/3; we will write this as $\bar{3}$, since it had its own special symbol
in the Egyptian form.

The papyrus therefore appropriately begins with a table giving values
of twice $\bar{n}$ for all odd n from 3 through 101. This is often referred to as
the $2/n$ table, since it gives the equivalents of 2 divided by n. Thus the
entry corresponding to 5 (that is, 2/5) is $\bar{3}\ \overline{15}$ (meaning 1/3 + 1/15;
the Egyptians did not use a symbol for addition here, but for ease of
reading the plus symbol will be inserted hereafter); 2/15 is given as
$\overline{10} + \overline{30}$; 2/31 as $\overline{20} + \overline{124} + \overline{155}$. Various theories have been offered
about the procedures used by the Egyptians to derive these values, but
no one principle works consistently for all values /VAN DER WAERDEN:
23–27/.

Multiplication as done by the Egyptians was a matter of repeated
duplications until appropriate values were found, whereupon a simple
addition would give the desired result. In Problem 79, the scribe demonstrates the multiplication of 2,801 by 7. The reader is to "do it thus":

	1	2,801
	2	5,602
	$\underline{4}$	$\underline{11,204}$
Total	7	19,607

Thus, 2,801 × 7 is 19,607.

(If the reader were asked to find 11 times some value, he would double, double, and double again, until he had values of once, twice, four
times, and eight times the number. Since 1 + 2 + 8 totals 11, the orig-

124

inal value and the doubles found at these numbers would be added to give the desired product. The student of today will notice an application of the distributive property here.)

The multiplication of fractions by whole numbers was handled in similar manner. In Problem 2, to divide 2 loaves of bread equally among 10 men (the first six problems consist of dividing 1, 2, 6, 7, 8, and 9 loaves equally among 10 men), the text merely states that each man is to receive 1/5 loaf, and then proceeds to verify the correctness of this value by multiplying 10 by 1/5.

$$1 \qquad \tfrac{1}{5}$$
$$\sqrt{2} \qquad \tfrac{1}{3} + \tfrac{1}{15}$$
$$4 \qquad \tfrac{2}{3} + \tfrac{1}{10} + \tfrac{1}{30}$$
$$\sqrt{8} \qquad 1\tfrac{1}{3} + \tfrac{1}{5} + \tfrac{1}{15}$$

"Total 2 loaves, which is correct." Although the scribe does not actually show this, we can check that $2 + 8 = 10$, and correspondingly $(1/3 + 1/15) + (1\ 1/3 + 1/5 + 1/15)$ is 2. In the above work the doubles of 1/5 and 1/15 are the values given in the introductory $2/n$ table. To double $\overline{n}$ when n is even, it is only necessary to divide n by 2.

The result of dividing 6 loaves among 10 men is given as $1/2 + 1/10$; of 7 loaves, $2/3 + 1/30$, illustrating the use of the 2/3 value when possible. /GILLINGS (a)./

One of the few generalizations of methods is given in Problem 61B: "To find 2/3 of $\overline{5}$ take thou its double and its sixfold, and do thou likewise for any fraction that may occur." Thus we are to find $\overline{2 \cdot 5} + \overline{6 \cdot 5}$, or $1/10 + 1/30$, which is 2/3 of 1/5; however, there is no proof that this method always gives the correct result.

The carrying out of the division process looks very similar to the method of multiplication. In Problem 69 it is necessary to divide 1,120 by 80, which yields the quotient 14. The instructions are to "multiply 80 [more literally, 'make thou the operation 80'] so as to get 1,120."

$$1 \qquad 80$$
$$10 \qquad 800\sqrt{}$$
$$2 \qquad 160$$
$$4 \qquad 320\sqrt{}$$

Sum is 1,120.

(By adding 10 and 4 we have the quotient 14, although the papyrus places the emphasis on obtaining the 1,120.)

If the division is not exact, halving is also used, with resulting fractions. Thus in Problem 24, in which an intermediate step requires dividing 19 by 8, we have:

1	8
2	16√
$\overline{2}$	4
$\overline{4}$	2√
$\overline{8}$	1√

Quotient is $2 + \overline{4} + \overline{8}$.

A series of similar problems that are essentially linear equations in one unknown are illustrated by Problem 24, "Aha, its whole, its seventh, it makes 19." ("Aha" is an expression for the unknown quantity, closely related to "a heap.") The resulting equation is $x + 1/7\, x = 19$. The solution is accomplished by the method of false position [90]. The numerical aspects of the solution are indicated by the intermediate requirement: As many times as 8 must be multiplied to give 19, so many times 7 must be multiplied to give the required number. The next problem is quite similar: "Aha, its whole, its fifth, together give 21."

Many of the problems are of a much more practical nature, a number of these involving the so-called *pesu* calculations. The technical term *pesu* (or *pesfu*) designates the number of loaves of bread or number of jugs of beer that can be made from a given quantity of grain. The nonequivalence of different grains in *pesu* value then leads to more complicated problems.

Problem 41 presents a challenge to the modern student: "Find the volume of a cylindrical granary of diameter 9 and height 10 cubits." (The Egyptian procedure to find the area of a circle is $A = (d - 1/9d)^2$; see [44]). After finding the volume in cubed cubits, the student should find the volume of *hekat* of grain contained. A *khar* is 2/3 of a cubed cubit, and 1/20 of the number of *khars* is the number of hundreds of quadruple *hekat* contained. (The result should be 48, or 4,800 quadruple *hekat;* a *hekat* was about 1/2 peck.)

A challenging and exciting discovery of a century ago gives fascinating exercise to the student of mathematics, both modern and ancient.

126

For Further Reading

CHACE

EVES (c): 37–43, 46–47
    [5th ed. 30–35, 40–43]

GILLINGS (a,)

———— (b)

———— (c)

GUGGENBUHL (b)

HOGBEN: 52–54, 63–64

MIDONICK: 706–32

NEUGEBAUER (a): 71–96

NEWMAN: I, 169–78.

RANSOM

VAN DER WAERDEN: 15–36

Capsule 31

THE MOSCOW PAPYRUS

THE Moscow papyrus, also referred to as the Golenischev papyrus for the man who owned it before its acquisition by the Moscow Museum of Fine Arts, was probably written about 1850 B.C. Although it contains only twenty-five problems and is smaller in size than the Rhind papyrus [**30**], being approximately 3 inches wide and 18 feet long, it is similar to that other important Egyptian work. The Moscow papyrus, also, is written in hieratic form, and it displays approximately the same types of mathematical problems and solutions.

An early translation of one of the geometrical problems in this work made it appear that the Egyptians were familiar with the formula for the area of a hemisphere, but this interpretation is no longer generally accepted /VAN DER WAERDEN: 33–34/. However, one problem does use the correct formula for the volume of a truncated square pyramid,

$$V = \frac{a^2 + ab + b^2}{3} \cdot h.$$

The solution is expressed only in terms of the necessary computational steps for the given numerical values: height of 6 and bases of sides 4 and 2. Various conjectures have been made about how the Egyptians might have derived this procedure, but the papyrus offers no help.

Eric Temple Bell has called this the Egyptians' "greatest pyramid."

127

For Further Reading

Boyer (j)   Gillings (d)
Eves (c): 47–48   Midonick: 507–21
[5th ed. 43]   Van der Waerden: 33–35

Capsule 32   William B. Wetherbee

KOREAN NUMBER RODS

The Chinese are known to have used bamboo rods as an aid to computation at least as early as 500 B.C. By the end of the sixth century A.D. the rods had come into use in Japan, by way of Korea. They continued in use in Korea long after various forms of the bead abacus had replaced them in China and Japan.

The early rods were round, sometimes as long as 12 to 18 inches, and a complete set consisted of 271 rods. They were sometimes made of other materials, such as bone or horn, and "to reckon with ivory rods" is still used as an allusion to wealth. The Korean rods were usually made from bamboo split into triangular or square-prism form to prevent rolling; their length was shortened to several inches. A set of about 150 rods, though incomplete, was sufficient for ordinary calculations.

I	II	III	IIII	X	XI	XII	XIII	XIIII	—	T	TT
I	2	3	4	5	6	7	8	9	10	II	12

Fig. [32]-1.—Korean Rod Numerals

In all cases, it is believed, calculations were carried out by the actual formation of the rod numerals on a counting board. The Korean form of the numerals is shown in Figure [32]-1. The counting board was divided into squares resembling our chessboard; zero was indicated simply by a blank space. The easy removal of rods from the board had some practical advantages over writing when it was time to cancel numbers that were no longer needed.

In later years both the Chinese and Japanese used the rod numerals on the counting board as a sort of coefficient array in the solving of

128

algebraic equations of the third and fourth degree. In laying down the numbers, red rods were used to represent positive numbers and black rods to represent negatives. In writing out such representations, however, the negative or subtractive concept was indicated by a mark placed obliquely across one of the numerals. By this time a circular symbol was used in writing to indicate the zero.

For Further Reading

MIKAMI: 27–31
NEEDHAM: III, 5–10, 126–33
D. E. SMITH (a): II, 169–74

SMITH and MIKAMI: 19–29, 47–58

Capsule 33

THE PERUVIAN QUIPU

THE use of knotted cords to record numbers has been a custom among many groups all over the world. Such cords have been found among some of the more developed civilizations, as well as among primitive tribes with no means of writing. The Peruvian quipu (knot record) is one of the best illustrations of this in the New World.

When the Spaniards first observed the use of the quipu among the Incas of the sixteenth century, they found a highly developed form of recording numbers by means of knots in strands of various thicknesses and colors. The decimal system was used, with both the types of knots and the number of knots forming various systems for recording numbers. Although some cords included a top strand that summed the numbers on a group of pendant strands to which it was attached, such devices appear to have been used for record keeping or census taking rather than for computational purposes.

The use of the quipu continued at least until the beginning of the twentieth century as a means of keeping track of sheep, rams, and lambs of the flocks of Peruvian shepherds.

For Further Reading

CAJORI (d): I, 38–41   MIDONICK: 643–59
DIANA   D. E. SMITH (a): II, 195–96

Capsule 34

MULTIPLICATION AND DIVISION PROCEDURES

PROCEDURES for carrying out basic mathematical computations such as multiplication and division have always been subject to the limitations that are implicitly imposed by the notational system in use. Similarly, the sorts of writing materials available—or not available—have influenced the written forms or algorithms. The extensive use of some form of abacus throughout much of the history of computation reflects both of these influences; similarly, it points to an apparent preference for some sort of mechanical procedure rather than a written or "thought" process.

The methods of the early Egyptians, which involved repeated doubling for multiplication (halving for division), followed by the summing of appropriate values, are described elsewhere [30]. A variation of this, called "duplation and mediation" (doubling and halving), was common in medieval Europe; it has been called "Russian multiplication," since it was supposedly used by Russian peasants until the time of World War I. Although the process of duplation and mediation is similar to the Egyptian doubling system, the selection of the proper summands now becomes automatic. To multiply 43 by 45, repeatedly halve the left side, disregarding any remainders, while doubling the right side (see Fig. [34]-1).

Then add the numbers in the right-hand column that correspond to odd numbers in the left-hand column, as shown in Figure [34]-2. A justification for this procedure may be found by expressing 43 in binary (base-two) notation.

At least as early as 2000 B.C., the Babylonians carried out operations of multiplication and division by the use of tables. Individual clay tablets have been found, each listing multiples of a given number,

130

43	45
21	90
10	180
5	360
2	720
1	1,440

FIGURE [34]-1

43	45	45
21	90	90
10	180	
5	360	360
2	720	
1	1,440	1,440
		1,935

FIGURE [34]-2

which we call the "principal value" of the table. Other tablets combine a number of such individual tables.

The use of a base-sixty system suggests that one might expect a complete set of sexagesimal multiplication tables to include tables for principal values of 2, 3, 4, . . . , 59, with the multiples for each ranging from 1 to 59. This is not the case. If the principal value of the table was p, one space-saving device was to give only the values of $1 \cdot p, 2 \cdot p, 3 \cdot p, \ldots , 19 \cdot p, 20 \cdot p, 30 \cdot p, 40 \cdot p$, and $50 \cdot p$. The addition of two values gives any basic product; for example, $43 \cdot p$ is the sum of $40 \cdot p$ and $3 \cdot p$.

The tables that have been found do not include all principal values from 2 to 59. But while such values as 11, 13, 14, 17, and 19 are missing, tables for the multiples of seemingly unusual numbers such as 3,45 and 44,26,40 have been found repeatedly (see [1] for explanation of Babylonian notation). The apparent mystery is resolved by the appearance of another type of table, which gives the reciprocals of various "regular" sexagesimal numbers, the reciprocals also expressed in sexagesimal form. The regular sexagesimal numbers are those whose reciprocals can be expressed in *finite* sexagesimal form. Thus the reciprocal of 8 is listed as 7,30. If this is interpreted as ;7,30, this means $7/60 + 30/3{,}600 = 450/3{,}600 = 1/8$.

An equivalent condition for a number to be regular is that it factors only into powers of the primes 2, 3, 5, since $60 = 2^2 \cdot 3 \cdot 5$. The listed reciprocal for 1,21 is 44,26,40. The actual reciprocal of 1,21 (i.e., 81_{ten}) expressed sexagesimally is 0;0,44,26,40, as may be readily verified. At first the lack of a Babylonian sexagesimal point seems to be a hindrance. But just as in base ten, where the reciprocal of 8 is .125 and the reciprocal of 80 is .0125, the important thing is the sequence of digits, and this is independent of the decimal point. Exactly the same thing holds true in the sexagesimal system.

The significance of the table of reciprocals now becomes clear. If one is to divide 43 by 8, one first uses this table to find the reciprocal of 8, which is 7,30. One then uses the multiplication table with principal value of 7,30. The operation of division is replaced by multiplication, once the proper table has been determined. Almost all of the multiplication tables known are those with principal values that are reciprocals of the sexagesimally regular numbers. Multiplication tables for 7 have also been preserved, thus providing sufficient means for carring out all basic multiplications.

The matter of dividing by a nonregular number presented some difficulty, since no finite reciprocal is available. The Babylonian tablets often remark, "11 does not divide." In such cases approximations were used, possibly arrived at by interpolation. A few tables that give approximate values for the reciprocals of some irregular numbers have been preserved.

Some of the tables of reciprocals that date from the later Seleucid period are quite extensive, giving regular numbers up to seven places with resulting reciprocals to as many as seventeen places. These apparently were used mainly for astronomical calculations. Additional computational work that could be carried out by the Babylonians is considered in /AABOE (b), NEUGEBAUER (a), and VAN DER WAERDEN/.

Some of the difficulties that are related to the written form of multiplication using the Greek Ionic (alphabetic) system become evident if one first writes the indicated products of 2 × 4, 20 × 40, 200 × 400, and then the actual products in this notation. The twenty-seven alphabetical symbols for numbers do not repeat a pattern of 2, 4, 8 that we have today. Specific details of procedures used by the Greeks are generally lacking, but at least some reliance on the abacus and tables for assistance seems to have been necessary. A number of examples of multiplications are given by Eutocius of Ascalon (c. A.D. 560) in his commentary on Archimedes, and one of these is given in Figure [34]-3. (The auxiliary symbols indicating that the letters are used here as numerals have been omitted, and the corresponding modern equivalents are given at the right.)

The Greeks developed their mathematics mainly within the framework of geometry rather than arithmetic or algebra because of a number of different factors, only one of which is the seeming notational handicap. The calculations used by Archimedes in *The Measurement of the Circle* (240 B.C.) involved considerable use of fractions

132

$$\begin{array}{lll}
\sigma\,\xi\,\epsilon & \text{---------} & 265 \\
\underline{\sigma\,\xi\,\epsilon} & \text{---------} & 265 \\
\end{array}$$

δ α			
M M,β,α ---------- 40,000	12,000	1,000	
α			
M ,β,γχτ ----------- 12,000	3,600	300	
,α τ κ ε ----------- 1,000	300	25	

ζ

M σ κ ε -----------70,225

FIG. [34]-3.—EXAMPLE OF MULTIPLICATION

and square roots of rather large numbers, so computational difficulties were not insurmountable. (Only the results of various steps are given in this work, not the corresponding procedures.) The French mathematician Paul Tannery, working within the past century, developed considerable skill in calculating in the Greek system and concluded that it has some advantages over our present one.

Although the origins of the presently used Hindu-Arabic notational system are usually placed prior to A.D. 600, it was not until about 1600 that written forms for carrying out calculations in the system were reasonably standardized. In multiplication, for example, the manner of writing the partial products underwent considerable experimentation; and when the Italian Luca Pacioli published his *Sūma* in 1494 he gave eight different forms or plans for multiplication.

The Hindu term for arithmetic after the seventh century A.D. was *patiginta*, a compound word meaning "science of calculation on the board." The scarcity of writing paper meant that figures were written in the dust or sand on a board or on the ground; possibly chalk or soapstone would be used on a board. The earliest procedures were influenced by the limited working space and also by the ease with which erasures could be accomplished. In later years and in other countries the basic procedure remained the same except that the digits were simply crossed out and their replacements written above them, giving rise to the so-called scratch methods.

In most cases the work proceeded from left to right, with the results written above the work rather than below. The successive steps in the problem of subtracting 839 from 5,625, as given in an Arabic manuscript, *Hindu Reckoning*, written by Kushyar ibn-Lebban about 1000, are the following. (Here we use scratches at first instead of erasures to indicate the order of steps.)

$$
\begin{array}{cccc}
48 & 79 & 86 \\
\not5\not625 & 4\not8\not25 & 47\not9\not5 & 4786 \\
\not839 & 8\not39 & 83\not9 & 839
\end{array}
$$

At the conclusion only the answer (4,786) and the subtrahend remained. A similar method for multiplication also worked from left to right. /D. E. Smith (a): II, 118./

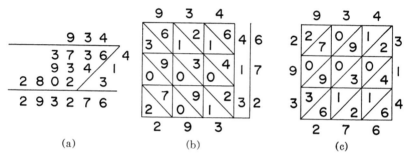

(a)   (b)   (c)

Fɪɢ. [34]-4.—Tʜʀᴇᴇ Mᴜʟᴛɪᴘʟɪᴄᴀᴛɪᴏɴ Mᴇᴛʜᴏᴅs

Three of the multiplication methods given in the Treviso arithmetic book and also described by Pacioli are used as shown in Figure [34]-4 for multiplication of 934 by 314. Method (a) is quite similar to that used today, the answer being read straight across the bottom of the computation. In Method (b) summing of the diagonal rows begins at the upper right-hand corner and continues in successive left-hand rows, moving diagonally down to the right. As can be seen, the answer is given beginning at the bottom and then reading up. In Method (c), similarly, the summing begins in the lower right-hand corner and continues in successive diagonal rows. This general method was often called the *gelosia, jalousie,* or "grating" method from its particular form.

A "scratch" method of division, possibly of Hindu origin, was the plan most commonly used until 1600. Several intermediate steps in the division of 65,284 by 594 are given below; the quotient is 109, with a remainder of 538. The outline of the final figure was thought to resemble a boat, and so it was often referred to by the terms *batello, galea,* or "galley method."

Calandri's arithmetic book of 1491 was the first to use the division form that is now standard. By the end of the seventeenth century

FIG. [34]-5.—GALLEY METHOD

it had become well established, and the galley method was considered only a curiosity.

For Further Reading

AABOE (b): 5–22   SCHRADER (a)
HEATH (c): 24–35   D. E. SMITH (a): II, 101–44
NEUGEBAUER (a): 3–52   VAN DER WAERDEN: 37–61
SANFORD (a)

Capsule 35   *Waldeck E. Mainville, Jr.*

FRACTIONS

FRACTIONS occur in the oldest mathematical records that have been found. The ancients, however, did not develop a generalized approach to fractions, so the special methods of handling them often imposed certain restrictions on their use.

Fractions were in use by the Babylonians as early as 2000 B.C. These were written in place-value form, essentially in the same fashion as our decimal fractions of today; however, the unwritten denominators were successive powers of sixty, and there was no indication of a separatix that would correspond to a decimal point [1].

The Rhind papyrus [30], sometimes called the Ahmes papyrus, con-

135

tains the first systematic treatment of unit fractions. Fractional values that could not be expressed by any one unit fraction were represented by the sum of two or more unit fractions, with a space instead of a plus sign. Thus 2/35 was written as 1/30 [+] 1/42. Unit fractions were written using a fraction symbol with the denominator underneath. In hieroglyphics (picture writing), the fraction 1/4 was written as ⌓, 1/13 as ⌓, and so forth. The fraction 2/3 had a special symbol, ⌓, and 1/2 was sometimes written ⌐. In cursive hieratic writing the unit fraction was indicated by a dot or symbol, called a *ro*, appearing above the denominator.

The Greeks also often relied on unit fractions. Their unit fractions were frequently represented by writing just the denominator with a single or double accent; thus $\lambda\beta''$ for 1/32. But Greek scholars did not confine themselves to unit fractions. General fractions were sometimes indicated by writing the numerator once with an accent and the denominator twice with a double accent. Thus $\varepsilon'\,\eta''\,\eta'' = 5/8$. In some cases the denominator was placed in the position of the modern exponent; in other cases the denominator was written directly above the numerator (just the opposite of modern notation, except that no bar was used between them).

In Rome the use of fractions occurred most frequently in money computation and in metrology. Each fraction had a special name, and the Romans usually kept the denominator a constant, 12, probably because their copper coin, *as*, weighing one pound, was divided into 12 *unciae*. Fractional computations were the chief part of arithmetical instruction in Roman schools.

The present method of writing fractions seems to have originated with the Hindus. Perhaps it was derived from Greece, where it had been used in the later period. In the Bakhshali manuscript (*c.* sixth century?) fractions were written with the numerator above the denominator without the dividing line. Integers were written as fractions with 1 as the denominator. As early as 1000 the Arabs had introduced the bar in the forms

$$a/b, \quad a-b, \quad \frac{a}{b},$$

but they did not make use of it in all cases. When the Jewish rabbi Abraham ben Ezra (1097–1167) adopted the Moorish forms he generally omitted the bar, but it was commonly found in manuscripts after that time.

The main impetus for the use of decimal fractions was given by

Simon Stevin in 1585; again there was delay in developing a suitable notation, and the decimal point was not universally used until the first quarter of the eighteenth century [**36**].

It was the practice of most arithmetical writers in Europe to postpone the discussion of common and decimal fractions to the end of their books. Apparently few students had expectations of ever reaching fractions.

The Arabic word for fraction, *al-kasr*, is derived from the stem of the verb meaning "to break." The Latin forms *fractio* and *minutum ruptus* were then given by early writers in English as "broken numbers." In early American arithmetics "vulgar" was often applied to common fractions to distinguish them from decimal fractions.

For Further Reading

CAJORI (d): I, 309–14
HEATH (c): 20–24
KARPINSKI: 121–27

SANFORD (d): 101–7
D. E. SMITH (a): II, 208–35

Capsule 36   Leland Miller and James Fey

DECIMAL FRACTIONS

ALTHOUGH the Hindu place-value numeration system was perfected in the fifth or sixth century, the natural extension of this system to decimal fractions did not take place until almost a thousand years later.

The idea of decimal fractions first arose in approximating roots of irrational numbers. About the twelfth century, John of Seville added $2n$ ciphers to the given number, calculated the square root, and took this root as the numerator of a fraction whose denominator was 1 followed by n ciphers. The German Adam Riese (1522) gave a table of square roots, stating that since the numbers had been multiplied by 1,000,000, the roots are 1,000 times too large. The square root of 2 thus appeared as 1 414, although the integral and fractional parts appeared in separate columns.

Another German, Christoff Rudolff, in setting up a compound-interest table for a book written in 1530, used a vertical bar exactly as we use a decimal point today.

It is at least possible that the idea of decimal fractions in Europe came through contact with the Orient. The Persian astronomer Jamshid al-Kashi (c. 1430) multiplied 25.07 by 14.3 to obtain 358.501, although he did not use a decimal point as such. Al-Kashi in turn may have been influenced by the Chinese and the Indians, among whom some systematic use of decimal fractions had been found.

In 1579 the Frenchman François Viète (also known as Vieta) published a work that included a systematic use of decimal fractions (using both the comma and the vertical bar as the separatrix) and a strong plea for their adoption throughout mathematics.

Despite these early suggestions, the invention of decimal fractions is most often credited to the Dutch scientist Simon Stevin. In 1585 Stevin published *La Disme*, a seven-page work in which decimal fractions were explained and rules were given for applying them to the operations of arithmetic. Stevin's idea was transmitted to England by a 1608 translation of *La Disme;* on the Continent, the Swiss Jobst Bürgi (1592) and the German Johann Hartmann Beyer (1603) published treatises on decimals. Beyer even claimed the invention as his own.

The only significant improvement in Stevin's formulation of decimal fractions has been in notation. Stevin wrote 5.912 as

$$\overset{0\ 1\ 2\ 3}{5912}$$

or 5 ⓪ 9 ① 1 ② 2 ③   Various suggestions were made for separating the integral and fractional parts of a numeral. Some writers wrote 75/321, others 75$\frac{321}{}$, and still others 75,321.

The greatest impetus to the use of decimal fractions resulted from the invention of logarithms. Although the first logarithms (published by John Napier in 1614) contained no decimal fractions, these did appear in a 1616 English translation with a point as decimal separatrix. In his Latin *Rabdologia* of 1617, Napier proposed the notation 1993,273 (with either point or comma suggested), although he also used 821, 2′5″ for our 821.25.

Even today, despite the broad use of decimal notation, there is no universally accepted form for writing the "decimal point." For 3.25 (in American notation) the English write 3·25 and the Germans and French write 3,25.

138

For Further Reading

BOYER (k)   D. E. SMITH (a): II, 235–47
CAJORI (d): I, 314–35   ——— (c): I, 20–24
MIDONICK: 733–50   STRUIK (d)
SANFORD (d): 107–14   ——— (e): 7–11

Capsule 37   Clarence H. Heinke

ORIGINS OF SYMBOLS FOR OPERATIONS

FIRST use in print of $+$ and $-$ symbols for the operations of addition and subtraction can be traced to a book by Johann Widmann in 1498. There is clear evidence that, as a lecturer at the University of Leipzig, Widmann had studied manuscripts in the Dresden library in which $+$ and $-$ signify operations, some of these having been written as early as 1486.

Many historians conjecture that the $+$ and $-$ symbols originated in mercantile practice, where they indicated excess and deficiency, respectively. Some attempt to trace the minus symbol as far back as Heron (or Hero) of Alexandria and Diophantus. The plus symbol as an abbreviation for the Latin word *et* ("and"), though appearing with the downward stroke not quite vertical, was found in a manuscript dated 1417.

Although two line segments in the form of the St. Andrew's cross were used earlier for other purposes, first use of the symbol $\times$ for multiplication is attributed to William Oughtred; it was used in his *Clavis mathematicae* (1631) and in an anonymous appendix to an earlier book (1618). The appendix is judged to have been written by Oughtred, even though it was unsigned. Leibniz objected to use of Oughtred's symbol because of possible confusion with the letter x. Therefore, in 1698 he suggested use of the dot as a symbol for multiplication. This suggestion is credited with influencing wide adoption of that symbol.

Around the year 1200 both the Arabic writer al-Hassar and

139

Fibonacci (Leonardo of Pisa) symbolized division in fraction form with use of the horizontal bar, but it is thought likely that Fibonacci adopted al-Hassar's introduction of this symbolization. In his *Arithmetica integra* (1544) Michael Stifel employed the arrangement 8)24 to mean 24 divided by 8. The ÷ symbol was first used to signify division by the Swiss Johann Heinrich Rahn, in an algebra published in 1659.

For Further Reading

CAJORI (d): I, 71–400   D. E. SMITH (a): II, 395–416

Capsule 38   William B. Wetherbee

CASTING OUT NINES

FEW historians agree on the origin of the process of "casting out nines." Its discovery has been credited to sources as diverse geographically as Arabia, India, and Rome, with dates ranging from the third to the twelfth century. A Roman, Hippolytos, is one of the first known to have used this process. He employed the excess of nines during the third century in connection with gematria.

Casting out nines was used in Arabian arithmetic, beginning with that of al-Khowarizmi in the ninth century, to check computations. However, much Arabian mathematics is known to have come from Greek and Indian sources, and some historians credit the Hindus with discovery of the process. Hindu astronomers definitely used it through the twelfth century.

The casting-out-nines process passed to the West from the Arabs. It appeared in the works of the Italian Fibonacci (1202), the German Widmann (1489), and the Englishman Recorde (1540), and it was brought to America by Isaac Greenwood (1729). It disappeared from American textbooks during the nineteenth century, but it was restored after 1900.

For Further Reading

CAJORI (e): 91   [5th ed. 174, 184]
DICKSON (c): I, 337   ORE (c): 225–30
EVES (c): 194, 204   D. E. SMITH (a): II, 151–54

Capsule 39   **Nathaniel Mann III**

NAPIER'S RODS

IN 1617, the year of his death, John Napier's *Rabdologia* was published. It describes three different methods of making calculations. One method of performing multiplication involves the use of rods with numerals imprinted on the four faces. *Rabdologia* attracted more general attention than Napier's work on logarithms, and the calculating rods came into widespread use both in Great Britain and on the Continent.

Napier's rods consist of ten rectangular blocks with multiples of a different digit on each of the four long faces. The multiples are listed in column form, each being set into a square divided by a diagonal so that the tens digit is at the top and the units digit below. The four sides of one of the rods are shown in Figure [**39**]–1.

Figure [**39**]–2 shows an example of the use of the rods in multiplying 7,259 by 364. In order to perform the multiplication, we select rods with columns headed by 7, 2, 5, and 9 and place them together in the proper order. In the manner of *gelosia* multiplication [**34**], the products are read from rows 3, 6, and 4 by adding the digits in each parallelogram. These results are recorded as shown in the figure, and simple addition then gives the final answer.

Napier's rods are often referred to as Napier's "bones" because of the title of a translation published in **1667** by William Leybourn; *The Art of Numbring By Speaking-Rods: Vulgarly termed Nepeir's Bones*. Leybourn had thought the latter part of the original title, *Rabdologia*, to be derived from the Greek word *logos*, speech, rather than *logia*, collection.

Napier also designed special rods for square roots and cube roots. Among the different variations of Napier's rods were Genaille's

141

FIGURE [39]-1

```
  7,259
    364
  21777
  43554
  29036
2642276
```

FIGURE [39]-2

rods, invented by Henri Genaille and "perfected" by Édouard Anatole Lucas in 1885. These eliminated one step in the use of the original rods, that of mentally adding the amount "carried" to the next rod at each step of the reading.

For Further Reading

JONES (h)   D. E. SMITH (c): I, 182–85
SANFORD (d): 339–40

Capsule 40   Bernard J. Yozwiak

LOGARITHMS

JOHN Napier (1550–1617), a Scotsman, is credited with being the inventor of logarithms. He published his discussion of logarithms in

1614 under the title *Mirifici logarithmorum canonis descriptio* ("A Description of the Wonderful Law of Logarithms"). The only rival to Napier as claimant to being the inventor of logarithms would be a Swiss watchmaker, Jobst Bürgi (**1552–1632**). Bürgi's table of logarithms appeared in **1620** under the title *Arithmetische und geometrische Progress-Tabulen*. It appears that the men worked independently, but the prior publication of Napier's work gives him the greater recognition.

In our modern terminology, if $a^x = y$, then the logarithm of y to the base a is x. Note that as x varies in an arithmetic progression y varies in a geometric progression. This fundamental correspondence between two series of numbers was arrived at by Napier in a geometric manner. Napier's definition of logarithms, however, had nothing to do with exponents—in fact, not even a standard or suitable notation for expressing exponents had been fully developed by his time.

The technique Napier used was to represent numbers and logarithms of these numbers by segments of lines cut off by two moving points. This can be illustrated as follows: Let a line segment AB be of fixed

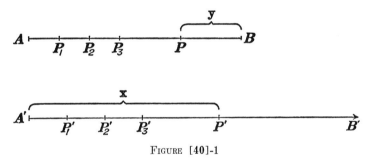

FIGURE [40]-1

length (Napier took the length to represent the number 10,000,000) and the line $A'B'$ be of infinite extent. Consider a point P to move on AB with a velocity numerically equal to the distance PB and a point P' to move on $A'B'$ with a constant velocity equal to the initial velocity of P, as in Figure [40]-1. Letting $P'_1, P'_2, \ldots$ correspond to $P_1, P_2, \ldots$, respectively, Napier defined the logarithm of PB to be the length $A'P'$. That is, the logarithm of $P_1B = A'P'_1$, the logarithm $P_2B = A'P'_2$, and so forth. Note that, though the lengths PB are decreasing, their logarithms are increasing. It turns out that, over a succession of equal periods of time, the lengths PB decrease in a geometric progression while the lengths AP increase in an arithmetic progression, thereby achieving the correspondence previously referred to.

143

In the construction of Napier's logarithms, the first member of the geometric series was taken to be 10,000,000 and assigned the logarithm 0. The common ratio of the geometric series was taken to be $1 - (1/10,000,000)$. The second member of the series is

$$10,000,000\,(1 - (1/10,000,000)) = 9,999,999;$$

to it was assigned the logarithm 1. The next member of the series, 9,999,998.0000001, was assigned the logarithm 2. The number whose logarithm was 100 was approximately 9,999,900. Napier also discovered that if $a:b = c:d$, then $\log a - \log b = \log c - \log d$, so that when the number whose logarithm was 100 was found, it was relatively easy to find the number whose logarithm was 200. For example, the number x satisfying the equation

$$9,999,900:10,000,000 = x:9,999,900$$

is the number whose logarithm is 200, since it must also satisfy

$$\log 9,999,900 - \log 10,000,000 = \log x - \log 9,999,000;$$

log 9,999,900 is 100, and log 10,000,000 is 0.

The number 10,000,000 for the length of the line AB was selected by Napier to be the sine of 90° with the decimal point deleted. The other numbers in his geometric progression corresponded (approximately to seven decimal places) to the sine of certain angles. He then constructed a table of logarithms of the sines of these angles and completed the table by interpolation.

The statement sometimes made that Napier's logarithms are natural logarithms (base e) is misleading. This can be demonstrated by using a little calculus. Referring to Figure [40]–1, let $PB = y$ and $A'P' = x$. Then $AP = 10,000,000 - y$. The velocity of point P is (d/dt) $(10,000,000 - y) = -(dy/dt)$, but the velocity is numerically equal to the distance $PB = y$, so $-(dy/dt) = y$. Separating variables and integrating yields $-\ln y = t + c$. At $t = 0$, $y = 10,000,000 = 10^7$ and so $c = -\ln 10^7$. The equation becomes

(1) $$-\ln y = t - \ln 10^7.$$

Since the velocity of P' is to equal the initial velocity of P, $dx/dt = 10^7$, and $x = 10^7 t + c$. But when $t = 0$, $x = 0$ and C must be 0. This implies that $t = x/10^7$. Substituting this result in Equation 1, $-\ln y = x/10^7 - \ln 10^7$, which simplifies to $x = 10^7 \ln(10^7/y)$. But x is Napier's logarithm of y. We then have the following relation between Napierian and natural logarithms:

144

$$(2) \qquad \text{Napierian log } y = 10^7 \log_e \frac{10^7}{y} = 10^7 \log_{1/e} \frac{y}{10^7}.$$

It is evident from this formula that Napier's logarithms are not the same as natural logarithms. Napier had no notion of base in his system.

Henry Briggs (1561–1631), an Englishman, visited Napier, and during this visit Napier and Briggs agreed that the tables of Napier would be more useful if they were altered so that the logarithm of 1 would be 0 and the logarithm of 10 would be 1. This change resulted in the invention of the "Briggsian" or "common" logarithms, which are so useful for computation.

The possibility of defining logarithms as exponents was recognized by John Wallis in 1685 and by Johann Bernoulli in 1694. A systematic exposition of logarithms based on this idea is attributable to William Jones and was given in Gardiner's *Tables of Logarithms* (1742). In 1749 Leonhard Euler published an article that gives the modern result that log n (n any nonzero real number) has an infinite number of values; these are all imaginary except when n is a positive number, in which case one logarithm out of this infinite set is real.

For Further Reading

CAJORI (e): 149–55, 235–37
EVES (c): 242–46
  [5th ed. 226–29]
HOOPER: 169–93

SANFORD (d): 191–97
SCHAAF (c)
D. E. SMITH (c): I, 149–55
STRUIK (e): 11–21

Capsule 41   James Fey

NOMOGRAPHY

THE use of graphic techniques for computation and solution of equations goes back to antiquity. In the time of Hipparchus (150 B.C.), graphic solution of sperical triangles was popular. During the Middle Ages, Arab mathematicians used geometric means to solve quadratic equations, and in the seventeenth century William Oughtred used graphic methods for solving spherical triangles. However, the key to general application of graphic methods to the solution of algebraic problems was analytic geometry, introduced by René Descartes in his *Discours de la méthode* (1637). The theory of nomograms rests largely on analytic geometry.

In 1842 Léon Lalanne pointed out that by altering the scales along the Cartesian axes it is often possible to simplify graphs of equations in two variables. Furthermore, he noted that if these changes are subject to certain minimal restrictions, the new graph is essentially equivalent to its Cartesian counterpart. Lalanne called his new theory "geometrical anamorphosis," and advances were made in this theory by J. Massau and Charles Lallemand during the 1880's.

These and other developments were foreshadowings. The real creator of nomography was the French mathematician Maurice d'Ocagne (1862–1938). D'Ocagne was the first to describe the "alignment chart" (1884), and he applied this chart to many engineering formulas. In 1899 he published *Traité de nomographie,* in which he brought together both the general theories and many applications of the subject. Since that time numerous texts on the subject have been published, and many nomograms have appeared in technical journals.

It is interesting that the original impetus for the study of nomography came from problems that arose during the construction of railroads in France. Thus most nineteenth-century contributors to the subject were engineers. In fact, nomography remains essentially a branch of applied mathematics with uses in engineering, industry, and the physical and natural sciences.

For Further Reading

D. E. Smith (c): I, 192–200

Capsule 42　　Harlen E. Amundson

PERCENT

Percent has been used since the end of the fifteenth century in business problems such as computing interest, profit and loss, and taxes. However, the idea had its origin much earlier. When the Roman emperor Augustus levied a tax on all goods sold at auction, *centesima rerum venalium,* the rate was 1/100. Other Roman taxes were 1/20 on every freed slave and 1/25 on every slave sold. Without recognizing percentages as such, they used fractions easily reduced to hundredths.

In the Middle Ages, as larger denominations of money came to be used, 100 became a common base for computation. Italian manuscripts of the fifteenth century contained such expressions as "20 p 100," "x p cento," and "vi p c°" to indicate 20 percent, 10 percent, and 6 percent. When commercial arithmetics appeared near the end of that century, use of percent was well established. For example, Giorgio Chiarino (1481) used "xx. per .c." for 20 percent and "viii in x perceto" for 8 to 10 percent. During the sixteenth and seventeenth century, percent was used freely for computing profit and loss and interest.

The percent sign, %, has probably evolved from a symbol introduced in an anonymous Italian manuscript of 1425. Instead of *"per* 100," *"P* 100," or *"P cento,"* which were common at that time, this author used "P $\stackrel{\text{o}}{\frown}$." By about 1650 the $\stackrel{\text{o}}{\frown}$ had become $\frac{\text{o}}{\text{o}}$, so "per $\frac{\text{o}}{\text{o}}$" was often used. Finally the "per" was dropped, leaving $\frac{\text{o}}{\text{o}}$ or %.

For Further Reading

Cajori (d): I, 312   D. E. Smith (a): II, 247–50

Capsule 43   Leland Miller

RADICAL SYMBOL

When mathematics was still in the rhetorical stage, the words for "root" or "side" were commonly written to represent what we now call the square root of the indicated number. The Arab writers thought of a square number as growing out of a root, and so works translated from the Arabic often used the word *radix;* we still speak of "extracting," or pulling out, the root. Latin writers thought of "finding" the *latus,* or the side of a geometric square.

Late medieval Latin writers contracted *radix* into a single symbol, ℞. This was used for more than a century, and the early French writer Nicolas Chuquet (1484) sometimes used ℞² for ℞; he also introduced ℞³ and ℞⁴ for cube and fourth roots.

The symbol √ first appeared in print in Christoff Rudolff's *Die Coss* in 1525, but without indices to indicate higher roots. The symbol √ may have been used because it resembled a manuscript form of the small *r* (*radix*), or it may have been an arbitrary invention. The cube and fourth roots were indicated by ₍c₎√ and √√. When Michael Stifel edited *Die Coss in* 1553, he indicated these roots by ₂√, ₃√, ₄√, and so on.

Rudolff's symbol did not gain immediate acceptance, even in his native Germany. The letter *ℓ* (*latus*, "side") was often used; thus *ℓ*4 was used for √4, and *ℓ*c 5 for ∛5. By the seventeenth century, however, the use of Rudolff's symbol for square root had become quite standard, although there were still many variations in the way the indices were written for the higher roots.

For Further Reading

SANFORD (d): 158–59   D. E. SMITH (a): II, 150–51, 407–10

Capsule 44   Hermann von Baravalle

THE NUMBER π

THE number π is the ratio of the circumference of a circle to its diameter. It is also the ratio of the area of a circle to the area of the square on its radius. Similarly, π appears as a ratio related to certain surface areas and volumes in solid geometry. Formulas of the ellipse and of various curves, such as the astroid, the cardioid, the limaçon, and the roses, also contain π. But the use of the number π is by no means restricted to geometrical situations. It appears in many branches of mathematics, including such seemingly unrelated fields as vibrational theory, the theory of numbers, statistics, and actuarial theory.

The gradual development of an understanding of this concept can be traced from the earliest historical records of mathematics to the present. This development includes an interplay of some of the most diverse aspects of mathematics, from empirical guesses to highly refined theoretical considerations.

One of man's earliest geometrical problems was to find a square equal in area to that of a given circle. (The expression "quadrature of the circle" is therefore associated with construction problems that involve π or an approximation to its value.) Many of the earliest references to this problem do not indicate that the concept of a constant ratio was clearly recognized at that time. Representative of early Egyptian mathematics is Problem 41 in the Rhind papyrus (*c.* 1650 B.C.) : "Example of working out a circular container of diameter 9 and height 10. You are to subtract a ninth of 9, namely 1; remainder 8. Multiply 8 eight times, result 64. You are to multiply 64 ten times, it becomes 640." If one generalizes this problem, the area of the base circle is found as the square of 8/9 of the diameter. The Egyptians, of course, did not generalize this into a stated formula, but the Rhind papyrus includes five such worked-out problems involving areas of circles, four using a diameter of 9 and the other a diameter of 10. If one converts the above results into the formula $A = k \cdot r^2$, the ratio k is approximately 3.1605. The often-found statement that the Egyptians' value of π was 3.16 must be interpreted in the above context. (A reproduction of the portion of the Rhind papyrus that contains this problem is found in /HOGBEN: 54/.)

In problems written on clay tablets by the Babylonians (probably 1800–1600 B.C.) the area of a circle is found from the circumference by means of the relation (expressed sexagesimally) $A = (0;5)(c^2)$. Since 0;5 (one-twelfth) thus becomes an approximation for $1/(4\pi)$, this yields an approximation of 3 for π. In some problems, however, a correction factor was introduced: the multiplication of the previous result by 0;57,36. The final result is the equivalent of using 3 1/8 for π /JONES (g): 164; NEUGEBAUER (a): 47/.

One sometimes finds the statement that the Bible uses a value of 3 for π. This is based on I Kings 7:23, which simply states the fact that a basin [built in connection with King Solomon's temple, *c.* 1000 B.C.] was ten cubits from one brim to the other, "and a line of 30 cubits did compass it round about."

One of the earliest Greek mathematicians to attempt to treat the problem of the "quadrature of the circle" in pure geometric form, with the specific restriction that only compass and straightedge should be used, was Hippocrates of Chios (*c.* 440 B.C.). He was able to show that the area of certain lunes (crescent-shaped figures formed by two intersecting arcs) could be represented exactly by triangular (and hence rectangular) areas. For example, if *AOB* is a quadrant of a circle and *AB* is the diameter of a semicircle lying outside the quadrant, then the lune bounded by the semicircle and the quadrant has the same area as triangle *AOB*. His success with such special cases led him to suppose that he could eventually draw a polygon and hence a square whose area is exactly that of a circle. /MIDONICK: 406–13; VAN DER WAERDEN: 131–36./

Another early Greek, Dinostratus (*c.* 350 B.C.) was able to use a curve called the "quadratrix," invented earlier by Hippias (*c.* 425 B.C.) to find the length of a quadrant. The curve, which divides the radius of the given circle at the ratio of $\pi/2$, is not determined exactly by ruler-and-compass construction, however; thus this method did not meet the restricted conditions set down by the Greeks/VAN DER WAERDEN: 191–92/.

The *Elements* of Euclid (300 B.C.) make no mention of the constant ratio between circumference and diameter, but Book II, Proposition 2, gives a formal proof that the areas of two circles are to each other as the squares on the diameters. This relation was already known to Hippocrates, as has been mentioned. Archimedes, in his work *The Measurement of the Circle* (240 B.C.), proved that the area of any circle is equal to the area of a right triangle with one leg equal to the radius and the other equal to the circumference of the circle.

A second important result given by Archimedes was that the ratio of the circumference of any circle to its diameter is less than 3 1/7 but greater than 3 10/71, and so 3 1/7 has often been called the "Archimedean value of π." His method depended on working with inscribed and circumscribed regular polygons, continually doubling the number of sides until the perimeters of polygons of 96 sides were obtained /ARCHIMEDES: 91–98; DÖRRIE: 184–88/.

Various approximations for π have been offered by mathematical writers in different countries. For example, an early Chinese worker in mechanics, Tsu Ch'ung-chih (*c.* A.D. 470) gave the rational value of 355/113 (3.1415929 . . .), which is correct to the sixth decimal place. Aryabhata in India (A.D. 510) gave for π: "Add 4 to 100, multiply by 8, and add 62,000. This is the approximate circumference of a circle, the diameter of which is 20,000." Written as a fraction, this is 62,832/ 20,000—as a decimal, 3.1416, which is less than .0001 above π.

Another Hindu writer, Brahmagupta (*c.* A.D. 628) gave 3 as the "practical value" and $\sqrt{10}$ (3.162...) as the "exact value." The latter value may be due to the commonly used approximation formula $a + (1/(2a + r))$ for $\sqrt{a^2 + r}$, since $\sqrt{10} = \sqrt{9 + 1} \approx 3 + 1/7$, the Archimedean value for π. The value $\sqrt{10}$ was used extensively during the Middle Ages.

Approximations for π by means of the "classical method" orginated by Archimedes can be carried out to as many decimal places as the computational ability and the perseverance of the worker permit. In 1579 the French mathematician François Viète used polygons having $6 \cdot (2^{16}) = 393,216$ sides to find π correct to nine decimal places. In 1610 Ludolph van Ceulen, of Germany, used polygons having 2^{62} sides to compute π to thirty-five places. (In Germany, π is still commonly called the Ludolphian number.)

The simplest approximation construction with straightedge and compass (it gives the value of π to 1/10,000 of the radius) was given by the Polish Jesuit Adamas Kochansky in 1685. The construction consists of a circle, its vertical diameter, and the two tangents at its end points. One tangent is extended to its intersection with the prolonged radius, which encloses an angle of 30° with the vertical diameter. The other tangent is extended to the length of three radii. The line segment between the end points of the two tangents has the approximate length of π times the radius. If a radius of 1 is used, the length is $\sqrt{40/3} - \sqrt{12} \approx 3.14153 \cdots$. /BARAVALLE (c): 345–46./

Despite the achievements in evaluating π up through the sixteenth century, it may be said that they were disappointing to those who accomplished them. It was expected that π, the ratio of the outstanding measurements of the most perfect geometrical form, would be a very special number. This expectation lived on as a firm conviction that its mathematical secret had not yet been revealed.

New light was shed on π approximately two thousand years after the Greek period. It came with the discovery of the previously mentioned French mathematician, Viète, who in 1592 found the formula

$$\pi = \frac{1}{\sqrt{\frac{1}{2}} \cdot \sqrt{\frac{1}{2} + \frac{1}{2}\sqrt{\frac{1}{2}}} \cdot \sqrt{\frac{1}{2} + \frac{1}{2}\sqrt{\frac{1}{2} + \frac{1}{2}\sqrt{\frac{1}{2}}} \cdots}} .$$

In 1658 William Brouncker expressed π in the form of an infinite continued fraction:

151

$$\pi = 4 \cdot \cfrac{1}{1 + \cfrac{1^2}{2 + \cfrac{3^2}{2 + \cfrac{5^2}{2 + \cfrac{7^2}{2 + \cdots}}}}}$$

About the same time an English mathematician, John Wallis, proved that

$$\frac{\pi}{2} = \frac{2}{1} \cdot \frac{2}{3} \cdot \frac{4}{3} \cdot \frac{4}{5} \cdot \frac{6}{5} \cdot \frac{6}{7} \cdot \frac{8}{7} \cdot \frac{8}{9} \cdots .$$

Finally, π was found as the limit of an infinite series. This was accomplished in 1674 by the German mathematician and philosopher Gottfried Wilhelm von Leibniz as

$$\pi = 4 \cdot (1 - \tfrac{1}{3} + \tfrac{1}{5} - \tfrac{1}{7} + \tfrac{1}{9} - \tfrac{1}{11} + \cdots).$$

An expression with simpler fractions and greater regularity could hardly be imagined. With the Leibniz series, the Greek expectations could not have received a more perfect answer. Other series have since been conceived with the aim of reaching a stronger convergence in order to ease the computing, but no other series has the same classical simplicity.

More concerning the exact nature of the number π was established in 1761 when Johann Heinrich Lambert showed that π is irrational (and hence cannot be expressed as a repeating decimal). Finally, in 1882, Ferdinand Lindemann proved that π is not an algebraic number and hence is classified as a transcendental number. (An algebraic number is one that is the root of a polynomial equation with rational coefficients.) This result conclusively established the fact that the quadrature of the circle by straightedge and compass alone is impossible.

The computation of π to a large number of places by means of the various series representations was aided by the use of such relations as $\pi/4 = 4 \arctan (1/5) - \arctan (1/239)$, given by John Machin in 1706. This relation was used by William Shanks of England to compute π to 707 places, a task completed in 1873 after fifteen years of work. After World War II, when computations of π were performed on electronic computers, a mistake was found in the 528th place of Shanks's work. Comparison of human versus electronic-computer speeds (as well as the increase in the speed of the computers) is vividly illustrated in the 1949 ENIAC calculation to 2,037 places in seventy hours; the calculation in 1958 to 10,000 places in one hour and forty minutes (forty seconds for the first 707 places!); and in 1961, the computation to 100,265 places in eight hours and forty-three minutes /WRENCH/.

In 1967 an approximation to π extending to 500,000 decimal places was attained in France on a CDC 6600 computer /EVES (c) [3d ed.]: 94/.

The adoption of the symbol π for the ratio is essentially due to the usage given it by Leonhard Euler from 1736 on. In the 1730's Euler first used p and c for the circumference-to-diameter ratio, then adopted the symbol π. However, he is not the originator of the symbol.

An actual ratio symbol, δ/π, had been used by William Oughtred (1647) and Isaac Barrow (1664) to indicate the ratio of the diameter to the circumference (periphery). David Gregory, nephew of James Gregory, used π/ρ for the ratio of circumference to radius (1697). In 1706 the English writer William Jones, in a work that gave the 100-place approximation of John Machin, first used the single symbol π.

The appearance of π in a different context is illustrated by Buffon's needle problem, which involves probability. In 1760 G. L. Leclerc, comte de Buffon, devised the problem of dropping at random a uniform rod (needle) of length l onto a plane ruled with parallel lines d distance apart ($l < d$). Buffon showed that the probability of the needle's falling across one of the ruled lines is $2l/\pi d$. If the length of the needle were 1/2 the distance between the lines, the probability of success would be $1/\pi$, and hence an approximation for π can be derived by experimentation. /EVES (c): 94–95./

The ranks of the circle squarers, those who have tried to find an exact value of π (and have often believed that they succeeded), have included men of all conditions and professions. One of the most famous was the English political philosopher Thomas Hobbes. In 1655, at the age of sixty-seven, he published the first of nearly a dozen different ways of squaring the circle. The next twenty-five years saw a continuing debate between Hobbes and the English mathematician John Wallis, who endeavored to make Hobbes's mathematical knowledge appear ridiculous and thereby discredit his religious and political opinions as well /GARDNER (d): 91–102/.

The expectations of the Greeks that the circle could be squared with compass and straightedge alone proved hopeless of fulfillment. But more important for the Greeks was their basic conviction that the order and beauty of the universe itself demands an outstanding mathematical solution for the ratio of its most perfect form. Seen in this light, the story of π ends no longer with a disappointment but, on the contrary, with its classical fulfillment in the Leibniz series.

153

For Further Reading

ARCHIMEDES: 91–98
BARAVALLE (c)
BARDIS
EVES (c): 89–97
  [5th ed. 83–90]
GARDNER (d): 91–102
HOBSON

JONES (g)
MIDONICK: 406–13
READ (c)
D. E. SMITH (a): II, 203–13
SMSG: VI, VII
STRUIK (e): 369–74
WRENCH

Capsule 45   James Fey

THE NUMBER *e*

THE invention of logarithms is generally attributed to the Scottish mathematician John Napier (1550–1617). Napier's logarithms, first described in 1614, were not dependent upon the idea of a base [40]. But in an appendix of the 1618 English translation of Napier's original Latin work, there is a table of logarithms that are actually natural logarithms. The table, which contains no decimal points, gives the logarithm of 10 as 2302584, whereas $\log_e 10 = 2.302584$.

The above table, probably attributable to William Oughtred, was expanded by John Speidell in his *New Logarithms* (1622) for the numbers 1–1,000. In modern notation, the logarithms of both Oughtred and Speidell are of the form $10^6 (\log_e x)$.

In 1667 another Scottish mathematician, James Gregory, showed how to compute logarithms by finding the areas of inscribed parallelograms between a hyperbola and its asymptotes, thus leading to the term "hyperbolic logarithms." By the end of the century it became apparent that such logarithms could be considered independently of the hyperbola and that logarithms could be considered as exponents with any number taken as the base of the system.

Building on the earlier developments of exponential series by Newton and Leibniz, Leonhard Euler published in 1748 his *Introductio in analysis infinitorum*, the most notable of all treatises dealing with the number e. Much of this well-developed account of his theory of the exponential function had been given earlier. At the age of twenty-one, while living at court in St. Petersburg, Russia, Euler wrote "Meditations upon Experiments Made Recently on the Firing of Cannon," in which he suggested, "For the number whose logarithm is unity, let e be written, which is **2.718281**. . . ." This number was given by the series

$$1 + \frac{1}{1} + \frac{1}{1 \cdot 2} + \frac{1}{1 \cdot 2 \cdot 3} + \frac{1}{1 \cdot 2 \cdot 3 \cdot 4} + \cdots .$$

The use of the symbol is original with Euler and marks the recognition by him of the existence of an exact number as the sum of a series and also as the base of the system of hyperbolic logarithms.

Euler derived two continued-fraction expansions for e:

$$e = 2 + \cfrac{1}{1 + \cfrac{1}{2 + \cfrac{1}{1 + \cfrac{1}{4 + \cfrac{1}{\ddots}}}}} \qquad\qquad e = 2 + \cfrac{1}{1 + \cfrac{1}{2 + \cfrac{2}{3 + \cfrac{3}{4 + \cfrac{4}{\ddots}}}}}$$

Using the relation

$$e = \lim_{n \to \infty} \left(1 + \frac{1}{n}\right)^n ,$$

he calculated e to twenty-three decimal places. Out of respect for these results and his establishment of the famous identity

$$e^{i\pi} + 1 = 0,$$

e is sometimes called Euler's number.

From his continued-fraction expansion for e Euler may have been the first to infer that e is irrational. After Joseph Liouville (1844) proved the existence of transcendental numbers [26], Charles Hermite in 1873 proved, moreover, that e is a transcendental number.

For Further Reading

BARAVALLE (b)   SCHAAF (c)
COOLIDGE (d)   D. E. SMITH (c): I, 95–106

PASCAL'S TRIANGLE

THE triangular table of numbers that is known as Pascal's triangle was named after Blaise Pascal (1623–1662) subsequent to the composition of his famous *Traité du triangle arithmétique,* published posthumously in 1665. However, this arithmetic triangle, which yields the binomial coefficients

$$
\begin{array}{ccccccc}
 & & & 1 & & & \\
 & & 1 & & 1 & & \\
 & & 1 & 2 & 1 & & \\
 & 1 & 3 & & 3 & 1 & \\
 1 & 4 & & 6 & & 4 & 1 \\
1 & 5 & 10 & & 10 & 5 & 1
\end{array}
$$

. . . .

appeared in China as early as 1303 in Chu Shih-chieh's *Precious Mirror of the Four Elements,* on algebra, and it is believed that Omar Khayyam (*c.* 1100) had some knowledge of the relationship between binomial coefficients. This device was also known to many other men preceding Pascal /BOYER (a): 388/.

In 1556 Niccolo Tartaglia gave it as his own invention! Nonetheless, Pascal proposed the triangle in a new form, shown below, and investigated its properties more deeply than his predecessors.

$$
\begin{array}{lllll}
1 & 1 & 1 & 1 & 1 & \cdots \\
1 & 2 & 3 & 4 & \cdots \\
1 & 3 & 6 & \cdots \\
1 & 4 & \cdots \\
1 & \cdots
\end{array}
$$

Some of the theorems in his work are proved by mathematical induction, this apparently being the first recorded use of this method [83]. A few examples of his results are the following (a figurative triangle being used to denote Pascal's triangle):

1. Any integer b in the triangle is equal to the sum of all preceding diagonal integers $r, s, t, \ldots, 1$. See Figure [46]-1.

2. Any integer b diminished by 1 is equal to the sum of all the integers above the diagonal lines shown (the shaded area in Fig. [46]-2).

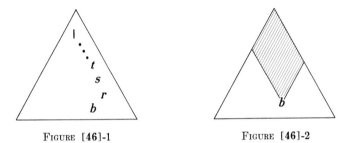

FIGURE [46]-1   FIGURE [46]-2

3. The sum of all the integers in a row equals twice the sum of all the integers in the preceding row.

It is not too well known that the Italian Girolamo Cardano (1501–1576) anticipated Pascal with respect to both form and study of the triangle /BOYER (a)/. But even Cardano's work was not entirely original. Tartaglia, in 1556, had published some work on the triangle and had used this array to determine coefficients of the binomial expansion to the twelfth power. The fact remains, however, that Cardano's work was quite masterful and inventive. He gave not only the rule

$$\binom{n}{r} = \frac{n - r + 1}{r} \binom{n}{r - 1},$$

in rather awkward nonsymbolic form, but also the triangle's application to progressions of higher order and to the study of musical theory. and harmony.

Pascal applied the triangular table to his fledgling theory of probability. It is fitting recognition of Pascal's significant contribution to the theory of this arithmetic triangle that it bears his name today.

Many papers have been written on this subject since Pascal's time. Rather recent results relate the coefficients of $(a + b)^{-n}$ to entries along the diagonals of the Pascal triangle /ROBINSON/. Even more recently, the association of the integers of the triangle with the Fibonacci sequence gives further evidence of the richness and importance of this simple array of numbers /RAAB (b)/.

157

For Further Reading

BOYER (a)

COOLIDGE (c): 92–95

EVES (c): 259–61, 277–78

  [5th ed. 244–45, 255–56]

HOGBEN: 160–67, 236–38

MIKAMI

RAAB (a)

—— (b)

ROBINSON

SANFORD (d): 177–79

D. E. SMITH (a): II, 508–10

—— (c): I, 67–69

SMITH and MIKAMI: 51

STRUIK (e): 21–26

Capsule 47   Jamey Fey

PROBABILITY

UNDOUBTEDLY questions of chance have engaged the attention of men since antiquity. However, the mathematical treatment of probability did not begin until the fifteenth century. One of the first printed discussions of games of chance is Luca Pacioli's *Sūma* (1494). In this work Pacioli proposed the now famous "problem of points" or "division problem" in which one is required to determine, on the basis of partial scores, a fair distribution of the stakes in an interrupted game of chance. In the sixteenth century Girolamo Cardano, another Italian, wrote a gambler's handbook in which he stated certain rules that allowed one to solve the dice problem (odds of getting a specified result) for one die. About fifty years later Galileo gave a complete table for the three-dice problem.

Despite these early attempts at mathematizing the laws of chance, the birth of probability theory as a mathematical discipline is identified with the year 1654. In that year George Brossin, chevalier de Méré, a courtier of Louis XIV, proposed both the dice problem and the division problem to Blaise Pascal. The chevalier had found that his theoretical reasoning on the problems did not agree with his observations. Pascal communicated the problem to Pierre de Fermat, and both soon arrived at solutions. Their work is considered the major breakthrough in founding the mathematical theory of probability.

The work of Pascal and Fermat generated widespread interest in probability. In 1655 Christiaan Huygens traveled to Paris to learn more of Pascal's ideas. Although he failed to see the famous Frenchman, he returned home and independently composed a treatise on the theory of probability that was the first of its kind to be published (*c.* 1658). Although in its infancy the study of probability centered about games of chance and was plagued by paradoxes resulting from an unsatisfactory definition of probability, after the year 1700 the theory began to advance at a rapid pace.

The second significant publication was an essay by Pierre-Rémond de Montmort (1708). In 1713 the first book devoted entirely to probability appeared. It was *Ars conjectandi* by Jakob (Jacques) Bernoulli—the first of a series of major contributions by the Bernoulli family. Abraham De Moivre published his *Doctrine of Chances* in 1718, and the theory of probability continued to grow throughout the eighteenth and nineteenth centuries with contributions from Euler, Laplace, Gauss, and many others.

In the last one hundred years probability theory has advanced significantly through the efforts of mathematicians to provide a firm mathematical base on which to build it. The culmination of this effort was publication, during the 1930's, of Kolmogorov's *Foundations of the Theory of Probability,* which developed probability theory on a rigorous axiomatic basis, thus completing the transition of the subject from a collection of suggestions for gamblers to a deductive mathematical system.

For Further Reading

Eves (c): 261, 294–96
  [5th ed. 245, 268–69, 272]
King and Read (b)

Ore (d)
D. E. Smith (a): II, 528–30
Todhunter

Capsule 48   James Fey and John W. Alexander, Jr.

CALCULATING PRODIGIES

The ability to execute mentally, at high speed, long and intricate calculations (such as multiplying numbers of ten or twenty digits and raising numbers to high powers) is a rare gift. It is a talent that might naturally be associated with general mathematical ability, and indeed the famous mathematicians Arthur Cayley and André Ampère demonstrated exceptional skill at reckoning with large numbers. However, the truly outstanding calculating prodigies have shown little other mathematical talent. W. W. Rouse Ball has compiled a fascinating account of the most prominent self-taught calculating prodigies /NEWMAN: I, 467–87/.

The first prodigy Ball mentions is Jedediah Buxton. In 1751 Buxton calculated mentally the number of cubic inches in a block of stone 23,145,789 yards long, 5,642,732 yards wide, and 54,965 yards thick. Thomas Fuller, brought to Virginia as a slave in 1724, could multiply two nine-digit numbers, state the number of seconds in a given period of time, and calculate the number of grains of corn in a given mass even though, like Buxton, he never learned to read or write.

In the nineteenth century Zerah Colburn earned the nickname "American Calculating Boy" by such computations as 8^{16} (281,474,-976,710,656), done in seconds, and the cube root of 268,336,125, given instantaneously. While on a tour of England in 1818, Colburn was bested by another self-taught boy, George Parker Bidder.

In addition to routine multiplication and finding roots, Bidder was adept at calculating compound interest. In one minute he figured the compound interest on £11,111 for 11,111 days at 5 percent. Unlike the other prodigies mentioned, Bidder later pursued formal studies and became a civil engineer of distinction. He retained his calculating powers throughout his life.

No doubt excellent memory and undivided devotion to mental calculation helped Buxton, Fuller, Colburn, Bidder, and other calculating prodigies perform their amazing feats. However, their strange power has never been fully explained, and it is rarely achieved.

For Further Reading

BELL (d): 380   NEWMAN: I, 467–87
BOWDEN: 311–16

THE MODERN DIGITAL COMPUTER

A DIGITAL computer is a device for performing mathematical operations with numbers expressed in the form of digits. Such devices stem from the abacus, a mechanical extension of the idea of finger counting [28, 29]. Computational aids that may be included in this family are Napier's rods (1617) and the calculating machines of Pascal (1642) and Leibniz (1684). The modern electronic digital computer is, of course, the most highly developed and useful member of this family.

The first suggestions for an automatic machine to do mathematical computation came from the mathematician Charles Babbage. He was born in Devonshire in 1792, the son of a banker from whom he inherited considerable private means.

When Babbage was twenty, he made his first suggestions for an automatic computer and started work on a scheme to prepare mathematical tables (such as tables of compound interest, logarithms, and trigonometric functions) without the help of a human operator except at the start of the computation. His idea was to start the tables by hand for a few entries, take differences out far enough to reach practically constant values for a long period, and then set the difference engine to extend this table.

Babbage proposed to use his engine for computing new tables and for checking tables that had already been published. The "difference engine" was designed to build up functions automatically by the use of high-order differences. Hence it needed to have a number of registers, one more than the degree of the polynomial, in which numbers could be stored; then it was to add successively the number in each register to that of an adjacent register.

Babbage persuaded the British government to finance the construction of a full-size difference machine. For the next ten years he worked on the design and the construction of his machine; he had to supervise the manufacture of various parts and invent hundreds of tools and machines. The government built him a workshop and a special fireproof vault to house the drawings that were prepared. Babbage

161

took much longer than he had anticipated; and in 1842, after the government had spent close to a million dollars, official support was withdrawn and the difference machine was abandoned.

While Babbage lost himself in a machine which, while simple in principle and sound in concept, was impossible to build because of the number of parts needed, George and Edward Scheutz built a small four-order "difference engine" in Sweden. Babbage deserves credit for inspiring this work, for it was an article on Babbage's machine in the July 1834 *Edinburgh Review* which inspired George Scheutz, a Stockholm printer, to tinker with difference machines. The machine, working properly, was completed in 1853 with financial assistance from the Swedish Diet. In 1856 the engine was purchased for the Dudley Observatory in Albany, New York, and it was of great value in the construction of tables.

In 1883 Babbage conceived the idea of an "analytical engine." This was to be an extension of the capabilities of the difference engine—to add, print, multiply, divide, and call for new data.

The analytical machine was never built, partly because of financial difficulties and partly because of engineering problems that were insurmountable at the time. Babbage did not publish many details of the mechanical design of the analytical engine, though some notes were included in a lecture given to the British Association. However, he left detailed drawings; and these, together with his notebook and a portion of the machine put together in 1906 by his son, H. P. Babbage, are now in the Science Museum, South Kensington, England.

An offshoot of Babbage's work is of great importance to business today. In order to insert data and instructions into the machine, he proposed to use a modification of the punched control cards used for weaving complex designs in the Jacquard loom. In 1780 Joseph Marie Jacquard in France had developed the idea of controlling a loom by means of holes, punched in cards, which served to control the selection of the threads used in the weaving process so that the loom would weave designs into the fabric. Babbage proposed to use holes in a similar fashion to control his machine's computational processes. Babbage's idea appears to have been two Jacquard mechanisms, each containing a string of cards to control the action of his machine.

The first programmed computer to operate successfully was built by Professor Howard H. Aiken of Harvard University. Begun in 1939, it was the first machine designed to use the principles of the analytical engine as they were conceived by Babbage. This Automatic Sequence Controlled Calculator (Harvard Mark I Calculator) was mechanical, although relays and clutches operated by electromagnets were used. The machine consisted of a panel 51 feet long and 8 feet high on which were mounted the tape readers, relays, and rotary systems that controlled the machine. As in Babbage's analytical engine, the numbers were stored in registers composed of wheels, each wheel capable of ten distinct positions; but each register was in itself a complete adding machine or an accumulator. There were 60 constant registers and 72 adding registers (accumulators), each capable of handling 23 decimals. The operations of the machine were controlled by 24-hole punched tape, which advanced about 200 steps per minute, so the basic time for addition was about 0.3 seconds. Built in were mechanisms for multiplication and for the computation of sin x, 10^x, and $\log_{10} x$. The mechanisms were called into action by groups of coding. The machine added, subtracted, multiplied, divided, compared quantities, consulted its memory of past operations when necessary, and referred to stored mathematical tables. It could be arranged to perform a series of steps necessary to solve logarithmic problems, compute various mathematical formulas, evaluate integrals, and solve differential equations.

Since the construction of the Mark I Calculator, further improved Harvard machines through Mark VI have been built. An outstanding feature of Mark V and Mark VI is the incorporation of elaborate error-detecting facilities.

The Electronic Numerical Integrator and Calculator (ENIAC) represented a considerable advance in computer-building technology, since it was entirely electronic in its internal operation. It was designed by Dr. J. P. Eckert and Dr. J. W. Mauchley at the Moore School of Electric Engineering of the University of Pennsylvania in cooperation with Major Herman H. Goldstine of Army Ordnance for the Ballistic Research Laboratory, Aberdeen (Maryland) Proving Ground, to compute firing tables for artillery. It was completed in 1946 and was much faster than any previous machine. The ENIAC in about seventy hours compiled π to 2,035 decimal places. The basic electronic device in this computer was the vacuum tube, which acted in the same manner as a relay. The vacuum tube was turned off or on by the electric current entering the tube. This machine used approximately 18,000 vacuum tubes as its operating components. The sequence control, to specify in advance a sequence of arithmetic operations and the numbers to be operated on, operated by means of many external wires running between plugboards and by external switches. To prepare ENIAC for a calculation, it was necessary to set up manually all the connections for the transmission of data from one unit to another.

The first theoretical advance in computer design resulted from the work of John von Neumann, who was influenced by early studies in pure logic made by A. M. Turing. While the ENIAC was being built, von Neumann was studying the logical design of computing machines. He suggested the use of a stored program, that is, the storing of instructions in the same manner as information. This gives a machine the ability to change and modify its instructions.

The EDSAC (Electronic Delayed Storage Automatic Computer), completed at the Mathematical Laboratory at the University of Cambridge in 1949 under the direction of M. V. Wilkes, was the first machine with a stored program to be completed and put into use.

No radically new ideas of the magnitude of the stored-program principle have appeared in the many computers designed since these early models. A great many advances, however, have been made in speed, reliability, and ease of use. Large-scale computers are made of internal memories that may contain 10 to 12 million ferrite "cores," each capable of storing one unit of information. The replacement of the vacuum tube by the transistor and now the replacement of the transistor by the microelectronic circuit have increased the speed of arithmetic and control circuits by reducing the distance an impulse has to travel.

Modern electronic digital computers possess many attributes in common. They are usually built in several units, only one being a computer or "processor." The other units are control, storage, and input-output devices. The modern machine is more often called a computing system. These systems use semiconductors and include magnetic-core and magnetic-tape storage. Almost every digital computer has been found capable of doing more than it was originally designed to do.

For Further Reading

BERNSTEIN

BOWDEN

EVES (c): 263–66 [5th ed. 484–89]

HAGA

LARRIVEE

MORRISON and MORRISON

NEWMAN: IV, 2066–123

WATSON and CALHOUN

IV

The History of GEOMETRY

an overview by

HOWARD EVES

T he story of the history of geometry, like that of many other growing and changing subjects, is composed of two intertwined strands. One strand narrates the growing content of the subject, and the other the changing nature of the subject. Everyone is aware that geometry must have started, probably far back in antiquity, from some very modest beginnings and then gradually grown to its present enormous size. On the other hand, not so many are aware that the nature, or inherent character, of the subject has had different connotations at different periods of its development. In the following brief history of geometry we shall endeavor to pay proper attention to both strands of the intriguing story.

SUBCONSCIOUS GEOMETRY

The first geometrical considerations of man are unquestionably very ancient. They would seem to have had their origin in simple observations stemming from human ability to recognize physical form and to compare shapes and sizes.

Much of the material of this overview is adapted, with appropriate permission, from the author's *An Introduction to the History of Mathematics* (Holt, Rinehart & Winston, 3d ed., 1969) and the historical introductions to some of the chapters of the author's *A Survey of Geometry* (Allyn & Bacon, Vol. I, 1963; Vol. II, 1964).

Innumerable circumstances in the life of even the most primitive man would lead to a certain amount of subconscious geometric discovery. The notion of distance was undoubtedly one of the first geometrical concepts to be developed. The need to bound land led to the notion of simple geometric figures such as rectangles, squares, and triangles. Other simple geometric concepts, such as the notion of vertical, of parallel, and of perpendicular, would have been suggested by the construction of walls and dwellings.

Many observations in the daily life of early man must have led to the conception of curves, surfaces, and solids. Instances of circles were numerous—for example, the periphery of the sun or the moon, the rainbow, the seed heads of many flowers, and the cross section of a log. A thrown stone describes a parabola; an unstretched cord hangs in a catenary curve; a wound rope lies in a spiral; spider webs illustrate regular polygons. The growth rings of a tree, the swelling circles caused by a pebble cast into a pond, and figures on certain shells suggest the idea of families of curves. Many fruits and pebbles are spherical, and bubbles on water are hemispherical; some bird eggs are approximately ellipsoids of revolution; a ring is a torus; tree trunks are circular cylinders; conical shapes are frequently seen in nature. Early potters made many surfaces and solids of revolution. The bodies of men and of animals, most leaves and flowers, and certain shells and crystals illustrate the notion of symmetry. The idea of volume arises immediately in the consideration of receptacles to hold liquids and other simple commodities.

Examples like the above can be multiplied almost indefinitely. Physical forms that possess an ordered character, contrasting as they do with the haphazard and unorganized shapes of most bodies, necessarily attract the attention of a reflective mind—and some elementary geometric concepts are thereby brought to light. Such geometry might, for want of a better name, be called "subconscious geometry." This subconscious geometry was employed by very early man in the making of decorative ornaments and patterns, and it is probably quite correct to say that early art did much to prepare the way for later geometric development. The evolution of subconscious geometry in little children is well known and easily observed.

Scientific Geometry

In the beginning, man considered only concrete geometrical problems, which presented themselves individually and with no observed interconnections. Later (but still before the dawn of recorded history), human intelligence evolved to the point where it was able to extract from a number of observations concerning the shapes, sizes, and spatial relations of specific physical objects certain general properties and relationships containing the former observations as particular cases. This introduced the advantage of ordering practical geometrical problems into sets such that the problems in a set can be solved by the same general procedure. One thus arrives at the notion of a geometrical law or rule. For example, comparing the lengths of circular courses with their diameters would lead, over a period of time, to the geometrical law that the ratio of circumference to diameter is a constant.

This higher stage in the development of the nature of geometry may be called "scientific geometry," in view of the fact that induction, trial and error, and empirical procedures were the tools of discovery. Geometry became a collection of general rule-of-thumb and laboratory results, some correct and some only approximate, concerning areas, volumes, and relationships of various figures suggested by physical objects.

No evidence permits us to estimate the number of centuries that passed before man was able to raise geometry to the status of a science, but writers of antiquity who concerned themselves with this matter unanimously agreed upon the Nile Valley of ancient Egypt as the place where subconscious geometry first became scientific geometry. The famous Greek historian Herodotus, of the fifth century B.C., stated the thesis in this wise:

> They said also that this king [Sesostris] divided the land among all Egyptians so as to give each one a quadrangle of equal size and to draw from each his revenues, by imposing a tax to be levied yearly. But every one from whose part the river tore away anything, had to go before him and notify what had happened. He then sent the overseers, who had to measure out by how much the land had become smaller, in order that the owner might pay on what was left, in proportion to the entire tax imposed. In this way, it appears to me, geometry originated, which passed thence to Hellas.

Thus the traditional account finds in early Egyptian surveying practices the beginnings of geometry as a science; indeed, the word "geometry" means "measurement of the earth." While we cannot be certain of this origin, it does seem safe to assume that scientific geometry arose from practical necessity, appearing several thousand years before our era in certain areas of the ancient Orient as a science to assist engineering and agricultural pursuits.[1] There is historical evidence that this occurred not only along the Nile River in Egypt but also in other great river basins, such as the Tigris and Euphrates of Mesopotamia, the Indus and Ganges of south-central Asia, and the Hwang Ho and the Yangtze of eastern Asia. These river basins cradled advanced forms of society known for their engineering prowess in marsh drainage, irrigation, flood control, and the erection of great edifices and structures. Such projects required the development of much practical geometry.

As far back as we can grope into the past, we still find present a sizable body of scientific geometry. Geometry appears to have remained of this type until the great Greek period of antiquity.

There is much to be said, at the elementary level of instruction, for the introduction of empirical, or experimental, geometry; many teachers feel it wise to precede a first course in demonstrative geometry with a few weeks of experimental geometry. The work of these weeks acquaints the student with many geometrical concepts, and it can be designed to emphasize both the values and the shortcomings of empirical geometry. Such instructional procedure follows the thesis that, in general, the learning program should parallel the historical development.

THE CONTENT OF PRE-HELLENIC GEOMETRY

The earliest existing records of man's activity in the field of geometry are some inscribed baked-clay tablets unearthed in Mesopotamia and believed to date, in part at least, from Sumerian times of about 3000 B.C. There are other generous supplies of Babylonian cuneiform tablets coming from later periods, such as the First Babylonian Dynasty of King Hammurabi's era, the New Babylonian Empire of Nebuchadnezzar II, and the following Persian and Seleucid eras. From these tablets we see that ancient Babylonian geometry is intimately related to practical mensuration. Numerous concrete examples show that the Babylonians of 2000–1600 B.C. were familiar with the general rules for computing the area of a rectangle, the areas of right and isosceles triangles (and perhaps the general triangle), the area of the special trapezoid having one side perpendicular to the parallel sides, the volume of a rectangular parallelepiped, and, more generally, the volume of a right prism with special trapezoidal base. The circumfer-

---

[1] An alternative thesis finds the origin of scientific geometry in religious ritual —agriculture, trade, and surveying being later contributions. See/SEIDENBERG (b)/.

ence of a circle was taken to be three times as long as the diameter, and the area of a circle was taken as one-twelfth the area of a square erected on a side having the length of the circumference of the circle (both measurements correct for $\pi = 3$): then the volume of a right circular cylinder was obtained by finding the product of the base and the altitude. The volume of a frustum of a cone or of a square pyramid appears incorrectly as the product of the altitude and half the sum of the bases. There also seems to be evidence that the ancient Babylonians used the incorrect formula

$$K = \frac{(a + c)(b + d)}{4}$$

for the area of a quadrilateral having a, b, c, d for consecutive sides. These peoples knew that corresponding sides of two similar right triangles are proportional, that the altitude through the vertex of an isosceles triangle bisects the base, and that an angle inscribed in a semicircle is a right angle. The Pythagorean theorem [58] was also known, even as far back as approximately 2000 B.C.

Our chief sources of information concerning ancient Egyptian geometry are the Moscow and Rhind papyri—mathematical texts containing 25 and 85 problems, respectively, and dating from approximately 1850 B.C. and 1650 B.C. [30, 31]. There is also, in the Berlin Museum, the oldest extant astronomical or surveying instrument—a combination plumb line and sight rod—which comes from the Egypt of about 1850 B.C. The Berlin Museum also possesses an Egyptian sundial that dates from about 1500 B.C. and is the oldest sundial known to be in existence. These instruments reveal, of course, a concurrent knowledge of some associated practical geometry. One should also point out that the great pyramid of Gizeh, whose very careful construction certainly involved some practical geometry, was erected about 2900 B.C.

Of the 110 problems in the Moscow and Rhind papyri, 26 are geometric. Most of these problems stem from mensuration formulas needed for computing land areas and granary volumes. The area of a circle is taken as equal to that of the square on 8/9 of the diameter, and the volume of a right circular cylinder as the product of the area of the base and the length of the altitude. Some of the problems seem to concern themselves with the cotangent of the dihedral angle between the base and a face of a pyramid. Although there is no documentary evidence that the ancient Egyptians were aware of the Pythagorean theorem early Egyptian surveyors realized that a triangle having sides of lengths 3, 4, and 5 units is a right triangle. It is curious that the incorrect formula

$$K = \frac{(a + c)(b + d)}{4}$$

for the area of an arbitrary quadrilateral with successive sides of lengths *a, b, c, d* appears in an inscription found in the tomb of Ptolemy XI, who died in 51 B.C.

Very remarkable is the existence in the Moscow papyrus of a numerical example of the correct formula for the volume of a frustum of a square pyramid,

$$V = \frac{h(a^2 + ab + b^2)}{3} \, ,$$

where *h* is the altitude and *a* and *b* are the lengths of the sides of the two square bases. No other unquestionably genuine example of this formula has been found in pre-Hellenic mathematics, and since its proof demands some form of integral calculus, its discovery by the Egyptians must certainly be regarded as an extraordinary piece of induction. Eric Temple Bell has aptly referred to this achievement as the "greatest Egyptian pyramid."

It is very likely that geometrical accomplishments similar to those of ancient Egypt and Babylonia occurred also in ancient India and China [**62**], but we know very little indeed with any degree of certainty about these accomplishments. The ancient Egyptians recorded their work on stone and papyrus, the latter fortunately resisting the ages because of Egypt's unusually dry climate; the Babylonians used imperishable baked-clay tablets. The early Indians and Chinese, however, used very perishable writing materials like bark bast and bamboo. Thus it has come to pass that we have a fair quantity of definite information, obtained from primary sources, about the geometry of ancient Egypt and Babylonia, while we know very little about the geometry of ancient India and China.

DEMONSTRATIVE GEOMETRY

The economic and political changes of the last centuries of the second millennium B.C. caused the power of Egypt and Babylonia to wane. New peoples came to the fore, and it happened that the further development of geometry passed over to the Greeks, who transformed the subject into something vastly different from the set of empirical conclusions worked out by their predecessors.

The Greeks insisted that geometric fact must be established, not by empirical procedures, but by deductive reasoning; geometrical truth was to be attained in the study room rather than in the laboratory. In short, the Greeks transformed the empirical, or scientific, geometry of the ancient Egyptians and Babylonians into what we might call "systematic," or "demonstrative," geometry.

It is disappointing that, unlike the situation with ancient Egyptian and Babylonian geometry, there exist virtually no primary sources for the study of very early Greek geometry. We are forced to rely on manuscripts and accounts that are dated several hundred years after the original treatments had been written.

Our principal source of information concerning very early Greek geometry is the so-called *Eudemian Summary* of Proclus. This summary constitutes several pages of Proclus' *Commentary on Euclid, Book I* and is a brief outline of the development of Greek geometry from the earliest times to Euclid. Although Proclus lived in the fifth century A.D., a good thousand years after the inception of Greek geometry, he still had access to a number of historical and critical works that are now lost to us except for the fragments and allusions preserved by him and others. Among these lost works is what was apparently a full history of Greek geometry, covering the period prior to 335 B.C., written by Eudemus, a pupil of Aristotle. The *Eudemian Summary* is so named because it is admittedly based on this earlier work.

According to the *Eudemian Summary*, Greek geometry appears to have started in an essential way with the work of Thales of Miletus in the first half of the sixth century B.C. This versatile genius, declared to be one of the "seven wise men" of antiquity, was a worthy founder of demonstrative geometry. He is the first known individual with whom the use of deductive methods in geometry is associated. Thales, the summary tells us, sojourned for a time in Egypt and brought back geometry with him to Greece, where he began to apply to the subject the deductive procedures of Greek philosophy. He is credited with a number of very elementary geometrical results [59], the value of which is not to be measured by their content but rather by the belief that he supported them with a certain amount of logical reasoning instead of intuition and experiment. For the first time a student of geometry was committed to a form of deductive reasoning, partial and incomplete though it may have been. Moreover, the fact that the first

171

deductive thinking was done in the field of geometry (instead of algebra, for instance) inaugurated a tradition in mathematics which was maintained until very recent times.

Early Greek Geometry and Material Axiomatics

The next outstanding Greek geometer mentioned in the *Eudemian Summary* is Pythagoras, who is claimed to have continued the systematization of geometry begun some fifty years earlier by Thales. Pythagoras was born about 572 B.C. on the island of Samos, one of the Aegean islands near Thales' home city of Miletus, and it is quite possible that he studied under the older man. It seems that Pythagoras then visited Egypt and perhaps traveled even more extensively about the Orient. When, on returning home, he found Ionia under Persian dominion, he decided to migrate to the Greek seaport of Crotona in southern Italy. There he founded the celebrated Pythagorean school, a brotherhood knit together with secret and cabalistic rites and observances and committed to the study of philosophy, mathematics, and natural science.

In spite of the mystical nature of much of Pythagorean study [9], the members of the society contributed, during the two hundred or so years following the founding of their organization, a good deal of sound mathematics. Thus, in geometry they developed the properties of parallel lines and used them to prove that the sum of the angles of any triangle is equal to two right angles. They contributed in a noteworthy manner to Greek geometrical algebra [54, 78], and they developed a fairly complete theory of proportion (though it was limited to commensurable magnitudes) and used it to deduce properties of similar figures. They were aware of the existence of at least three of the regular polyhedral solids [60], and they discovered the incommensurability [18] of a side and a diagonal of a square. Although much of this information was already known to the Babylonians of earlier times, the deductive aspect of geometry is thought to have been considerably exploited and advanced in this work of the Pythagoreans. Chains of propositions in which successive propositions were derived from earlier ones began to emerge. As the chains lengthened and some were tied to others, the bold idea of developing all of geometry in one long chain suggested itself. It is claimed in the *Eudemian Summary* that a Pythagorean, Hippocrates of Chios, was the first to attempt, with at least partial success, a logical presentation of geometry in the form of a single chain of propositions based on a few initial definitions and assumptions. Better attempts were made by

172

Leon, Theudius, and others. And then, about 300 B.C., Euclid produced his epoch-making effort, the *Elements,* a single deductive chain of 465 propositions neatly and beautifully comprising plane and solid geometry, number theory, and Greek geometrical algebra. From its very first appearance this work was accorded the highest respect, and it so quickly and so completely superseded all previous efforts of the same nature that now no trace remains of the earlier efforts. The effect of this single work on the future development of geometry has been enormous and is difficult to overstate.

Much was accomplished by the Greeks during the three hundred years between Thales and Euclid. Not only did the Pythagoreans and others develop the material that ultimately was organized into the *Elements* of Euclid, but they developed notions concerning infinitesimals and limit and summation processes (notions that did not attain final clarification until the invention of the calculus in modern times) and also considerable higher geometry, or the geometry of curves other than the circle and the straight line and of surfaces other than the sphere and the plane. Curiously enough, much of this higher geometry originated in continued attempts to solve the three famous construction [50] problems of antiquity: the duplication of a cube, the trisection of an arbitrary angle, and the quadrature of a circle [51, 52, 53].

Also during the first three hundred years of Greek mathematics, there developed the Greek notion of a logical discourse as a sequence of statements obtained by deductive reasoning from an accepted set of initial statements. Now both the initial and the derived statements of the discourse are statements about the technical matter of the discourse and hence involve special or technical terms. The meanings of these terms must be clear to the reader, and so, the Greeks felt, the discourse should start with a list of explanations and definitions of these technical terms. After these explanations and definitions have been given, the initial statements, called "axioms" and/or "postulates" of the discourse, are to be listed. These initial statements, according to the Greeks, should be so carefully chosen that their truths are quite acceptable to the reader in view of the explanations and definitions already cited.

A discourse conducted according to the above plan is today said to be developed by "material axiomatics." Certainly the most outstanding contribution of the early Greeks to mathematics was the formulation of the pattern of material axiomatics and the insistence that geometry be systematized according to this pattern. Euclid's *Elements*

173

is the earliest extensively developed example of the use of the pattern that has come down to us. In more recent years, as we shall see, the pattern of material axiomatics has been significantly generalized to yield a more abstract form of discourse known as "formal axiomatics."

LATER GREEK GEOMETRY

The three most outstanding Greek geometers of antiquity were Euclid (*c.* 300 B.C.), Archimedes (287–212 B.C.), and Apollonius (*c.* 225 B.C.); and it is no exaggeration to say that almost every significant subsequent geometrical development, right up to and including the present time, finds its seeds of origin in some work of these three great scholars.

All three of these men were prolific writers. Thus, though the *Elements* of Euclid is by far his most influential work—indeed, the most influential single work in geometry in the entire history of the subject—he wrote a number of other geometrical treatises of which some knowledge of eight or so has come down to us.

Some ten mathematical treatises of Archimedes have survived to the present day, and there are traces of various lost works. Of his extant works, three are devoted to plane geometry and two to solid geometry. These works are not compilations of achievements of predecessors but are highly original creations, marking Archimedes as one of the very greatest mathematicians of all time, and certainly the greatest of antiquity. In one of his works devoted to plane geometry Archimedes inaugurated the classical method of perimeters for computing π [**44**], and found that π lies between 223/71 and 22/7, or that, to two decimal places, π is given by 3.14. This procedure of Archimedes was the start in the long history of securing ever more accurate approximations for the value of π, reaching, in 1967, the fantastic accuracy of 500,000 decimal places. In his other works on plane geometry Archimedes anticipated some of the methods of the integral calculus [**98, 109**]. In one of his works on solid geometry we find, for the first time, the correct formulas for the areas of a sphere and of a zone of one base, and for the volumes of a sphere and of a spherical segment of one base.

There is a geometrical assumption explicitly stated by Archimedes in his work *On the Sphere and Cylinder* which deserves special mention; it is one of the five geometrical postulates assumed at the start of the work, and it has become known as "the postulate of Archimedes." A simple statement of the postulate is: *Given two unequal line segments, there is always some finite multiple of the shorter one which is longer than the other.* In some modern treatments of geometry

174

this postulate serves as part of the postulational basis for introducing the concept of continuity. It is a matter of interest that in the late nineteenth and early twentieth centuries geometric systems were constructed which deny the Archimedean postulate, thus giving rise to so-called non-Archimedean geometries.

Although Apollonius was an astronomer of merit, and although he wrote on a variety of mathematical subjects, his chief bid to fame rests on his extraordinary and monumental *Conic Sections,* a work that earned him the title, among his contemporaries, of "the Great Geometer." Apollonius' *Conic Sections* is a thorough investigation of these curves, and it completely superseded all earlier works on the subject. It was Apollonius who supplied the terms "ellipse," "parabola," and "hyperbola" [61]. Because of subsequent commentary, we are aware of the contents of six other geometrical works of Apollonius. One of these concerned itself with the construction, with straightedge and compasses, of a circle tangent to three given circles; this attractive problem is today known as "the problem of Apollonius." In another work we encounter the so-called circle of Apollonius of present-day college geometry courses.

With the passing of Apollonius the golden age of Greek geometry came to an end. The lesser geometers who followed did little more than fill in details and perhaps independently develop certain theories the germs of which were already contained in the works of the three great predecessors. In particular a number of newer plane curves of higher degree were discovered, and the applications of geometry were exploited. Among these later geometers special mention should be made of Heron (or Hero) of Alexandria (*c.* A.D. 75), Menelaus (*c.* 100), Claudius Ptolemy (*c.* 85–*c.* 165), and Pappus (*c.* 320). In geometry Heron concerned himself with plane and solid mensuration, and Menelaus and Ptolemy contributed to trigonometry as a handmaiden of astronomy. Pappus, the last of the creative Greek geometers, who lived some five hundred years after Apollonius, strove vainly, although with enthusiasm, to breathe fresh life into languishing Greek geometry. His great work, the *Collection,* most of which has come down to us, is a combined commentary and guidebook of the existing geometrical works of his time. It is sown with numerous original propositions, improvements, extensions, and valuable historical comment. The *Collection* proved to be the requiem of Greek geometry; for after Pappus, Greek geometry ceased to be a living study: merely its memory was perpetuated by minor writers and commentators.

In ancient Greek geometry, both in its form and in its content, we

175

find the fountainhead of the subject. One can scarcely overemphasize the importance to all subsequent geometry of this remarkable bequest.

The Detour Through India and Arabia

The closing period of ancient times was dominated by Rome. One Greek center after another fell before the power of the Roman armies, and in 146 B.C. Greece became a province of the Roman Empire. Conditions proved more and more stifling to original scientific work, and a gradual decline in creative thinking set in. The arrival of the barbarians in the West and the eventual collapse of the slave market, with their disastrous effects on Roman economy, found science and mathematics reduced to a mediocre level.

The period starting with the fall of the Roman Empire in the middle of the fifth century and extending into the eleventh century is known as Europe's Dark Ages, for during this period civilization in western Europe reached a very low ebb. Schooling became almost nonexistent, Greek learning all but disappeared, and many of the arts and crafts bequeathed by the ancient world were forgotten.

During this bleak period of learning the people of the East, especially the Hindus and the Arabs, became the major custodians of mathematics. However, the Greek concept of rigorous thinking—in fact, the very idea of deductive proof—seemed distasteful to the Hindu way of doing things. Although the Hindus excelled in computation, contributed to the devices of algebra, and played an important role in the development of our present positional numeral system, they produced almost nothing of importance in geometry or in basic mathematical methodology. Perhaps the best piece of Hindu geometry of the period, and somewhat solitary in its excellence, was the arithmetico-geometric work of Brahmagupta (c. 628) on cyclic quadrilaterals, each with rational sides, diagonals, and area.

The spectacular episode of the rise and decline of the Arabian Empire occurred during the period of Europe's Dark Ages. Within a decade after Mohammed's flight from Mecca to Medina in 622, the scattered and disunited tribes of the Arabian Peninsula were consolidated by a strong religious fervor into a powerful nation. Within a century, force of arms had extended the Moslem rule and influence over a territory reaching from India through Persia, Mesopotamia, and northern Africa, clear into Spain. Of considerable importance for the preservation of much of world culture was the manner in which the Arabs seized upon Greek and Hindu erudition. Numerous Hindu

176

and Greek works in astronomy, medicine, and mathematics were industriously translated into the Arabic tongue and thus saved until later European scholars were able to retranslate them into Latin and other languages. But for the work of the Arabian scholars, a great part of Greek and Hindu science would have been irretrievably lost over the long period of the Dark Ages.

In addition, the Arabian mathematicians made some small contributions of their own. In geometry one can mention the work done by Abu'l-Wefa (940–998) with "rusty" compasses, or compasses of fixed opening [50], the geometrical solution of cubic equations by Omar Khayyam [71] (c. 1044–c. 1123), and the researches of Nasir eddin (c. 1250) on Euclid's parallel postulate. Like the Hindus, the Arabian mathematicians generally regarded themselves primarily as astronomers and thus showed considerable interest in trigonometry. They may be credited with using all six of the trigonometric functions [95, 96] and with improving upon the derivation of the formulas of spherical trigonometry.

THE RETURN OF GEOMETRY TO WESTERN EUROPE

It was not until the latter part of the eleventh century that Greek classics in science and mathematics began once again to filter into Europe. There followed a period of transmission during which the ancient learning preserved by Moslem culture was passed on to the western Europeans through Latin translations made by Christian scholars traveling to Moslem centers of learning, and through the opening of western European commercial relations with the Levant and the Arabian world. The loss of Toledo by the Moors to the Christians in 1085 was followed by an influx of Christian scholars to that city to acquire Moslem learning. Other Moorish centers in Spain were infiltrated, and the twelfth century became, in the history of mathematics, a century of translators.

The thirteenth century saw the rise of the universities at Paris, Oxford, Cambridge, Padua, and Naples. Universities were to become potent factors in the development of mathematics, since many mathematicians associated themselves with one or more such institutions. During this century, about 1260, Johannes Campanus made a Latin translation of Euclid's *Elements* which later, in 1482, became the first printed version of Euclid's great work.

The fourteenth century was a mathematically barren one. It was the century of the Black Death, which swept away more than a third of the population of Europe; during this century the Hundred Years War, with its political and economic upheavals in northern Europe, got well under way.

The fifteenth century, an early period of the Renaissance, witnessed the rebirth of art and learning in Europe. With the collapse of the Byzantine Empire, culminating in the fall of Constantinople to the Turks in 1453, refugees flowed into Italy, bringing with them treasures of Greek civilization. Many Greek classics, known up to that time only through Arabic translations that were often inadequate, could now be studied from original sources. Also, about the middle of the century, the invention of printing from movable type revolutionized the book trade and enabled knowledge to be disseminated at an unprecedented rate. Mathematical activity in this century was largely centered in the Italian cities and in the central European cities of Nuremberg, Vienna, and Prague. It concentrated on arithmetic, algebra, and trigonometry, under the practical influence of trade, navigation, astronomy, and surveying.

In the sixteenth century the development of arithmetic and algebra continued, the most spectacular mathematical achievement of the century being the discovery, by Italian mathematicians, of the algebraic solution of cubic and quartic equations. This continued development of algebra, during which the subject changed from a rhetorical to a symbolic form, later had a marked effect, as we shall see, on the expansion of geometry. A more immediate stimulus to the further development of geometry was the translation, in 1533, of Proclus' *Commentary on Euclid, Book I.* The first important translation into Latin of Books I–IV of Apollonius' *Conic Sections* was made by Federigo Commandino in 1566; Books V–VII did not appear in Latin translation until 1661. In 1572 Commandino made a very important translation of Euclid's *Elements* from the Greek. This translation served as a basis for many subsequent translations, including a very influential work by Robert Simson (1687–1768), from which, in turn, so many English editions were derived. By this time a number of the works of Archimedes had also been translated into Latin. With so many of the great Greek works in geometry available, it was inevitable that sooner or later some aspects of the subject should once again claim the attention of researchers.

PROJECTIVE GEOMETRY

In an effort to produce more realistic pictures, many of the Renaissance artists and architects became deeply interested in discovering the formal laws controlling the construction of projections of objects on a screen, and as early as the fifteenth century a number of these men created the elements of an underlying geometrical theory of perspective. (Some aspects of this subject had been considered by the ancient Greeks.) The theory was considerably extended in the early seventeenth century by a small group of French mathematicians, the motivator of whom was Gérard Desargues, an engineer and architect. Influenced by the growing needs of artists and architects for a deeper theory of perspective, Desargues published, in Paris in 1639, a remarkably original treatise on the conic sections which exploited the idea of projection. But this work was so neglected by most other mathematicians of the time that it was soon forgotten and copies of the publication disappeared until, over two centuries later (in 1845), it was resurrected by the French geometer and historian of geometry Michel Chasles.

There are several reasons for the initial neglect of Desargues's little volume. It was overshadowed by the more supple analytic geometry introduced by René Descartes two years earlier. Geometers were generally either developing this new powerful tool or trying to apply infinitesimals to geometry. Also, Desargues unfortunately adopted a style and a terminology so eccentric that they beclouded his work and discouraged others from attempting properly to evaluate his accomplishments.

The reintroduction of projective considerations into geometry did not occur until the late eighteenth century, when the great French geometer Gaspard Monge created his descriptive geometry. This subject, which concerns a way of representing and analyzing three-dimensional objects by means of their projections on certain planes, had its origin in the design of fortifications. Monge was a very inspiring teacher, and there gathered about him a group of brilliant students of geometry, among whom were Lazare Carnot, Charles J. Brianchon, and Jean Victor Poncelet.

The real revival of projective geometry was launched by Poncelet. As a prisoner of war, captured by the Russians during Napoleon's retreat from Moscow, Poncelet, with no books at hand, planned his great work on projective geometry, which, after his release and return to France, he published in Paris in 1822. This work gave tremendous impetus to the study of the subject and inaugurated the so-called great period in the history of projective geometry. There followed into the field a host of mathematicians, among them Gergonne, Brianchon, Chasles, Plücker, Steiner, Staudt, Reye, and Cremona—

men whose names are great in the history of geometry and in the history of projective geometry in particular. The remarkable "principle of duality" of projective geometry seems to have been independently discovered by Gergonne and Poncelet.

ANALYTIC GEOMETRY

While Desargues and Blaise Pascal were opening the new field of projective geometry, Descartes and Pierre de Fermat were conceiving ideas of modern analytic geometry. There is a fundamental distinction between the two studies, for the former is a *branch* of geometry whereas the latter is a *method* of geometry.

Few academic experiences can be more thrilling to the student of advanced high school or beginning college mathematics than an introduction to this new and powerful method of attacking geometrical problems. The task of establishing a theorem in geometry is cleverly shifted to that of establishing a corresponding theorem in algebra. Since many students are considerably more able as algebraists than as geometers, analytic geometry has been described as the "royal road" in geometry that Euclid thought did not exist. (Proclus tells that Ptolemy once asked Euclid whether there was any shorter way to a knowledge of geometry than by a study of the *Elements,* whereupon Euclid answered that there was "no royal road to geometry.")

There is no unanimity among historians of mathematics concerning who invented analytic geometry, nor even concerning what age should be credited with the invention. Much of this difference of opinion stems from a lack of agreement regarding just what constitutes analytic geometry.

Those who favor antiquity as the era of the invention point out the well-known fact that the concept of fixing the position of a point by means of suitable coordinates was employed in the ancient world by the Egyptians and the Romans in surveying and by the Greeks in map making. And, if analytic geometry implies not only the use of coordinates but also the geometric interpretation of relations among coordinates, then a particularly strong argument in favor of crediting the Greeks is the fact that Apollonius derived the bulk of his geometry of the conic sections from the geometrical equivalents of certain Cartesian equations of these curves, an idea that seems to have originated with Menaechmus about 350 B.C.

Others claim that the invention of analytic geometry should be credited to Nicole Oresme, who was born in Normandy about 1323. Oresme, in one of his mathematical tracts, anticipated another aspect of analytic geometry when he represented certain laws by graphing the dependent variable against the independent one as the latter variable was permitted to take on small increments. A century after Oresme's tract was written it enjoyed several printings, and in this way it may possibly have influenced later mathematicians.

However, before analytic geometry could assume its present highly practical form, it had to await the development of algebraic symbolism. Accordingly, it may be more nearly correct to agree with the majority of historians, who regard the decisive contributions made in the seventeenth century by the two French mathematicians Descartes and Fermat as the essential origin of at least the modern spirit of the subject. Not until after the impetus given to the subject by these two men do we find analytic geometry in a form with which we are familiar.

Descartes's claim to the invention of analytic geometry rests on one of the three appendixes to his famous philosophical treatise on universal science, *Discours de la méthode pour bien conduire sa raison et chercher la vérité dans les sciences* ("Discourse on the Method of Rightly Conducting Reason and Seeking Truth in the Sciences"), which was published in 1637. It is the last, or third, appendix, entitled *La géométrie*, which contains Descartes's contributions to analytic geometry. Fermat's claim to priority rests on a letter written to Gilles Persone de Roberval in September 1636 in which it is stated that the ideas of the writer were even then seven years old.

Although Descartes had mentioned solid analytic geometry, he did not elaborate it. Its first systematic development was made by Antoine Parent in 1700. Alexis Claude Clairaut, in 1731, was the first to write analytically on nonplanar curves in space, and somewhat later Leonhard Euler advanced the whole subject well beyond its elementary stages.

Much terminology, like our classification of curves (and surfaces) into linear, quadratic, cubic, and so forth, stems from our use of Cartesian-coordinate systems. Some curves, however, such as many spirals, have intractable equations when referred to a Cartesian frame, whereas they enjoy relatively simple equations when referred to some other skillfully designed coordinate system. Particularly useful in the case of spirals is the polar-coordinate system [64], which was considered in 1691 by Jakob Bernoulli (also known as Jacques Bernoulli). Further coordinate systems were not investigated until toward the close of the nineteenth century, when geometers were led to break away from the Cartesian systems in situations where the peculiar necessities of a problem indicated that some other algebraic apparatus would be more suitable.

An interesting development in coordinate systems was inaugurated by Julius Plücker in 1829, when he noted that our fundamental element need not be the point but can be any geometric entity. This, in turn, led Plücker to the concept of the dimension of a manifold of geometric entities as simply the essential number of coordinates needed to determine one of the entities in the manifold.

The first nebulous notions of a hyperspace which is n-dimensional ($n > 3$) in points are lost in the dimness of the past and were confused by metaphysical considerations [57]. A bold plunge into the study of such spaces was not made until the middle of the nineteenth century —by Arthur Cayley in 1843, Hermann Grassmann in 1844, and G. F. Bernhard Riemann in 1854.

Differential Geometry

Many new and extensive fields of mathematical investigation were opened up in the seventeenth century, making that era an outstandingly productive one in the development of mathematics. Unquestionably the most remarkable mathematical achievement of the period was the invention of the calculus, toward the end of the century, by Isaac Newton and Gottfried Wilhelm von Leibniz. The new tool proved to be almost unbelievably powerful in its astonishingly successful disposal of hosts of problems that had been baffling and quite unassailable in earlier days.

A fair share of this remarkable applicability lies in the field of geometry, and it has come to pass that there is an exceedingly vast body of geometry wherein one studies properties of curves and surfaces, and their generalizations, by means of the calculus. This body of geometry is known as "differential geometry." For the most part, differential geometry investigates curves and surfaces only in the immediate neighborhood of any one of their points. This aspect of differential geometry is known as "local differential geometry," or "differential geometry in the small." However, sometimes properties of the total structure of a geometrical figure are implied by certain local properties of the figure that hold at every point of the figure. This leads to what is known as "integral geometry," or "global differential geometry," or "differential geometry in the large."

Though one can find geometrical theorems deduced from a study of evanescent figures in Archimedes' determination of areas and volumes and in Apollonius' treatment of the normals to conic sections—and, indeed, during the seventeenth century, in Bonaventura Cavalieri's method of indivisibles and Christiaan Huygens' beautiful work on curvature and evolutes—it is probably quite correct to say that differential geometry, at least in its modern dress, started in the early part of the eighteenth century with the interapplications of the calculus and analytic geometry. But the first real stimulus to the subject, beyond planar situations, was furnished by Gaspard Monge, who can be considered the father of the differential geometry of curves and surfaces of space. Monge and his students thus initiated what is usually referred to as the first period in the history of differential geometry.

The second period was inaugurated by Carl Friedrich Gauss (1777–1855), who introduced the singularly fruitful method of studying the differential geometry of curves and surfaces by means of parametric representations of these objects.

The third great period in the history of differential geometry was introduced by Bernhard Riemann. Here we find an assertion of that tendency of mathematics of recent times to strive for the greatest possible generalization. Two things were found necessary for this further development: an improved notation and a procedure independent of any particular coordinate system employed. The tensor calculus was accordingly devised and developed. Generalized differential geometries, known as Riemannian geometries, were explored intensively; and these in turn led to non-Riemannian, and other, geometries. Much of this material has found significant application in relativity theory and other parts of modern physics.

NON-EUCLIDEAN GEOMETRY

There is evidence that a logical development of the theory of parallels gave the early Greeks considerable trouble [55]. Euclid met the difficulties by defining parallel lines as coplanar straight lines that do not meet one another however far they may be produced in either direction, and by adopting as an initial assumption his now famous parallel postulate: *If a straight line intersects two straight lines so as to make the interior angles on one side of it together less than two right angles, the two straight lines will intersect, if indefinitely produced, on the side on which are the angles which are together less than two right angles.* This postulate lacks the terseness of Euclid's other postulates, and it does not seem to possess the quality, demanded by Greek material axiomatics, of self-evidence or ready acceptability on the part of the reader. Actually, the postulate is the converse of Proposition 17 of Euclid's Book I, and all in all it seemed more like a proposition than a postulate. Moreover, Euclid himself made no use of the parallel postulate until he reached Proposition 29 of Book I. It was natural to wonder if the postulate was really needed at all, and to think that perhaps it could be derived as a theorem from the remaining postulates, or, at least, that it could be replaced by a more acceptable equivalent.

Of the many substitutes that have been devised to replace Euclid's parallel postulate, the one most commonly used is that made well known in modern times by the Scottish physicist and mathematician John Playfair, although this particular alternative had been used by others and had been stated as early as the fifth century by Proclus. Playfair, whose *Elements of Geometry* was published in 1795, gave the substitute usually encountered in present-day high school texts: *Through a given point not on a given line can be drawn only one line parallel to the given line.*

The attempts to derive the parallel postulate as a theorem from the remaining postulates of Euclid's *Elements* occupied geometers for over two thousand years and culminated in some of the most far-reaching developments of modern mathematics. Many "proofs" of the postulate were offered, but each was sooner or later shown to rest upon a tacit assumption equivalent to the postulate itself. Of these many attempts and investigations, particularly outstanding were those made by Girolamo Saccheri in 1733, by Johann Heinrich Lambert in a work posthumously published by his friends in 1788, and by Adrien Marie Legendre in an appendix of successive editions of his very popular *Éléments de géométrie*, which first appeared in 1794. Each of these three men tried to institute a *reductio ad absurdum* by producing a contradiction under the assumption of a denial of some equivalent of the parallel postulate. Though they all failed in their efforts, they did bring to light a number of consequences that are recognized today as important theorems in a non-Euclidean geometry.

The first to suspect, and even proclaim, the impossibility of obtaining a contradiction under one of the denials of the parallel postulate were Gauss of Germany, Janos Bolyai (1802–1860) of Hungary, and Nikolai Ivanovich Lobachevsky (1793–1856) of Russia. These three men approached the subject through the Playfair form of the parallel postulate by considering three possibilities: Through a given point can be drawn *more than one,* or *just one,* or *no* line parallel to a given line. By tacitly assuming, as in fact did Euclid, the infinitude of a straight line, the third case was easily eliminated; but, in spite of prolonged

184

investigations, none of these men could obtain a contradiction under the first possibility. Each began, in time, to suspect that no contradiction could be found and that the resulting geometry, though quite different from the Euclidean geometry, was just as consistent as the Euclidean geometry. Gauss was the first of the three to reach these advanced conclusions, but since throughout his life he failed to publish anything on the matter, the honor of discovering this particular non-Euclidean geometry must be shared with Bolyai and Lobachevsky. Bolyai published his findings in 1832 in an appendix to a mathematical work of his father. Later it was learned that Lobachevsky had published similar findings as early as 1829–30; but, because of language barriers and the slowness with which information of new discoveries traveled in those days, Lobachevsky's work did not become known in western Europe for some years.

The actual independence of the parallel postulate from the other postulates of Euclidean geometry was not unquestionably established until consistency proofs of the non-Euclidean geometry were furnished. These were now not long in coming and were supplied by Eugenio Beltrami, Felix Klein, Henri Poincaré, and others. The method was to set up a model wherein the non-Euclidean geometry has an interpretation in a part of Euclidean space; then any inconsistency in the non-Euclidean geometry would imply a corresponding inconsistency in Euclidean geometry.

In 1854 Riemann showed that if the infinitude of a straight line be discarded and merely its boundlessness be assumed, then, with some other slight adjustments of the remaining postulates, another consistent non-Euclidean geometry can be developed from the third case above.

In 1871 Klein gave to the three geometries—that of Bolyai and Lobachevsky, that of Euclid, and that of Riemann—the names "hyperbolic geometry," "parabolic geometry," and "elliptic geometry."

It took unusual imagination to entertain the possibility of a geometry different from Euclid's, for the human mind had for two millennia been bound by the prejudice of tradition to the firm belief that Euclid's system was most certainly the only way geometrically to describe physical space, and that any contrary geometric system simply could not be consistent.

TOPOLOGY

Topology started as a branch of geometry, but during the second quarter of the twentieth century it underwent such generalization and became involved with so many other branches of mathematics that it is now more properly considered, along with geometry, algebra, and analysis, a fundamental division of mathematics. Today topology may roughly be defined as the mathematical study of continuity. In this overview, however, we shall restrict ourselves to those aspects of the subject that reflect its geometric origin. We shall, accordingly, here regard topology as the study of those properties of geometric figures which remain invariant under so-called topological transformations—that is, under single-valued continuous mappings possessing single-valued continuous inverses.

Topology, as a self-connected study, scarcely predates the middle of the nineteenth century, but one can find some earlier isolated topological investigations. Toward the end of the seventeenth century Leibniz used the term *geometria situs* to describe a sort of qualitative mathematics that today would be thought of as topology, and he predicted important studies in this field. An early-discovered topological property of a simple closed polyhedron is the relation

$$V - E + F = 2,$$

where V, E, and F denote the number of vertices, edges, and faces, respectively, of the polyhedron. This relation was known to Descartes in 1640, but the first proof of the formula was given by Euler in 1751. Euler had earlier, in 1736, considered some topology of linear graphs in his treatment of the Königsberg bridge problem. Gauss made several contributions to topology. Of the several proofs he furnished of the fundamental theorem of algebra, two are explicitly topological. Later Gauss briefly considered the theory of knots, which today is an important subject in topology. In 1840 A. F. Möbius enunciated the four-color problem, which was soon taken up by others. At this time the subject of topology was known as *analysis situs*.

The term "topology" was introduced in 1847 by J. B. Listing, one of Gauss's students, in the title, *Vorstudien zur Topologie*, of the first book devoted to the subject. The German word *Topologie* was later anglicized to "topology" by Professor Solomon Lefschetz of Princeton University. Gustav Robert Kirchhoff, another of Gauss's students, in 1847 employed the topology of linear graphs in his study of electrical networks. But of all of Gauss's students, the one who contributed by far the most to topology was G. F. Bernhard Riemann, who, in his doctoral thesis of 1851, introduced topological concepts into the study of complex-function theory. The chief stimulus to topology furnished by Riemann was the notion of "Riemann surface," a topological device for rendering multiple-valued complex functions into single-valued functions. Also of importance in topology is Riemann's probationary lecture of 1854, concerning the hypotheses that lie at the foundations of geometry; this lecture furnished the breakthrough to higher dimensions.

About 1865 Möbius wrote a paper in which a polyhedron was viewed simply as a collection of joined polygons. This introduced the concept of "2-complexes" into topology. In his systematic development of 2-complexes, Möbius was led to the surface now referred to as a "Möbius strip." In 1873 James Clerk Maxwell used the topological theory of connectivity in his study of electromagnetic fields.

Poincaré ranks high among the early contributors to topology. A paper of his, written in 1895 and entitled *Analysis Situs*, introduced the important homology theory of n dimensions. With Poincaré's work, the subject of topology was well under way, and an increasing number of mathematicians entered the field. Especially important names in topology since Poincaré are Oswald Veblen, J. W. Alexander, Solomon Lefschetz, L. E. J. Brouwer, and Maurice Fréchet.

The notion of a geometric figure as made up of a finite set of joined fundamental pieces, as was emphasized by Möbius, Riemann, and Poincaré, gradually gave way to the Cantorian concept of an arbitrary set of points, and it was recognized that any collection of things—be it a set of numbers, or of algebraic functions, or of nonmathematical objects—can constitute a topological space in some sense or other. Studies governed by this latter, and very general, view of topology have become known as "set topology," whereas studies more intimately connected with the earlier view have become known as "combinatorial topology" or "algebraic topology"—though it must be confessed that this division is perhaps one of convenience rather than logic.

The "Erlanger Programm"

By the middle of the nineteenth century a number of different geometries had come into existence, and the time was ripe for some sort of codification and synthesizing device to give a sense of order and classification to these geometries. Such a scheme was announced in 1872 by Felix Klein, in his inaugural address upon appointment to the Philosophical Faculty and the Senate of the University of Erlangen. This address, based on work he and Sophus Lie had done in group theory, set forth a remarkable definition of "a geometry"— one that served to codify essentially all the existing geometries of the time and pointed the way to new and fruitful avenues of research in geometry. This address, with the program of geometrical study advocated by it, has become known as the *Erlanger Programm*.

Somewhat oversimply stated, the *Erlanger Programm* claims that a geometry is the investigation of those properties of figures which remain unchanged when the figures are subjected to a group of transformations. For plane Euclidean metric geometry, the group of transformations is the set of all rotations and translations in the plane; for plane projective geometry, the group of transformations is the set of all so-called planar projective transformations; for topology, the group of transformations is the set of all topological transformations. Each geometry has its underlying controlling transformation group. In building up a geometry, then, one is at liberty to choose, first of all, the fundamental element of the geometry (point, line, circle, sphere, etc.); next, the totality or manifold of these elements (plane of points, ordinary space of points, spherical surface of points, plane of lines, pencil of circles, etc.); and, finally, the group of transformations to which the manifold of elements is to be subjected. The *Erlanger Programm* advocated the classification of existing geometries, and the creation and study of new geometries, according to this scheme. In particular, one should study the geometries characterized by the various proper subgroups of the transformation group of a given geometry, thereby obtaining geometries that embrace others.

ABSTRACT SPACES

In 1906 Maurice Fréchet inaugurated the study of "abstract spaces," and some very general geometries came into being which no longer necessarily fit into the neat Kleinian codification. A "space" became merely a set of objects, for convenience called "points," together with a set of relations in which these points are involved, and a geometry became simply the theory of such a space. The set of relations to which the points are subjected is called the "structure" of the space, and this structure may or may not be explainable in terms of the invariant theory of a transformation group. Through set theory, then, geometry received a further generalization or metamorphosis.

Although abstract spaces were first formally introduced in 1906, the idea of a geometry as the study of a set of "points" with some superimposed structure was really already contained in remarks made by Riemann in his great lecture of 1854. These new geometries have found invaluable application in the modern development of analysis. Important among abstract spaces are the so-called metric spaces (these are

188

the abstract spaces introduced by Fréchet in 1906), Hausdorff spaces, topological spaces, Hilbert spaces, and vector spaces.

HILBERT'S "GRUNDLAGEN" AND FORMAL AXIOMATICS

The discovery by Gauss, Bolyai, and Lobachevsky of a self-consistent geometry different from the geometry of Euclid liberated geometry from its traditional mold. A deep-rooted and centuries-old conviction that there could be only the one possible geometry was shattered, and the way was opened for the creation of many different systems of geometry. With the possibility of creating such purely "artificial" geometries, it became apparent that geometry is not necessarily tied to actual physical space. The postulates of geometry became, for the mathematician, mere hypotheses whose physical truth or falsity need not concern him. The mathematician may take his postulates to suit his pleasure, just so long as they are consistent with one another. Whereas it had been customary, in material axiomatics, to think of the objects that represent the primitive terms of an axiomatic discourse as being known prior to the postulates, now the postulates became regarded as prior to the specification of primitive terms. This new viewpoint of the axiomatic method has become known as "formal axiomatics" in contrast to the earlier "material axiomatics." In a formal axiomatic treatment the primitive terms have no meaning whatever except that implied by the postulates, and the postulates have nothing to do with "self-evidence" or "truth"—they are merely assumed statements about the undefined primitive terms.

Many mathematicians have come to regard any discourse conducted by formal axiomatics as a "branch of pure mathematics." If for the primitive terms in such a postulational discourse we substitute terms of definite meaning which convert the postulates into true statements about those terms, then we have an "interpretation" of the postulate system. Such an interpretation will also, if the reasoning has been valid, convert the derived statements of the discourse into true statements. Such an evaluation of a branch of pure mathematics has been called a "branch of applied mathematics." Clearly, a given branch of pure mathematics may possess many interpretations and may thus lead to many branches of applied mathematics. From this point of view, we see that material axiomatics is the independent axiomatic development of some branch of applied mathematics. In a formal axiomatic treatment, one strips the discourse of all concrete content and goes to the abstract development that lies behind any specific application.

189

Formal axiomatics was first systematically developed by David Hilbert in his famous *Grundlagen der Geometrie* (*The Foundations of Geometry*) of 1899. This little work, which has run through nine editions, is today a classic in its field. Next to Euclid's *Elements,* it may be regarded as perhaps the most influential work so far written in the field of geometry. Backed by the author's great mathematical authority, the work firmly implanted the postulational method of formal axiomatics not only in the field of geometry but also in nearly every branch of mathematics of the twentieth century. The book, by plugging up the tacit assumptions made by Euclid, offers a completely acceptable postulate set for Euclidean geometry, and it can be read, in great part, by any intelligent student of high school geometry. It has been used by a number of authors and writing groups of the second half of the twentieth century as a basis for a rigorous high school treatment of geometry.

A Modern View of Geometry

For a long time geometry was intimately tied to physical space, actually beginning as a gradual accumulation of subconscious notions about physical space and about forms, content, and spatial relations of specific objects in that space. We have called this very early geometry "subconscious geometry." Later, human intelligence evolved to the point where it was able consciously to consolidate some of the early geometrical notions into a collection of somewhat general geometrical laws or rules. We have called this laboratory phase in the development of geometry "scientific geometry." About 600 B.C. the Greeks began to inject deduction into geometry, giving rise to what we have called "demonstrative geometry."

In time demonstrative geometry became a material-axiomatic study of idealized physical space and of the shapes, sizes, and relations of idealized physical objects in that space. For the Greeks there was only one space and one geometry; these were absolute concepts. The space was not thought of as a collection of points but rather as a realm, or locus, in which objects could be freely moved about and compared with one another. From this point of view, the basic relation in geometry was that of congruence or superposability.

With the elaboration of analytic geometry in the first half of the seventeenth century, space came to be regarded as a collection of points; and with the invention, about two hundred years later, of the classical non-Euclidean geometries, mathematicians accepted the situation that there is more than one conceivable space and hence more

190

than one geometry. But space was still regarded as a locus in which figures could be compared with one another. The central idea became that of a group of congruent transformations of a space into itself, and geometry came to be regarded as the study of those properties of configurations of points which remain unchanged when the enclosing space is subjected to these transformations. We have seen how this point of view was expanded by Klein in his *Erlanger Programm* of 1872. In the *Erlanger Programm* a geometry was defined as the invariant theory of a transformation group. Since, from this point of view, a geometry is defined by a set of any objects and a group of transformations to which the set of objects may be subjected, geometry came to be rather far removed from its former intimate tie-up with physical space, and it became a relatively simple matter to invent new and perhaps bizarre geometries.

At the end of the last century, Hilbert and others formulated the concept of formal axiomatics, and there developed the idea of a branch of mathematics as an abstract body of theorems deduced from a set of postulates. Each geometry became, from this point of view, a particular branch of mathematics. Postulate sets for a large variety of geometries were studied; but the *Erlanger Programm* was in no way upset, for a geometry could be regarded as a branch of mathematics which is the invariant theory of a transformation group.

In 1906, however, Fréchet inaugurated the study of abstract spaces, and some very general studies came into being which did not fit into the Kleinian scheme but which mathematicians still wished to call geometries. A space became merely a set of objects together with a set of relations in which the objects are involved, and a geometry became the theory of such a space. It must be confessed that this latter notion of a geometry is so embracive that the boundary lines between geometry and other areas of mathematics have become very blurred, if not entirely obliterated. It is essentially only the terminology and the mode of thinking involved that makes the subject "geometric."

There are many areas of mathematics where an injection of geometrical terminology and procedure greatly simplifies both the understanding and the presentation of some concept or development. This is becoming increasingly evident in so much of mathematics that some mathematicians of the second half of the twentieth century feel that perhaps the best way to describe geometry today is not as some separate and prescribed body of knowledge but as a *point of view*—a particular way of looking at a subject. Not only is the language of geometry often much simpler and more elegant than the language of algebra and analysis, but it is at times possible to carry through rigorous trains of reasoning in geometrical terms without translating them into algebra or analysis. There results a considerable economy both in thought and in the communication of thought. Moreover, and perhaps most important, the suggested geometrical imagery frequently leads to further results and studies, thus furnishing us with a powerful tool of inductive or creative reasoning.

A great deal of modern analysis has become singularly compact and unified through the employment of geometrical language and imagery. There seems little doubt that much of this will filter down into the elementary courses in analysis, and that the presently overfat calculus texts may become slimmer again, and also easier for the student to comprehend, by the use of a geometrical viewpoint. For example, the whole epsilon-delta procedure of classical analysis will, in all likelihood, be supplanted by some sort of geometrical-topological approach.

Capsule 50   Merlyn Retz and Meta Darlene Keihn

COMPASS AND STRAIGHTEDGE CONSTRUCTIONS

IN THE first three postulates of the *Elements*, Euclid states the three "constructions" permitted in his geometry: (1) to draw a straight line from any point to any other point; (2) to produce a finite straight line continuously in a straight line; and (3) to describe a circle with any center and any distance. Euclid does not use the word "compass" in his *Elements*—he never describes how the constructions are to be performed. The restriction that these constructions are to be carried out by the use of only an unmarked straightedge and a compass has traditionally been ascribed to Plato (*c.* 390 B.C.).

The third statement above and the way in which this postulate is used in his construction problems result in a limitation that is usually expressed by saying that Euclid used a "collapsible compass," which in effect closed as soon as one of its points was removed from the paper. This seems to imply the inability to transfer a given line segment. However, the first three propositions in Book I show how (by a simple construction of an equilateral triangle with a given segment as one of the sides) it is possible to construct any given segment on a given line through a given point. The collapsible compass and straightedge are thus equivalent in use to the modern compass and straightedge.

Although the Greeks came very close to performing all the constructions that are permissible using only these two geometric tools, they were also aware of a number of relatively simple problems that they were not able to solve by just these means: (1) to inscribe in a circle any regular polygon, (2) to trisect any given angle, (3) to find the side of a cube that has twice the volume of a given cube, and (4) to construct a square having an area equal to that of a given circle.

While it has been customary to emphasize the futile search of the Greeks for such solutions (perhaps because amateur mathematicians at all periods of time since then have exercised their ingenuity on these problems), a more accurate appraisal would seem to be that even the early Greek geometers must have surmised that the allowable means were inadequate. They set to work to find other means to solve these problems, and here they were successful. By making use of certain curves, not circles, supposedly already completely drawn, they were able to solve many of these problems. The discovery of the conic sections and the use of such curves as the conchoid and the quadratrix to effect solutions attest to the ingenuity of the Greek geometers. The fact that they lacked the necessary mathematical tools of analytic geometry and algebraic theory to describe the possibilities of various geometrical instruments (and thereby also to show what is impossible) cannot be held against the Greeks. A valid proof that the fourth problem above cannot be solved by compass and straightedge alone was not given until 1882 [**44, 53**].

Another point of interest for the study of geometrical constructions has been that of "constructions with limited means." The phrase "limited means" refers to additional restrictions on the choice of geometrical tools to be used and/or to the manner in which they are to be used. Among the best known of these constructions are the "Mascheroni constructions," more properly called "Mohr-Mascheroni constructions."

In 1797 the Italian mathematician Lorenzo Mascheroni published *Geometria del compasso*, in which he showed that the construction problems that can be solved by use of straightedge and compass can be solved by means of the compass alone. Mascheroni's work, dedicated to Napoleon Bonaparte, was translated into French a year later. A German translation and a second French edition thirty years later aided in associating Mascheroni's name with this type of construction. However, in 1928 a booklet entitled *Euclides Danicus* came to light in a secondhand bookshop in Denmark. This work, written by Georg Mohr and published originally in 1672 in both Danish and Dutch editions, contained the same basic result that the "compass only" is equivalent to compass and straightedge. Since it is clearly evident that Mascheroni was unaware of the work Mohr had done over one hundred years earlier, the dual designation is justified.

How does one show the equivalence of compass only (or of any other "limited means") to compass and straightedge? Formal criteria in general were given by Jean Victor Poncelet in 1822. He noted that additional points in compass and straightedge constructions are found as the intersection of two lines, a line and a circle, or two circles. If one can show how such points can be obtained by any restricted use of geometric tools, the equivalence is thereby established. (It is obvious that not all of the points of a continuous straight line can be drawn by compass only, but as many points of the line as are necessary can be found by this means.) Mascheroni gave constructions for such points of intersection, together with solutions for various construction problems; but he did not explicitly state that the equivalence is established by such "intersections criteria."

The earlier approach of Mohr was in the form of an answer to the question, "What is possible using only the compass?" His answer was, "All of Euclid," and he validated this answer by actually giving compass-only solutions to all of the construction problems of Euclid!

Mascheroni apparently thought that his work might be of some practical value in the construction of astronomical instruments because of the greater accuracy of the compass over the straightedge; the intersection of arcs yields a more clearly defined point than that of two lines, in many cases, and straightedges are not necessarily "straight." His constructions were largely based on the idea of reflection of points; a somewhat more orderly development can be based on the idea of inversion, discovered by Jakob Steiner in 1824.

An even more extensive history is associated with another restriction on these traditional tools—the straightedge is again permitted, but the compass is restricted to one and the same opening throughout the construction. The first explicit consideration of such fixed-compass (or "rusty compass") constructions was apparently given by an Arab of the tenth century, Abu'l-Wefa. In his work on geometrical constructions, five problems specifically require that the construction is to be accomplished using only one opening of the compass. Two of these problems involve the construction of a regular pentagon, in one case with the fixed opening equal to the given side and in the other with the opening equal to the radius of the circle in which the pentagon is to be inscribed. Over a dozen other constructions are carried out using only one opening, although in these the restriction is not explicitly called for. These include the basic constructions of bisecting angles, segments, and arcs and of constructing right angles and perpendiculars through points both on and off a given line.

The fixed compass with straightedge seems to have been one of the instruments of the artists of the fifteenth and sixteenth centuries—both Leonardo da Vinci and Albrecht Dürer described constructions based on just one such opening. Many of these were related to the construction of regular polygons, useful to the artist in decorations and architecture; after the invention of firearms, they were used in drawing plans for fortifications.

Mathematically speaking, the period of the middle 1500's is most commonly associated with the discovery by Italian mathematicians of general solutions of the cubic and quartic equations. The mathematical controversy that arose between Tartaglia and Cardano as a result of this activity [71] resulted in a mathematical duel between Ferrari, Cardano's pupil, and Tartaglia. In response to seventeen problems that called for constructions using a fixed opening of the compass, Ferrari demonstrated "all of Euclid" (at least the first six books) under this restriction. Within ten years additional sets of solutions of "all of Euclid," using fixed compass and straightedge, were given by Cardano, Tartaglia, and the latter's pupil, Benedetti.

During the next century these solutions disappeared from view, although there were some hazy references to them in the "practical geometries" of the period. In 1673, a year after the *Euclides Danicus* had been published, a similar booklet, *Euclidis Curiosi*, appeared. This treated "all of Euclid" by means of "one single opening of the compasses and a ruler." Although this booklet was published anonymously in Dutch, the author was again Georg Mohr, motivated to find such solutions by the fact that he had heard that a set of solutions did exist, although he had never set eyes on a manuscript concerning it.

The Poncelet-Steiner theorem, so named for Jean Victor Poncelet, who first stated it in 1822, and for Jakob Steiner, who gave a detailed proof of it in 1833, gives a new form to the fixed-compass problem.

It states that if a single circle and its center are once drawn in the plane, every construction possible with ruler and compass can be carried through with the ruler alone. Thus the fixed compass needs to be used only once to draw the original circle. More sophisticated geometrical properties of the circle (including harmonic properties, centers of similitude, and the radical axis) are used in the proof of this theorem, and the properties are then used to establish the "intersections criteria."

In 1904 Francesco Severi successfully answered the question: "If the pair of compasses broke before the entire Poncelet-Steiner circle had been completed, what would one do in this 'catastrophe'?" The Severi theorem states that any small arc of the circle, together with the center, is sufficient to resolve all of the problems of regular compass and straightedge.

Problems in construction can thus be solved by various uses of the basic geometrical tools, and in most cases in more than one way with each. A natural question is "Which way is best?" One set of criteria was established by Émile Lemoine in 1907: The simplicity of the construction is the sum of the numbers of the various simple operations used in the construction /EVES (c): 100/.

The early Greeks may have given special attention to geometric constructions because each served as a sort of existence theorem for the figure or concept involved. The establishing of the various equivalence theorems (e.g., that the compass alone is equivalent to straightedge and modern compass) reverses the approach—one is interested in showing that theoretically, at least, the results are attainable, even if one would not bother to carry out the actual work. Jakob Steiner made the contrast rather clear when he said in reference to geometrical constructions in general, "It is a very different matter actually to carry out the constructions, i.e., with the instruments in the hand, than it is to carry them through, if I may use the expression, simply by means of the tongue!"

For Further Reading

BECKER	HALLERBERG (a)
CHENEY	———(b)
COOLIDGE (b): 43–46, 51–58	HLAVATY
COURANT and ROBBINS: 147–52	F. KLEIN (b)
COURT	KOSTOVSKII
EVES (c): 98–99 [5th ed. 90–91]	SMSG: IV, X

196

DUPLICATION OF THE CUBE

ONE OF the "three famous problems of antiquity" was to find a geometrical construction for the edge of a cube having twice the volume of a given cube. It probably dates to the time of the Pythagoreans (*c.* 540 B.C.). The Pythagorean theorem suggests a simple means for finding a square with twice the area of a given square—it is the "square on the diagonal." If the side of the square is of unit length, we have thus solved the problem of finding a line segment of length $\sqrt{2}$. The corresponding problem of finding a segment of length $\sqrt[3]{2}$ was stated in a much more interesting form by the Greeks.

The commentator Eutocius (*c.* A.D. 560) tells of a letter supposedly written by Eratosthenes to Ptolemy I (not to be confused with the mathematician of the same name) concerning King Minos, who had a cubical tomb constructed for his son. The king was displeased with the size of the monument, however, and so ordered it doubled in size —by doubling the side. Eratosthenes points out that this was an error as the tomb would thereby be increased fourfold in area and eightfold in volume; but, he says, the geometers then tried to solve the problem.

A second and better-known story is also told of the source of the problem. It is said that the gods sent a plague to the people of Athens. The people sent a delegation to the oracle at Delos to ask what could be done to appease the gods. They were told to double the size of the cubical altar to Apollo, and the plague would cease. They built a new altar, each edge of which was twice as long as each edge of the old altar. But since the gods' demand had not been fulfilled, the plague continued. The story fails to relate what was finally done to appease the gods, but evidently the plague eventually left the city.

The search for solutions to this problem, to be carried out if possible under the restriction of using only straightedge and compass, was to lead the Greeks to many mathematical discoveries during the next several centuries. A compass-and-straightedge construction for this problem was not one of their discoveries, however; it can be proven that this cannot be done under these restrictions.

197

Hippocrates (*c.* 440 B.C.) was the first to make progress on the problem. He showed that the duplication problem was the same as the problem of finding two mean proportionals between two given line segments of lengths A and $2A$. In our terminology (not that of Hippocrates), one must find x and y such that $A/x = x/y = y/2A$. Since $x^2 = Ay$ and $y^2 = 2Ax$, elmination of y leads to $x^3 = 2A^2$. Thus a cube of side x has a volume twice that of a cube of side A. At this stage in Greek mathematics, however, the simultaneous solution of the two parabolas was not possible.

Archytas of Tarentum (*c.* 400 B.C.) later solved the problem by the intersection of three surfaces of revolution. His ingenious space construction involved the intersection of a right cone, a cylinder, and an anchor ring with inner diameter of zero. /GRAESSER; VAN DER WAERDEN: 150–52./

Menaechmus (*c.* 350 B.C.) is given credit for discovering the conic sections in the process of trying to find a solution for this problem. He gave two solutions, one involving the intersection of two parabolas, and the other the intersection of a hyperbola and a parabola. (It can easily be seen by analytic geometry that when the equations $y = x^2$ and $xy = 2$ are solved simultaneously, then $x = \sqrt[3]{2}$.) It should be emphasized that these were perfectly legitimate solutions, but they did not satisfy the Greek criterion of restricting the tools used to straightedge and compass. Plato (340 B.C.) discovered a mechanical solution, and during the third century B.C. Nicomedes used the curve called the conchoid. Diocles (*c.* 180 B.C.) used the cissoid to effect a duplication. /HEATH (c): 164–69./

Viète, in 1593, proved that every cubic equation not otherwise solvable leads to either a duplication or a trisection problem. It remained for Descartes in 1637 to prove the impossibility of solution by means of lines and circles. He showed that a parabola and a circle can be used to find the roots of a cubic equation if the second-degree term is missing. Since every cubic may be reduced to one with no second-degree term, every cubic may be solved by means of a circle and a parabola. But the parabola may not be constructed with straightedge and compass, hence neither the duplication of the cube nor the trisection of the angle may be so performed.

For Further Reading

COOLIDGE (b): 46
GRAESSER

HEATH (c): 154–70
F. KLEIN (b): 1–47

SANFORD (d): 265–68   VAN DER WAERDEN: 139–41, 150–

D. E. SMITH (a): II, 313–16   65, 230–31, 236–37

Capsule 52   *Philip Habegger*

THE TRISECTION PROBLEM

THE ease with which any angle can be bisected and any line segment can be divided into any number of equal parts, using only straightedge and compass, readily gives motivation to attempt the multisection of any angle under similar restrictions. It is possible also that the problem of constructing a regular polygon of nine sides, which requires the trisection of a 60° angle, was the motivation the early Greeks had for attempting this problem and thus adding it to their "famous problems of antiquity."

In the nineteenth century algebraic theory supplied the basis for the proof that in general an arbitrary angle cannot be trisected with only compass and straightedge. (Many special angles, e.g., an angle of 90°, can of course be trisected.) The Greeks did devise various methods for trisecting any angle, but these were dependent on less restrictive procedures or on special geometric curves assumed to be already drawn.

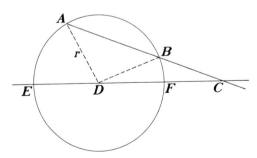

FIG. [52]-1.—TRISECTION OF AN ANGLE

A typical construction of this sort can be based on a proposition given by Archimedes. This is illustrated in Figure [52]-1. If one extends a chord AB of an arbitrary circle by a segment BC that is equal to the radius, and if C is then joined to the center D to give diameter EDF, $\angle ADE$ is three times the size of $\angle BDF$. (The proof follows easily from the properties of isosceles triangles and the exterior angle of a triangle.) Now in order to trisect the arc AE (or the corresponding $\angle ADE$) one can extend the diameter EDF and then lay off segment BC equal to r in such a way that the extension of BC passes through point A. This can be accomplished by using a straightedge on which the distance r has been marked off; the straightedge is then slid so that it passes through point A while the end points of the r segment simultaneously lie on the circle and the diameter EDF extended. The line through points A, B, and C is then drawn with the help of the straightedge.

This latter procedure is an example of a *neusis* ("verging," "inclination," "insertion") construction frequently used by the early Greeks: a segment of given length is placed with its end points on two given curves in such a way that the segment extended passes through a given point. The use of such a "marked" straightedge violates the customary compass and (unmarked) straightedge restriction, but it does effect a solution.

A similar *neusis* construction for trisecting an angle can be accomplished using a special curve, the conchoid (or cochloid) invented by Nicomedes (*c.* 240 B.C.). Let the ℓ of Figure [**52**]-2 be any line; O a point not on ℓ; Q any point on ℓ; and P the point on line OQ such that the segment QP is of given length k. The locus of point P as Q moves along ℓ is one branch of the conchoid of ℓ for the pole O and the constant k. (The second branch would be determined by the locus of P' at distance k in the opposite direction from Q.)

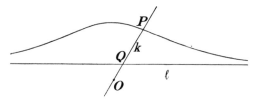

FIG. [**52**]-2.—CONCHOID OF NICOMEDES

Pappus (*c.* A.D. 320) states that Nicomedes described an apparatus that would draw conchoids; with such an instrument one can easily trisect any angle /EVES (c) : 86–87/.

Pappus also gave a construction for trisecting an angle with the aid of a hyperbola. This construction used the focus and directrix property of the conic /Eves (c): 105/.

One of the earliest constructions using a so-called higher curve was that of Hippias (425 B.C.). This curve was the quadratrix [53]; it could be used to divide an angle in any given ratio.

Many centuries later René Descartes gave a method for trisecting an angle using a parabola and a circle /Tietze: 53–55/.

Other methods use various mechanical solutions, including linkages, a carpenter's square, special compasses, and many others. There are also constructions that produce good approximations to the trisection of an angle, although they are not theoretically exact /Yates/.

The first rigorous proof of the impossibility of the trisection of any given angle by compass and straightedge was given by Pierre Wantzel in 1837. The needed criteria for constructibility are essentially algebraic in nature /Eves (c): 97–99/; involved are considerations of such concepts as domain of rationality, algebraic numbers, and group theory.

For Further Reading

Eves (c): 85–89; 97–99
    [5th ed. 80–83, 406–7]
Heath (c): 147–54
Sanford (d): 261–65

Schaaf (d): 102–5
Tietze: 46–63
Yates

Capsule 53   Lee Anderson

THE QUADRATURE OF THE CIRCLE

The quadrature of the circle, sometimes called "squaring the circle," is a plane geometry construction problem: to construct a square having an area equal to that of a given circle. If one introduces the restriction that only compass and straightedge are to be used in the construction, no solution is possible; more than two thousand years passed before this was proved. It can be solved by higher algebraic and transcendental curves, as we shall see below.

The quadrature problem is a very old problem. The Egyptian Rhind papyrus (*c.* 1650 B.C.) relates several problems in which the area of a circle is expressed as the area of a square with side ⅑ less than the diameter of the circle [44]; there, however, the problem arises as an area problem and not as a construction problem.

The Greeks were aware of the problem before 400 B.C., for Aristophanes refers to this problem in his play *Birds*, written about 414 B.C. (Aristophanes has Meton, an astronomer, bring with him a ruler and compass and make certain constructions "in order that your circle may become square." This "solution" is just a play on words, for Meton merely inscribed a square in a circle.) However, the problem was seriously worked on by Anaxagoras (*c.* 440 B.C.), Hippocrates (*c.* 440 B.C.), Antiphon (*c.* 430 B.C.), Hippias (*c.* 425 B.C.), and Archimedes (*c.* 287–212 B.C.) Hippocrates succeeded in squaring certain lunes, and Antiphon thought he had solved the problem by inscribing regular polygons in a circle. Hippias developed the "quadratrix" to solve this problem, and Archimedes used the "spiral of Archimedes" to obtain a solution.

The quadratrix of Hippias illustrates such a higher curve (this one probably invented to aid in trisecting an angle) which can be used to obtain a solution of the problem. Suppose (Fig. [53]-1) that *ABCD* is a square and arc *BED* is a quadrant of a circle with *A* as center and

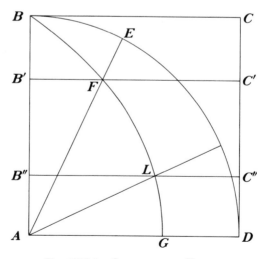

Fig. [53]-1.—Quadratrix of Hippias

segment AB as radius. Suppose a radius moves uniformly from position AB to position AD, while at the same time segment BC moves uniformly from position BC to position AD, always parallel to itself.

The relative motions of the radius and segment are so timed that they begin moving from original positions at the same instant and reach position AD at the same instant. The corresponding points of intersection of the radius and the segment at each moment form the desired locus, and so $AFLG$ is the quadratrix.

Pappus proved by *reductio ad absurdum* /SANFORD (d): 260/ that the following proportion holds:

arc of quadrant BED/segment AB = segment AB/segment AG.

Consequently, a line segment whose length is $\frac{1}{4}$ of the circumference may be constructed as the third proportional to segments AG and AB. Let us call this line segment q. In modern notation, the area of the circle is $\frac{1}{2}rC = \frac{1}{2}r(4q) = 2rq$. Construct the mean proportional s between $2r$ and q (i.e., find s such that $2r/s = s/q$). Hence s^2, which is the area of a square with side s, is equal to $2rq$, which is the area of the circle. It must be pointed out that the solution depends upon the previous existence of point G, which can be determined only approximately, since the radius and segment coincide completely at the last instant.

The spiral of Archimedes can also be defined dynamically. Let ray OA revolve uniformly about fixed point O, starting from initial posi-

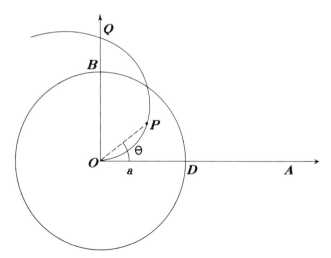

FIG. [53]-2.—SPIRAL OF ARCHIMEDES

tion *OD*. At the same time let point *P* also move uniformly along *OA*, starting from *O* at the same time the ray is in initial position. The locus of point *P* is the spiral. (The equation is easily represented in modern polar-coordinate form as $r = a\theta$, where a is some constant.)

When *OA* has rotated to position *OB* perpendicular to *OD*, point *P* will have reached point *Q*, and arc *DB* (with radius a) will equal *OQ*, since each is equal to $a\theta$. Hence we have found a line segment whose length equals ¼ of the circumference; the solution can be completed as in the quadratrix problem above.

The impossibility of performing the construction with compass and straightedge alone depends upon the fact that π is a transcendental number, proved by Ferdinand Lindemann in 1822. (See [**44**], "The Number π," which is closely related to this topic.)

For Further Reading

Eves (c): 89–99
[5th ed. 83–90]
Heath (c): 139–47
Hobson: 1–57

Sanford (d): 257–61
Schaaf (d): 106–8
Tietze: 90–105

Capsule 54   Cynthia Schenck and Samuel Selby

THE GOLDEN SECTION

When the Greek commentator Proclus said that Eudoxus (*c.* 370 B.C.) continued the researches on the section begun by Plato, he was referring to what has become the second most familiar ratio known to mathematicians (π stands in undisputed first place). The "golden section" (ratio, mean), as it has come to be known, was thus studied by the Greeks before the time of Euclid. Euclid describes this section in his Proposition VI, 30: "to cut a line segment in extreme and mean ratio."

The point *B* (see Fig. [**54**]-1) is said to divide the segment *AC* in mean and extreme ratio if the ratio of the shorter segment to the longer

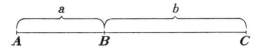

Fig. [54]-1.—Golden Section

is equal to the ratio of the longer to the entire line segment, that is, $AB/BC = BC/AC$. If we use modern notation, writing this as $a/b = b/(a + b)$, with $a < b$, and then set $b/a = x$, we have the equation $x^2 - x - 1 = 0$, or $x = (1 \pm \sqrt{5})/2$. The positive root, $1.618034\cdots$, is often designated by the symbol ϕ, phi (sometimes by τ, tau). Since the equations $x^2 = 1 + x$ and $1/x = x - 1$ are equivalent to the original equation, it immediately follows that $\phi^2 = 2.618\cdots$ and $1/\phi = .618\cdots$. (The ratio a/b above therefore is $.618\cdots$, given by some writers as the golden ratio.)

Euclid, of course, did not work with numbers or algebra in the forms just given; he gave a geometrical construction that determined the point where the segment was cut (sectioned) in the desired ratio. Actually, his Proposition II, 11 resolves an equivalent problem with an easier solution: "to cut a given segment so that the rectangle contained by the whole and one of the segments is equal to the square on the remaining segment" [78]. Described informally, the construction is carried out as follows: If $ABCD$ (see Fig. [54]-2) is a square with

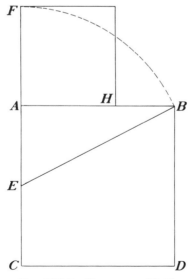

Fig. [54]-2.—Proposition II, 11

205

AB as the given segment, find E, the midpoint of AC. With E as center and EB as radius, find the intersection of this arc with the line CA extended, giving point F. Completing the square on side AF gives the desired point H.

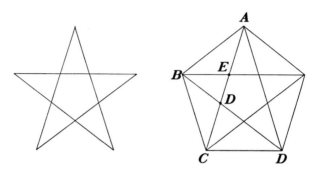

FIG. [54]-3.—STAR PENTAGRAM AND PENTAGON

The sectioning of a segment in mean and extreme ratio is found in the star pentagram, the symbol of health and the distinguishing mark among the Pythagoreans. Each of the five segments divides others in mean and extreme ratio. In fact, if one considers a regular pentagon and the corresponding star pentagram, one can readily show that

$$\phi = AC/AB = AC/AD = AD/AE = AE/DE,$$

and so forth. Euclid also includes a number of propositions on the golden section in Book XIII, which is devoted to the theory of the regular polyhedra.

The golden ratio may have been known even before the time of the Greeks. The Greek historian Herodotus relates that the Egyptian priests had told him that the proportions of the great pyramid of Gizeh were so chosen that the area of a square whose side is the height of the great pyramid equals the area of a face triangle. Rather simple algebra can be used to show that the ratio of the altitude of a face triangle to half the base length is ϕ /BARAVALLE (a)/. Actual measurements on the pyramid seem to give a close approximation to this ratio.

The aesthetic and artistic properties of this ratio are exhibited in the "golden rectangle"—a rectangle with sides in the ratio of 1 to ϕ or ϕ to 1. Such a rectangle, among all possible ones, is considered by some to be the most pleasing to the eye. (File cards in either of the two standard proportions, 3-by-5 or 5-by-8, are close approximations to the golden rectangle.)

206

Many famous works of architecture and art, such as the Greek Parthenon and some of the paintings of Leonardo da Vinci, appear to be framed in the golden rectangle, although of course this does not prove that the creator necessarily began with this specific ratio in mind. In 1509 Luca Pacioli wrote the treatise *De divina proportione* ("Of the Divine Proportion"), which was illustrated by Leonardo da Vinci and which treats of this ratio. Leonardo referred to the *sectio aurea* ("the golden section"), while Kepler referred to it as *sectio divina* ("divine section").

To anyone familiar with the Fibonacci sequence [22], the numbers of the ratio 3-to-5 or 5-to-8 are recognizable as adjacent terms in the sequence 1, 1, 2, 3, 5, 8, 13, 21, 34, . . . , where any term after the second is the sum of the two preceding terms. This is not coincidence; it can be proved that the ratio of any two successive terms of this sequence approaches ϕ as a limit as one goes farther and farther out in the sequence.

Even nature seems to show a peculiar interest in displaying patterns involving these relations. The seeds in a sunflower or the tiny florets in the core of a daisy blossom are arranged in two overlapping sets of spirals, radiating in clockwise and counterclockwise directions. A counting of the number of spirals in each case almost invariably yields two consecutive terms in the Fibonacci sequence, such as 21 and 34 or 34 and 55 /BERGAMINI: 92–97/. Similar relations are often found in various plants with a spiral leaf-growth pattern.

For Further Reading

BERGAMINI: 92–97
BARAVALLE (a)
GARDNER (c): 89–103

SCHAAF (d): 139–42
SMSG: IX

Capsule 55   Alice I. Robold

NON-EUCLIDEAN GEOMETRY

By the term "non-Euclidean geometry" we mean a system of geometry built up without the aid of the Euclidean parallel

207

hypothesis and containing an assumption as to parallels incompatible with that of Euclid.

EUCLID, about 300 B.C., collected and arranged the propositions of plane geometry, based upon a set of five postulates. He defined parallel straight lines as "straight lines which being in the same plane and being produced indefinitely in both directions, do not meet one another in either direction." The fifth postulate, referred to as the "parallel postulate," states:

> *If a straight line falling on two straight lines make the interior angles on the same side less than two right angles, the two straight lines, if produced indefinitely, meet on that side on which are the angles less than the two right angles.*

From the beginning, this postulate was criticized. It lacked the terseness of the first four postulates and seemed not sufficiently evident to be accepted without proof. Proclus considered it "alien to the special character of postulates." Euclid proved twenty-eight propositions before using the fifth postulate in a proof. The converse of the parallel postulate, *"The sum of two angles of a triangle is less than two right angles,"* was proved by Euclid as a theorem, therefore it was thought that the parallel postulate should also be capable of proof.

Heath, in his translation and commentary /EUCLID: I, 202–20/, relates several noteworthy attempts to prove the postulate—by Ptolemy, Proclus, Nasir eddin, John Wallis, Girolamo Saccheri, Johann Heinrich Lambert, and Adrien Marie Legendre. As a result of such attempts many equivalent axioms were discovered. More important, however, was the work leading up to the discovery of non-Euclidean geometry.

Of special importance here is the work of Saccheri, a Jesuit priest in Italy, who attempted to prove the postulate by *reductio ad absurdum.* Using a two-right-angled isosceles quadrilateral, Saccheri proved /SISTER FITZPATRICK/ that

> interior angles at the straight joining the extremities of equal perpendiculars erected . . . from two points of another straight, as base, not merely are equal to each other, but besides are either right or obtuse or acute according as that join is equal to, or less, or greater than the aforesaid base; and inversely.

He mistakenly believed that he was successful in reaching a contradiction in the cases of the hypotheses of both the obtuse and acute angles. Of the latter he said:

> For then we have two lines which produced must run together into the same line and have at one and the same infinitely distant point a common perpendicular. . . . The hypothesis of the acute angle is absolutely false, because repugnant to the nature of a straight line.

Even though he reached a faulty conclusion, his work is significant in that he developed a series of propositions that constitute an important part of non-Euclidean geometry, the possible truth of which he was unable to perceive.

The actual development of non-Euclidean geometry, however, was not based on Saccheri's work and did not come about until the early nineteenth century, over two thousand years after Euclid. Amazingly, it was then developed independently by three people, Lobachevsky, Bolyai, and Gauss. The first to publish was Nikolai Lobachevsky, mathematics professor at the University of Kasan. Janos Bolyai, a Hungarian, published his development as an appendix to a work of his father, Farkas (or Wolfgang) Bolyai. When the proud father sent a copy of the book to his friend Carl Friedrich Gauss, the great German mathematician replied that he could hardly praise Janos' work because it coincided with work and results Gauss himself had developed, but not yet published, over a period of thirty to thirty-five years.

Little attention was paid then to the subject until 1866, when G. F. Bernhard Riemann suggested a geometry in which no two lines are parallel and the sum of the angles of a triangle is greater than two right angles. The geometry of his predecessors was synthetic. Riemann's geometry was closely related to the theory of surfaces.

The development of non-Euclidean geometries has been of special significance in showing *why* attempts to prove Euclid's parallel postulate failed. The successful development of a consistent geometry using the first four postulates of Euclid but replacing the fifth with another that is incompatible with it proves that the fifth postulate is indeed independent of the other four; thus it could not be proved. This realization has helped to lead to a more careful examination of the foundations of mathematics.

One may ask whether a geometry is based on a set of consistent postulates, whether these postulates are independent of one another, or whether this geometry will serve better than another geometry for a given application. But the question of whether a geometry is "true" has no place in pure science.

For Further Reading

BONOLA
CARSLAW
EUCLID: I, 202–20
NAGEL *et al.*: 613–21

SISTER FITZPATRICK
D. E. SMITH (c): II, 351–58
WOLFE: 1–64

Capsule 56   Roger Lowe

THE WITCH OF AGNESI

PIERRE de Fermat (1601–1665) studied the quadrature (determination of area) of several types of curves. Among these was one that he wrote as $e(a^2 + b^2) = b^3$. In modern terminology this would be written as $y(x^2 + a^2) = a^3$, and in English it is known as the "witch of Agnesi."

Fermat himself did not name the curve. Guido Grandi (1671–1742) studied it later and gave it the name *versoria* (a Latin word meaning a rope that guides a sail). Why he gave it this name has never been clear—there is a similar obsolete Italian word *versorio* that means

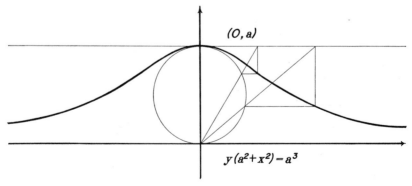

$$y(a^2 + x^2) = a^3$$

FIG. [56]-1.—THE WITCH OF AGNESI

free to move in every direction; but there is no particular reason to associate the curve with this concept, either. When the Italian mathematician Maria Gaetana Agnesi referred to this curve in *Instituzioni analitche* (published in 1748, when she was thirty years old), she confused the word *versoria* with a different Latin word, *versiera*, which could be translated as "devil's grandmother," "female goblin," or—as John Colson in fact did translate it—"witch." Agnesi's work on the analysis of finite and of infinitesimal quantities was widely used for many years, and the name has become standard in English.

Maria Agnesi was a brilliant girl whose father was professor of mathematics at the University of Bologna. By the time she had reached the age of thirteen she was proficient in Hebrew, French, Spanish, German, and several other languages in addition to her native Italian. Her father enjoyed hosting gatherings of the intelligentsia at which Maria would converse with the learned professors on any topic they might pick and in their native languages. She disliked this type of publicity, however, and from the age of twenty led a rather secluded life apart from her mathematical activities. In 1752, when her father became terminally ill, she was appointed to his chair of mathematics at the University of Bologna.

For Further Reading

LARSEN   STRUIK (e): 178–80

Capsule 57   *Duane E. Deal*

FOUR-DIMENSIONAL GEOMETRY

UNTIL almost the beginning of the present century the attitude of mathematicians and lay people alike toward geometries of more than three dimensions—if notice was given to them at all—was one of skepticism. It was generally taken that reference to physical considerations alone was sufficient to preclude the existence of more than three dimensions.

Aristotle, in his *Heaven,* said that a solid has magnitude "in three ways, and beyond these there is no other magnitude because the three are all." In *Physics* he talked of six dimensions, but these were up and down, before and behind, and right and left.

Ptolemy, as quoted by Simplicius, said, "It is possible to take only three lines that are mutually perpendicular; two by which the plane is defined and a third measuring depth." To the Greeks no further explanation was necessary.

In the fourteenth century Nicole Oresme sought a graphic representation of Aristotelian forms as heat, velocity, sweetness, and so on by laying down a line as a basis, called *longitudo,* and taking one of the forms to be represented by lines perpendicular to this as a *latitudo.* The form was then represented by a surface. Taking a surface as a basis, with *latitudo* perpendicular at each point, a solid was formed. He even went on to take a solid as a basis and at each point considered an increment. But he rejected a fourth-dimensional figure out of hand and considered instead the solid as consisting of infinitely many planes. Taking perpendiculars at each point of each plane, the result was an infinite set of intersecting solids. He did use the phrase "fourth dimension," however, in Latin as 4^{am} *dimensionem.*

Girolamo Cardano in his *Ars magna* ("great art") of 1545 referred to powers of numbers in geometrical terms, saying, "The first power [of a number] refers to a line, the square to a surface, the cube to a solid, and it would be fatuous indeed for us to progress beyond for the reason that it is contrary to nature."

Some progress is noted by the sixteenth century, for Michael Stifel in 1553 said that in arithmetic "we set down corporeal lines and surfaces, and pass beyond the cube as if there were more than three dimensions, although this is contrary to nature." At about the same time Christopher Clavius tried a detailed proof that not more than three concurrent lines can be drawn, each perpendicular to each of the others. This proof was not published until its appearance in a German encyclopedia in 1802.

René Descartes tried to find a graphic representation of the motion of freely falling bodies. He said that if a body is acted on by one accelerating force, motion is represented by a triangle. If it is affected by two forces, it is represented by a triangular pyramid. But if acted on by three forces, it is represented "by other figures." What these figures are Descartes did not try to point out.

Blaise Pascal at about the same time used a summation (or, in effect, an integration) which we may write in our numeration system as

$$\sum_{m=0}^{a} x^m \cdot \Delta x = \frac{a^{m+1}}{m+1}.$$

This yields a surface or a solid, but for $m = 3$ he said it yields a "plane-plane, composed of . . . solids each of which is multiplied by a small division of the axis forming . . . small plane-planes of the same altitude . . . and one need not be disturbed by this fourth dimension, because . . . on taking planes in place of sides, solids, or even lines . . . the sum of the lines gives a plane which takes the place of this 'plane-plane.' " He avoided the difficulty of four dimensions by letting a line represent (numerically at least) a solid.

Henry More in *The Immortality of the Soul* (1659) wrote of a fourth dimension that he called the "fourth spissitude" or thickness. He thought of a spirit as "redoubling into itself," as a string may be doubled and redoubled, and as a spiritual substance that might "double and redouble" into a fourth dimension. He went no further, as "the fourth mode or dimension is the state when a spirit will not easily admit of further redoubling," being in "saturation." This of course is not really mathematical thinking, nor was it intended to be.

John Keill (1698), of Oxford, cited More, among others, as proof that philosophers maintain "opinions more absurd than can be found in any of the most Fabulous Poets."

John Wallis' fascinating *Algebra* of 1685 includes this passage:

> A Line drawn into a Line shall make a Plane or Surface; this drawn into a Line shall make a Solid; But if this Solid be drawn into a Line, or this Plane into a Plane, what shall it make? A Plane-Plane? That is a Monster in Nature, and less possible than a Chimaera or Centaure. For Length, Breadth, and Thickness take up the whole of Space. Nor can our Fansie imagine how there should be a Fourth Local Dimension beyond those Three.

Jacques Ozanam (1698) rejected four dimensions because "in nature we do not know of any quantity which has more than three dimensions." Immanuel Kant said that because material bodies are influenced by the law of inverse squares, three dimensions arise. The soul, he reasoned, is subject to the same law. Though he did not give a formal proof, he said if God had wanted He might have given us a law of inverse cubes and hence a four-dimensional space.

A more modern note was struck by Jean Le Rond d'Alembert (1754):

> I stated above that it is impossible to conceive of more than three dimensions. A man of parts, of my acquaintance, holds that one may however, look upon duration as a fourth dimension. This idea may be challenged but it seems to me to have some merit other than that of mere novelty.

Joseph Louis Lagrange, in 1797, said:

> Since the position of a point in space depends upon three rectangular coordinates, these coordinates in the problems of mechanics are conceived as being functions of t. Thus we may regard mechanics as a geometry of four dimensions, and mechanical analysis as an extension of geometrical analysis.

In 1885 an anonymous letter to the British magazine *Nature* noted that there could be *many* fourth dimensions, and called time "a" fourth dimension, not "the" fourth dimension. This letter refers to "time-space," and says a "cube and the whole of the three-dimensional space in which it is situated are floating away in time-space." There are several other references in the nineteenth century to four dimensions in occult, mystical, and theological writing.

In 1891 W. W. Rouse Ball tried to help the physicists who were bothered by the "luminiferous ether" that was to be an elastic solid yet offer no resistance to the passage of the planets through it. He suggested that the ether was in a fourth dimension by itself but made contact at its various borders with the small particles of bodies in our three-dimensional world.

Florian Cajori spoke of a "literary man in Berkeley" who asks, "Is it not nonsense for mathematicians to talk about a fourth dimension, and could they not employ their time in some way useful to mankind?"

With the advent of Einstein's theory of relativity and subsequent discussions of time-space and the possible curvature of our three-space in a four-dimensional space, four-dimensional concepts are spoken of quite calmly, and we now realize that the physical existence or nonexistence of a four-dimensional body has nothing to do with its existence as a mathematical entity.

For Further Reading

SCHAAF (d): 131–34

THE PYTHAGOREAN THEOREM

THE "Pythagorean" theorem is one of the most important propositions in the entire realm of geometry. Despite the strong Greek tradition that associates the name of Pythagoras with the statement that
*"the square on the hypotenuse of the right-angled triangle is equal to
the squares on the sides containing the right angle,"* there is no doubt
that this result was known prior to the time of Pythagoras. Apollodorus
comments on the "noble sacrifice" offered by Pythagoras upon proof
of this theorem. (Pythagoras is said to have sacrificed a hecatomb, a
herd of one hundred oxen, in keeping with the practices of thanks-
giving at that time.) The facts that such a sacrifice was contrary to
the tenets of the Pythagoreans and that the same thing is said of
Thales in connection with his supposed discovery that the angle
in a semicircle is a right angle both cast suspicion on the authenticity
of the story. Nevertheless, it is the sort of story appropriate to the
significance of the event. The Pythagorean society may have had the
first actual proof of the statement, but this can only be conjecture.

Neugebauer /(a): 35–40/ speaks of the Babylonian study and discovery of the diagonal of a square, given the measure of the side,
as "sufficient proof that the 'Pythagorean' theorem was known more
than a thousand years before Pythagoras." Further evidence may be
found in the clay tablet text, Plimpton 322, which contains columns
of figures related to Pythagorean triples [16].

The frequent textbook reference to Egyptian "rope-stretchers" and
their knotted surveying ropes as proof that these ancients knew the
theorem is erroneous. While it is known (from a twelfth-dynasty
papyrus fragment) that the Egyptians realized as early as 2000 B.C.
that $4^2 + 3^2 = 5^2$, there is no evidence that the Egyptians knew or
could prove the right-angle property of the figure involved. The
proposition of proving that a 3–4–5 triangle has a right angle is a
challenge to good students, when they are required to do so in the
"Egyptian way," that is, without using the Pythagorean theorem or
its converse.

The conception of this theorem may not be entirely Western. The

215

Chinese *Chou Pei Suan Ching* (dating back to the Han period, 202 B.C.–A.D. 220, and possibly considerably earlier) includes an excerpt in which the speaker directs the listener to "break the line and make the breadth 3, the length 4; then the distance between the corners is 5." In this same document are found wood-block printings (see Fig. [58]-1) of several diagrams that are now associated with proofs of the theorem offered in many elementary geometry texts. No actual proof is given, however.

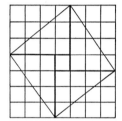 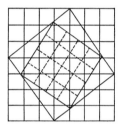

FIG. [58]-1.—DIAGRAMS FOR THEOREM

The *Sulvasutras* of Hindu mathematics, dating from the last five centuries preceding the Christian era, give rules relating to the proportions of altars and imply a statement of the theorem. There is no reason for believing, however, that the Hindus had any idea of the nature of a geometric proof.

It is possible that Pythagoras may have given proof of the theorem based on the proportionality of similar figures. With the later realization that all lines are not necessarily commensurable, this proof became invalid. Thus, at the time of Euclid's *Elements* (*c.* 300 B.C.) there was need for a more adequate proof. (Proclus' speculation was simply that Euclid rewrote the proof in order that he might put the proposition in his first book to complete it. There is also considerable evidence that the first book was written to lead to the climax of this theorem and its converse.)

Euclid's Proposition I, 47, is the Pythagorean theorem, with a proof universally credited to Euclid himself. When, in 1907, Elisha Scott Loomis prepared the initial manuscript for *The Pythagorean Proposition*, a work that eventually (in its second edition) grew to contain 370 proofs of this theorem, he lamented, "I have noticed lately two or three American texts on geometry in which the above proof [Euclid's] does not appear. I suppose the author wishes to

216

show his originality or independence—possibly up-to-dateness. He shows something else. The leaving out of Euclid's proof is like the play of *Hamlet* with Hamlet left out."

Euclid's proof follows, in somewhat abbreviated form.

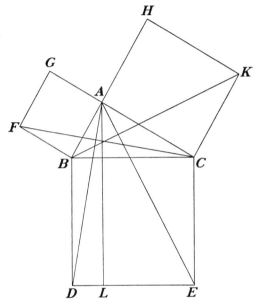

FIG. [58]-2.—EUCLID'S PROOF

Let angle *BAC* of Figure [58]-2 be the right angle of triangle *ABC*. The squares *BG*, *BE*, and *CH* are formed on the respective sides, and *AL* is drawn parallel to *BD* (or *CE*). Points *C*, *A*, *G*, and also points *B*, *A*, *H* are shown to be collinear. Then triangle *ABD* is shown to be congruent to triangle *FBC* (Euclid calls them "equal") by Proposition I, 4, which is Euclid's statement of "s.a.s." Parallelogram *BL* is twice triangle *ABD*, and square *BG* is twice triangle *FBC*, hence parallelogram *BL* is equal to the square *BG*.

Similarly, parallelogram *CL* can be proved equal to the square *CH*. Therefore square *BDEC*, made up of the two parallelograms *BL* and *CL*, is equal to the two squares *BG* and *CH*. Q.E.D.

The characteristic diagram immediately identifies Proposition I, 47, regardless of the language into which the translation of Euclid has been made. Bergamini /78/ gives reproductions from the Greek (*c.* 800), Arabic (*c.* 1250), Latin (1120), French (1564), English (1570), and Chinese (1607). The figure is sometimes referred to as the "Bride's Chair," supposedly because it resembles the chair, carried on the back of a slave, in which an Eastern bride was sometimes transported to the wedding ceremony.

The most abbreviated proof of the theorem was given by the Hindu scholar Bhaskara (*c.* 1150). He offers the diagram given in Figure [58]-3 with no explanation—only the word "Behold!" A little algebra supplies the proof /Eves (c): 189/.

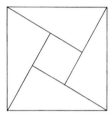

Fig. [58]-3.—"Behold!"   Fig. [58]-4.—Known to Tabit ibn-Qorra

Another dissection proof, developed by H. Perigal in 1873 (Fig. [58]-4), is a rediscovery of one known to Tabit ibn-Qorra in the ninth century /Eves (c): 73/. Knowing the formulas for areas of the triangle and square, one may complete this proof by addition of areas.

Probably no other theorem in mathematics can be demonstrated by such a wide variety of algebraic and geometric proofs.

For Further Reading

Bergamini: 78
Euclid: I, 349–68, 417–18
Eves (c): 59–61, 73–74
  [5th ed. 53–54, 66–67, 169–70]
Heath (c): 95–100
Jones (f)

Loomis
Neugebauer (a): 35–40
Shanks: 121–30
D. E. Smith (a): I, 30–31; II, 288–90
Van der Waerden: *See index*

PONS ASINORUM

"Pons asinorum" is a name commonly used for Proposition 5 of Book I of Euclid's *Elements:*

> *In isosceles triangles the angles at the base are equal to one another, and if the equal straight lines be produced further, the angles under the base will be equal.*

The term is an anglicized Latin phrase meaning "bridge of asses"; but who originated the term, or why, is unknown.

Some suppose that the term arose during the Dark Ages, when the Greek tradition in mathematics was forgotten. The standard mathematical reference at this time was a work by Boethius (*c.* 475–524), consisting of some arithmetic and the statement of some of the propositions of Euclid. No proofs of these were given beyond Euclid, Book I, 5, which thus represented a sort of high-water mark in demonstrative mathematics at this time. It is thought that in the monastic schools where Boethius' work was long used, the term "pons asinorum" referred to the implication that anyone who could not cross this bridge was a fool. An alternative explanation is that those who balked at going further than this in geometry were like asses who refuse to cross a bridge.

Still another explanation comes from the resemblance of the diagram for Euclid, Book I, 5, to a trestled bridge, which could be crossed only by a surefooted animal such as the ass.

Accordingly, only the surefooted student could proceed beyond this point in geometry.

Thales (*c.* 600 B.C.) is credited with the discovery of Euclid I, 5; he may have attempted a formal proof of it. In addition to Euclid's proof there are two demonstrations given by Proclus (*c.* 410–485), one of which is attributed to Pappus (*c.* 320). In another proof, apparently related to that of Pappus, the triangle is supposed to be taken up, turned over, and placed on itself. The difficulty of supposing that a triangle can be taken up, and at the same time in some sense remain where it is, was recognized at an early date.

C. L. Dodgson /48/ comments humorously on this proof:

Minos: It is proposed to prove I, 5 by taking up the isosceles Triangle, turning it over, and then laying it down on itself.

Euclid: Surely this has too much of the Irish Bull about it, and reminds one a little too vividly of the man who walked down his own throat, to deserve a place in a strictly philosophical treatise?

Minos: I suppose its defenders would say that it is conceived to leave a trace of itself behind, and that the reversed Triangle is laid down upon the trace so left.

For Further Reading

Euclid: I, 251–55, 415–16

Capsule 60   Daniel L. Klaasen

REGULAR POLYHEDRA

The regular polyhedra have been a part of geometric study ever since such study began. They have a symmetrical beauty that has fascinated men of all ages. Some regular polyhedra were known to the ancient Egyptians, who used them in their architecture.

The Pythagoreans (c. 500 B.C.) probably discovered three of the five regular polyhedra and made them an important part of the study of geometry. The Greeks believed that the five solids corresponded to the elements of the universe—the tetrahedron to fire, the cube to earth, the octahedron to air, the icosohedron to water, and the dodecahedron to the universe. Soon after the Pythagoreans, Plato (c. 350 B.C.) and his followers studied these solids to such an extent that they became known as the "Platonic solids."

The regular polyhedra do occur in nature: the first three in crystal form and the last two as skeletons of microscopic sea animals. However, it is their beauty and symmetry that have kept man interested in them through the centuries. There is no distinct mathematical discipline based on the five solids, but much important mathematics has been discovered as a by-product of the study of these figures. Theaetetus wrote a treatise on the five solids about 380 B.C., and is said to have been the first to prove that there are exactly five regular polyhedra. Later Euclid (*c.* 300 B.C.) devoted most of his thirteenth book to theorems dealing with the solids.

After the Greeks, interest in the subject diminished, and the solids have never again risen to the same degree of popularity or relative importance that they enjoyed during the Greek period. Current considerations of the five solids tend to be topological, as can be seen by a modern definition, that is, that a solid is a regular convex polyhedron if its faces are all congruent regular polygons, if its vertices are convex, and if the same number of faces meet at each vertex. The Swiss mathematician Ludwig Schlafli (1814–1895) devised the current symbol for the regular polyhedron $\{p, q\}$, where p stands for the number of sides around each regular polygon and q stands for the number of polygons meeting at each vertex.

It is not difficult to show that there are no regular polyhedra other than the five already named.

Let $\{p, q\}$ be any regular polyhedron. The size (in degrees) of each angle of the regular polygons forming its sides can be expressed as $180 - (360/p)$. Since $\{p, q\}$ is convex, the sum of the angles at one vertex is less than 360 degrees. Therefore we can set up the following inequality:

$$\left(180 - \frac{360}{p}\right)q < 360,$$

$$180\left(1 - \frac{2}{p}\right)q < 360,$$

$$(p - 2)(q - 2) < 4.$$

Now p and q are both larger than 2. If $p = 3$ we can have from the inequality $\{3, 3\}$, $\{3, 4\}$, $\{3, 5\}$; if $p = 4$ we have $\{4, 3\}$, and if $p = 5$ we have $\{5, 3\}$. Since there is no allowable value for q when $p > 5$ there are no other regular polyhedra.

For Further Reading

BECK *et al.:* 3–78
BOYER (g): 94–97, 129–31

HEATH (c): 106–9, 175–78, 250–54
VAN DER WAERDEN: 100, 173–75

HISTORY OF THE TERMS "ELLIPSE," "HYPERBOLA," AND "PARABOLA"

THE evolution of our present-day meanings of the terms "ellipse," "hyperbola," and "parabola" may be understood by studying the discoveries of history's great mathematicians. As with many other words now in use, the original application was very different from the modern.

Pythagoras (*c.* 540 B.C.), or members of his society, first used these terms in connection with a method called the "application of areas." In the course of a solution (often a geometric solution of what is equivalent to a quadratic equation) one of three things would happen: the base of the constructed figure would either fall short of, exceed, or fit the length of a given segment. (Actually, additional restrictions would be imposed on certain of the geometric figures involved.) These three conditions were designated as ἔλλειψις, *elleipsis*, "defect"; ὑπερβολή, *hyperbole*, "excess"; and παραβολή, *parabole*, "a placing beside." It should be noted that the Pythagoreans were not using these terms in reference to the conic sections.

In the history of the conic sections Menaechmus (350 B.C.), a pupil of Eudoxus, is credited with the first treatment of the conic sections. Menaechmus was led to the discovery of the curves of the conic sections by following the path Archytas (*c.* 400 B.C.) had suggested: that is, he sought to solve the Delian problem by a consideration of sections of geometrical solids. Proclus, the Greek commentator, reported that the three curves were discovered by Menaechmus; consequently they were called the "Menaechmian triads." It is thought that Menaechmus discovered the curves now known as the ellipse, parabola, and hyperbola by cutting cones with planes perpendicular to an element and with the vertex angle of the cone being acute, right, or obtuse, respectively.

The fame of Apollonius of Perga (*c.* 225 B.C.) rests mainly on his extraordinary *Conic Sections*. This work was written in eight books, seven of which have been preserved. The work of Apollonius on the conic sections differed from that of his predecessors in that he obtained all of the conic sections from one right double cone by varying the angle at which the intersecting plane would cut the element.

All of Apollonius' work was presented in regular geometric form, without the aid of the algebraic notation of present-day analytic geometry. However, his work can be described more easily here by using modern terminology and symbolism. Let A be the vertex of a conic (as in Fig. [61]-1), AB the principal axis of the conic, P any point on the conic, and Q the foot of the perpendicular from P onto AB. At A draw the perpendicular to AB, and on this mark off the distance AR equal to what is now called the "latus rectum," or "parameter p" (the length of the chord that passes through a focus of the conic perpendicular to the principal axis).

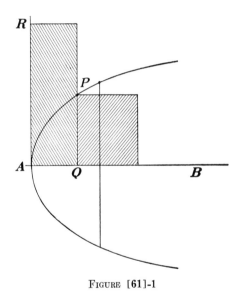

Figure [61]-1

Now apply to the segment AR a rectangle of area $(PQ)^2$, having AQ for one side. Depending on whether the application falls short of, exactly equals, or exceeds the segment AR, Apollonius called the conic an ellipse, parabola, or hyperbola.

If the conic is referred to a rectangular coordinate system in the usual manner, (Fig. [61]-2), with point A as the origin and with (x, y) as coordinates of any point P on the conic, the standard equation of the parabola, $y^2 = px$ (where p is the length of the latus rectum) is immediately verified. Similarly, if the ellipse or hyperbola

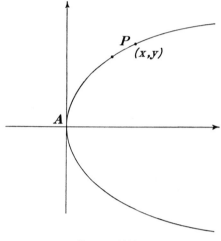

Figure [61]-2

is referred to a coordinate system with vertex at the origin, it can be shown that $y^2 < px$ or $y^2 > px$, respectively /Boyer (g): 163/.

The three adjectives "hyperbolic," "parabolic," and "elliptic" are encountered in many places in mathematics, including projective geometry and non-Euclidean geometries. Often they are associated with the existence of exactly two, one, or none of something of particular relevance /Eves (d)/. The relationship arises from the fact that the number of points in common with the so-called line at infinity in the plane for the hyperbola, parabola, and ellipse is two, one, and zero, respectively.

For Further Reading

Boyer (g): 162–63   Heath (c): 355–59
Eves (d)   Van der Waerden: 241–48

GEOMETRY IN CHINA

THE Westerner is often impressed by the achievements of the Chinese in the practical arts. It is from the practical realm that most Chinese mathematics evolved. The earliest mathematical interests were astronomy and calendar making; this situation led to the development of a usable arithmetic. Geometry arose through the need to find distances, areas, volumes, and so forth, and it was arithmetical in nature. Unlike the Greeks, the Chinese never developed geometry in an abstract and systematic manner—arithmetic and the concept of number were always necessary.

The first surviving Chinese works that involve geometry were written between the third and first century B.C., but some authorities consider these to be commentaries on or compilations of earlier works. At the least, it does seem likely that the subject matter is representative of an earlier age. Some historians place the origins of parts of these works in the twelfth century B.C. The period of greatest productivity in Chinese geometry was from about 200 B.C. to A.D. 500.

In one early work, the *Chou Pei Suan Ching*, there is a brief study of the 3-4-5 right triangle. A figure is included, but there is no formal demonstration of the Pythagorean theorem [58]. Here can be seen a specific case of the dependence of Chinese geometry on the number concept. There is also a passage dealing with shadows of the gnomon that may be interpreted to show that the early Chinese recognized the ratios of corresponding sides of similar right triangles.

Another work, the *Chiu Chang Suan Shu* ("Nine Chapters on the Mathematical Art"), deals in part with areas of plane figures and the volumes of a number of solids. Statements of problems are followed by answers given in prose form, many answers being equivalent or approximately equivalent to formulas developed at a much later date in Western mathematics.

Not generally known is the existence of a vestige of a theoretical geometry, one that originated around 330 B.C., in the *Mohist Canon*. This appeared as a group of geometric definitions that do not employ

225

arithmetic concepts. Such geometrical notions as point, lines of equal length, bounded space, and parallel lines are defined there /NEEDHAM: 91–95/. The *Canon* touches on almost all branches of physical science and seems to be an attempt to pass from the practical to the philosophical. Unfortunately, the work is quite fragmentary and gives no hint that the Mohists went beyond this elementary stage. This attempt to develop a more formal system of geometry seems to have had little or no influence, however, on succeeding generations of Chinese mathematicians.

In circular mensuration many approximations for π were made, 3 being the value commonly used. However, in the third century A.D. an approximation of 3.14159 was made, using a regular polygon of 3,072 sides. Two centuries later π was found to lie between an "excess value" of 3.1415927 and a "deficit value" of 3.1415926, a record calculation that stood for a thousand years.

The period from A.D. 200 to 1500 saw increased contact with the world to the west, first through the introduction of Buddhism from India and then through trading contacts with Middle Eastern countries and, finally, with Europe. After about 500 few if any advances in native geometry were seen, the main advances in Chinese mathematics being in the art of calculating and in "rhetorical" algebra.

There is a possibility that in the thirteenth century the books of Euclid were translated from the Arabic into Chinese; but any such translations are now lost, and they had no apparent effect. It was not until the seventeenth century that Euclid became a permanent part of China's mathematical knowledge—Matteo Ricci, a Jesuit, translated the first six books of Euclid in the period between 1603 and 1607. (Ricci also introduced modern trigonometry into China.) The remaining books of Euclid were translated in 1857.

As far as is known, the Chinese did not develop any geometrical concepts relating to conics except in connection with a few problems about tangent circles inscribed in figures such as fans. No interest in polyhedra was shown. Trisection of an angle and duplication of the cube, two of the three famous problems of classical geometry, were ignored, and the quadrature of a circle was dealt with only as a variant in finding the numerous approximations of π.

In general, the early mathematics of the Chinese is easily comparable with that of other pre-Renaissance cultures. Their developments in the art of calculation, algebra, and related practical fields such as surveying and engineering are quite noteworthy. Only the Greeks, however, were able to develop geometry as a systematic body of knowledge.

For Further Reading

BOYER (g): 217–28   NEEDHAM: 19–53, 91–108

MIKAMI   STRUIK (b)

Capsule 63   *Tom Foltz*

THE CYCLOID

IF A circle rolls along a fixed line, any point of the circle will trace out a cycloid. This curve possesses many interesting mathematical and physical properties; it has engendered so many quarrels that it has been called "the Helen of geometry."

Much of the history of this most interesting curve is obscure. Galileo tried to find the area under one arch of the cycloid and suggested that it would make an attractive arch for a bridge. By 1634 the Frenchman Gilles Persone de Roberval was able to show that the area under one arch of the curve is exactly three times the area of the generating circle. He did not publish these results, but when the Italian Évangelista Torricelli did publish them in 1644, Roberval accused him of plagiarism. Both men solved the problem of constructing tangents to the curve at any point, as did Fermat and Descartes.

In 1658 Christopher Wren computed the length of a cycloidal arch as four times the diameter of the generating circle. The same year Pascal briefly returned to the study of mathematics and found certain areas, volumes, and centers of gravity associated with the curve. (Pascal gave the curve the name of "roulette," while Roberval had called it the "trochoid.")

In 1696, Johann Bernoulli set the following problem before the scholars of his time: Given two points A and B in a vertical plane, find for the movable particle M the path AMB that M would follow, descending by its own gravity, to reach the point B from the point A in the shortest time. Bernoulli had previously worked out the solution to this problem, utilizing a generalization of Snell's law that Fermat had developed, in his work in geometrical optics, to find the path of a ray of light in a nonhomogeneous media.

227

Bernoulli discovered that the path that this falling body traverses is a cycloid. Nowadays the development of the curve of least descent (brachistochrone) is usually shown in texts on the calculus of variations, a powerful tool that was not available to either Fermat or Bernoulli. It is most interesting that, with the limited tools available to the mathematicians of the seventeenth century, concepts such as that of the brachistochrone could have been discovered and developed to the extent that they were.

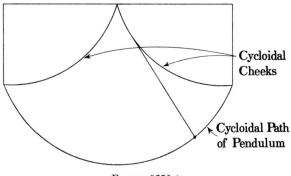

Cycloidal Cheeks

Cycloidal Path of Pendulum

FIGURE [63]-1

Christiaan Huygens made use of the cycloid in his great work *Horologium oscillatorium* (1673). He designed a pendulum to swing between two cheeks having the form of an inverted cycloid, following his discovery that the evolute of a cycloid is another cycloid. Accordingly the pendulum, by being wrapped against the cycloidal cheeks, could be made to swing in the arc of a cycloid instead of a circle. This device made a theoretically perfect timepiece, since the cycloid also possesses the property of being a tautochrone—the time required for a particle to fall down the cycloid to its lowest point is the same for any starting point on the curve.

For Further Reading

BOYER (g): 389–90, 400, 410–15
EVES (c): 261–62, 296, 357
　[5th ed. 245–46, 272–73, 323]

PHILLIPS
STRUIK (e): 232–38, 263–69

POLAR COORDINATES

POINTS are most commonly described today by ordered pairs (x, y) in the Cartesian system, where x is the directed distance from the vertical axis and y is the directed distance from the horizontal axis. For certain kinds of curves, however, a more convenient and useful form of representation is that of polar coordinates. The polar ordered pair is (r, θ), where θ is the angle the vector makes with the reference line or polar axis and r is the length of the vector.

Isaac Newton was the first to think of using polar coordinates. In a treatise *Method of Fluxions* (written about 1671), which dealt with curves defined analytically, Newton showed ten types of coordinate systems that could be used; one of these ten was the system of polar coordinates. However, this work by Newton was not published until 1736; in 1691 Jakob Bernoulli derived and made public the concept of polar coordinates in the *Acta eruditorum*. The polar system used for reference a point on a line rather than two intersecting lines. The line was called the "polar axis," and the point on the line was called the "pole." The position of any point in a plane was then described first by the length of a vector from the pole to the point and second by the angle the vector made with the polar axis.

After Bernoulli, Jacob Hermann, in a paper of 1729, asserted that polar coordinates were just as useful for studying geometric loci as were the Cartesian coordinates. However, Hermann's work was not well known, and it remained for Euler, about twenty years later, to make the polar coordinate system really popular.

For Further Reading

BOYER (e): *See index*   ——— (g): 448–49, 458, 474–75

THE NINE-POINT CIRCLE

The circle which passes through the feet of the perpendiculars, dropped from the vertices of any triangle on the sides opposite them, passes also through the midpoints of these sides as well as through the midpoints of the segments which join the vertices to the point of intersection of the perpendiculars.

This theorem, known as the "nine-point circle theorem," was discovered by Charles J. Brianchon and Jean Victor Poncelet, who published a proof in a joint paper in 1820.

In 1822 a German, Karl Feuerbach, proved not only that the nine points lie on a circle but also that the circle is tangent to the inscribed circle and the three escribed circles of the given triangle. Feuerbach's work with the nine-point circle resulted in its being called the "Feuerbach circle" in Germany.

Since the circle had been attributed, although erroneously, to Leonhard Euler, it is sometimes referred to as Euler's circle. In 1765 Euler proved that the circumcenter, the centroid, and the orthocenter are all collinear, and this line is called the "Euler line" of the triangle.

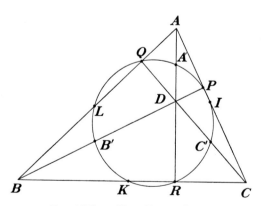

Fig. [65]-1.—Nine-Point Circle

It can also be proved that the nine-point center bisects the segment determined by the circumcenter and the orthocenter of the triangle; the centroid divides the segment in the ratio of one to two.

For Further Reading

BOYER (g): 573–74   D. E. SMITH (c): II, 337–45
EVES (c): 138
   [5th ed. 437–38]

Mainstreams
in the Flow of Algebra

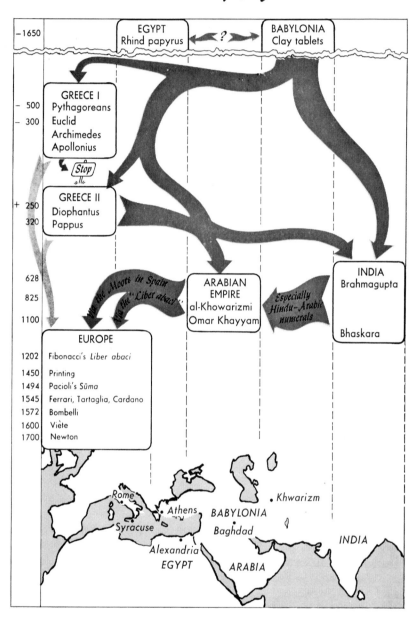

V

The History of ALGEBRA

an overview by

JOHN K. BAUMGART

E xotic and intriguing is the origin of the word "algebra." It does not submit to a neat etymology as does, for example, the word "arithmetic," which is derived from the Greek *arithmos* ("number").

Algebra is a Latin variant of the Arabic word *al-jabr* (sometimes transliterated *al-jebr*) as employed in the title of a book, *Hisab al-jabr w'al-muqabalah*, written in Baghdad about A.D. 825 by the Arab mathematician Mohammed ibn-Musa al-Khowarizmi (Mohammed, son of Moses, the Khowarezmite). This treatise on algebra is commonly referred to, in shortened form, as *Al-jabr*.

A literal translation of the book's full title is "science of restoration (or reunion) and opposition," but a more mathematical phrasing would be "science of transposition and cancellation"—or, as Carl Boyer puts it /(g): 252-53/, "the transposition of subtracted terms to the other side of an equation" and "the cancellation of like [equal] terms on opposite sides of the equation." Thus, given the equation

$$x^2 + 5x + 4 = 4 - 2x + 5x^3,$$

al-jabr gives

$$x^2 + 7x + 4 = 4 + 5x^3,$$

and *al-muqabalah* gives

$$x^2 + 7x = 5x^3.$$

Perhaps the best translation would be simply "the science of equations."

While speaking of etymologies and al-Khowarizmi it is interesting to note that the word "algorism" (or algorithm), which means any special process of calculating, is derived from the name of this same author, al-Khowarizmi, because he described processes for calculating with Hindu-Arabic numerals in a book whose Latin translation is usually referred to as *Liber algorismi* ("Book of al-Khowarizmi").

Perhaps a final philological comment on the lighter side is worthwhile. The Moroccan Arabs introduced the word *algebrista* ("restorer [that is, reuniter] of broken bones, bonesetter") into Moorish Spain. Since bonesetting and bloodletting were additional fringe benefits available at the barbershop, the local barber was known as an *algebrista*. Hence, also, the bloody barber poles!

Although originally "algebra" referred to equations, the word today has a much broader meaning, and a satisfactory definition requires a two-phase approach: (1) Early (elementary) algebra is the study of equations and methods for solving them. (2) Modern (abstract) algebra is the study of mathematical structures such as groups, rings, and fields—to mention only a few. Indeed, it is convenient to trace the development of algebra in terms of these two phases, since the division is both chronological and conceptual.

ALGEBRAIC EQUATIONS AND NOTATION

The early (elementary) phase, which spanned the period from about 1700 B.C. to A.D. 1700, was characterized by the gradual invention of symbolism and the solving of equations (usually with numerical coefficients) by various methods which showed only minor improvements until the "general" solutions of cubic and quartic equations (c. 1545) and the enlightened treatment of polynomial equations in general by François Viète, also known as Vieta (1540–1603).

The development of algebraic notation progressed through three stages: the *rhetorical* (or verbal), the *syncopated* (in which abbreviated words were used), and the *symbolic*. In the third stage, notation went through many modifications and changes until it became fairly stable by the time of Isaac Newton (c. 1700). It is interesting to note that even today there is some lack of uniformity in the use of symbols. For example, Americans write "3.1416" as an approximation for π, and many Europeans write "3,1416." The symbol "$\doteq$" is sometimes used for "approaches a limit" and sometimes for "is approximately equal to." In some European countries "$\div$" means "minus."

Babylonian Algebra—Rhetorical Style

Since algebra probably originated in Babylonia, it seems appropriate to illustrate the rhetorical style by an example from that country. The following problem shows the relatively sophisticated level of Babylonian algebra. It is typical of the problems found on cuneiform clay tablets dating back to the reign of King Hammurabi (*c.* 1700 B.C.). The explanation, of course, is expressed in English; and the Hindu-Arabic decimal notation is used in place of cuneiform sexagesimal notation. The parallel column on the right gives the corresponding steps in modern notation.

[1] Length, width. I have multiplied length and width, thus obtaining area: 252. I have added length and width: 32. Required: length and width.

[2] [Given] 32 the sum;
           252 the area.

$$\left.\begin{array}{c} x + y = k. \\ xy = P. \end{array}\right\} \cdots (A)$$

[3] [Answer] 18 length, 14 width.

[4] One follows this method:
    Take half of 32 [this gives 16].

$$\frac{k}{2}$$

  16 × 16 = 256.

$$\left(\frac{k}{2}\right)^2$$

  256 − 252 = 4.

$$\left.\left(\frac{k}{2}\right)^2 - P = t^2.\right\} \cdots (B)$$

  The square root of 4 is 2.

$$\sqrt{\left(\frac{k}{2}\right)^2 - P} = t.$$

  16 + 2 = 18 length.

$$\frac{k}{2} + t = x.$$

  16 − 2 = 14 width.

$$\frac{k}{2} - t = y.$$

[5] [Check] I have multiplied 18 length by 14 width.

  18 × 14 = 252 area.

$$\left(\frac{k}{2} + t\right)\left(\frac{k}{2} - t\right)$$
$$= \frac{k^2}{4} - t^2 = P = xy.$$

Notice that (1) the problem is stated, (2) the given data are listed, (3) the answer is given, (4) the method of solution is explained *with numbers*, and finally (5) the answer is checked.

The above "recipe" (as B. L. Van der Waerden /65/ calls the method of solution) is used repeatedly in similar problems. It has historical significance and current interest for several reasons.

First of all, it is not the way we would solve system (A) today. The standard procedure in our college-algebra texts is to solve, say, the first equation for y (in terms of x), to substitute in the second equation, and then to solve the resulting quadratic equation for x; that is, we would use the method of elimination. The Babylonians also knew how to solve systems by elimination but preferred often to use their parametric method. That is—and we use modern notation here—they thought of x and y in terms of a new unknown (or parameter) t by letting $x = (k/2) + t$ and $y = (k/2) - t$. Then the product

$$xy = \left(\frac{k}{2} + t\right)\left(\frac{k}{2} - t\right) = \left(\frac{k}{2}\right)^2 - t^2 = P$$

led them to relation (B):

$$\left(\frac{k}{2}\right)^2 - P = t^2.$$

Secondly, the above problem has historical significance because the Greek (geometric) algebra of the Pythagoreans and Euclid (who in his *Elements* organized most of the existing mathematics of his time) followed the same method of solution—phrased, however, in terms of line segments and areas and illustrated by geometric figures. Some centuries later another Greek, Diophantus, also used the parametric approach in his work with "Diophantine" equations; and he made a start toward modern symbolism by introducing abbreviated words and avoiding the rather cumbersome style of geometric algebra.

Thirdly, the Arab mathematicians (including al-Khowarizmi) did *not* use the method of the above problem; they preferred to eliminate one of the unknowns by substitution and to express all in terms of words and numbers.

Before leaving the Babylonians and their algebra we note that they were able to solve a rather surprising variety of equations, including certain special types of cubics and quartics—all with numerical coefficients, of course.

236

ALGEBRA IN EGYPT

Algebra appeared in Egypt almost as soon as in Babylonia; but Egyptian algebra lacked the sophistication in method shown by Babylonian algebra, as well as its variety in types of equations solved, if we are to judge by the Moscow papyrus and the Rhind papyrus—Egyptian documents dating from about 1850 B.C. and 1650 B.C. respectively but reflecting mathematical methods derived from an earlier period. For linear equations the Egyptians used a method of solution consisting of an initial estimate followed by a final correction—a method to which Europeans later gave the rather abstruse name "rule of false position." The algebra of Egypt, like that of Babylonia, was rhetorical.

The numeration system of the Egyptians, relatively primitive in comparison with that of the Babylonians, helps to explain the lack of sophistication in Egyptian algebra. As we shall see later, European mathematicians of the sixteenth century had to extend the Hindu-Arabic notion of number before they could progress significantly beyond the Babylonian results in solving equations.

GREEK GEOMETRIC ALGEBRA

Greek algebra as formulated by the Pythagoreans (*c.* 540 B.C.) and Euclid (*c.* 300 B.C.) was geometric. For example, what we write as

$$(a + b)^2 = a^2 + 2ab + b^2$$

was thought of by the Greeks in terms of the diagram shown in Figure V-1 and was quaintly stated in the following way by Euclid in his *Elements*, Book II, Proposition 4:

> *If a straight line be divided into any two parts, the square on the whole line is equal to the squares on the two parts, together with twice the rectangle contained by the parts.* [That is, $(a + b)^2 = a^2 + b^2 + 2ab$.]

One is tempted to say that for the Greeks of Euclid's day a^2 was really a square!

There is no doubt that the Pythagoreans were quite familiar with Babylonian algebra and that, in fact, they followed the standard Babylonian methods of solving equations. Euclid has recorded these Pythagorean results, and we illustrate by choosing the theorem that corresponds to the Babylonian problem considered above. (We give

237

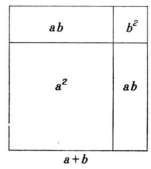

FIGURE V-1

here a slightly simplified statement; Euclid's wording is more general and also more ponderous.)

From Book VI of the *Elements* we have Proposition 28:

> *Given the straight line AB* [that is, $x + y = k$], *construct along this line a rectangle equal to a given area* [$xy = P$], *assuming that the rectangle "falls short" of AB by an amount "filled out" by another rectangle* [the square *BF* in Fig. V-2], *similar to a given rectangle* [which here we take to be any square].

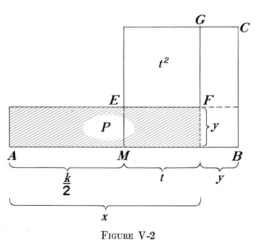

FIGURE V-2

In the solution of this required construction (see Fig. V-2) Euclid's work parallels almost exactly the Babylonian solution of the equivalent problem. As indicated by T. L. Heath /EUCLID: II, 263/, the steps are as follows:

238

Bisect AB at M:

$$\frac{k}{2}.$$

Draw square $MBCD$:

$$\left(\frac{k}{2}\right)^2.$$

Using VI, 25,
construct square $DEFG$
with area equal to the
excess of $MBCD$ over the
given area P:

$$t^2 = \left(\frac{k}{2}\right)^2 - P.$$

Then it is clear that

$$y = \frac{k}{2} - t.$$

As was often his custom, Euclid left the other case to the student—in this case, $x = (k/2) + t$, which Euclid certainly realized but did not state.

It is truly remarkable that most of the standard Babylonian problems are "redone" this way by Euclid!

But why? What caused the Greeks to give their algebra this unwieldy formulation? The answer (which Van der Waerden /125/ suggests) is quite basic: They had conceptual difficulties with fractions and irrational numbers.

Even though the Greek mathematicians were able to circumvent fractions by treating them as ratios of integers, they had insurmountable difficulties with such numbers as   . One recalls the "logical scandal" of the Pythagoreans when they discovered that the diagonal of the unit square is incommensurable with the side (i.e., diag/side $\neq$ ratio of two integers).

So it was their strict mathematical rigor that forced them to use a *set of line segments* as a suitable domain of elements. Thus, although   cannot be expressed in terms of integers or their ratios, it *can* be represented as a line segment that is precisely the diagonal of the unit square.

Perhaps it is not entirely facetious to say that their linear continuum was literally linear!

In passing we must mention Apollonius ($c.$ 225 B.c), who applied geometric methods to the study of conic sections. In fact, his great treatise *Conic Sections* contains more "analytic geometry" of conics—all phrased in geometric terminology—than the standard college course today.

Then Greek mathematics came to a screeching halt. The Roman

occupation had begun, and it did not foster mathematical scholarship, although it did encourage some other branches of Greek culture. Because of the ponderous style of geometric algebra (just try reading Apollonius!), it could not survive solely by a written tradition; it needed a living, oral, means of communication. One could follow the flow of ideas as long as an instructor pointed to diagrams and explained them; but schools of direct instruction did not survive.

SYNCOPATED ALGEBRAIC
NOTATION

However, some centuries later the Greek mathematician Diophantus made a fresh start in algebra along the lines of the older Babylonian methods. Diophantus introduced the *syncopated* style of writing equations. He studied and worked at the University of Alexandria, Egypt, where Euclid had once taught. Very little else is known about his life except what is claimed in the following algebraic puzzle rhyme from the *Anthologia Palatina:*

> "Here lies Diophantus." The wonder behold—
> Through art algebraic, the stone tells how old:
> "God gave him his boyhood one-sixth of his life,
> One-twelfth more as youth while whiskers grew rife;
> And then yet one-seventh ere marriage begun;
> In five years there came a bouncing new son.
> Alas, the dear child of master and sage
> — Met fate at just half his dad's final age.
> Four years yet his studies gave solace from grief;
> Then leaving scenes earthly he, too, found relief."

(What answer do you get for Diophantus' age? Eighty-four?)

Diophantus' claim to fame rests on his *Arithmetica,* in which he gives an ingenious treatment of indeterminate equations—usually two or more equations in several variables that have an infinite number of rational solutions. These are often called Diophantine equations, although he was not the first to solve such systems. His approach is clever, but he does not develop a systematic method for finding general solutions. His approach is along Babylonian lines in the sense that he expresses all unknowns in terms of a parameter and then obtains an equation containing only the parameter.

In leaving Diophantus we give an example of his syncopated algebra:

240

that is,

$$\mathrm{K}^{\mathrm{T}}\beta \quad s\eta \bigwedge \Delta^{\mathrm{T}}\epsilon \quad \overset{\mathrm{o}}{\mathrm{M}}\delta \quad \dot{\epsilon}\sigma\tau\iota \quad \mu\delta;$$

or

$$x^3 2 \quad x8 - x^2 5 \quad 1 \cdot 4 \quad = \quad 44,$$

$$2x^3 + 8x - (5x^2 + 4) \quad = \quad 44.$$

Note the minus sign, $\bigwedge$. (For complete explanation of the symbols and their meanings, see [66]).

HINDU AND ARABIC ALGEBRA

The scene now shifts briefly to India and the Arabic civilization. Little is known about Hindu mathematics before the fourth or fifth century A.D. because few records of the ancient period have been found. India was subjected to numerous invasions, followed by the Pax Romana, which facilitated the exchange of ideas. Babylonian and Greek accomplishments, in particular, were apparently known to Hindu mathematicians. Brahmagupta worked in a syncopated style in which $5xy + \sqrt{35} - 12$ would be written in the following way (the meaning is given below—see also [66]:

ya	ka	5	bha	k(a) 35	ru	$\overset{\mathrm{o}}{12}$
x	y	5	product	irrational 35	"pure" number	-12

Brahmagupta (c. 628) and Bhaskara (c. 1150) were the most prominent of the Hindu algebraists. The Hindus solved quadratic equations by completing the square, and they accepted negative and irrational roots; they also realized that a quadratic equation (with real roots) has two roots.

Hindu work on indeterminate equations was superior to that of Diophantus; the Hindus attempted to find *all possible integer solutions* and were perhaps the first to give general methods of solution. The work of Brahmagupta and Bhaskara on the so-called Pell equation,

$$y^2 = ax^2 + 1 \text{ (where } a \text{ is a nonsquare integer)},$$

is extremely good; and they showed how to obtain an infinite number of solutions from a given solution x, y (provided $xy \neq 0$).

The advent of Mohammedanism provided the impetus that soon (c. 700) led to the Arab conquest of India, Persia, Mesopotamia, North Africa, and Spain. Thus the Arabs acquired (most eagerly) the Greek and Hindu scientific writings, which they translated into Arabic and preserved through the Dark Ages of Europe. One of their most glitter-

241

ing acquisitions was the system of Hindu (often called Arabic) numerals, as we shall see.

Our main interest during the Arabic period centers on al-Khowarizmi because his books (the *Al-jabr* and the *Liber algorismi*, both of which were mentioned earlier) were later translated into Latin (*c.* 1200) and thus greatly influenced European mathematics. But his work was not up to previous standards; as Van der Waerden points out, al-Khowarizmi rejected "Greek erudition" and ignored other available results. His aim was to write a practical book on solving equations. His style was rhetorical, and his work was not as good as that of the Babylonians or the Hindus.

Some readers may be delightfully surprised to learn that the author of the *Rubaiyat,* Omar Khayyam (*c.* 1100), was not only a Persian poet but a first-class mathematician of the Arab world—one who made a contribution to algebra by using intersecting conics to provide a geometric solution for certain types of cubic equations.

When Toledo, in Moorish Spain, fell to the Christians (1085), al-Khowarizmi's *Al-jabr,* Ptolemy's *Almagest* (a work on astronomy), Euclid's *Elements,* and many other Greek and Arabic texts were translated into Latin by Christian scholars who now entered Spain. However, even more important for Europe, especially Italy, was the *Liber abaci* (1202) of Fibonacci (Leonardo of Pisa), in which he solved equations in the rhetorical and general style of al-Khowarizmi and strongly advocated the use of Hindu-Arabic numerals, which he discovered on his journeys to many lands as a merchant and tradesman.

How fortunate for mathematics that our traveling salesman Fibonacci had these intellectual interests!

It is not surprising that at first the local chambers of commerce (in Pisa and neighboring city-states of Italy) resisted the adoption of the "new" Hindu-Arabic numerals and in fact viewed them with suspicion; but they were gradually adopted, and the old abacus was stored in the attic. There is no evidence that Fibonacci had to appear before the Unnumerical Activities Committee!

In passing, we note that the Arab term for the unknown was *shai* ("thing"), which was translated into Latin as *res* and into Italian as *cosa*; hence algebra was known for some time in England as the "cossic art" and in Germany as *"die Coss."*

Algebra in Europe

The algebra that entered Europe (via Fibonacci's *Liber abaci* and translations) had retrogressed both in style and in content. The semi-

242

symbolism (syncopation) of Diophantus and Brahmagupta and their relatively advanced accomplishments were not destined to contribute to the eventual breakthrough in European algebra. The renaissance and rapid growth of algebra in Europe were due to the following factors:

1. The facility in handling numerical work by the Hindu-Arabic numeral system, which was vastly superior to systems (such as that of Roman numerals) requiring use of the abacus
2. The invention of printing with movable type (*c.* 1450), which hastened standardization of symbolism through improved communications based on wide distribution
3. The resurgence of the economy, which supported scholarly activity; and the resumption of trade and travel, which facilitated the exchange of ideas as well as goods

Powerful commercial cities arose first in Italy (1200–1300), and it was here that the algebraic renaissance in Europe really began.

Symbolic Algebraic
Notation

Modern symbolism began to emerge around 1500. Perhaps the best way to convey the process of development is to give some examples that show not only the initial poverty and later diversity of symbols but also the gradual improvement and standardization of notation. (Additional examples will be found in [**66**] and [**89**].) Modern notation is given beneath each of the older forms.

Cardano (1545): cubus $\bar{p}$ 6 rebus aequalis 20.

$$x^3 + 6x = 20.$$

Bombelli (1572): $\overset{6}{\underbrace{\text{I. p.}}} \overset{3}{\underbrace{8.}}$ Eguale à 20.

$$x^6 + 8x^3 = 20.$$

Viète (1591): I QC − 15 QQ + 85 C − 225 Q + 274 N aequatur 120.

$$x^5 - 15x^4 + 85x^3 - 225x^2 + 274x = 120.$$

Harriot (1631): aaa − 3bba ====== + 2.ccc.

$$x^3 - 3b^2x = 2c^3.$$

Descartes (1637): $x^3 - 6xx + 13x - 10 \, \infty \, 0.$

Wallis (1693): $x^4 + bx^3 + cxx + dx + e = 0$.

Who invented a particular symbol is a question requiring detailed research. Often it cannot be determined with certainty. Two symbols and their inventors may be mentioned in passing:

1. The $=\!\!=$ sign introduced by Robert Recorde in his *The Whetstone of Witte* (1557)
   (He used this symbol "bicause noe.2. thynges, can be moare equalle.")
2. The $\sqrt{}$ sign, possibly a modification of r for *radix* ("root," "radish"), introduced in print by Christoff Rudolff in his algebra book *Die Coss* (1525)

The Algebra of Cardano and Viète

A banner year was 1545: in that year Girolamo Cardano, an Italian scholar, published his *Ars magna,* containing Scipione del Ferro's solution of the cubic and Ludovico Ferrari's solution of the quartic. These solutions represented the first really new material since antiquity, even though these essentially general solutions were achieved by "ingenious devices" rather than advances in insight and theory.

Although the symbolic style of *Ars magna* is disappointing, Cardano, besides publishing the results of others—

1. Was the first to exhibit three roots for a particular cubic.
   (He suspected that there should be three roots for every cubic, but he was puzzled by negative roots and imaginary roots.)
2. Recognized, at least in some sense, negative roots, which he called "fictitious."
3. Had the intellectual curiosity to see what would happen if one operated with such numbers as $5 \pm \sqrt{-15}$, which we now call "complex" or "imaginary."
4. Recognized the "irreducible case" in the solution of a cubic.
5. Removed the x^2 term from a cubic equation.
   (This reduction was more "theoretically general" than he realized.)
6. Stated that the sum of the three roots of a cubic is the negative of the coefficient of x^2.

The watershed of algebraic thought (separating the early shallow flow of "manipulative solution of equations" from the deeper modern stream which began with the theoretical properties of equations) is personified in the Frenchman François Viète, who was the first, in his *logistica speciosa*, to introduce letters as general (positive) coefficients and to put some other finishing touches on symbolism, which was finally up to date by the time of Isaac Newton. But Viète's most significant contributions were contained in *De aequationum recognitione et emendatione*, published posthumously in 1615. In this work he—

1. Gave transformations for increasing or multiplying the roots of an equation by a constant.
2. Indicated awareness of relations between roots and coefficients of a polynomial equation.
3. Stated a transformation that rids a polynomial of its next-to-highest-degree term.

The Development of Complex Numbers

Before tracing the mainstream in the development of algebraic structure, we shall briefly follow a parallel and concurrent stream in the development of the complex-number system and see how each of these nourished the other. Viète's inability to accept negative numbers (not to mention imaginary numbers) prevented him from attaining the generality he sought (and partly comprehended) in giving, for example, relations between the roots and the coefficients of a polynomial equation.

Cardano, also, would probably have made more progress if imaginary numbers had not been so puzzling to him; yet he had the intellectual curiosity and courage to investigate the problem "Divide 10 into two parts such that the product of one times the remainder is 40," even though he called it "manifestly impossible." Unable to restrain his curiosity, he then boldly proclaimed, "Nevertheless we will operate," and obtained the two solutions $5 + \sqrt{-15}$ and $5 - \sqrt{-15}$ (as we would write them today). He checked the answers by showing that their product is indeed $25 - (-15)$, or 40, remarked that these answers (numbers) are "truly sophisticated," and finally concluded (alas!) that to continue working with these numbers would involve one in arithmetic "as subtle as it would be useless."

One is tempted to say that the operation was a success but the patient died! Yet it is quite true that Cardano was in advance of his time.

It was Albert Girard (1629) who approached both negative and imaginary numbers with great boldness; he used negative numbers in solving geometric problems and suggested that by also accepting imaginary numbers as roots it would be possible to assert that an equation has as many roots as its degree. He also stated relations between roots and coefficients of a polynomial equation and suggested that imaginary roots are helpful in making these relations general. As an example, for the equation

$$x^4 - 4x + 3 = 0$$

he gave the roots 1, 1, $-1 + \sqrt{-2}$, and $-1 - \sqrt{-2}$. (We use modern notation here.)

René Descartes's geometric representation of negative numbers (1637) helped to make them more acceptable; and in 1659 Johann Hudde took the significant step of using a letter in a formula for *any* real number, whether positive or negative.

Imaginary numbers were successfully used by Leonhard Euler in the eighteenth century; it is to him that we owe the formula $e^{2\pi i} = 1$. To many mathematicians this looked like higher magic—but it worked. However, imaginary numbers attained wider acceptance after geometric representations were given, first by Caspar Wessel (1797), then by Jean Robert Argand (1806), and then by Carl Friedrich Gauss, who, in a treatise of 1831, published the geometric representation that he had already, in 1811, described in a letter to F. W. Bessel. Another, and purely arithmetical, definition of a complex number $a + bi$ as an ordered pair (a, b) of real numbers subject to the standard rules of combining these pairs was given by William Rowan Hamilton in 1837. We now accept complex numbers rather glibly, but Gauss found occasion to remark that "the true metaphysics of $\sqrt{-1}$" is difficult. His results and reputation settled the acceptance of imaginary numbers.

In this connection it is amusing to quote Gottfried Wilhelm von Leibniz (1702), who, perhaps because of his philosophic bent, spoke of the imaginary number as "that wonderful creature of an ideal world, almost an amphibian between things that are and things that are not."

Descartes (1637) originated the terms "real" and "imaginary," and Euler (1777) introduced the letter i for $\sqrt{-1}$. Augustin Louis Cauchy (1821) contributed the terms "conjugate" and "modulus," and Gauss (1831) introduced the term "complex."

The foundations begun by Viète for the modern structural formulation of algebra had to wait some two hundred years before Niels Henrik Abel (1824) and especially Évariste Galois (1831) introduced the idea of a group, in their independent proofs that a polynomial equation of degree greater than four has no general algebraic solution. When we say that the quadratic equation $ax^2 + bx + c = 0$ (a second-degree polynomial equation) has a general algebraic solution, we mean that each of the two roots can be expressed in terms of a finite number of additions, subtractions, multiplications, divisions, and root extractions performed on the coefficients a, b, c. Thus the two roots are $(-b + \sqrt{b^2 - 4ac})/2a$ and $(-b - \sqrt{b^2 - 4ac})/2a$. In a somewhat similar way the three roots of the cubic equation

$$ax^3 + bx^2 + cx + d = 0$$

can be expressed in terms of the coefficients a, b, c, d. A quartic equation, also, submits to this type of treatment. However, when the degree of the polynomial equation is greater than four, no general algebraic solution is possible.

THE DEVELOPMENT OF THE GROUP CONCEPT

During the two hundred years from Viète to Abel and Galois, mathematicians were not idle; the group concept, of course, did not emerge suddenly with Abel and Galois, as we shall see.

The period from 1600 to 1800 saw such developments as the invention of logarithms by John Napier (1614) and Henry Briggs (1615); the work of Pierre de Fermat in probability and in number theory (1635) and, contemporaneously with Descartes, in analytic geometry (1637); the beginning of probability theory by Blaise Pascal (1653) in correspondence with Fermat; work on the binomial series by James Gregory and Isaac Newton (*c.* 1670); the founding of the calculus by Newton and Leibniz (*c.* 1680); initial work in actuarial mathematics by Abraham De Moivre (1720); the prolific work in analysis by Euler (1750); and the versatile and considerable mathematics produced by Joseph Louis Lagrange (1780), Pierre Simon Laplace (1805), Adrien Marie Legendre (1805), Cauchy (1820), and Gauss (1820).

In the work of some of these men an implicit grasp of the group concept was already to be found. It will be convenient at this point to define a group, and we shall illustrate the definition by giving an example of a permutation group, since permutation groups were historically important in developing the group concept. Consider the set $G = \{I, a, b, c, d, e\}$ where the six elements of G are, respectively, the six permutations 123, 132, 213, 231, 312, 321. That is,

$I = (1, 2, 3)$ and replaces 1 by 1, 2 by 2, and 3 by 3.
$a = (1, 3, 2)$ and replaces 1 by 1, 2 by 3, and 3 by 2.
$b = (2, 1, 3)$ and replaces 1 by 2, 2 by 1, and 3 by 3.
$c = (2, 3, 1)$ and replaces 1 by 2, 2 by 3, and 3 by 1.

$d = (3, 1, 2)$ and replaces 1 by 3, 2 by 1, and 3 by 2.

$e = (3, 2, 1)$ and replaces 1 by 3, 2 by 2, and 3 by 1.

The "product" ab is defined to mean that first permutation a is performed and then permutation b is performed on the result of a. Thus, a produces $(1, 3, 2)$ and, applying b to $(1, 3, 2)$, we get the result $(2, 3, 1)$, which is c. So we say that $ab = c$. Notice that $ba = d$, and so our multiplication in this example is not commutative. All possible products are given in Figure V-3. (To find the product ab from the table, use the left and top margins [in that order] for a and b; the entry is c, indicated by the dotted lines.) By referring to the table the following defining properties of a group are easily checked:

	I	a	b	c	d	e
I	I	a	b	c	d	e
a	a	I	c	b	e	d
b	b	d	I	e	a	c
c	c	e	a	d	I	b
d	d	b	e	I	c	a
e	e	c	d	a	b	I

FIGURE V-3

1. The set G is closed with respect to the defined multiplication; that is, the product of any two elements of G is again an element of G. Thus, $bc = e$, and e is an element of set G.
2. The associative law holds for any three elements of G. Thus,

$$(ab)c = a(bc),$$

$$(c)c = a(e),$$

$$d = d.$$

3. The set G has an identity element (in this example, I) with the property that $aI = Ia = a$, $bI = Ib = b$, etc.
4. Each element of G has an inverse; for example, we can find—

   An x so that $ax = I$ (in fact, $x = a$, since $aa = I$).

   A y so that $by = I$ (in fact, $y = b$, since $bb = I$).

   A z so that $cz = I$ (in fact, $z = d$, since $cd = I$).

The set G is called a group if and only if the above four properties hold. Some groups have the additional property of commutativity, and they are called Abelian (or commutative) groups.

The relation of the group concept to that of a field may be briefly indicated by noting that the set F of real numbers is an example of a field because the elements of F satisfy the five axioms for a commutative group with respect to addition (where the identity element is represented by 0); the nonzero elements of F satisfy five more axioms for a commutative group with respect to multiplication (where the identity element is represented by 1); and the elements of F satisfy a final axiom called the distributive law, namely,

$$a \cdot (b + c) = a \cdot b + a \cdot c,$$

which "distributes" the multiplication "over" the addition.

Some of the components of the group concept (i.e., those essential properties that were later abstracted and formulated as axioms), and also of the field concept, were discernible as early as 1650 B.C., when the Egyptians showed a curious awareness that something was involved in assuming that $ab = ba$ (to use modern notation). In his commentary on the Rhind papyrus, Arnold B. Chace /I, 6/ remarks that the Egyptian method of multiplication emphasized a distinction between multiplicand and multiplier but adds that it was known that if the two were interchanged, the product would be the same. An example of this interchange of factors occurs in Problem 26 of the Rhind papyrus /CHACE: I, 68/:

Problem 26

A quantity and its ¼ added together become 15. What is the quantity? Assume 4.

$$\sqrt{\quad} \quad 1 \qquad 4$$

$$\sqrt{\quad} \quad 1/4 \qquad 1$$

$$\text{Total,} \qquad 5.$$

As many times as 5 must be multiplied to give 15, so many times must 4 be multiplied to give the required number. Multiply 5 so as to get 15.

$$\sqrt{\quad} \quad 1 \qquad 5$$

$$\sqrt{\quad} \quad 2 \qquad 10$$

$$\text{Total,} \quad 3.$$

Multiply 3 by 4.

249

	1	3
	2	6
√	4	12.

The quantity is 12.

	1/4	3
Total,		15.

We notice that in the above problem the Egyptian scribe determines that 4 must be multiplied by 3. Then to get the product he multiplies 3 by 4, the reverse order. The probable reason for this is that the Egyptian did his multiplication by doubling; consequently it was easier to multiply by powers of 2 than by other numbers. If the Egyptian had multiplied 4 by 3 he would have had to double *and* add:

	1	4
√	1	4
√	2	8

| Totals, | 3 | 12. |

The Egyptian also freely used the distributive law, namely, $a(b + c) = ab + ac$ (as we write it), but without comment. An illustration of this occurs in Problem 68 of the Rhind papyrus /CHACE: I, 104/. To double 3 21/64, which the scribe writes as $3 + 1/4 + 1/16 + 1/64$ (as it would appear in modern notation), he simply doubles each term and gets $6 + 1/2 + 1/8 + 1/32$, or 6 21/32.

The Babylonians (*c.* 1700 B.C.) also used the commutative and distributive laws. These laws were tacitly assumed in their rhetorical algebra when, in effect, they used such formulas as $(a + b)^2 = a^2 + 2ab + b^2$.

Looking at Greek mathematics, we see that Euclid was more aware of the explicit nature of the distributive law. In the geometric algebra of Book II he states and proves Proposition 1:

> *If there be two straight lines, and one of them be cut into any number of segments whatever, the rectangle contained by the two straight lines is equal to the rectangles contained by the uncut straight line and each of its segments.*

That is (see Fig. V-4),

250

or

area $BGHC$ = area $BGKD$ + area $DKLE$ + area $ELHC$,

$$a(b + c + d) = ab + ac + ad.$$

```
B           D    E         C
 ┌──────────┬────┬─────────┐
a│          │    │         │
 │    b     │  c │    d    │
 └──────────┴────┴─────────┘
G           K    L         H
```

FIGURE V-4

Somewhat later Diophantus /4/, exhibited interesting insights regarding multiplicative inverses and the unity element when he wrote the following:

> Every number multiplied by a [unit] fraction having this number as denominator gives unity [i.e., $a(1/a) = 1$].
> Unity being invariable and always constant, its expression multiplied by itself remains the same expression.

But additive inverses were more elusive for him, it seems; and we find him saying that the equation $4x + 20 = 4$, written here in modern style, "*is absurd* because the 4 ought to be some number greater than 20." He was not prepared to accept the negative number -4 as a root.

The group *concept* was not recognized so explicitly as were some of its components (axioms); but even so, it was implicitly sensed and used before Abel and Galois brought it into focus, and before Arthur Cayley (1854) defined a general abstract group.

One might perhaps claim, as does G. A. Miller /(a): I, 430/, that the concept of a cyclic group is prehistoric in the sense that the ancients measured a circle by using equal divisions of its circumference, or that the 24-hour clocks of the Babylonians and Egyptians were (implicitly) examples of finite additive groups with 24 used as a zero element; and it is of some interest that Andreas Speiser suggests that Euclid's work contains at the implicit level what we would now classify as algebraic number theory and group theory /G. A. MILLER (a): I, 431/.

Less speculation, however, is required to assess the developments from Viète (1600) to Galois (1831).

During the seventeenth century it was clear to those working with the nth roots of unity that these n elements formed a multiplicative cyclic group and that the primitive nth roots could be used as generators of the group.

Another example of the use of group theory at the subformal level—and a striking one—is found in Euler's proof (1760–61) of a generalization of Fermat's "little theorem."

We illustrate the ideas with the following example. Let the composite number $m = pq$ be the product of two primes. We take $m = 35 = 5 \cdot 7 = pq$. Now arrange all the integers from 1 to 35 in a rectangular array, deleting those that are multiples of 5 or 7, as shown in Figure V-5, and rewrite as shown in Figure V-6.

1	2	3	4	~~5~~
6	~~7~~	8	9	~~10~~
11	12	13	~~14~~	~~15~~
16	17	18	19	~~20~~
~~21~~	22	23	24	~~25~~
26	27	~~28~~	29	~~30~~
31	32	33	34	~~35~~

FIGURE V-5

1	2	3	4	
6	12	8	9	
11	17	13	19	$q - 1 = 6$ rows
16	22	18	24	
26	27	23	29	
31	32	33	34	
$p - 1 = 4$ columns				

FIGURE V-6

The so-called Euler ϕ-function, $\phi(m) = \phi(35)$ in this case, is defined to be the number of positive integers less than m and relatively prime to m; in this case, $\phi(m) = \phi(35) = 4 \cdot 6 = (p - 1)(q - 1) = 24$, which is the number of integers in the second array above, which has $p - 1 = 4$ columns and $q - 1 = 6$ rows. Now the theorem that Euler stated asserts that $a^{\phi(m)} - 1$ is exactly divisible by m provided a and m are relatively prime. In the present example this means that $a^{24} - 1$ is exactly divisible by 35 provided we choose an a that has no factor in common with $m = 35$. If we take $a = 2$, then Euler's theorem assures us that $2^{24} - 1$ is exactly divisible by 35; indeed, $(2^{24} - 1)/35 = 479{,}349$.

In proving the theorem Euler made an array, like the ones in Figures V-5 and V-6, that displayed the elements of a multiplicative group (modulo m). In our present example, the modulus is $m = 35$; and to multiply 11 by 17 we first get the usual product, 187, and then divide by $m = 35$ to get a remainder of 12, which is the so-called product. Hence we say that $11 \cdot 17 \equiv 12 \pmod{35}$; similarly $4 \cdot 9 \equiv 1 \pmod{35}$.

The most interesting feature of the array in Figure V-6 is that the first column is itself a group and thus a subgroup of the whole array.

G. A. Miller /(a): I, 427/ suggests that by using the above array Euler was actually, at a subformal level, using an idea later formulated by Lagrange (1770) and now known as Lagrange's theorem, which says that the number of elements, 6, in the first-column subgroup divides the number of elements, 24, in the whole array.

Lagrange gave this idea an explicit and general formulation and showed that the number of elements in a symmetric group is divisible by the number of elements in any subgroup (which is, of course, a permutation group). Hence his result was valid for non-Abelian permutation groups as well as for Abelian groups.

In 1799 Paolo Ruffini showed that the converse of Lagrange's theorem is not true; that is, a group of order n does not necessarily have a subgroup of order s just because s divides n. More remarkable is Ruffini's proof (along the same lines as Abel's later and independent proof of 1824) that it is (algebraically) impossible to solve the general polynomial equation of degree greater than four.

Just preceding Abel and Galois, Cauchy (1815) published his first article on group theory, dealing with permutation groups; and Gauss (1820) formalized modular systems (which are additive cyclic groups) and was the first to use the special symbol $\equiv$ for the concept of congruence—although, of course, the concept itself was not original with Gauss. Hans Wussing suggests that Gauss's *Disquisitiones arithmeticae* (1801) can be considered a virtual source of implicit group theory; that the determination of the so-called Gaussian periods of the cyclotomic function is essentially equivalent to the determination of the subgroups of the Galois group of the cyclotomic equation; and that Gauss, in his work on the theory of composition of quadratic forms, derived a complete set of properties which, taken axiomatically, define an Abelian group.

Search for the "General Solution"

Before describing the momentous work of Abel and Galois, we note some of the events immediately preceding and directly influencing the remarkable achievements of these gifted young mathematicians, both of whom died in their twenties.

In 1770 Euler devised a new method (differing from Ferrari's) for solving the quartic equation, but his optimistic hope that some similar method would solve the general polynomial equation was ill-fated. In the same year Lagrange considered the problem of solving the general polynomial equation by comparing the known solutions of quadratic, cubic, and quartic equations and noting that in each of these three cases a certain reduction transformed the equation to one of lower degree; but, unhappily, when Lagrange tried this "reduction" on a quintic equation, the degree of the resulting equation was increased rather than decreased. Although Lagrange did not succeed in his main objective, his attack on the problem made use of permutations of the roots of the equation; and he discovered the key to the theory of permutation groups, including the property mentioned earlier and now called Lagrange's theorem.

Both Abel and Galois built on Lagrange's work.

It is not surprising that Abel, too, approached the general problem of trying to solve the polynomial equation of degree n by trying to solve the general quintic equation. In fact, he thought he had succeeded, and the "solution" was sent to a leading mathematician of Denmark. While waiting for a reply Abel fortunately discovered his mistake, and this misadventure caused him to wonder whether a general algebraic solution was possible. In his own words /BELL (d): 310–12/:

> One of the most interesting problems of algebra is that of the algebraic solution of equations. Thus we find that nearly all mathematicians of distinguished rank have treated this subject. We arrive without difficulty at the general expression of the roots of equations of the first four degrees. A uniform method for solving these equations was discovered and it was believed to be applicable to an equation of any degree; but in spite of all the efforts of Lagrange and other distinguished mathematicians the proposed end was not reached. That led to the presumption that the solution of general equations was impossible algebraically; but this is what could not be decided, since the method followed could lead to decisive conclusions only in the case where the equations were solvable. In effect they proposed to solve equations without knowing whether it was possible. In this way one might indeed arrive at a solution, although that was by no means certain; but if by ill luck the solution was impossible, one might seek it for an eternity, without finding it. To arrive infallibly at something in this matter, we must therefore follow another road. We can give the problem such a form that it shall always be possible to solve it. . . . Instead of asking for a relation of which it is not known whether it exists or not, we must ask whether such a relation is indeed possible. . . . When a problem is posed in this way, the very statement contains the germ of the solution and indicates what road must be taken; and I believe there will be few instances where we shall fail to arrive at

propositions of more or less importance, even when the complication of the calculations precludes a complete answer to the problem.

Although Abel succeeded in showing that for n greater than four the general polynomial equation could not be solved algebraically, he did not claim to have completely achieved the objectives he set for himself:

1. To find all the equations of any given degree which are solvable algebraically
2. To determine whether a given equation is or is not solvable algebraically

It was fortunate that Abel's proof, in which he used permutation groups to some extent, received early publication in the first volume of August Leopold Crelle's *Journal*. This proof caught the imagination of Galois, who gave complete answers to the questions proposed by Abel. In 1831 Galois showed that a polynomial equation is solvable if and only if its group, over the coefficient field, is solvable. The concepts associated with this result are usually characterized as Galois's theory. In his work he used the idea of isomorphic groups, and was the first to demonstrate the importance of invariant (or normal) subgroups and factor groups. The term "group" is due to Galois, and the term "factor group" is due to Otto Hölder.

Although Galois's accomplishments were mathematical landmarks of the greatest significance and originality, they did not immediately make their full impact on his contemporaries because these men were slow to understand, appreciate, and publish Galois's work, admittedly *"à peu près inintelligible"* ("almost unintelligible"), to quote a committee of the French Academy of Sciences.

After Galois was killed in a duel in 1832 at the age of twenty-one, the development of group theory was substantially advanced by Cauchy. During the years 1840–60 he published many articles on the subject, totaling some 280 pages, in the Paris *Comptes Rendus*.

In 1854 Cayley published an article entitled "On the Theory of Groups as Depending on the Symbolic Equation $\theta^n = 1$," which is noteworthy because it gives what is probably the earliest definition of a finite abstract group. It also gives the result now known as Cayley's theorem, that *"every finite group is isomorphic to a regular permutation group."*

In 1870 Camille Jordan published his *Traité des substitutions*—a

255

masterly presentation of permutation groups—covering the results of Lagrange, Ruffini, Abel, Galois, Cauchy, and Serret, as well as his own contributions to the subject.

In the same year Leopold Kronecker gave a set of axioms defining finite Abelian groups. It is different enough from the modern definition to arouse some curiosity:

> Let θ', θ'', θ''', . . . be elements of a finite set, and of such a nature that two of these at a time determine (with the help of a certain procedure) a third one. Then, letting f denote the result of this procedure, θ''' is said to exist for any two elements, θ' and θ'', (and these two elements may be equal). This is expressed by $f(\theta', \theta'')$. Moreover,

$$f(\theta', \theta'') = f(\theta'', \theta')$$

$$f(\theta', f(\theta'', \theta''')) = f(f(\theta', \theta''), \theta''').$$

> However, if θ'' and θ''' are distinct, then $f(\theta', \theta'')$ and $f(\theta', \theta''')$ are also distinct.

Out of this complete axiomatic system for (finite) Abelian groups, Kronecker—still working with a completely arbitrary, abstract set of elements—derived the customary group properties, such as the existence of the unity element for the set, inverses, etc.

INTRODUCTION AND INFLUENCE OF QUATERNIONS

In 1837, six years after Gauss invented his treatment of complex numbers, Hamilton arrived at his own independent discovery of the same ideas, which he applied to rotations and vectors in the plane, as others had done. In a second paper on this subject (1843) he generalized from ordered pairs to n-tuples with emphasis on quadruples (or "quaternions"), which extended the algebra of vectors in the plane to vectors in space. Thus the concept of a complex number, $a + bi$, was extended to the form $a + bi + cj + dk$ (a, b, c, d real) where $i^2 = j^2 = k^2 = -1 = ijk$. The most remarkable property of these quaternions was that the commutative law of multiplication did not hold. It had taken fifteen years of labor before it dawned on Hamilton that it was possible to create a consistent and useful mathematical system that contradicted the time-honored law that $AB = BA$. This flash of insight occurred one October day when he was out strolling with his wife along the Royal Canal in Dublin, and he carved the basic formulas on a stone in the Brougham Bridge.

Hamilton also found occasion to comment on the associative law in his *Lectures on Quaternions* (1853). In the Preface he says, "To this associative property or principle I attach much importance." Of course, the associative law was used long before it was named, and it was already explicitly noted in 1830 when Legendre called attention to it in his *Théorie des Nombres*. Legendre wrote: "One ordinarily supposes that in multiplying a given number C by another number N which is the product of two factors A and B, one gets the same result whether he multiplies C by N all at once, or C by A and then by B." Symbolically, Legendre wrote,

$$C \times \overline{AB} = \overline{CA} \times B.$$

The commutative and distributive laws were given their present names, interestingly enough, by F. J. Servois (1814) in a discussion involving functions. Servois comments that if $f\phi z = \phi f z$, where f and ϕ are functions and z an independent variable, then the functions are called "commutative." He also says that if $f(x + y + \dots) = fx + fy + \dots$, then the function is called "distributive." (He used parentheses more sparingly than we do today.)

It is only fair to mention in passing that Hermann Grassmann simultaneously and independently created an even more general theory of n-tuples than Hamilton; but Grassmann's ponderous and philosophically phrased publications were not read by mathematicians in time for his ideas to have the dramatic impact that Hamilton's quaternions had.

As it turned out, quaternions were not as practical as Hamilton had believed, and they were soon eclipsed by later inventions that were easier to apply; but they began to do for algebra what non-Euclidean concepts were doing for geometry. Once it was realized that $AB = BA$ was not an irrevocable axiom, mathematicians began to experiment with new systems in which *other* axioms were also changed.

Boolean Algebra and Matrix Algebra

Algebras were created in which, for example, it was possible that $xy = 0$ with both $x \neq 0$ and $y \neq 0$. An interesting application of this idea is found in George Boole's algebra of logic where he lets x represent the class of "men" and y the class of "not-men" (which he denotes by $1 - x$) and calls 0 the "class 'nothing.'" "Hence," says Boole, "$x(1 - x)$ will represent the class whose members are at once 'men' and 'not-men,' and the equation $x(1 - x) = 0$ thus expresses the principle *that a class whose members are at the same time men and not-men does not exist. . . .* But this is identically that 'principle of

257

contradiction' which Aristotle has described as the fundamental axiom of all philosophy."

The principles that Boole established in his *Laws of Thought* (1854) were along the lines of Leibniz' "universal characteristic"; and their development by Gottlob Frege (1884) and others culminated in the *Principia mathematica* (1910–13) of Bertrand Russell and A. N. Whitehead and in the whole of present-day mathematical logic.

It is almost unnecessary to mention the important application of Boolean algebra to computer design because of the present popularity of computers in most of the modern mathematics curricula.

Computers are widely used today in linear programming, a technique directly related to the idea of a matrix, which was implicit in the "open" or "indeterminate" product of Grassmann's hypercomplex numbers—as was also the later vector analysis of Josiah Willard Gibbs. However, Cayley, who originated the explicit theory of matrices, stated that he got the notion "either directly from that of a determinant; or as a convenient mode of expression" of the following equations:

$$x' = ax + by.$$

$$y' = cx + dy.$$

He represented this transformation by

$$\begin{bmatrix} a & b \\ c & d \end{bmatrix}$$

and developed an algebra of matrices by observing properties of transformations on linear equations.

Cayley also showed (1858) that a quaternion could be represented in matrix form as shown above where a, b, c, d are suitable complex numbers. For example, if we let the quaternion units 1, i, j, k be represented by

$$\begin{bmatrix} 1 & 0 \\ 0 & 1 \end{bmatrix}, \quad \begin{bmatrix} i & 0 \\ 0 & -i \end{bmatrix}, \quad \begin{bmatrix} 0 & 1 \\ -1 & 0 \end{bmatrix}, \quad \text{and} \quad \begin{bmatrix} 0 & i \\ i & 0 \end{bmatrix},$$

the quaternion $4 + 5i + 6j + 7k$ can be written as shown below.

$$\begin{bmatrix} 4 + 5i & 6 + 7i \\ -6 + 7i & 4 - 5i \end{bmatrix}$$

This led Peter Guthrie Tait, a disciple of Hamilton, to conclude erroneously that Cayley had used quaternions as his motivation for matrices.

In 1925 August Heisenberg discovered that the algebra of matrices is just right for the noncommutative mathematics describing phenomena in quantum mechanics.

In collaboration with James Joseph Sylvester, Cayley (c. 1846) also began work on the theory of algebraic invariants, which had been in the air for some time and which, like matrices, received some of its motivation from determinants.

We have seen how a desire for a general theory of structure (beginning with the solution of polynomial equations and the relations between their roots and coefficients) led to the "completion" of the complex-number system; and how, in turn, an extension of the complex numbers to hypercomplex numbers created new structures.

Mathematics had proliferated to such an extent that a most welcome event was Felix Klein's address (in 1872) called the *Erlanger Programm*, in which he showed that the group concept would help in classifying many branches of mathematics. He applied this concept with brilliant success in showing that many different kinds of geometry are all interrelated by virtue of their group structures. The group concept also proved useful in synthesizing wide areas of algebra and geometry, especially in the work of Gaspard Monge, Jean Victor Poncelet, Arthur Cayley, Alfred Clebsch, Hermann Grassmann, and Bernhard Riemann.

Since Klein's time there has been even more proliferation, and Howard Eves estimates that more than two hundred algebraic structures have been studied. Besides groups, a few of the more familiar structures are rings, integral domains, and fields.

The set of real numbers, under ordinary addition and multiplication, is the most familiar example of a field. In ordinary algebra, in which the letters represent real numbers, the field axioms are assumed. One of the most interesting field properties usually assumed in ordinary algebra (actually it is not an axiom but a theorem) is the "nonexistence of zero divisors." This is used in solving quadratic equations by the factoring method and guarantees that if a product like $(x - 2)(x - 3)$ is zero, at least one of the factors must be zero.

The concept of a field was used by both Abel and Galois at an intuitive, subformal level in their work on polynomial equations. In 1871 Richard Dedekind gave a concrete formulation, and the earliest expositions of the theory of fields from the general point of view were given independently by Heinrich Weber and Eliakim H. Moore in 1893.

Today's student of algebra is psychologically in a position very similar to that of the innovators we have met in this overview. Stories about their work—such as Cardano's exploration of imaginary numbers or Cayley's invention of matrices by noticing coefficient patterns in equations—may well serve to excite the curiosity and spirit of adventure of the modern mind.

Capsule 66   Kenneth Cummins

EQUATIONS
AND THE WAYS THEY WERE WRITTEN

IF A student of the time of Diophantus had been confronted with an expression of the now-common form illustrated by $x^2 - 7x + 12 = 0$, he would have been utterly baffled; this modern symbolic style is of relatively recent invention.

There is no complete agreement on "the time of Diophantus"; some authorities believe that he lived in the third century A.D., but some place him as early as the first century. It is known, however, that he was a Greek mathematician working "in residence" at the University of Alexandria, Egypt, and that he made a start on the use of algebraic symbolism, which eventually supplanted the writing of algebra in a verbal or prose style called "rhetorical algebra."

To illustrate rhetorical algebra we choose an example from an Arab mathematician of a later period: al-Khowarizmi, whose book *Al-jabr* (*c.* 825) both named and greatly influenced European algebra. (It is curious that even al-Khowarizmi used words for numbers, since it was his book *Liber algorismi* [to use the Latin] that introduced Hindu-Arabic numerals into Europe.) He states and solves, as follows, the problem given in modern notation as $x^2 + 21 = 10x$:

What must be the amount of a square, which, when twenty-one dirhems are added to it, becomes equal to the equivalent of ten roots of that square? Solution: Halve the number of the roots; the half is five. Multiply this by itself; the product is twenty-five. Subtract from this the twenty-one which are connected with the square; the remainder is four. Extract its root; it is two. Subtract this from the half of the roots, which is five; the remainder is three. This is the root of the square which you required and the square is nine. Or you may add the root to the half of the roots; the sum is seven; this is the root of the square which you sought for, and the square itself is forty-nine.

Of course, his solution amounts to our writing

$$x = \frac{10}{2} \pm \sqrt{\left(\frac{10}{2}\right)^2 - 21} = 5 \pm \sqrt{4} = 3, 7.$$

If al-Khowarizmi's algebra seems prosaic, it might be worthwhile to comment that ideas often precede notation; symbolism is invented as needed.

"Syncopated algebra," the use of abbreviated words, was introduced by Diophantus; and somewhat later, in India, Brahmagupta (*c.* 628) invented his own system of abbreviations. Unfortunately, other writers often chose to ignore (or were unaware of) existing progress in notation; thus al-Khowarizmi used the rhetorical style of the preceding example.

The original of Diophantus' thirteen-volume work, the *Arithmetica*, has been lost, and the earliest existing copy of any part of the work was made more than a thousand years after it was written.

Here is an example from one of the earlier manuscripts, followed by interpretations in modern form and an explanation of the Greek:

$$K^T\beta \quad s\eta \bigwedge \Delta^T\epsilon \quad \overset{o}{M}\delta \quad \acute{\epsilon}\sigma\tau\acute{\iota} \quad \mu\delta;$$

that is,

$$x^3 2 \quad x8 - x^2 5 \quad 1\cdot 4 \quad = \quad 44,$$

or

$$2x^3 + 8x - (5x^2 + 4) \quad = \quad 44.$$

K^T is an abbreviation for KYBOΣ (*KUBOS*, "cube").

s is an abbreviation for $\alpha\rho\iota\theta\mu os$ (*arithmos*, "number").

$\bigwedge$ is a combination of Λ and I in ΛEIΨΣIΣ (*LEIPSIS*, "lacking").

Δ^T is an abbreviation for ΔYNAMIΣ (*DUNAMIS*, "power").

$\overset{o}{M}$ is an abbreviation for MONAΔEΣ (*MONADES*, "units").

261

Equality is expressed by ἐστί ("is equal to") and also by ισ for ἴσος (*isos*, "equal").

The first nine letters of the Greek alphabet, α, β, γ, δ, ϵ, ς, ζ, η, and θ, stand for 1, 2, 3, 4, 5, 6, 7, 8, and 9; and ι, κ, λ, μ, ν, ξ, o, π, and ϙ (obsolete koppa), stand for 10, 20, 30, 40, 50, 60, 70, 80, and 90.

The example given above uses some capital letters, some lowercase. Later manuscripts use only lowercase letters.

To illustrate the syncopated style of Brahmagupta, we give the following example, with an interpretation into modern notation:

ya ka 7 *bha k(a)*	$7xy + \sqrt{12} - 8$
ya v(a) 3 *ya* 10	$= 3x^2 + 10x.$

It will be quickly noted that equality is expressed by writing the left-hand member of the equation above the right-hand member (to use modern terminology). The shortened form *ya* stands for *yavattavat*, the first unknown; *ka* for *kalaka* ("black"), a second unknown; *bha* for *bhavita*, ("product"); *k(a)* for *karana*, ("irrational" or "root"). The dot placed above a number, as it is here placed over the 8, indicates a negative number; *ru* stands for *rupa*, ("pure" or "plain" number); *v(a)* for *varga*, ("square number"). Additional unknowns would have been expressed by using abbreviations for additional colors, thus: *ni* for *nilaca* ("blue"), *pi* for *pitaca* ("yellow"), *pa* for *pandu* ("white"), and *lo* for *lohita* ("red").

The accompanying list of examples will give some indication of the ways in which algebraic notation gradually progressed from the rhetorical stage to the symbolic. (See also the examples in the overview for this chapter and in [89].) To help the reader decipher some of the abbreviations we make the following brief introductory comments.

A pure number is often followed by *N*, *numeri*, or ϕ (analogous to our writing $7x^0$ for 7). Abbreviations for x are many, including *Pri.* for *primo* ("first"), n° for *numero* ("number," "unknown"), ρ for *res* ("thing," "unknown"), and *N* for *numerus* ("number," "unknown"). The square (of x) is written in many ways, including *Se.* for *secundo* ("second"). Addition and subtraction are often indicated by $\bar{p}$ for *piu* ("more") and $\bar{m}$ for *meno* ("less").

1494	Trouame .I.n°. che giōto al suo q̄drat° facia .12.			
Pacioli	x	$+$	x^2	$= 12.$

1514	4Se. − 51Pri. − 30N dit is ghelijc 45.			
Vander Hoecke	$4x^2 - 51x$	$- 30$	$=$	45.

1521   I □ e $32C°$ — 320 numeri.
Ghaligai $x^2 + 32x = 320.$

1525   Sit I₃ aequatus 12 𝒳 — 36.
Rudolff x^2 $=$ $12x - 36.$

1545   cubus p̄   6   rebus aequalis 20.
Cardano $x^3 + 6x$ $=$ 20.

1553   2 𝒳 A + 2₃. aequata. 4335.
Stifel $2x$ $A +$ $2x^2$ $=$ 4,335.

1557   14.𝓮. + .15.𝓺 ══ 71.𝓺.
Recorde $14x$ +   15   =   71.

1559   I ◇ P 6ρ P 9 [I ◇ P 3ρ P 24.
Buteo x^2 $+ 6x + 9 = x^2$ $+ 3x + 24.$

1572 $\overset{6}{\text{I}}$. p. $\overset{3}{8}$. Eguale à 20.
Bombelli $x^6 + 8x^3$ $=$ 20.

1585   3② + 4 egales à 2① + 4.
Stevin $3x^2 + 4$ $=$ $2x$ + 4.

1591   I QC — 15   QQ + 85C — 225Q + 274N aequatur 120.
Viète $- 15x^4$ $+ 85x^3 - 225x^2 + 274x$ $=$ 120.

1631 $aaa - 3\,bba$ ══ $+2\,ccc.$
Harriot $x^3 - 3b^2x$ $=$ $2c^3.$

1637 $yy \propto cy - \dfrac{cx}{b}\,y + ay - ac.$
Descartes

1693 $x^4 + bx^3 + cxx + dx + e = 0.$
Wallis

For Further Reading

CAJORI (d): I, 71–400   D. E. SMITH (a): II, 421–35
SANFORD (d) 153–59

THE BINOMIAL THEOREM

THE "arithmetic triangle" is often associated with the name of Blaise Pascal, who in 1653 discussed many of its properties and applied it to the expansion of $(a + b)^n$, with n a positive integer. He did not claim to have invented the "triangle" or the binomial theorem, but he was probably unaware that the Hindus and Arabs had worked with these ideas as early as the beginning of the twelfth century, when Omar Khayyam claimed to know the binomial expansion for degrees four, five, six, and higher (and for particulars referred the reader to another of his works—which has since been lost).

The Hindus and Arabs used the expansions of $(a + b)^2$ and of $(a + b)^3$ in finding square roots and cube roots. If they were given a positive number N and required to find its square root, they would choose a nearby perfect square number, say s^2, and let d be another number such that $s^2 + d = N$. The correction on s so that $(s + \text{correction})^2 = N$ was all that was required. Continuing to use modern notation here, we can describe their square-root process as follows: Let x be the required correction; then

$$N = s^2 + d = (s + x)^2 = s^2 + 2sx + x^2$$
$$d = x(2s + x)$$
$$x = \frac{d}{2s + x}.$$

By discarding the x on the right we obtain a first approximation to $\sqrt{N}$:

$$\sqrt{N} \approx s_1 = s + \frac{d}{2s}.$$

A machine program for this will be easy to write. Using $N = 5$ as an illustration, we have

$$5 = 2^2 + d = (2 + x)^2$$

$$x = \frac{d}{2s} = \frac{1}{2(2)} = 0.25$$

$$s_1 = 2 + 0.25 = 2.25.$$

264

Repeating this process, we have

$$5 = (2.25)^2 + d_1 = (2.25 + x_1)^2$$

$$x_1 = \frac{d_1}{2s_1} = \frac{-0.0625}{2(2.25)} = -0.01389$$

$$s_2 = s_1 + x_1 = 2.23611$$

$$d_2 = 5 - (2.23611)^2 = 0.00019;$$

and so on, until d_n (which does, in fact, converge to zero) is as close to zero as required.

For cube roots a similar iterative procedure results from letting $N = s^3 + d$. Then

$$x = \frac{d}{3s^2 + 3sx + x^2}$$

$$\approx \frac{d}{3s^2}.$$

It is interesting to observe that the correction term, x, which gave $d/2s$ and $d/3s^2$ for square and cube roots, reminds one of Newton's method in the calculus.

In 1676 Isaac Newton wrote two letters to Henry Oldenburg in which Newton stated without proof the binomial formula

$$(P + PQ)^{\frac{m}{n}} = P^{\frac{m}{n}} + \frac{m}{n} AQ + \frac{m - n}{2n} BQ$$

$$+ \frac{m - 2n}{3n} CQ + \frac{m - 3n}{4n} DQ + \cdots ,$$

where $A = $ first term $= P^{\frac{m}{n}}$, $B = $ second term $= \frac{m}{n} AQ$, and so forth, and the exponent $\frac{m}{n}$ was a rational fraction (positive or negative). The form of the theorem more familiar to the modern reader is obtained if one makes the indicated substitution for $A, B, C, \ldots$:

$$(P + PQ)^{\frac{m}{n}} = P^{\frac{m}{n}} + \frac{m}{n} P^{\frac{m}{n}-1}(PQ) + \frac{\left(\frac{m}{n}\right)\left(\frac{m}{n} - 1\right)}{1 \cdot 2} P^{\frac{m}{n}-2}(PQ)^2$$

$$+ \frac{\left(\frac{m}{n}\right)\left(\frac{m}{n} - 1\right)\left(\frac{m}{n} - 2\right)}{1 \cdot 2 \cdot 3} P^{\frac{m}{n}-3}(PQ)^3 + \cdots .$$

The first proof (not up to modern standards of rigor) for arbitrary positive integral power (i.e., $m/n =$ positive integer) seems to be that given by Jakob (or Jacques) Bernoulli in his *Ars conjectandi,* which was published in 1713, eight years after his death. In 1826 the twenty-four-year-old Niels Henrik Abel, poverty-stricken and suffering from lumbar tuberculosis but already a famous mathematician, published the first general proof of the formula for arbitrary complex exponents. This appeared in the *Journal für die reine und angewandte Mathematik,* customarily referred to as *Crelle's Journal.*

It might be stated that in the expansion of $(1 + x)^\alpha$, the successive terms form a sequence that is finite only if α is a nonnegative integer. In case α is fractional or negative, the question of convergence—both of the sequence of successive terms and of the series, which is itself the general binomial expansion—immediately arises.

We do not often think of the binomial theorem, even in its general form, as opening doors to more advanced mathematics; yet a discussion of the two following expressions,

$$\lim_{n \to \infty} \left(1 + \frac{1}{n}\right)^n$$

and

$$\lim_{\substack{y \to 0 \\ x \text{ fixed}}} (1 + xy)^{1/y},$$

leads to the definition of the transcendental number e and the transcendental function e^x. With this in mind, it is no longer mysterious that the Maclaurin series expansion for e^x looks like a modification of the binomial expansion of $(1 + xy)^{1/y}$.

For Further Reading

ORE (b)

CONTINUED FRACTIONS

THE equality
$$\frac{318}{76} = 4 + \cfrac{1}{5 + \cfrac{1}{2 + \frac{1}{3}}}$$

shows that the common fraction 318/76 can be written as a *continued fraction*. If all the numerators in a continued fraction are 1's (as in the above example) it is called a *simple* continued fraction.

Perhaps the most interesting elementary property of continued fractions is their close relationship with the Euclidean algorithm for finding the greatest common divisor of two integers:

$$318 = 76(4) + 14$$

$$76 = 14(5) + 6$$

$$14 = 6(2) + 2$$

$$6 = 2(3) + 0$$

$$\frac{318}{76} = 4 + \cfrac{1}{5 + \cfrac{1}{2 + \frac{1}{3}}}$$

Remarks:

318 ÷ 76 gives a quotient of 4 and a remainder of 14, and so forth.

The last nonzero remainder, 2, is the G.C.D. of 318 and 76.

Remark:

To obtain the above, write

$$\frac{318}{76} = 4 + \cfrac{1}{\cfrac{76}{14}}$$

and then replace $\dfrac{76}{14}$ by $5 + \cfrac{1}{\dfrac{14}{6}}$, and

so forth.

The striking similarity of the expressions in the parallel columns above (especially with respect to the digits 4, 5, 2, and 3) leads some writers to say that continued fractions were already known to the Greeks, "though not in our present notation."

Rafael Bombelli seems to have been the first to make explicit use of (infinite) continued fractions when he wrote the following in 1572 (modern notation is used here) :

$$\sqrt{13} = 3 + \cfrac{4}{6 + \cfrac{4}{6 + \cfrac{4}{6 + \cdots}}} \; ;$$

and he probably recognized the above as a special case of

$$\sqrt{a^2 + b} = a + \cfrac{b}{2a + \cfrac{b}{2a + \cfrac{b}{2a + \cdots}}} \; .$$

The above expression for $\sqrt{13}$ is called an "infinite continued fraction" and can be obtained by equating it to $3 + 1/x$; then

$$\frac{1}{x} = \frac{\sqrt{13} - 3}{1} = \frac{4}{\sqrt{13} + 3} = \frac{4}{3 + \sqrt{13}} = \cfrac{4}{3 + \left(3 + \cfrac{1}{x}\right)} = \cfrac{4}{6 + \cfrac{1}{x}} \; ;$$

hence

$$\sqrt{13} = 3 + \frac{1}{x} = 3 + \cfrac{4}{6 + \cfrac{1}{x}} \; ;$$

now just keep replacing $1/x$ by $4/(6 + (1/x))$.

This process for finding an infinite sequence of successive approximations for $\sqrt{13}$ gives the first three convergents as follows:

$$C_1 = 3 \qquad\qquad = 3.$$

$$C_2 = 3 + \tfrac{4}{6} \qquad = 3\tfrac{2}{3}.$$

$$C_3 = 3 + \cfrac{4}{6 + \tfrac{4}{6}} = 3\tfrac{6}{10}.$$

These converge to $\sqrt{13}$, oscillating back and forth across $\sqrt{13}$ as shown in Figure [68]-1.

FIGURE [68]-1

John Wallis (*c.* 1685) found many properties of these convergents, including recurrence (or recursion) formulas that express a particular convergent, $C_k = N_k/D_k$, in terms of the preceding two sets of N's and D's. One of the interesting examples discussed by Wallis is the one discovered by William Brouncker (1658):

$$\frac{4}{\pi} = 1 + \cfrac{1^2}{2 + \cfrac{3^2}{2 + \cfrac{5^2}{2 + \cfrac{7^2}{2 + \ldots}}}}$$

A modern form of symbolism was introduced by Christiaan Huygens (1629–1695), who expressed the ratio 77,708,431/2,640,858 in this form:

$$29 + \frac{1}{2} + \frac{1}{2} + \frac{1}{1} + \frac{1}{5} + \frac{1}{1} + \frac{1}{4} \ldots.$$

This ratio actually arose in the solution of a practical problem which Huygens attacked in 1680, in designing the toothed wheels of his planetarium. In 365 days the annual movement of the earth is $359°45'40''31'''$, while that of Saturn is $12°13'34''18'''$. Converting to units of sixtieths of a second, 77,708,431 is to 2,640,858 as the period of Saturn is to the period of time during which the earth makes its revolution around the sun. The corresponding simple continued fraction given above is sometimes expressed today in the more convenient form (29; 2, 2, 1, 5, 1, 4, . . .), introduced by Dirichlet in 1854.

Huygens wished to find two smaller integers with almost the same ratio, so that no pair of smaller integers would yield a closer approximation. Denoting the simple continued fraction in the modern form $(a_0; a_1, a_2, a_3, \ldots)$, Huygens' approximation was made by attempting to determine a_k so that both $|\,a_k - a_{k+1}\,|$ and $|\,a_k - a_{k-1}\,|$ were maximized. He then used $(a_0; a_1, a_2, \ldots a_{k-1})$ as his approximation. Hence he chose (29; 2, 2, 1) = 206/7; his wheel of Saturn had 206 teeth while its motor wheel had 7 teeth. Using these numbers made it necessary to advance the wheel of Saturn by one tooth every 1,346 years.

It was Pietro Cataldi (1613) who began working on the theory of continued fractions and who also introduced—in his treatise published in Bologna on finding the square roots of numbers—the motivation for the notation that was to be used later by Huygens.

Leonhard Euler (1737) secured the foundation of the modern theory and showed that any quadratic irrational (like $\sqrt{13}$, above) can be represented by a simple repeating (or periodic) continued fraction; thus $\sqrt{13}$ can also be written in the following form:

$$3 + \cfrac{1}{1 + \cfrac{1}{1 + \cfrac{1}{1 + \cfrac{1}{1 + \cfrac{1}{6 + \ldots}}}}} \; .$$

More compactly, $\sqrt{13} = 3;11116\ 11116\ \ldots$.

Johann Heinrich Lambert (1761) showed that the following simple continued fraction for π,

$$3 + \cfrac{1}{7 + \cfrac{1}{15 + \cfrac{1}{1 + \cfrac{1}{292 + \ldots}}}} \; ,$$

was not periodic and hence not a quadratic irrational ($a + \sqrt{b}$, a, b rational).

Joseph Louis Lagrange (1798) proved that periodic simple continued fractions represent solutions of quadratic equations with rational coefficients. Thus $\sqrt{13} - 1 = 2;11116\ 11116\ \ldots$ is a root of $x^2 + 2x - 12 = 0$. Lagrange also gave the first complete exposition of convergence of convergents. He showed that in general (see Fig. [68]-1) every odd convergent is less than all following convergents (in the sequence C_1, C_2, C_3, C_4, C_5, $\ldots$) and every even convergent is greater than all following convergents. From this (and the fact that the C's approach $\sqrt{13}$) it follows that, for example, C_4 differs from $\sqrt{13}$ by less than

Adrien Marie Legendre (1794) proved that every infinite continued fraction is irrational.

Thomas Joannes Stieltjes (1894) found a relationship between divergent series and convergent continued fractions which made it possible to define integration for the series; Stieltjes' integrals were to some extent a result of his work with continued fractions.

270

For Further Reading

BELL (a): 298–99, 476–78

CAJORI (d): II, 48–57

COURANT and ROBBINS: 49–51, 301–3

DANTZIG (b): 155–57, 312–16

NIVEN (a): 51–67

D. E. SMITH (a): II, 418–21

——— (c): I, 80–84

STRUIK (e): 111–15

Capsule 69   *Richard M. Park*

OUGHTRED AND THE SLIDE RULE

WILLIAM Oughtred (1574–1660), the vicar of Shalford and rector of Albury, Surrey, was one of the most influential mathematicians of his time. He was in great demand as a teacher, since the universities of that time offered little instruction in mathematics. A systematic treatment of much of the then-known work in arithmetic and algebra was published in his *Clavis mathematicae* (1631), which ran through six editions.

Oughtred placed unusual emphasis on mathematical symbols, developing or fostering many symbols in use today. Major examples are $\times$ for multiplication, :: for proportion, and $-$ for difference.

Today, however, Oughtred is best remembered for his invention of both the circular and the rectilinear slide rules. His circular slide rule is described in his *Circles of Proportion* (1632) as eight fixed circles on one side of the instrument with an index operating much like a compass (Fig. [69]-1). Calling the outermost (largest) circle the first and the innermost circle the eighth, the scales on each of the eight circles are as shown below.

1. Sines from 5°45' to 90°
2. Tangents from 5°45' to 45°
3. Tangents from 45° to 84°15'
4. Logarithmically spaced integers 2, 3, 4, 5, 6, 7, 8, 9, 1
5. Equally spaced integers 1, 2, 3, 4, 5, 6, 7, 8, 9, 0
6. Tangents from 84° to 89°24'

271

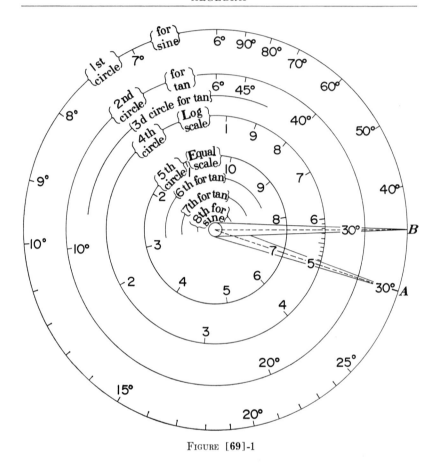

FIGURE [69]-1

7. Tangents from 35′ to 6°
8. Sines from 35′ to 6°

An example of its use: To find the value for sin 30° (see point *A* in Fig. [69]-1) one leg of the index (or compass) is placed at 30° on the first circle; the corresponding number on the fourth circle, 5, gives 0.5000 as the sine of 30°. Similarly, to find tan 30° refer to point *B*, 30° on the second circle, and read the corresponding answer, 0.5774, on the fourth circle. (Oughtred could get accuracy to four places.)

The fourth circle is used for multiplication. For 2 · 3 (see Figs. [69]-2 and -3) open and turn the two legs of the index so that they point to 1 and 2; then, with the angle *α* between the two legs held

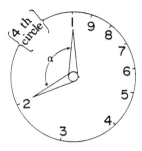

FIGURE [69]-2

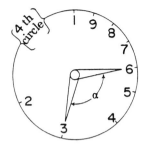

FIGURE [69]-3

constant, rotate the index so that one leg points to 3. Then the other leg points to 6, the desired product.

Around 1622 Oughtred invented his rectilinear slide rule, which consisted of two logarithmically calibrated rulers, one sliding along the other (without a fixed track or groove). He gave full credit to Edmund Gunter for the latter's invention in 1620 of a single rectilinear logarithmic scale, used to multiply numbers by adding the corresponding segments mechanically with the aid of a pair of dividers.

In 1630, two years before Oughtred published his *Circles of Proportion,* one of his former students, Richard Delamain, published *Grammelogia.* This, also, contained a description of a circular slide rule. Each man accused the other of having stolen his invention, but Cajori /(e): 158/ and D. E. Smith /(c): I, 160/ think it probable that each man invented the circular slide rule independently.

For Further Reading

CAJORI (c)
——— (d): I, 187–99
——— (e): 158–59

SANFORD (d): 343–47
D. E. SMITH (c): I, 160–64

273

HORNER'S METHOD

WHAT we know today as Horner's method (for approximating real roots of polynomial equations with real, numerical coefficients) was known in an equivalent form by the Chinese for many centuries before it was published by Chhin Chiu-shao in 1247. It was called the "celestial element method"; and it appears also, though in more primitive form, in the *Nine Chapters*, written before the Christian era.

It is quite likely that in his travels Fibonacci (Leonardo of Pisa) learned of this method, which in 1225 he described rather well up to a certain point, after which he stopped explaining the method and merely gave the answer, to an excellent degree of accuracy. To solve (we use modern notation here)

$$x^3 + 2x^2 + 10x = 20,$$

he writes the equation in the form

$$x + \tfrac{1}{10}x^3 + \tfrac{1}{5}x^2 = 2,$$

from which it is clear that $x < 2$. The original equation shows that $x > 1$, since $1 + 2 + 10 < 20$. Then he shows that x cannot equal a rational fraction, a/b, because $(a/b) + (a^3/10b^3) + (a^2/5b^2)$ is not an integer; hence x is irrational. Further, x is not the square root of a positive integer, a, because the given equation implies that

$$x = \frac{20 - 2x^2}{10 + x^2},$$

which for $x = \sqrt{a}$ becomes the impossible statement that

$$\sqrt{a} = \frac{20 - 2a}{10 + a}.$$

Then Fibonacci abruptly gives the answer (in base sixty) as

$$x = 1°22'7''42'''33^{iv}4^{v}40^{vi},$$

that is, as

274

$$1 + \frac{22}{60} + \frac{7}{60^2} + \frac{42}{60^3} + \frac{33}{60^4} + \frac{4}{60^5} + \frac{40}{60^6} \, ,$$

as it had probably been given to him during his travels.

François Viète (1600), apparently unaware of earlier results, gave a systematic process that showed a new insight into the general theory of equations, but the process becomes very laborious for equations of high degree.

Isaac Newton (1669) simplified Viète's method, the simplification being essentially like that found in texts in college algebra or theory of equations (not the Newton's method found in books on the calculus).

Paolo Ruffini (1803) and William George Horner (1819) independently worked out and published very similar methods for finding approximations of real roots of numerical polynomial equations. They both thought of their methods as better ways to find cube roots, fourth roots, and so on. At first they explained their methods in terms of the calculus, but later each of them was able to use only elementary algebra.

Ruffini's later method is actually closer than is Horner's to what present-day texts call "Horner's method."

Although Horner did not attend a university, he became a master in the Kingswood School of Bristol at the age of nineteen. He was not, however, a great mathematician. It was a stroke of good fortune that this mathematical accomplishment—his only one—was published in the *Philosophical Transactions* of the Royal Society (although not without some objections because of the elementary nature of his paper); the intricate style of his exposition made the work seem more impressive than it really was.

For Further Reading

BELL (a): 108–14   D. E. SMITH (a): II, 471–72
COOLIDGE (c): 186–94   ———— (c): 232–52

SOLUTION OF POLYNOMIAL EQUATIONS
OF THIRD AND HIGHER DEGREES

THE first records of man's interest in cubic equations date from the time of the Old Babylonian civilization, about 1800–1600 B.C. Among the mathematical materials that survive are tables of cubes and cube roots, as well as tables of values of $n^2 + n^3$. Such tables could be used to solve cubics of special types.

For example, to solve the equation $2x^3 + 3x^2 = 540$, the Babylonians might have first multiplied by 4 and made the substitution $y = 2x$, giving $y^3 + 3y^2 = 2{,}160$. Letting $y = 3z$, this becomes $z^3 + z^2 = 80$. From the tables, one solution is found to be $z = 4$, and hence 6 is a root of the original equation.

In the Greek period concern with volumes of geometrical solids led easily to problems that in modern form involve cubic equations. The well-known problem of duplicating the cube is essentially one of solving the equation $x^3 = 2$. This problem, impossible of solution by ruler and compasses alone, was solved in an ingenious manner by Archytas of Tarentum ($c.$ 400 B.C.), using the intersections of a cone, a cylinder, and a degenerate torus (obtained by revolving a circle about its tangent) /GRAESSER/.

The well-known Persian poet and mathematician Omar Khayyam (A.D. 1100) advanced the study of the cubic by essentially Greek methods. He found solutions through the use of conics. It is typical of the state of algebra in his day that he distinguished thirteen special types of cubics that have positive roots. For example, he solved equations of the type $x^3 + b^2x = b^2c$ (where b and c are positive numbers) by finding intersections of the parabola $x^2 = by$ and the circle $y^2 = x(c - x)$, where the circle is tangent to the axis of the parabola at its vertex. The positive root of Omar Khayyam's equation is represented by the distance from the axis of the parabola to a point of intersection of the curves.

The next major advance was the algebraic solution of the cubic. This discovery, a product of the Italian Renaissance, is surrounded by an atmosphere of mystery; the story is still not entirely clear /CARDANO: ix–xii; FELDMAN (a)/. The method appeared in print in 1545 in the *Ars magna* of Girolamo Cardano of Milan, a physician, astrologer, mathematician, prolific writer, and suspected heretic, altogether one of the most colorful figures of his time.

The method has gained currency as "Cardan's formula," Cardan being the English form of the name. According to Cardano himself, however, the credit is due to Scipione del Ferro, a professor of mathematics at the University of Bologna, who in 1515 discovered how to solve cubics of the type $x^3 + bx = c$. As was customary among mathematicians of that time, he kept his methods secret in order to use them for personal advantage in mathematical duels and tournaments. When he died in 1526, the only persons familiar with his work were a son-in-law and one of his students, Antonio Maria Fior of Venice.

In 1535 Fior challenged the prominent mathematician Niccolo Tartaglia of Brescia (then teaching in Venice) to a contest because Fior did not believe Tartaglia's claim of having found a solution for cubics of the type $x^3 + bx^2 = c$. A few days before the contest Tartaglia managed to discover also how to solve cubics of the type $x^3 + ax = c$, a discovery (so he relates) that came to him in a flash during the night of February 12/13, 1535. Needless to say, since Tartaglia could solve two types of cubics whereas Fior could solve only one type, Tartaglia won the contest.

Cardano, hearing of Tartaglia's victory, was eager to learn his method. Tartaglia kept putting him off, however, and it was not until four years later that a meeting was arranged between them. At this meeting Tartaglia divulged his methods, swearing Cardano to secrecy and particularly forbidding him to publish it. This oath must have been galling to Cardano. On a visit to Bologna several years later he met Ferro's son-in-law and learned of Ferro's prior solution. Feeling, perhaps, that this knowledge released him from his oath to Tartaglia, Cardano published his version of the method in *Ars magna*. This action evoked bitter attack from Tartaglia, who claimed that he had been betrayed.

Although couched in geometrical language, the method itself is algebraic and the style syncopated. Cardano gives as an example the equation $x^3 + 6x = 20$ and seeks two unknown quantities, p and q, whose difference is the constant term 20 and whose product is the cube of 1/3 the coefficient of x, 8. A solution is then furnished by the difference of the cube roots of p and q. For this example the solution is

$$\sqrt[3]{\sqrt{108} + 10} - \sqrt[3]{\sqrt{108} - 10}.$$

This procedure easily applies to the general cubic after being transformed to remove the term in x^2.

This discovery left unanswered such questions as these: What should be done with negative and imaginary roots, and (a related question) do three roots always exist? What should be done (in the so-called irreducible case) when Cardano's method produced apparently imaginary expressions like

$$\sqrt[3]{81 + 30\sqrt{-3}} + \sqrt[3]{81 - 30\sqrt{-3}}$$

for the real root, -6, of the cubic $x^3 - 63x - 162 = 0$? These questions were not fully settled until 1732, when Leonhard Euler found a solution.

The general quartic equation yielded to methods of similar character; and its solution, also, appeared in *Ars magna*. Cardano's pupil Ludovico Ferrari was responsible for this result. Ferrari, while still in his teens (1540), solved a challenging problem that his teacher could not solve.

His solution can be described as follows: First reduce the general quartic to one in which the x^3 term is missing, then rearrange the terms and add a suitable quantity (with undetermined coefficient) to both sides so that the left-hand member is a perfect square. The undetermined coefficients are then determined so that the right-hand member is also a square, by requiring that its determinant be zero. This condition leads to a cubic, which can now be solved—the quartic can then be easily handled.

Later efforts to solve the quintic and other equations were foredoomed to failure, but not until the nineteenth century was this finally recognized. Carl Friedrich Gauss had proved in 1799 that every algebraic equation of degree n over the real field has a root (and hence n roots) in the complex field. The problem was to express these roots in terms of the coefficients by radicals. Paolo Ruffini, an Italian teacher of mathematics and medicine at Modena, is considered to have given (in 1813) an essentially satisfactory proof of the impossibility of doing this for equations of degree higher than four. Better known is the work of a brilliant young Norwegian mathematician, Niels Henrik Abel. After first thinking he had solved the general quintic, Abel found his error; and in 1824 he published at his own expense (in Christiania, now Oslo) his proof of its impossibility. His result appeared also, two years later, in the first volume of *Crelle's Journal* (Berlin), thus helping to inaugurate at a high level one of the great mathematical periodicals of the world. Abel's work in turn stimulated the young Frenchman Évariste Galois (1811–1832), who before his early death in a duel showed that every equation could be associated with a characteristic group and that the properties of this group could be used to determine whether the equation could be solved by radicals.

For Further Reading

BOYER (g): 310–17
CARDANO: vii–xxii
COOLIDGE (c): 19–29
EVES (c)
    [5th ed. 200–204]
FELDMAN (a)
GRAESSER

MIDONICK: 583–98
NEUGEBAUER (a): 44, 51
ORE (a)
———— (b)
D. E. SMITH (a): II, 454–70
STRUIK (e): 62–73
JACOB YOUNG: 213–21

Capsule 72   Leonard E. Fuller

VECTORS

THE roots of vector algebra go back to the geometric concept of directed line segments in space. The composition of forces by the parallelogram law led to the idea of addition of vectors. Their representation as ordered sets of real numbers occurred only after the extension of number systems beyond the complex numbers.

Hermann Grassmann, in his monumental *Ausdehnungslehre*, published in 1844, freed his thinking from three-dimensional Euclidean space. He discussed manifolds of n dimensions and developed algebras for these systems. This enabled him to consider an extension of complex numbers to hypercomplex numbers. He made a significant stride when he found that he had to give up the property of commutativity of multiplication. This was the major stumbling block in the extension. His work included also the theory of tensor calculus, which was destined to play a key role in the theory of relativity. Unfortunately, Grassmann's work was not properly understood by others, so its true significance had to wait for the passage of time. In 1862 he published a second edition of his work, in which he attempted to clarify the first and to add to it; but again he met with little success.

The year before Grassmann published the first edition of his *Ausdehnungslehre*, William Rowan Hamilton discovered the basic idea for quaternions. He, too, was bold enough to sacrifice the commutative property of multiplication. In 1853 he published *Lectures on Quaternions*, a work that was better understood and appreciated than Grassmann's, perhaps because it was not so general. Hamilton devoted the rest of his career to developing the theory of quaternions. He seemed convinced that this theory held the key to many ideas.

There was opposition to Hamilton's ideas, perhaps because of the complexity of the algebra involved. As a result, others tried to develop their own substitutes for it.

A disciple of Hamilton, Peter Guthrie Tait, devoted his life to quaternions. He stirred up a fight between mathematicians that extended over fifty years. His chief opponent was Josiah Willard Gibbs, who developed an excellent departure from quaternions with his vector analysis. A student of Gibbs, Edwin Bidwell Wilson, put the theory of vector analysis in book form in 1901. It is ironic that the idea that could have resolved the conflict much earlier was in Grassmann's *Ausdehnungslehre*. Actually, it was resolved by Grassmann's tensor calculus, which was further developed by C. G. Ricci, who published a work on it in 1888. At first, little attention was paid to this work; it was only after Einstein used it in his theory of relativity that it gained general acceptance. This theory of relativity vindicated the work of Grassmann and showed that he had been more than fifty years ahead in his thinking.

Today vectors are studied from the geometric point of view as directed line segments in three dimensions, largely as a result of Gibbs's work, and from the algebraic point of view as n-dimensional manifolds, largely as a result of Grassmann's.

For Further Reading

BELL (a): 182–211	NEWMAN: I, 162–63, 697–98
CAJORI (e): 334–45	D. E. SMITH (c): II, 677–96

DETERMINANTS AND MATRICES

THE Japanese mathematician Seki Kowa (1683) systematized an old Chinese method of solving simultaneous linear equations whose coefficients were represented by calculating sticks—bamboo rods placed in squares on a table, with the positions of the different squares corresponding to the coefficients. In the process of working out his system, Kowa rearranged the rods in a way similar to that used in our simplification of determinants; thus it is thought that he had the *idea* of a determinant.

Ten years later in Europe Gottfried Wilhelm von Leibniz formally originated determinants and gave a written notation for them. In a letter to Marquis de L'Hospital Leibniz gave a discussion of a system of three linear equations in two unknowns /D. E. SMITH (c): I, 268–69/. A translation appears in the left-hand column, below, with a more modern version in the right-hand column.

I suppose that

$$10 + 11x + 12y = 0.$$
$$20 + 21x + 22y = 0.$$
$$30 + 31x + 32y = 0.$$

$$a_{10} + a_{11}x + a_{12}y = 0.$$
$$a_{20} + a_{21}x + a_{22}y = 0. \qquad (1)$$
$$a_{30} + a_{31}x + a_{32}y = 0.$$

where . . . eliminating y first from the first and second equations, we shall have

$$\frac{10.22 + 11.22x}{-12.20 - 12.21\ldots} = 0,$$

$$(a_{10}a_{22} - a_{12}a_{20})$$
$$+ (a_{11}a_{22} - a_{12}a_{21})x = 0.$$

and from the first and third

$$\frac{10.32 + 11.32x}{-12.30 - 12.31\ldots} = 0.$$

$$(a_{10}a_{32} - a_{12}a_{30})$$
$$+ (a_{11}a_{32} - a_{12}a_{31})x = 0.$$

It remains now to eliminate the letter x . . . and as the result we shall have

$$1_0.2_1.3_2 \quad 1_0.2_2.3_1$$

$$1_1.2_2.3_0 = 1_1.2_0.3_2.$$

$$1_2.2_0.3_1 \quad 1_2.2_1.3_0$$

$$a_{10}a_{21}a_{32} \quad\quad a_{10}a_{22}a_{31}$$

$$+a_{11}a_{22}a_{30} = +a_{11}a_{20}a_{32}$$

$$+a_{12}a_{20}a_{31} \quad\quad +a_{12}a_{21}a_{30}$$

or, moving all terms to the left side of the equation,

$$a_{10}a_{21}a_{32} - a_{10}a_{22}a_{31}$$

$$+a_{11}a_{22}a_{30} - a_{11}a_{20}a_{32} = 0,$$

$$+a_{12}a_{20}a_{31} - a_{12}a_{21}a_{30}$$

or

$$\begin{vmatrix} a_{10} & a_{11} & a_{12} \\ a_{20} & a_{21} & a_{22} \\ a_{30} & a_{31} & a_{32} \end{vmatrix} = 0. \quad\quad (2)$$

(The reader may recall, or easily verify, that (2) is the condition for the three straight lines represented by (1) to pass through a common point.) The now-standard "vertical line notation" used in (2) above was given in 1841 by Arthur Cayley.

Determinants were invented independently by Gabriel Cramer, whose now well-known rule for solving linear systems was published in 1750, although not in present-day notation.

Many other mathematicians also made contributions to determinant theory—among them Alexandre Théophile Vandermonde, Pierre Simon Laplace, Josef Maria Wronski, and Augustin Louis Cauchy. It is Cauchy who applied the word "determinant" to the subject; in 1812 he introduced the multiplication theorem.

Although the idea of a matrix was implicit in the quaternions (4-tuples) of William Rowan Hamilton and also in the "extended magnitudes" (n-tuples) of Hermann Grassmann [72], the credit for inventing matrices is usually given to Cayley, with a date of 1857, even though Hamilton obtained one or two isolated results in 1852. Cayley says that he got the idea of a matrix "either directly from that of a determinant; or as a convenient mode of expression of the equations $x' = ax + by,\ y' = cx + dy$."

It was shown by Hamilton in his theory of quaternions [**77**] that one could have a logical system in which the multiplication is not commutative. This result was undoubtedly of great help to Cayley in working out his matrix calculus because matrix multiplication, also, is noncommutative.

Cayley's theory of matrices grew out of his interest in linear transformations and algebraic invariants, an interest he shared with James Joseph Sylvester. They investigated algebraic expressions that remained invariant (unchanged except, possibly, for a constant factor) when the variables were transformed by substitutions representing translations, rotations, dilatations ("stretchings" from the origin), reflections about an axis, and so forth. Thus, for example, if one transforms the conic

$$(1) \qquad\qquad Ax^2 + Bxy + Cy^2 = K$$

by applying the substitution

$$x = \frac{1}{\sqrt{2}}\, x' - \frac{1}{\sqrt{2}}\, y'$$

$$y = \frac{1}{\sqrt{2}}\, x' + \frac{1}{\sqrt{2}}\, y',$$

which is a linear transformation representing a rotation of axes through 45°, this becomes

$$(2) \qquad\qquad A'x'^2 + B'x'y' + C'y'^2 = K,$$

where

$$A' = -A + C, \qquad B' = \tfrac{1}{2}(A + B + C), \qquad C' = \tfrac{1}{2}(A - B + C).$$

It is easily checked that the "discriminant" $B^2 - 4AC$ of (1) is equal to the discriminant $B'^2 - 4\,A'C'$ of (2), no matter what values are used for A, B, C. Hence this discriminant, $B^2 - 4AC$, is called an invariant (under the rotation). Under the 45° rotation, $3x^2 + 2xy + 3y^2 = 5$ becomes $4x'^2 + 0x'y' + 2y'^2 = 5$. The discriminants are, respectively, $2^2 - 4 \cdot 3 \cdot 3$ and $0^2 - 4 \cdot 4 \cdot 2$ (both equal to -32).

Today, matrix theory is usually considered part of the broader subject of linear algebra, and it is a mathematical tool of the social scientist, geneticist, statistician, engineer, and physical scientist.

For Further Reading

BELL (a): 182–89, 424–27   FELDMAN (b)
——— (d): 378–405   MIDONICK: 196–211
CAJORI (e): 332–45   NEWMAN: I, 341–65

Capsule 74   Anice Seybold

BOOLEAN ALGEBRA

THE idea of laying down postulates for the manipulation of abstract symbols (not necessarily numbers) seems to have occurred first in England and at about the time of George Boole (1815–1864). Boole published his basic ideas in 1847 in a pamphlet entitled *Mathematical Analysis of Logic*. In 1854, in *An Investigation of the Laws of Thought on Which Are Founded the Mathematical Theories of Logic and Probabilities*, he presented a more thorough exposition of his work. "Boolean algebra" is a term often applied to the algebra of sets, although it can also be interpreted so as to yield what we now call "the propositional calculus" or "truth-function logic," which is studied largely by means of truth tables.

Boole used lowercase letters such as x, y, z, to denote sets, whereas we often use uppercase A, B, C, and so on. It is assumed that we can tell whether a given thing does or does not belong to a given set. A set can be described by saying it consists of all items having a given property or characteristic. The set containing no elements is called the null set—in Boole's notation written as the number 0, in modern notation written as $\emptyset$ or O, uppercase letter oh. The set of all elements under consideration (containing all sets under consideration and perhaps more, too) is the universal set—1 in Boole's notation and now frequently I, uppercase letter eye. If we take the set of all human beings for the universal set, then all human males, all people over fifty years old, all blue-eyed people, and all brown-eyed people are four differents sets that are subsets of the universal one. The set of all two-headed people is the null set (we hope).

Sets can be combined to form new sets in two basic ways. The logical

284

product or intersection of two sets x and y (or A and B)—denoted by Boole as xy or $x \cdot y$ (now frequently as $A \cap B$, called "A cap B")—consists of all elements that are in both sets. If A is the set of all human males and B is the set of all blue-eyed people, then $A \cap B$ is the set of all blue-eyed men. If C is the set of all brown-eyed people, then we have $B \cap C = \emptyset$, where $=$ is used to connect two different symbols for the same set (in this case, the null set). By the meaning of logical product we must have $X \cap X = X$. Boole wrote this as $x^2 = x$. When this equation is regarded as a condition on unknown numbers rather than as a set-theoretic statement, it has only 0 and 1 as roots. This led Boole to search out his set-theoretic interpretations for 0 and 1 which we have already observed.

By logical sum of two sets A and B—denoted $A \cup B$ and called "A cup B"—we mean the set whose members are members of the set A or the set B or both. Using A and B as in the last paragraph, the set $A \cup B$ would consist of all people who are males or who have blue eyes, including, of course, all blue-eyed men.

Boole's "logical sum" was a little different. His logical sum of sets x and y—denoted $x + y$, read "x plus y"—consisted of elements in x or y but not in both. Just as we agreed with Boole that $x^2 = x$, we might have expected him to agree with us that $1 + 1 = 1$ and $x + x = x$. But his logical sum $x + x$ is difficult to interpret. Whenever it occurred he gave it the formal designation $2x$; this caused him complications that need not concern us here.

The analogues of certain laws in ordinary algebra are seen to hold in Boolean algebra. For instance, $A \cap B = B \cap A$ is the commutative law for logical products. Also,

$$A \cap (B \cup C) = (A \cap B) \cup (A \cap C)$$

is an analogue of the distributive law. The correspondence of this law to the distributive law of ordinary algebra is especially obvious when we use Boole's symbols:

$$x(y + z) = xy + xz.$$

Another way of constructing a new set comes, not from combining two sets, but from complementation. If we remove from the universal set I all members of the set A, the remaining elements constitute a set called the complement of A and variously denoted by $I - A$, $-A$, A', and $\bar{A}$. By definition, $A \cup A' = I$, and $A \cap A' = 0$.

In the propositional calculus, letters stand for statements that may be true or false instead of for numbers (as in high school and college algebra) or for sets (as in Boolean algebra). To give some indication of the relation between Boolean algebra and the propositional calculus, we mention only that if A is "Roses are red" and B is "Violets are blue," then $A \wedge B$ is "Roses are red and violets are blue," and $A \vee B$ is "Either roses are red or violets are blue or both statements are true."

John Venn (1834–1923), a contemporary of Boole's and also an Englishman, invented a way of representing clearly such Boolean expressions as the right and left members of the distributive law. Similar diagrams had been invented independently by Leonhard Euler (these were called Euler circles) and by Augustus De Morgan and others /Sister Stephanie/. In Venn diagrams we draw a fence around all members of a set so as to exclude all nonmembers. Then the "area" common to the regions representing the two sets, the shaded area in Figure [**74**]-1, represents their logical product. Figure [**74**]-2 represents the case where $A \cap B = \emptyset$.

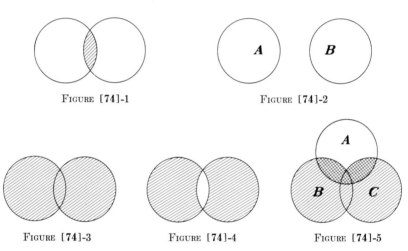

Figure [**74**]-1   Figure [**74**]-2

Figure [**74**]-3   Figure [**74**]-4   Figure [**74**]-5

The logical sum as defined by modern mathematicians would be represented by the shaded area in Figure [**74**]-3. However, according to Boole, the logical sum would be represented by the shaded area in Figure [**74**]-4. In order to find the Venn representation of $A \cap (B \cup C)$, the left member of our distributive law, we shade first the logical sum $B \cup C$, then its logical product with A, obtaining the doubly shaded area in Figure [**74**]-5.

Similar analysis of the right member of the distributive law yields the same set. Hence the two members are merely different names for the same set. (The reader might like to apply Venn diagrams to the other distributive law of Boolean algebra, $A \cup (B \cap C) = (A \cup B) \cap (A \cup C)$.)

The most interesting recent development in connection with Boolean algebra is its application to the design of electronic computers through the interpretation of Boolean combinations of sets as switching circuits. The logical product of two sets corresponds to a circuit with two switches in series. Electricity flows in such a circuit only if both the first and second switches are closed. The logical sum of two sets corresponds to a circuit with two switches in parallel. Electricity flows in such a circuit if either one or the other or both switches are closed.

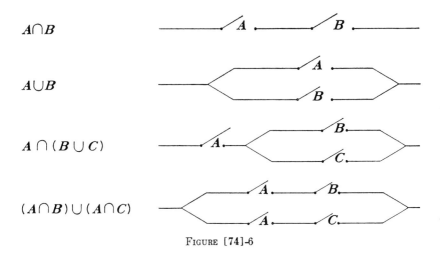

$A \cap B$

$A \cup B$

$A \cap (B \cup C)$

$(A \cap B) \cup (A \cap C)$

FIGURE [74]-6

In the last diagram of FIGURE [74]-6 the two A switches must be linked mechanically so that they are always both open or both closed. The last two circuits are equivalent (they correspond to identical sets by the distributive law); but the hardware for the first of these, $A \cap (B \cup C)$, is simpler.

For Further Reading

BELL (d): 433–47
BOOLE
CAJORI (d): II, 290

MIDONICK: 147–65, 774–85
NEWMAN: III, 1852–1931
SISTER STEPHANIE

CONGRUENCE (Mod m)

LET m be a fixed, positive integer. For arbitrary integers x and y we write $x \equiv y$ (mod m), read "x is congruent to y, modulo m," in case the integer $x - y$ is divisible by the integer m. The concept and notation were introduced by Carl Friedrich Gauss in 1801, when he was twenty-four years old. The integer m is called the modulus.

The property described above means that there exists an integer q such that $x - y = qm$, or (what is the same) $x = y + qm$. For every integer x, the long-division process guarantees the existence of integers q and r such that $x = qm + r$, $0 \leq r < m$. Since x is thus congruent to r, modulo m, it follows that (modulo m) each integer x is congruent to one and only one of the integers $0, 1, \cdots, m - 1$; this integer is called the "least residue" of x, modulo m.

From the definition one can readily prove that—

1. If $x \equiv y$ (mod m), $y \equiv z$ (mod m), then $x \equiv z$ (mod m).
2. If $x \equiv y$ (mod m), then $y \equiv x$ (mod m).
3. If $x \equiv y$ (mod m), $a \equiv b$ (mod m), then
   a) $x + a \equiv y + b$ (mod m).
   b) $x - a \equiv y - b$ (mod m).
   c) $xa \equiv yb$ (mod m).
   d) $x^k \equiv y^k$ (mod m), k any positive integer.
   e) $kx \equiv ky$ (mod m), k any integer.

It follows that if

$$f(x) = a_n x^n + a_{n-1} x^{n-1} + \cdots + a_0,$$

where x and all the coefficients a_i are integers, and if $x \equiv y$ and every $a_i \equiv b_i$, modulo m, then

$$f(x) \equiv b_n y^n + \cdots + b_0 \pmod{m}.$$

Although congruences form a vital tool in the theory of integers, Gauss recognized their utility, also, in showing certain polynomial equations to have no rational roots. Consider the equation

288

$$f(x) = x^n + a_{n-1} x^{n-1} + \cdots + a_0 = 0,$$

where all a_i are integers. All rational roots of $f(x)$ are known to be integers dividing the constant term, a_0; call the integral divisors of a_0 "potential roots" of $f(x) = 0$.

If r is actually an integral root of $f(x) = 0$, then $f(r) = 0$, whence $f(r) \equiv 0 \pmod{m}$ for every choice of the modulus m. In considering a potential root r, if in some manner we find a positive integer m such that $f(r) \not\equiv 0 \pmod{m}$, then we are assured that r is not a root of $f(x) = 0$. The value of this method for eliminating potential roots lies in the fact that calculating $f(r)$ to determine whether $f(r) = 0$ is often far more difficult than "calculating $f(r)$, modulo m." The latter phrase refers to the determination of the least residue of $f(r)$, modulo m; if this residue is not 0, then $f(r) \not\equiv 0 \pmod{m}$ and $f(r) \neq 0$.

It is convenient to use the same modulus m in checking all potential roots r, but this is not essential. In selecting m, one will never gain any knowledge from an m that is a factor of a_0 and of r, for then we always find $f(r) \equiv 0 \pmod{m}$.

An example is shown below.

$$f(x) = x^{14} + x^3 - x^2 + x + 6 = 0.$$

Potential roots are ± 6, ± 3, ± 2, and ± 1. We try the modulus $m = 5$, since this is the smallest positive integer not dividing 6. Note that in any congruence modulo 5, the term 6 may be replaced by 1. Thus

$$f(1) \equiv 1 + 1 - 1 + 1 + 1 \equiv 3 \pmod{5}$$

$$f(-1) \equiv 1 - 1 - 1 - 1 + 1 \equiv -1 \equiv 4 \pmod{5}$$

$$f(2) \equiv 2^{14} + 3 - 4 + 2 + 1 \equiv 2^{14} + 2 \pmod{5}.$$

Since

$$2^4 \equiv 1 \pmod{5},$$

it follows that

$$2^{14} \equiv 2^{10} \cdot 2^4 \equiv 2^{10} \equiv 2^6 \equiv 2^2 \pmod{5}$$

$$f(2) \equiv 4 + 2 \equiv 1 \pmod{5}$$

$$f(-2) \equiv 2^{14} - 8 - 4 - 2 + 1 \pmod{5}$$

$$\equiv 4 - 3 - 4 - 2 + 1 \equiv 1 \pmod{5}$$

$$f(3) \equiv 3^{14} + 2 - 4 + 3 + 1 \equiv 3^{14} + 2 \pmod{5}.$$

Then

$$3^4 \equiv 9 \cdot 9 \equiv 4 \cdot 4 \equiv 1 \pmod 5$$

$$3^{14} \equiv 3^{10} \equiv 3^6 \equiv 3^2 \equiv 4 \pmod 5$$

$$f(3) \equiv 4 + 2 \equiv 1 \pmod 5$$

$$f(-3) \equiv 3^{14} - 2 - 4 - 3 + 1 \pmod 5$$

$$\equiv 3^{14} - 3 \equiv 4 - 3 \equiv 1 \pmod 5.$$

Since

$$6 \equiv 1 \pmod 5$$

it follows that

$$6^i \equiv 1^i \pmod 5;$$

$$f(6) \equiv f(1) \equiv 3 \pmod 5.$$

Similarly,

$$-6 \equiv -1 \pmod 5;$$

$$f(-6) \equiv f(-1) \equiv 4 \pmod 5.$$

In every case the least residue fails to be 0. Thus no potential root is an actual root, whence $f(x) = 0$ has no rational roots.

For Further Reading

MIDONICK: 380–86   STRUIK (e): 49–54

ORE (c): 209–33

Capsule 76   Eugene W. Hellmich

COMPLEX NUMBERS
(THE STORY OF $\sqrt{-1}$)

HISTORY shows the necessity for the invention of new numbers in the orderly progress of civilization and in the evolution of mathematics.

The story of $\sqrt{-1}$, the imaginary unit, and of $x + yi$, the complex number, originates in the logical development of algebraic theory.

Deploring the use of the word "imaginary" by calling it "the great algebraic calamity" but "too well established for mathematicians to eradicate" is quite proper from the modern point of view; but the use of this word reflects the elusive nature of the concept for distinguished mathematicians who lived centuries ago.

Early consideration of the square root of a negative number brought unvarying rejection. It seemed obvious that a negative number is not a square, and hence it was concluded that such square roots had no meaning. This attitude prevailed for a long time.

Perhaps the earliest encounter with the square root of a negative number is in the expression $\sqrt{81 - 144}$, which appears in the *Stereometrica* of Heron of Alexandria (*c.* A.D. 50); the next known encounter is in Diophantus' attempt to solve the equation $336x^2 + 24 = 172x$ (as we would now write it), in whose solution the quantity $\sqrt{1{,}849 - 2{,}016}$ appears (again using modern notation).

The first clear statement of difficulty with the square root of a negative number was given in India by Mahavira (*c.* 850), who wrote: "As in the nature of things, a negative is not a square, it has no square root." Nicolas Chuquet (1484) and Luca Pacioli (1494) in Europe were among those who continued to reject imaginaries.

Girolamo Cardano (1545), who is also known as Jerome Cardan, is credited with some progress in introducing complex numbers in his solution of the cubic equation, even though he regarded them as "fictitious." He is credited also with the first use of the square root of a negative number in solving the now-famous problem, "Divide 10 into two parts such that the product $\cdots$ is 40," which Cardano first says is "manifestly impossible"; but then he goes on to say, in a properly adventurous spirit, "Nevertheless, we will operate." (This was due, no doubt, to his medical training!) Thus he found $5 + \sqrt{-15}$ and $5 - \sqrt{-15}$ and showed that they did indeed have a sum of 10 and a product of 40.

Cardano concludes by saying that these quantities are "truly sophisticated" and that to continue working with them would be "as subtle as it would be useless."

Cardano did not use the symbol $\sqrt{-15}$. His designation was "R_x. m," that is, "radix minus," for the square root of a negative number. Rafael Bombelli (*c.* 1550) used "$d.m$" for our $\sqrt{-1}$. Albert Girard (1629) included symbolism such as "$\sqrt{-2}$." René Descartes (1637) contributed the terms "real" and "imaginary." Leonhard Euler (1748) used "i" for $\sqrt{-1}$. Caspar Wessel (1797) used "$\sqrt{-1} = \epsilon$." Carl Friedrich

Gauss (1832) introduced the term "complex number." William Rowan Hamilton (1832) expressed the complex number in the form of a number-couple.

Bombelli continued Cardano's work. From the equation $x^2 + a = 0$, he spoke of "$+ \sqrt{-a}$" and "$- \sqrt{-a}$." The special case of this equation, $x^2 + 1 = 0$, affords an excellent approach to i and i^2, as follows: If $x^2 + 1 = 0$, then $x^2 = -1$ and $x = \pm \sqrt{-1}$. Now, if $i = \sqrt{-1}$, then $i^2 + 1 = 0$ when x is replaced by i, and $i^2 = -1$. From this it follows as a good exercise that $i^3 = -\sqrt{-1}$, $i^4 = 1$, $i^5 = \sqrt{-1}$, $\cdots$, $i^{243} = -\sqrt{-1}$, and so forth.

In his *Algebra* (1673, republished in 1693 in *Opera mathematica*; see /D. E. SMITH (c): I, 48/) John Wallis associated "—1600 square perches" with a loss and then supposed this to be in the form of a square with a side [160 square perches = 1 English acre]:

> What shall this side be? We cannot say it is 40, nor that it is −40. (Because either of these multiplied into itself, will make +1600; not −1600). But thus rather, that it is $\sqrt{-1600}$, (the Supposed Root of a Negative Square:) or (which is equivalent thereunto) $10\sqrt{-16}$, or $20\sqrt{-4}$, or $40\sqrt{-1}$.

Wallis, Wessel (1798), Jean Robert Argand (1806), Gauss (1813), and others made significant contributions to the understanding of complex numbers through graphical representation, and in 1831 Gauss defined complex numbers as ordered pairs of real numbers for which $(a, b) \cdot (c, d) = (ac - bd, ad + bc)$, and so forth. Wessel's representation is given as follows /D. E. SMITH (c): I, 60/:

> Let +1 designate the positive rectilinear unit and $+\epsilon$ a certain other unit perpendicular to the positive unit and having the same origin; then the direction angle of +1 will be equal to 0°, that of −1 to 180°, that of $+\epsilon$ to 90°, and that of $-\epsilon$ to −90° or 270°. By the rule that the direction angle of the product shall equal the sum of the angles of the factors, we have: $(+1)(+1) = +1$; $(+1)(-1) = -1$; $(-1)(-1) = +1$; $(+1)(+\epsilon) = +\epsilon$; $(+1)(-\epsilon) = -\epsilon$; $(-1)(-\epsilon) = +\epsilon$; $(+\epsilon)(+\epsilon) = -1$; $(+\epsilon)(-\epsilon) = +1$; $(-\epsilon)(-\epsilon) = -1$. From this it is seen that ϵ is equal to $\sqrt{-1}$, and the divergence of the product is determined such that not any of the common rules of operation are contravened.

Of a similar representation it has been said /BELL (d): 234/:

> All this of course proves nothing. *There is nothing to be proved*; we *assign*

to the symbols and operations of algebra *any meanings whatever* that will lead to consistency. Although the *interpretation* ··· *proves* nothing, it may suggest that there is no occasion for anyone to muddle himself into a state of mystic wonderment over nothing about the grossly misnamed "imaginaries."

A geometric representation credited to Wessel and Argand independently is based on the geometric principle that the altitude to the hypotenuse of a right triangle is a mean proportional between the segments into which the altitude divides the hypotenuse. In Figure [76]-1, $OD_1 = d_1 = +1, OD_2 = d_2 = -1$. $\angle D_1RD_2$ is a right angle, and $OR = d$. Then $d_1 : d = d : d_2$. Now $d = \sqrt{d_1 d_2} = \sqrt{+1 \cdot -1} = \sqrt{-1} = i$.

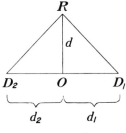

FIGURE [76]-1

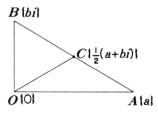

FIGURE [76]-2

Some interesting geometric proofs can result from the representation of the complex number $a + bi$ by the point in the plane with rectangular coordinates a and b. An example is the proof that the midpoint of the hypotenuse of a right triangle is equidistant from the three vertices. In Figure [76]-2 O is the vertex of the right angle of right triangle AOB, and C is the midpoint of the hypotenuse AB. Using the coordinates in the figure,

$$OC = |\tfrac{1}{2}(a + bi) - 0| = \tfrac{1}{2}|(a + bi)| = \tfrac{1}{2}\sqrt{a^2 + b^2},$$

and

$$AB = |a - bi| = \sqrt{a^2 + b^2}.$$

Hence

$$OC = \tfrac{1}{2} AB = BC = CA.$$

Lastly, among the more valuable relations involving the imaginary, is that suggested by Abraham De Moivre (1730):

$$(\cos \theta + i \sin \theta)^n = \cos n\theta + i \sin n\theta.$$

An illustration of De Moivre's relation in the development of trigonometric identities follows /JONES (c)/:

(1) $$(\cos \theta + i \sin \theta)^3 = \cos 3\theta + i \sin 3\theta.$$

But by the binomial theorem we have

(2) $$(\cos \theta + i \sin \theta)^3$$
$$= \cos^3 \theta + 3i \cos^2 \theta \sin \theta + 3i^2 \cos \theta \sin^2 \theta + i^3 \sin^3 \theta.$$

Equating the right-hand members of (1) and (2), we have

$$\cos 3\theta + i \sin 3\theta = \cos^3 \theta - 3 \cos \theta \sin^2 \theta + i(3 \cos^2 \theta \sin \theta - \sin^3 \theta).$$

Equating the real parts gives

$$\cos 3\theta = \cos^3 \theta - 3 \cos \theta \sin^2 \theta$$
$$= 4 \cos^3 \theta - 3 \cos \theta.$$

Equating the imaginary parts gives

$$\sin 3\theta = 3 \cos^2 \theta \sin \theta - \sin^3 \theta$$
$$= -4 \sin^3 \theta + 3 \sin \theta.$$

For Further Reading

BELL (d): 233–34
JONES (c)
MIDONICK: 804–14

D. E. SMITH (a): II, 261–67
——— (c): I, 46–66

294

QUATERNIONS

VECTORS are objects that can be added or subtracted, and multiplied amongst themselves; they can also be multiplied by real numbers. In each case the result is another vector.

William Rowan Hamilton (who was born in Dublin in 1805 and was appointed professor of astronomy at Trinity College, Dublin, in 1827) was disturbed by the lack of any concept of *quotient* of vectors. That is, for any two vectors u and v, with $v \neq 0$, he wanted to find a unique vector q such that the vector product qv was equal to u. His investigations showed the system of vectors to be too small for this purpose and led him to an enlarged system whose members he called "quaternions." His work stirred up considerable disputation throughout the Western world on the question whether quaternions should replace vectors as an everyday tool in physics and mathematics, and it resulted in the formulation of an international association to study the question. We shall look briefly at the way in which Hamilton was led to quaternions.

Consider a rectangular coordinate system with axes X, Y, and Z and with unit vectors i, j, and k drawn on these respective axes. All vectors used herein emanate from the origin, whence the vector terminating in the point (x, y, z) is $xi + yj + zk$. Let u and v be two vectors; $u = ai + bj + ck$, and $v = di + ej + fk \neq 0$. We shall consider a process for converting v into u, meanwhile counting how many real numbers are needed to specify the process completely in the general case.

First, vectors v and u determine a plane π. Imagine a movable vector v_0, which initially lies on top of v. In the plane π we rotate v_0 until it lies on the ray containing vector u, the angle of rotation being designated as δ.

This number δ does not determine our rotation, since we are not content to rotate v_0 through *any* angle equal to δ, but only through an angle δ lying in the appropriate plane. We therefore consider what numbers may serve to specify the particular plane π through the origin. If a movable plane π_0 is pictured initially as coinciding with

295

the XZ plane, we may rotate π_0 about the Z-axis until it includes v; next, we rotate π_0 about v until it contains the vector u. In this final position the plane π_0 coincides with π; and if γ and β are the two angles employed, the triple (β, γ, δ) determines the contemplated rotation carrying the ray of vector v to that of u.

The length of v may very well differ from that of u, so we now multiply the former length by some constant α to convert it to that of u. Altogether four constants, α, β, γ, and δ, serve to convert v to u. To express this "fourness" Hamilton coined the name "quaternion" for whatever algebraic object he could find to accomplish the desired conversion.

It turned out that his purposes were served admirably by the notation

$$(1) \qquad w = \alpha_0 + \alpha_1 i + \alpha_2 j + \alpha_3 k,$$

where α_0, α_1, α_2, and α_3 are arbitrary real numbers. These symbols are to be combined under addition and subtraction by the usual rules. For example, if w' is given by

$$(2) \qquad w' = \beta_0 + \beta_1 i + \beta_2 j + \beta_3 k,$$

then both $w + w'$ and $w' + w$ are equal to

$$(\alpha_0 + \beta_0) + (\alpha_1 + \beta_1)i + (\alpha_2 + \beta_2)j + (\alpha_3 + \beta_3)k.$$

The product ww' is defined by use of the usual distributive laws of algebra together with the following stipulations: $ij = k$; $ji = -k$; $jk = i$; $kj = -i$; $ki = j$; $ik = -j$; and $i^2 = j^2 = k^2 = -1$. Thus, for

$$(3) \qquad w = 1 + 2i + 3j + 4k; \qquad w' = 2 + i + 5k,$$

we find that

$$ww' = 2 + 4i + 6j + 8k$$
$$+ i - 2 + 3ji + 4ki$$
$$+ 5k + 10ik + 15jk - 20$$
$$= -20 + 20i + 10k.$$

A similar computation shows that $w'w = -20 - 10i + 12j + 16k$.

Since $ww' \neq w'w$ in the computations above, the commutative law of multiplication is not valid for quaternions. Another instance is given by the equations $ij = k$, $ji = -k$. For special pairs w and w', however, the product may be commutative. This is the case, for example, if w is arbitrary and $w' = \beta_0 + 0i + 0j + 0k = \beta_0$; $ww' = w'w = \beta_0 w$.

The system of quaternions so constructed includes the familiar vectors $ai + bj + ck$; and when the laws of quaternion addition and multiplication are applied to these vectors, the usual results are obtained except that products now consist of a real term plus the usual vector product. But now we shall demonstrate that an additional feature is present: Every nonzero vector—also every nonzero quaternion —has an inverse in the system of quaternions.

The quaternion w in (1) is 0 if and only if all of its coefficients are 0. Let $w \neq 0$, whence the number

$$(4) \qquad p = \alpha_0^2 + \alpha_1^2 + \alpha_2^2 + \alpha_3^2$$

is a real, positive number. If we write w as

$$w = \alpha_0 + v, \qquad v = \alpha_1 i + \alpha_2 j + \alpha_3 k,$$

then v is called the "vector part" of w; and $\bar{w} = \alpha_0 - v$ is called the "conjugate" of w. Note that the conjugate of $\bar{w}$ is w. The norm of w is defined to be $w\bar{w}$. A short computation shows that both w and $\bar{w}$ have norm equal to the number p in (4): $w\bar{w} = p = \bar{w}w$. It follows that $w((1/p)\bar{w}) = 1 = ((1/p)\bar{w})w$, whence $(1/p)\bar{w}$ is the inverse of w. For the quaternion w in (3) the inverse is $(1/30)(1 - 2i - 3j - 4k)$.

As a standard device for everyday use in physics, quaternions have disappeared entirely. They are, however, very much alive now with a different raison d'être. Today mathematicians are interested in studying number systems in their entirety, in learning their properties, and in learning how to construct new ones. One prominent type is called an associative division algebra over a field. It is known that there are only three such algebras over the real field: (1) the real number system, (2) the complex number system, and (3) the system of quaternions. Thus the system of quaternions may be designated as the only noncommutative associative division algebra over the real field.

The noncommutativity of quaternion multiplication gives rise to a curious property. An equation of degree n can no longer be said to have at most n distinct roots, at least not if quaternion solutions are admitted. For example, the quadratic equation $w^2 + 1 = 0$ has three obvious quaternion solutions: $w = i$, $w = j$, and $w = k$. In actuality there are infinitely many. It is easy to verify that $w = \alpha_0 + \alpha_1 i + \alpha_2 j + \alpha_3 k$ satisfies the condition $w^2 + 1 = 0$ if and only if $\alpha_0 = 0$ and $\alpha_1^2 + \alpha_2^2 + \alpha_3^2 = 1$.

For Further Reading

Bell (a): 182–211   Boyer (g): 624–26
——— (d): 340–61   D. E. Smith (c): II, 677–83

Capsule 78   Gertrude V. Pratt

EARLY GREEK ALGEBRA

The algebra of the early Greeks (of the Pythagoreans and Euclid, Archimedes, and Apollonius, 500–200 B.C.) was geometric because of their *logical* difficulties with irrational and even fractional numbers and their *practical* difficulties with Greek numerals [4], which were somewhat similar to Roman numerals and just as clumsy. It was natural for the Greek mathematicians of this period to use a geometric style for which they had both taste and skill.

The Greeks of Euclid's day thought of the product ab (as we write it) as a rectangle of base b and altitude a, and they referred to it as "the rectangle contained by CD and DE" (Fig. [78]-1).

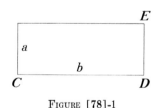

FIGURE [78]-1

To illustrate the style and method of Greek geometric algebra we show how they solved a particular kind of quadratic equation. The theorem—in this case, really a problem to be solved—is given in Euclid's own words /I, 402/; and the "proof" (a construction of the positive root of the equation, followed by a verification) is almost step by step the same as that given by Euclid. Book II, Proposition 11, is as follows:

298

To cut a given straight line so that the rectangle contained by the whole and one of the segments is equal to the square on the remaining segment. [Find H so that $a(a - x) = x^2$; in other words, find the positive root x (or AH) of the quadratic equation $x^2 + ax - a^2 = 0$.]

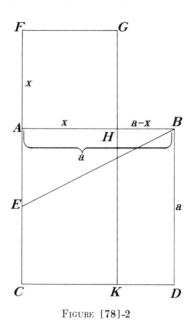

FIGURE [78]-2

AB, or a, is the given segment (Fig. [78]-2). Construct square $ABDC$. Bisect AC at E. Draw EB. Extend CA to F so that $EF = EB$. Construct square $FGHA$. Then H is the required point (so that $x = AH$ is the positive root of $x^2 + ax - a^2 = 0$).

Verification follows, modern notation being used in the right-hand column.

By an earlier proposition (II, 6)	Prop. II, 6 is a form of the identity
$$CF \cdot FG + AE^2 = EF^2.$$	$$(\alpha + \beta)(\alpha - \beta) + \beta^2 \equiv \alpha^2 \qquad (1)$$
	or
	$$(\alpha + \beta)(\alpha - \beta) \equiv \alpha^2 - \beta^2,$$
	where, in the present context

$$\alpha = x + \frac{a}{2} \quad \text{and} \quad \beta = \frac{a}{2}$$

so that

$$\alpha + \beta = a + x \quad \text{and} \quad \alpha - \beta = x.$$

By construction $EF = EB$; hence

Hence (1) gives

$$CF \cdot FG + AE^2 = EB^2.$$

$$(a + x)(x) + \left(\frac{a}{2}\right)^2 = \left(x + \frac{a}{2}\right)^2.$$

By the Pythagorean theorem,

By the Pythagorean theorem,

$$CF \cdot FG + AE^2 = AB^2 + AE^2$$

$$(a + x)(x) + \left(\frac{a}{2}\right)^2 = a^2 + \left(\frac{a}{2}\right)^2$$

$$- AE^2 = \qquad - AE^2$$

$$- \left(\frac{a}{2}\right)^2 = \qquad - \left(\frac{a}{2}\right)^2$$

$$\overline{CF \cdot FG \qquad = AB^2}$$

$$\overline{(a + x)(x) \qquad = a^2}$$

$$- AHKC = \qquad - AHKC$$

$$- ax \quad = \qquad - ax$$

$$\overline{AH^2 \qquad = DB \cdot HB}$$

$$\overline{x^2 \qquad = a(a - x).}$$

or

$$AH^2 \qquad = AB \cdot HB. \qquad (2)$$

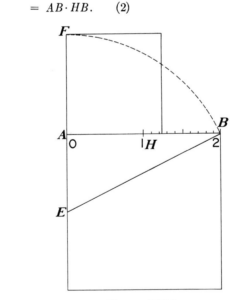

FIGURE [78]-3

$AH^2 = AB \cdot HB.$ (2)

Hence H is the required point (so that AH, or x, satisfies the condition (2)).

As an example, see Figure [78]-3.

Let $AB = a = 2$ to get the quadratic equation $x^2 + 2x - 4 = 0$. Carrying through the above construction we find that $AH = x \approx 1.236$, which agrees with the positive root obtained from the quadratic formula, $x = -1 + \sqrt{5}$.

For Further Reading

AABOE (b): 61–65   EVES (c): 64–69

EUCLID   [5th ed. 57–62]

EVANS   VAN DER WAERDEN: 118–26

Capsule 79   **Ferna E. Wrestler**

HINDU ALGEBRA

THE Hindu work on astronomy *Surya Siddhanta* ("Knowledge from the Sun"), written around A.D. 500, provided the motivation for a remarkable development of arithmetic and algebra in India as shown by the works of Aryabhata (*c.* 525), Brahmagupta (**628**), Mahavira (*c.* 850), and Bhaskara (1150). After Bhaskara, Hindu mathematics showed no progress until modern times.

Brahmagupta gave an interesting rule for finding one of the two positive roots of the quadratic equation $x^2 - 10x = -9$ (using modern notation), which in the original is written as shown here:

$$ya \; v \; 1 \; ya \; \overset{.}{1}0$$

$$ru \; \overset{.}{9}$$

In this, ya is the unknown; v means "squared"; the dot above a number indicates that it is a negative number. The left-hand member of the equation (as we would describe it) is written on one line and the

301

right-hand member beneath; *ru* means "absolute" ("plain") number.

The three columns below give the solution as translated /D. E. SMITH (a): II, 445/, then in modern notation and with a generalization for $ax^2 + bx = c$.

$$x^2 - 10x = -9. \qquad ax^2 + bx = c.$$

Here absolute number (9) multiplied by (1) the [coefficient of the] square [is] (9)

$$(-9)(1) = -9 \qquad (c)(a) = ca$$

and added to the square of half the [coefficient of the] middle term, 25, makes 16;

$$-9 + \left(\frac{-10}{2}\right)^2 = 16 \qquad ca + \left(\frac{b}{2}\right)^2$$

of which the square root 4, less half the [coefficient of the] unknown (5), is 9;

$$\sqrt{16} - \left(\frac{-10}{2}\right) = 9 \qquad \sqrt{ca + \left(\frac{b}{2}\right)^2} - \frac{b}{2}$$

and divided by the [coefficient of the] square (1) yields the value of the unknown 9.

$$\frac{9}{1} = 9. \qquad \frac{\sqrt{ca + \left(\frac{b}{2}\right)^2} - \frac{b}{2}}{a} = x$$

or

$$x = \frac{-b + \sqrt{b^2 + 4ac}}{2a}.$$

The method used in the above example is essentially the same as our present method of "completing the square" and consists of adding the shaded area $(b/2)^2$ of Figure [79]-1 to the unshaded area

$$(a^2x^2 + abx) + \left(\frac{b}{2}\right)^2,$$

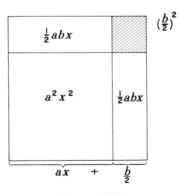

FIGURE [79]-1

which gives the whole area:

$$\underbrace{(a^2x^2 + abx)}_{ca} + \left(\frac{b}{2}\right)^2.$$

Since it was given that $ax^2 + bx = c$, put ca for $a^2x^2 + abx$; add $(b/2)^2$ to equal

$$\left(ax + \frac{b}{2}\right)^2.$$

Hence the side, $ax + (b/2)$, of the large (completed) square is

$$\sqrt{ca + \left(\frac{b}{2}\right)^2};$$

and ax is $b/2$ less than

$$\sqrt{ca + \left(\frac{b}{2}\right)^2}.$$

Finally, divide by a to obtain x.

The example given shows that Hindu algebra was largely verbal (rhetorical), although in the statement of the problem use is made of abbreviations, illustrating the so-called syncopated style. Especially noteworthy is the correct use of negative numbers, written by placing a dot above the number. Imaginary numbers escaped the Hindus, who, however, at least recognized them as rating a comment: "as in the nature of things, a negative is not a square, it has therefore no square root" (Mahavira). They operated freely with irrational numbers and used the identity that would be written in modern notation as

$$\sqrt{a \pm \sqrt{b}} = \sqrt{\frac{a + \sqrt{a^2 - b}}{2}} \pm \sqrt{\frac{a - \sqrt{a^2 - b}}{2}}.$$

They realized that a quadratic equation with real roots would have two roots, but they did not always bother to find both roots, as we have seen. Negative roots were discarded as "inadequate."

The Hindus worked with arithmetic and geometric progressions, permutations, and linear equations; and they could solve some equations of degree higher than two.

The Hindus made their greatest progress in indeterminate analysis. For an equation $ax + by = c$ (a, b, and c integers) with an integral solution, they could determine the solution by continued fractions, a method that is still used. After finding one solution, $x = p$, $y = q$, they found others by using $x = p + bt$, $y = q - at$ for any integer t. Likewise, if one pair of integers p and q could be found to satisfy a so-called Pell equation, $y^2 = ax^2 + 1$ (a an integer that is not a square), they could find more by using the following property /CAJORI (e): 95/: "If p and q is one set of values of x and y, and p' and q' is the same or a different set, then $qp' + pq'$ and $app' + qq'$ is another solution." A problem from Bhaskara's works is this: "What square number multiplied by 8 and having 1 added shall be a square?" One solution of the equation $8x^2 + 1 = y^2$ is $x = 6$, $y = 17$, from which it is readily seen that $x = 204$, $y = 577$ is another.

Of interest is Bhaskara's solution of a problem on right triangles. The problem is given as follows /SCOTT: 73/: "The hypotenuse being 85, say, learned men, what upright sides will be rational?" (In modern symbolism, "Find rational values of x and y if $x^2 + y^2 = h^2$.") The solution is given below, with modern symbolism at the right.

Double the hypotenuse.	170	$2h$
Multiply by an arbitrary number, say 2.	340	$2ah$
Divide by the square of the arbitrary number increased by 1.	$\dfrac{340}{5}$	$\dfrac{2ah}{a^2 + 1}$
This gives one side.	68	
Multiply by the arbitrary number, 2.	136	$\dfrac{2a^2h}{a^2 + 1}$
Subtract the hypotenuse.	$136 - 85$	$\dfrac{2a^2h}{a^2 + 1} - h$
This gives the other side.	51	$\dfrac{h(a^2 - 1)}{a^2 + 1}$

This is equivalent to saying that the three sides of a right triangle are proportional to $a^2 + 1$, $2a$, $a^2 - 1$; and the values are not unique but depend on the choice of the arbitrary number a.

The following problem is typical of this period: "The horses belonging to four persons are 5, 3, 6, 8, respectively. The camels pertaining to the same are 2, 7, 4, 1. The mules belonging to them are 8, 2, 1, 3, and the oxen 7, 1, 2, 1. All four persons being equally rich, tell me the price of each horse and the rest." (One solution is horses, 85; camels, 76; mules, 31; oxen, 4.)

For Further Reading

CAJORI (e): 83–98
EVES (c): 181–91
  [5th ed. 161–71]
MIDONICK: 116–40, 166–80

SCOTT: 66–80
D. E. SMITH (a): I, 152–64, 274–81

Capsule 80   Cecil B. Read

ARABIC ALGEBRA, 820-1250

LITTLE is known about Arabian history before the time of Mohammed (570–632). Mohammed was instrumental in forming a powerful nation that eventually extended over parts of India, Persia, Africa, and Spain. Baghdad was the Eastern intellectual center, and Cordova, in Spain, the Western.

The rulers, called caliphs, supported scientific research. The Arabs, conquering Egypt, acquired some Greek masterpieces from the Alexandrian library. Conquering part of India, they came in contact with the Hindus. Works of Hindu mathematicians were translated, and Hindu numerals entered Arabia. Greek mathematical works, including Euclid's *Elements* and the writings of Archimedes, Heron, Ptolemy, Apollonius, and Diophantus, were also translated into Arabic. Often Arabic translations of Hindu and Greek works are the only known copies.

Arabic algebra came from both the Hindus and the Greeks. The Arabs treated algebra numerically like the Hindus and geometrically like the Greeks.

The early Arabs wrote out problems entirely in words. After contact with other peoples, symbols and Hindu numerals were gradually introduced; but later Arabian writers reverted to writing out problems completely, showing perhaps the influence of Greek methods.

305

Possibly the greatest of the Arabic mathematical writers was al-Khowarizmi (*c.* 825), although some think his algebra shows little originality. He used a type of "transposition" that is not found in Hindu or Greek works, and he seems to have been the first to collect like powers of the unknown. These may be his original ideas. He solved linear and quadratic equations, both numerically and geometrically. He recognized the existence of negative roots (as the Hindus also did) but consciously rejected them.

The original Arabic edition of al-Khowarizmi's best-known work, *Hisab al-jabr w'al muqabalah,* is lost, but a Latin translation exists (dating from the twelfth century). One translation of the title is "The Science of Transposition and Cancellation." The book became known as *Al-jabr,* from which we get our word "algebra." Subsequent Arabic and medieval algebras were based on al-Khowarizmi's work.

The following example shows, in al-Khowarizmi's own words (as translated /D. E. SMITH (a): II, 447/), how he found the positive root of the quadratic equation that we would write as $x^2 + 10x = 39$. The second column shows this in numerical values, and the third gives a generalization for $x^2 + px = q$.

$$x^2 + 10x = 39. \qquad x^2 + px = y.$$

You halve the number of roots, which in the present instance yields five.	$\frac{1}{2}(10) = 5$	$\frac{p}{2}$
This you multiply by itself; the product is twenty-five.	$5 \cdot 5 = 25$	$\left(\frac{p}{2}\right)^2$
Add this to thirty-nine; the sum is sixty-four.	$25 + 39 = 64$	$\left(\frac{p}{2}\right)^2 + q$
Now take the root of this, which is eight,	$\sqrt{64} = 8$	$\sqrt{\left(\frac{p}{2}\right)^2 + q}$
and subtract from it half the number of the roots, which is five; the remainder is three.	$8 - \frac{10}{2} = 3$	$\sqrt{\left(\frac{p}{2}\right)^2 + q} - \frac{p}{2} = x$
This is the root of the square which you sought for; the square itself is nine.		$x = \dfrac{-p + \sqrt{p^2 + 4q}}{2}.$

or

The method used is essentially the same as our present-day method of "completing the square" and consists literally of adding the shaded

306

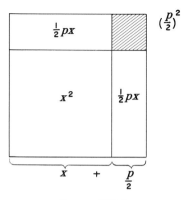

FIGURE [80]-1

area $(p/2)^2$ of Figure [80]-1 to the unshaded area $(x^2 + px)$ to complete the square of side $(x + (p/2))$. That is,

$$\left(\frac{p}{2}\right)^2 + (x^2 + px) = \left(x + \frac{p}{2}\right)^2;$$

but since it was given that $(x^2 + px) = q$, $(p/2)^2 + q = (x + p/2)^2$. Hence the side $(x + (p/2))$ of the completed square is equal to

$$\sqrt{\left(\frac{p}{2}\right)^2 + q};$$

and x is $p/2$ less than that quantity.

Notice that the *other* root, -13, of the equation $x^2 + 10x = 39$ was ignored because it is negative. If both roots had been positive they would probably have both been found.

Abu Kamil (*c.* 900) wrote a more extensive treatise on algebra. It was so good that later writers used much of it, although without mentioning his name. The methods were well known and considered to be common property. He used both the terms "square" and "root." The Greeks thought of 5 as the side of a square with area 25; the Arabs, following the Hindus, thought of 25 as growing, like a tree, out of the number 5 as a root. Both concepts appear in "square root." The Latin word for "root" is *radix*; from it comes our word "radical."

Like others, Abu Kamil solved equations algebraically and geometrically. He classified quadratic equations into six types, presenting no general method. To give a single example indicating that he did work of more than elementary difficulty: he showed, without using the modern notation employed here, the equality

$$\sqrt{12 \pm 2\sqrt{20}} = 10 \pm \sqrt{2}.$$

One of the best Arabic algebraists was Omar Khayyam (*c.* 1100), known usually only as the author of the *Rubaiyat*. He used geometric algebra, solving cubic equations by finding the intersections of conics. Some think that this was the greatest achievement in Arabic algebra. Omar Khayyam thought the cubic was insolvable by purely algebraic means.

His method of solving $x^2 + 10x = 39$ (as we write it) was really the same as al-Khowarizmi's, but we state it because of its historical interest: "Multiply half of the root by itself; add the product to the number and from the square root of this sum subtract half the root. The remainder is the root [side] of the square." /D. E. SMITH (a): II, 447./ It is not a coincidence that the same numbers (10 and 39) appear in the two examples. This particular problem was a favorite in the Arab schools of that time.

Note especially that Arab mathematicians would not have thought of the above example in our customary form, $x^2 + 10x - 39 = 0$, because they simply did not grasp negative numbers; this difficulty with negative numbers and the subtleties of zero products probably explains why solution by factoring came rather late (in the time of Thomas Harriot, 1631).

Some work was done with indeterminate equations by al-Karkhi (*c.* 1020), who tended to follow the style of the Greek mathematician Diophantus. As one problem he proposes this: "Find rational numbers x, y, and z such that $x^3 + y^3 = z^2$."

Arabic algebra used the rules of false position and of double false position [**90**]. They explained the rule of three, which today we call proportion. The Hindu mathematicians had used the terms, and the Arabs translated directly.

Some historians think the Arabs added little that was new, but all agree that throughout the Dark Ages the Arabs preserved the Greek and Hindu works for posterity. Without their translations, most of this prior work would be lost.

It was principally through the Arabs that algebra entered Europe. Hindu influence dominated, hence algebra came to Europe with little axiomatic foundation. Perhaps this explains why, until quite recently, geometry was based on postulates and theorems while elementary algebra emphasized method rather than logical foundations.

For Further Reading

BOYER (g): 249–69   ——— (e)
CAJORI (e): 99–112   KHOWARIZMI
CARNAHAN (a)   MIDONICK: 418–34
COOLIDGE (c): 19–29   OMAR KHAYYAM
EVES (c): 191–95   D. E. SMITH (a): II, 446–48
[5th ed. 171–78]   STRUIK (e): 55–60

Capsule 81   Sister M. Stephanie Sloyen

ALGEBRA IN EUROPE, 1200-1850

EUROPEAN algebra was based directly upon Arabic algebra and developed rather slowly from what might be termed its beginning,
around 1200, until the nineteenth century, when discoveries followed
closely upon one another.

Much of the early work was done in Italy. There Fibonacci
(Leonardo of Pisa) did a great deal to popularize Hindu-Arabic numerals with his book on arithmetic and algebra, *Liber abaci* ("Book
of Calculation"), written in 1202. This book also contains the famous
Fibonacci sequence: $1, 1, 2, 3, 5, 8, 13, \ldots$ [22].

For the next few centuries there was very little algebraic activity
in Europe; however, during the period from 1515 to 1545 it was again
Italy that produced the algebraists. During that time many mathematics books were published in Italy, although mathematicians
did not send their disoveries to journals for publication. They
preferred to use their new knowledge in order to shine in public con-
tests, challenging one another in problem solving. Scipione del Ferro,
a professor at the University of Bologna, in 1515 devised a method of
solving the cubic equation $x^3 + bx = c$, but he did not circulate his
work. Niccolo Tartaglia solved the cubic equation $x^3 + ax^2 = c$ and
then also the cubic $x^3 + bx = c$ (about 1535) and used his information
in order to vanquish challengers [71]. Girolamo Cardano, a physician
and mathematician who was called the "gambling scholar" by Oystein
Ore /(a)/, obtained the solution from Tartaglia and made many improvements on Tartaglia's solution, solving (at least for positive roots)

all possible cases except the "irreducible" one. He then published the complete solution of all varieties of the cubic equation (except the irreducible case involving "imaginaries") in his *Ars magna,* giving full credit to Tartaglia. It was Ludovico Ferrari who successfully solved the general quartic equation. (Rafael Bombelli, a sixteenth-century Bolognese mathematician, made progress on the irreducible case of the cubic by recognizing in 1572 that *apparently* imaginary expressions like

$$\sqrt[3]{81 + 30\sqrt{-3}} + \sqrt[3]{81 - 30\sqrt{-3}}$$

were real—in this case, -6.)

Algebraists of the seventeenth century include Thomas Harriot, an Englishman who introduced the signs $<$ and $>$ /C. SMITH; EVES (a)/ and the use of aa for what we call a^2 and aaa for a^3. While we may think this awkward, it is an improvement over the A *cubum* of François Viète or even the *res cubum* of earlier times. William Oughtred, another Englishman, was responsible for the slide rule, the multiplication sign $\times$, and the sign $::$ for proportion.

René Descartes, a Frenchman, was one of the greatest mathematicians of this century and a prolific writer. His outstanding contribution was, of course, his work on plane analytic geometry, but he also improved the symbolism of algebra and introduced our present system of positive, integral exponents. A large part of Descartes's *La géométrie* consists of what we now call "theory of equations," and it contains Descartes's rule of signs for determining the number of positive and negative roots of an equation. Descartes used the letters at the end of the alphabet, . . . , x, y, z, for variables, and the early letters, a, b, c, . . ., for constants. Viète, in the sixteenth century, had used vowels for variables and consonants for constants.

Pierre de Fermat's work in the seventeenth century in France was chiefly in number theory; theorems in Diophantine analysis (of which he left no proof) are due to him. Isaac Newton, genius in many fields and inventor of the calculus, discovered the binomial theorem in 1664 when he was twenty-two. The theory of symmetric functions of the roots of an equation, first perceived by Viète, was firmly established by Newton, who also gave a method for finding approximations to the roots of numerical equations.

In the nineteenth century mathematicians began to work in specialized fields, but Carl Friedrich Gauss was an exception to this rule. In his doctoral dissertation, written when he was twenty and published in **1799**, he gave the first rigorous proof of the fundamental theorem of algebra: *Every algebraic equation of degree n has a root* [and hence *n* roots]. Later he published three more proofs of the same theorem. It was he who called it "fundamental." Much of the work on complex number theory is Gauss's. He was one of the first to represent complex numbers as points in a plane. From **1807** until his death in **1855**, Gauss was director and professor of astronomy at the observatory in Göttingen, Germany, where he had graduated from the university.

Évariste Galois, killed in a duel in **1832** at the age of twenty-one, was a genius never recognized in France during his lifetime. On the eve of the duel he wrote to a friend /D. E. SMITH (c): **285**/:

> Ask Jacobi or Gauss publicly to give their opinion, not as to the truth, but as to the importance of the theorems [see below]. Subsequently there will be, I hope, some people who will find it to their profit to decipher all this mess.

This note was attached to what Galois thought were some new theorems in the theory of equations; these turned out to contain the essence of the theory of groups, so important today. At about the same time Niels Henrik Abel, in Norway, thought he had found a method of solving the general quintic equation, but later he corrected himself and proved that a solution by means of radicals was impossible.

Finally, we take note of two English algebraists, Arthur Cayley and James Joseph Sylvester. As a young man Cayley practiced law in London, and it was there that he met Sylvester, an actuary. For the rest of their lives they worked together on the theory of algebraic invariants.

Although he spent most of his life in England, Sylvester brought his work to America (he taught briefly at the University of Virginia in **1841/42** and returned to the States to teach at Johns Hopkins University from **1877** to **1883**). He established graduate study in mathematics in this country, and "American algebra" might be said to begin with him.

For Further Reading

BOYER (g): 333-38, 367-81, 544-49, 629-32

EVES (a)

ORE (a)

C. SMITH

STRUIK (e): 74-111, 115-22

FUNCTION

Definition 1.—A function is a set of ordered pairs whose first elements are all different.

Definition 2.—When the value of one variable depends on another, the first is said to be a function of the second.

Definition 3.—If to each permissible value of x there corresponds one or more values of y, then y is a function of x.

Definition 4.—If y is a function of x, then it is equal to an algebraic expression in x.

TWENTY elementary algebra texts were examined for definitions of "function"; eleven of these texts were published before 1959, nine after 1959. The older texts used Definitions 2, 3, 4, and others; six of the newer ones used Definition 1.

Fifteen college algebra texts were examined, seven published before 1959 and eight after 1959. None of the older texts used Definition 1; four of the eight newer ones did.

This quite recent history of "function" has additional significance in the context of the earlier history of both the idea and the word.

Eric Temple Bell suggests /(a): 32/ that the Babylonians of c. 2000 B.C. might be credited with a working definition of "function" because of their use of tables like the one for $n^3 + n^2$, $n = 1, 2, \ldots , 30$, suggesting the definition that a function is a table or correspondence (between n in the left colunm and $n^3 + n^2$ in the right column).

More explicit ideas of function seem to have begun about the time of René Descartes (1637), who may have been the first to use the term; he defined a function to mean any positive integral power of x, such as x^2, x^3, ...

Gottfried Wilhelm von Leibniz (1692) thought of a function as any quantity associated with a curve, such as the coordinates of a point on a curve, the length of a tangent to the curve, and so on.

Johann Bernoulli (1718) defined a function to be any expression involving one variable and any constants.

Leonhard Euler (1750) called functions in the sense of Bernoulli's definition "analytic functions" and used also a second definition, according to which a function was not required to have an analytic expression but could be represented by a curve, for example. Euler also introduced the now standard notation $f(x)$.

Joseph Louis Lagrange (1800) restricted the meaning of function to a power series representation. Jean Joseph Fourier (1822) stated that an arbitrary function can be represented by a trigonometric series. P. G. Lejeune Dirichlet (1829) said that y is a function of x if y possesses one or more definite values for each of certain values that x may take in a given interval, x_0 to x_1.

More recently, the study of point sets by Georg Cantor and others has led to a definition of function in terms of ordered pairs of elements, not necessarily numbers.

For Further Reading

Bell (a): *See index*
Boyer (f): 243, 276–77
——— (g): 290–92
Cajori (d): II, 267–70
Eves (c): 371–72
  [5th ed. 462–63]
Growth of Mathematical Ideas:
  65–110, 445–49

Insights into Modern Mathematics: 55–58, 220, 241–72, 409–11
F. Klein (a): I, 200–207
G. A. Miller (b)
Read (e)
Selected Topics: 42–56
John Young: 192–200

Capsule 83

MATHEMATICAL INDUCTION

From the mathematical experiment

$$1 + 3 \qquad\quad = 2^2,$$

$$1 + 3 + 5 \quad\ = 3^2,$$

$$1 + 3 + 5 + 7 = 4^2,$$

etc.,

one is led to the formula

$$1 + 3 + 5 + \cdots + (2n - 1) = n^2;$$

and then this conjecture is proved deductively by using the principle of mathematical induction.

B. L. Van der Waerden /126/ points out that "in essence" the principle of mathematical induction was known to the Pythagoreans but that Francesco Maurolico was the first to make fairly explicit use of it (in his *Arithmetic*, 1575). Blaise Pascal (c. 1653) was the next person to use the idea, as he did repeatedly in his work on the so-called Pascal triangle, which he called the "arithmetic triangle."

The induction proofs of Maurolico are given in a rather sketchy style not easily followed. Pascal's style is more nearly along modern lines, and we present in modern notation a translation of his induction proof that

$$\frac{{}_nC_r}{{}_nC_r + 1} = \frac{r + 1}{n - r},$$

where

$$_nC_r = \frac{n!}{(n - r)!\, r!}$$

and r is any "cell" from the 0th to the $(n - 1)$th in Figure [83]-1.

Consequence XII: *In every arithmetic triangle, two adjoining cells on the same line* [have the property that] *the lower is to the higher as the number of cells below (and including) the lower is to the number of cells above (and including) the higher.*

Let E and C be any two adjoining cells on the same line; I say that

E is to	C	as	2	to	3
lower	higher		because there are two cells from E to the bottom, namely, E, H.		because there are three cells from C to the top, namely, C, R, μ.

Although this proposition has an infinity of cases, I shall give a very short demonstration based on two lemmas:

The first, which is self-evident, that this proportion is true on the second line [of the triangle]; because it is easily seen that ϕ is to σ as 1 is to 1 [let $n = 1$; then ${}_1C_0/{}_1C_1 = (0 + 1)/(1 - 0)$].

314

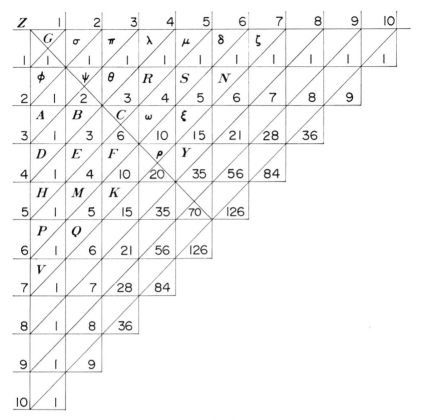

Figure [83]-1

The second, that if this proportion is true on any line it will necessarily be true on the following line. [Let $n = k$. Then

$$_kC_r / {_kC_{r+1}} = (r + 1)/(k - r)$$

implies

$$_{k+1}C_r / {_{k+1}C_{r+1}} = (r + 1)/((k + 1) - r),$$

and hence the theorem is true for $n = k + 1$ *if* it is true for $n = k$.] From which it is apparent that it is necessarily true on all the lines: for it is true on the second line by the first lemma; therefore by the second [lemma] it is true on the third line; therefore on the fourth, and so on.

It is necessary therefore only to prove the second lemma in this way: If the proportion is true on any line, as on the fourth $D \lambda$; for example, if D is to B as 1 to 3, and B to θ as 2 to 2, and θ to λ as 3 to 1, and so forth, I say that this same proportion will be true on the following line $H\mu$ and that, for example, E is to C as 2 to 3.

For D is to B as 1 to 3 by hypothesis.
Therefore $D + B$ is to B as $1 + 3$ to 3

$$E \quad \text{to } B \text{ as } \quad 4 \quad \text{to } 3.$$

Likewise B is to θ as 2 to 2 by hypothesis.
Therefore $B + \theta$ is to B as $2 + 2$ to 2

$$C \quad \text{to } B \quad 4 \quad \text{to } 2.$$

But E to B as 4 to 3 [and B to C as 2 to 4] (as was shown). Then [by multiplying these last two proportions] E is to C as 2 to 3. Which it was required to show.

One can show the same on all the rest [of the lines], since this proof is based only on that proportion found for the preceding [line], and [the property] that each cell is equal to its preceding [one on the left] plus the one above it, which is true everywhere [in the triangle].

The "property" referred to is, for example,

$$E = D + B;$$

or, in general,

$$_nC_r = {_{n-1}C_{r-1}} + {_{n-1}C_r},$$

which is Pascal's rule of formation (definition) of the arithmetic triangle.

For Further Reading

MESCHKOWSKI (b): 36–43   STRUIK (e): 21–26
D. E. SMITH (c): I, 67–79

Capsule 84

FUNDAMENTAL THEOREM OF ALGEBRA

CARL Friedrich Gauss, at the age of twenty (1797), gave the first satisfactory proof of the theorem which he called fundamental and which was the topic for his doctoral dissertation at the University of

Helmstädt, *A New Proof that Every Rational Integral Function of One Variable Can Be Resolved into Real Factors of the First or Second Degree.* (Equivalent statements are "Every algebraic equation of degree n has n roots," and "Every algebraic equation of degree n has a root of the form $a + bi$, where a and b are real.") Actually, Gauss gave four proofs for the theorem, the last when he was seventy; in the first three proofs he assumes the coefficients of the polynomial equation are real, but in the fourth proof the coefficients are any complex numbers.

The words "new proof" in Gauss's title indicate that the ideas summarized in the statement of the theorem had been considered by earlier mathematicians. The Hindus (by 1100 at the latest) realized that quadratic equations (with real roots) had two roots. Girolamo Cardano realized in 1545, though somewhat vaguely because negative and imaginary numbers were not clearly defined at this time, that cubics should have three roots; and he exhibited three roots for some cubics. Similar ideas were held with respect to quartic equations by Cardano and other Italian algebraists of this period.

François Viète (*c.* 1600) considered the possibility of factoring the left member of the polynomial equation $f(x) = 0$ (with positive coefficients) into linear factors but was foredoomed to only partial success because of his marked aversion to negative and imaginary numbers.

Peter Roth seems to have been the first writer to say definitely that a polynomial equation of degree n has n roots. This was in 1608. Albert Girard stated in 1629 that every algebraic equation has as many roots as the degree of its highest power.

The insights of René Descartes on this matter are of special interest because they are related to his famous "rule of signs." We quote from his *La géométrie* (1637) /(b): 159–60/:

> Every equation can have as many distinct roots (values of the unknown quantity) as the number of dimensions [i.e., degree] of the unknown quantity in the equation. . . .
>
> It often happens, however, that some of the roots are false or less than nothing. . . .
>
> We can determine also the number of true [positive] and false [negative] roots that any equation can have, as follows: An equation can have as many true roots as it contains changes of sign . . . and as many false roots as the number of times two + or two − signs are found in succession.

The first attempt at a proof seems to have been made by Jean Le Rond d'Alembert, in 1746, and for this reason the theorem is sometimes called d'Alembert's theorem, especially in France. Leonhard Euler (1749) and Joseph Louis Lagrange also tried to prove the theorem.

A correct proof was not given until Gauss wrote his doctoral dissertation, which was published in 1799. This included "geometrically obvious" assumptions for which later standards of rigor required proof, which A. Ostrowski gave in 1920.

For Further Reading

BELL (a): 178
───── (d): 218–69
COURANT and ROBBINS: 269–71
DUNNINGTON

F. KLEIN (a): I, 101–4
D. E. SMITH (c): I, 292–306
STRUIK (e): 81–87, 99–102, 115–22

Capsule 85   Donald W. Western

DESCARTES'S RULE OF SIGNS

IN 1637 the French philosopher René Descartes (1596–1650) published a book with a lengthy title commonly abbreviated to *Discours de la méthode,* a full translation being "Discourse on the Method of Rightly Conducting One's Reason and Seeking Truth in the Sciences." Three appendixes were included: *La dioptrique* ("Optics"), *Les météores* ("Meteorology"), and *La géométrie* ("Geometry"). The third part of the third appendix is entitled, in translation, "On the Construction of Solid and Supersolid Problems." It deals with many basic ideas for solving equations that arise in connection with geometric problems (primarily the study of conic sections by algebraic methods).

After posing some problems on mean proportions, Descartes proceeds to construct a fourth-degree polynomial equation by multiplying together the linear factors $(x - 2)$, $(x - 3)$, $(x - 4)$, and $(x + 5)$ to obtain

$$x^4 - 4x^3 - 19x^2 + 106x - 120 = 0.$$

He remarks that the polynomial is divisible by no other binomial factors and that the equation has "only the four roots 2, 3, 4, and 5." The fact that the fourth root is −5 rather than 5 is recognized by speaking of 5 as a "false" root, in contrast to the positive numbers, which are called "true" roots. (The minus sign is not used by Descartes to designate negative numbers.) Then comes the statement of the celebrated rule of signs /DESCARTES (b) : 160/:

> We can determine also the number of true and false roots that any equation can have, as follows: An equation can have as many true roots as it contains changes of sign, from + to − or from − to +; and as many false roots as the number of times two + signs or two − signs are found in succession.

Following this general comment, Descartes points out the three changes of sign and the one succession (permanence) of sign in his example and concludes, "On connoit qu'il y a trois vraies racines; et une fausse"; that is "We know there are three true roots and one false root."

As is often the case with the promulgation of a significant mathematical result, this first statement of the relation between changes in signs of the successive terms of the polynomial and the nature of the roots was not complete. Neither was any attempt made at proof, other than the illustrative example that accompanied it.

There is some disagreement in the literature whether the rule of signs was generally known before Descartes's publication of *La géométrie*. Smith and Latham state in a footnote of their translation /DESCARTES (b): 160/ that [Thomas] Harriot had given it in his *Artis analyticae praxis*, published in London in 1631. However, Moritz Cantor denies this possibility, since Harriot did not admit negative roots. Girolamo Cardano (1501–1576) had stated a relation between one or two variations in sign and the occurrence of positive roots.

The process of refining the rule of signs continued over a period of two centuries. In this process two points, specifically, were clarified: (1) the fact that variations in sign determine only upper bounds for the number of positive roots because of the possibility of imaginary roots and (2) the fact that the permanences of sign determine bounds for the number of negative roots only for a complete polynomial— that is, one with no coefficients equal to zero.

Isaac Newton, in his work *Arithmetica universalis* (published in 1707 but written some thirty years earlier), gave an accurate statement

319

of the rule of signs and presented without proof a procedure for determining the number of imaginary roots. At about the same time Gottfried Wilhelm von Leibniz pointed out a line of proof, although he did not give it in detail. In 1675 Jean Prestet published an insufficient proof; Johann Andreas Segner published one proof in 1725 or 1728 and in 1756 a more complete one. In 1741 Jean Paul de Gua de Malves gave a demonstration, introducing the argument that is basic to modern proofs. (This type of argument was employed more clearly by Segner in 1756.) Several other proofs were given in the period from 1745 to 1828. In 1828 Carl Friedrich Gauss added the significant contribution to the statement of the rule that if the number of positive roots falls short of the number of variations, it does so by an even integer.

The complete statement of Descartes's rule of signs is as follows:

> Let $P_n(x) = a_0 x^n + a_1 x^{n-1} + \cdots + a_n$, where the coefficients a_0, a_1, $\cdots$, a_n are real numbers, $a_0 \neq 0$. Then the number of positive real roots of the equation $P_n(x) = 0$ [a root of multiplicity m being counted m times] is either equal to the number of variations in signs or less than that number by a positive even integer.

The matter of negative roots of $P_n(x) = 0$ is handled simply by considering the positive roots of $P_n(-x) = 0$. Thus the matter of permanence of sign is avoided.

The crux of the proof stems from the work of Gua de Malves and Segner. It consists in showing that if

$$P_n(x) = (x - r)P_{n-1}(x),$$

where $P_{n-1}(x)$ has real coefficients and r is positive, then $P_n(x)$ has at least one more variation in sign than does $P_{n-1}(x)$—for the general case, an odd number more.

For Further Reading

BELL (d): 35–55

CAJORI (e): 178–79, 248

DESCARTES (b)

STRUIK (e): 89–99

SYMMETRIC FUNCTIONS

A SYMMETRIC function of two or more variables is a function that is not affected if any two of the variables are interchanged. Perhaps the most familiar symmetric functions are those met in elementary theory of equations where for the cubic equation

$$x^3 + C_1 x^2 + C_2 x + C_3 = 0$$

we have

$$r_1 + r_2 + r_3 = -C_1, \qquad r_1 r_2 + r_1 r_3 + r_2 r_3 = C_2, \qquad r_1 r_2 r_3 = -C_3.$$

These last three equalities express the coefficients of the cubic equation as symmetric functions of the roots r_1, r_2, r_3.

When François Viète made his first tentative discoveries concerning symmetric functions in the late sixteenth century, the very notion of the roots of an algebraic equation was incomplete, in large measure because of an incomplete understanding of negative and imaginary numbers. Viète himself worked only with positive roots. He noticed that if the equation $x^3 + b = ax$ ($a > 0, b > 0$) has two *positive* roots, r_1 and r_2, then

$$(1) \qquad\qquad r_1^2 + r_2^2 + r_1 r_2 = a,$$

and

$$(2) \qquad\qquad r_1 r_2 (r_1 + r_2) = b.$$

Also, as Cajori says /(b): 230/:

> His nearest approach to complete recognition of the facts is contained in the statement that the equation
>
> $$x^3 - (u + v + w)x^2 + (uv + vw + wu)x - uvw = 0$$
>
> has three roots, u, v, w. For cubics, this statement is perfect, if u, v, w are allowed to represent any numbers. But Viète is in the habit of assigning to letters only positive values, so that the passage really means less than at first sight it appears to do.

Albert Girard was interested in extending Viète's result. He considered all roots—those he called "impossible" (i.e., imaginary) as well as negative and positive roots. He studied the sums of their products taken two at a time (analogous to Viète's (1), above), then three at a time (analogous to Viète's (2), above), and so on.

But Girard was also interested in obtaining expressions for the sums of given powers of the roots; these sums constituted a different set of symmetric functions than the one Viète had essentially pioneered. Girard published his results in Amsterdam in 1629 in a pamphlet *Invention nouvelle en l'algèbre,* which contained the statement /FUNK-HOUSER: 361/ that if

$$x^n - Ax^{n-1} + Bx^{n-2} - Cx^{n-3} + \cdots = 0,$$

then

$$
\left.\begin{array}{l}
A \\
A^2 - 2B \\
A^3 - 3AB + 3C \\
A^4 - 4A^2B + 4AC + 2B^2 - 4D
\end{array}\right\}
\text{ will be the sum of }
\left\{\begin{array}{l}
\text{solutions} \\
\text{squares} \\
\text{cubes} \\
\text{biquadrates.}
\end{array}\right.
$$

Girard stated this result rather casually. Perhaps because of this and perhaps also because seventeenth-century mathematicians were not ready, Girard's remark went unnoticed until it reappeared, without proof, in Isaac Newton's *Arithmetica universalis* (1707) and became famous. It also became one of several theorems that are called "Newton's theorem."

For a hundred years after Newton many mathematicians, including Colin Maclaurin, Leonhard Euler, and Joseph Louis Lagrange, concerned themselves with proofs and generalizations of this theorem.

For Further Reading

CAJORI (b): 230–31 STRUIK (e): 81–87
FUNKHOUSER

DISCRIMINANT

As A result of the historical development of ideas leading to the term "discriminant," there is today a slight inconsistency in the use of the word. Texts dealing with the equation

$$Ax^2 + Bx + C = 0$$

call $B^2 - 4AC$ the discriminant of the equation. Other texts, discussing the binary quadratic form

$$Q(x, y) = Ax^2 + 2Bxy + Cy^2,$$

call $AC - B^2$ the discriminant of Q. Similar though these expressions are, the first is negative four times what we would expect it to be if notations were uniform. And even if this were corrected, it would not be immediately obvious that we are justified in using the same name—that we have the same mathematical entity.

By the middle of the eighteenth century it was well known that a necessary and sufficient condition for the equation $Ax^2 + Bx + C = 0$ to have two identical roots was $B^2 - 4AC = 0$. The expression was known; mathematicians knew what it signified and how to work with it; but it was not yet recognized as a mathematical entity.

During the next hundred years mathematicians studied several expressions related to the quadratic form. In 1748 Leonhard Euler used conditions involving expressions like those above to determine whether a quadric surface is contained in finite space; but Euler did not give a name to these expressions.

The expression that was not yet an entity reappeared in 1773. Joseph Louis Lagrange was studying the binary quadratic form given above. He proved that if $x + \lambda y$ were substituted for x, leading to a new form

$$A(x + \lambda y)^2 + 2B(x + \lambda y)y + Cy^2,$$

then if the new expression is simplified to

$$A'x^2 + 2B'xy + C'y^2,$$

323

we must have

$$A'C' - B'^2 = AC - B^2.$$

Other mathematicians turned to the study of such invariants, and similar expressions kept reappearing. Carl Friedrich Gauss called such an expression a "determinant" of the function. It remained for the tempestuous James Joseph Sylvester, who called himself the "mathematical Adam" because of his habit of giving names to mathematical creatures, to name this one. In 1851 he was studying invariants in reducing certain sixth-degree functions of two variables to simpler forms. What he found was what he called (and what we now recognize as) the "discriminant of a cubic."

His explanation in a long, testy, and somewhat defensive footnote is amusing and enlightening:

> "Discriminant," because it affords the *discrimen* or test for ascertaining whether or not equal factors enter into a function of two variables, or more generally of the existence or otherwise of multiple points in the locus represented or characterized by any algebraical function, the most obvious and first observed species of singularity in such function or locus. Progress in these researches is impossible without the aid of clear expression; and the first condition of a good nomenclature is that different things should be called by different names. The innovations in mathematical language here and elsewhere (not without high sanction) introduced by the author, have been never adopted except under actual experience of the embarrassment arising from the want of them, and will require no vindication to those who have reached that point where the necessity of some such addition becomes felt.

Both our cases satisfy Sylvester's definition. The discriminant is a combination of constants which vanishes if at least two factors of a function are the same. If

$$B^2 - 4AC = 0, \qquad A \neq 0,$$

then

$$Ax^2 + Bx + C = A(x + B/2A)^2;$$

under the same conditions (or equivalently, with changed notation, if $AC - B^2 = 0, A \neq 0$),

$$Ax^2 + 2Bxy + Cy^2 = \frac{1}{A}(Ax + By)^2.$$

For Further Reading

BELL (d): 378–405   BOYER (g): 253, 258

Capsule 88   L. S. Shively

INTEREST AND ANNUITIES

IN THE *Liber abaci* of Fibonacci (Leonardo of Pisa), written in 1202, the following problem appears /EVES (c): 210/:

> A certain man puts one denarius at [compound] interest at such a rate that in five years he has two denarii, and in every five years thereafter the money doubles. I ask how many denarii he would gain from this one denarius in one hundred years?

The answer, $(2^{20} - 1)$ denarii, is easily obtained, since exactly 20 doublings are involved. The implied interest rate of $16\frac{1}{3}$ percent compounded annually is possibly a commentary on the rather high rates charged in medieval Europe in spite of certain restrictions by the Church.

The custom of charging interest is found as early as 2000 B.C., as recorded on ancient Babylonian clay tablets. We give one example: /D. E. SMITH (a): II, 560/:

> Twenty manehs of silver, the price of wool, the property of Belshazzar, the son of the king. . . . All the property of Nadin-Merodach in town and country shall be the security of Belshazzar, the son of the king, until Belshazzar shall receive in full the money as well as the interest upon it.

Interest rates in Babylonia ran as high as 33 percent. In Rome during Cicero's day 48 percent was allowed; Justinian later set the maximum allowable rate at 0.5 percent per month, which gave rise to the common rate of 6 percent a year. In India, however, during the twelfth century, rates as high as 60 percent are recorded.

The origin of the word "interest" is related to church policy, which forbade usury, payment for the use of money. The moneylender got

325

around this restriction of canon law by collecting a fee only if the money was repaid tardily (which happened often enough even in those days!). The lender argued that the fee compensated him for the monetary difference between his poorer financial standing, because of late payment, and what would have been the standing under prompt repayment. This difference was referred to as *id quod interest* ("that which is between").

Annuities were known as early as 1556, the year in which Niccolo Tartaglia, in his *General trattato,* gives the following problem, which he said was brought to him by gentlemen from Barri who said that the transaction had actually taken place /SANFORD (d): 136/:

> A merchant gave a university 2,814 ducats on the understanding that he was to be paid 618 ducats a year for nine years, at the end of which the 2,814 ducats should be considered as paid. What interest was he getting on his money?

The answer to the problem is that the interest rate was slightly more than 19 percent; but without logarithms and annuity tables, it was not considered easy.

In 1693 Edmund Halley, who is best known for his work as an astronomer, contributed to the study of life insurance annuities with the publication of *Degrees of Mortality of Mankind . . . with an Attempt to Ascertain the Price of Annuities upon Lives.* This included the following formula /CAJORI (e): 171/:

> To find the value of an annuity, multiply the chance that the individual concerned will be alive after n years by the present value of the annual payment due at the end of n years; then sum the results thus obtained for all values of n from 1 to the extreme possible age for the life of that individual.

Halley probably used the mortality table published in 1662 by John Graunt of London in his *Natural and Political Observations . . . Made upon the Bills of Mortality,* which was based on records of deaths that were kept in London beginning in 1592. (These records were originally intended to keep track of deaths due to the plague.)

For Further Reading

SANFORD (d): 127–31   D. E. SMITH (a): II, 559–65

EXPONENTIAL NOTATION

THE great French mathematician René Descartes is credited with first introducing, in about 1637, the use of Hindu-Arabic numerals as exponents on a given base. To any modern schoolboy the idea of writing $x \cdot x \cdot x$ as x^3 or $x \cdot x \cdot x \cdot x$ as x^4 seems so obvious that it is quite natural for one to feel that Descartes probably hit upon this idea without help from his many predecessors in mathematics. But that was not the case! Ingenious inventions very often result from the insights of men who have learned from the trials and errors of others; such was the case with Descartes's use of exponents.

In this short capsule we shall look at some examples of early exponential symbolism and shall see that the idea of an exponent was available when Descartes took the very significant step of using Hindu-Arabic numerals placed to the upper right of the base.

Sometime around 1552 an Italian mathematician, Rafael Bombelli, worked very diligently on a manuscript that he published in 1572 as an algebra book called *L'Àlgebra*. In this volume he wrote the solution to a problem, beginning it as shown below:

$$\text{4.p.R.q.} \underset{\displaystyle \lfloor 24.\text{m}.20, \rfloor}{} \text{Eguale à2.}$$

A first glance at this line of symbols might lead one to think that Bombelli was using a very complicated secret code, as in a sense he was; he was writing the equation we represent by writing

$$4 + \sqrt{24 - 20x} = 2x.$$

Let us pause for a moment and compare Bombelli's equation with our present-day form. It is easy to see that "Eguale à" probably means "equals." Continuing, we can see that "p." probably stands for "plus" and "m." for "minus." The symbol "R.q." represents "square root"; the two angular symbols mean the same as parentheses in modern symbolism. Thus "R.q.⌊ ⌋" means the square root of the polynomial

327

written within the symbols. To write positive integral powers of a variable, x, Bombelli wrote the exponent in a small circular arc above a numeral, so that

$$\overset{\smile 1}{2}, \overset{\smile 2}{2}, \overset{\smile 3}{2}$$

meant the same as $2x$, $2x^2$, and $2x^3$ in modern notation. Thus

$$\text{R.q.} \lfloor 24.\text{m}.\overset{\smile 1}{20} \rfloor$$

means $\sqrt{24 - 20x}$.

At first thought, it may seem that Bombelli's method is much better than our present symbolism because he did not need to write the letter x. But suppose a mathematician wished to represent $x^2 - y^2$. Could he do this by writing

$$\overset{\smile 2}{1}.\text{m}.\overset{\smile 2}{1} ?$$

No! For this reason, Bombelli's exponents were short-lived.

The complete solution as Bombelli included it in *L'Algebra* is given in the left-hand column below. Cover the modern version in the right-hand column if you want to test your skill in translating.

4. p. R. q. $\lfloor 24.$ m. $\overset{\smile 1}{20},\rfloor$
 Eguale à $\overset{\smile 1}{2}.$ $4 + \sqrt{24 - 20x} = 2x.$

R. q. $\lfloor 24.$ m. $\overset{\smile 1}{20} \rfloor$
 Eguale à $\overset{\smile 1}{2}.$ m. 4. $\sqrt{24 - 20x} = 2x - 4.$

24. m. $\overset{\smile 1}{20}$
 Eguale à $\overset{\smile 2}{4}.$ m. $\overset{\smile 1}{16}.$ p. 16. $24 - 20x = 4x^2 - 16x + 16.$

24. p. $\overset{\smile 1}{16}$
 Eguale à $\overset{\smile 2}{4}.$ p. $\overset{\smile 1}{20}.$ p. 16. $24 + 16x = 4x^2 + 20x + 16.$

24
 Eguale à $\overset{\smile 2}{4}.$ p. $\overset{\smile 1}{4}.$ p. 16. $24 = 4x^2 + 4x + 16.$

8
 Eguale à $\overset{\smile 2}{4}.$ p. $\overset{\smile 1}{4}.$ $8 = 4x^2 + 4x.$

2

328

Eguale à $\overset{\smile 2}{1}$. p. $\overset{\smile 1}{1}$.

2¼

    Eguale à $\overset{\smile 2}{1}$. p. $\overset{\smile 1}{1}$. p. ¼.

1½

    Eguale à $\overset{\smile 1}{1}$. p. ½

1

    Eguale à $\overset{\smile 1}{1}$.

$2 = x^2 + x.$

$2\tfrac{1}{4} = x^2 + x + \tfrac{1}{4}.$

$1\tfrac{1}{2} = x + \tfrac{1}{2}.$

$1 = x.$

Bombelli was not the only mathematician before Descartes to write a numeral above the coefficient to indicate the power of the variable. Nicolas Chuquet, a physician in Lyons, France, wrote 12^0, 12^1, 12^2, and 12^3 to designate 12, $12x$, $12x^2$, and $12x^3$ in his *Le triparty en la science des nombres*, written about 1484. He also used

$$12.^{1.\tilde{m}}$$

to designate $12x^{-1}$. Later, about 1610, Pietro Cataldi wrote

$$\cancel{0},\; \cancel{2},\; \cancel{3},\; \cancel{4},$$

to stand for x^0, x^2, x^3, and x^4; and in 1593 the Dutch writer Adrianus Romanus used

$$1(\overline{45})$$

for x^{45}. In 1619 the Swiss mathematician Jobst Bürgi used Roman numerals as exponents. He wrote

$$\overset{\text{vi}}{8} + \overset{\text{v}}{12} - \overset{\text{iv}}{9} + \overset{\text{iii}}{10}$$

to indicate the polynomial

$$8x^6 + 12x^5 - 9x^4 + 10x^3.$$

The accompanying table summarizes the main items in the historical development of exponents, including negative and fractional exponents. Cardano's verbal notation and J. Buteo's pictorial notation illustrate styles otherwise omitted because they did not contribute to the development of our present system.

The history of the development of exponential notation is a credit to man's genius in finding facile symbolisms for expressing mathematical concepts.

TABLE [89]-1

HISTORICAL DEVELOPMENT OF EXPONENTS

Modern Notation	$7x$	$7x^2$	$7x^3$	Commentary
1360 Oresme				Different manuscripts show different notations. For $2^{1/2}$ he wrote $\frac{1}{2}2^p$; also $\boxed{\frac{1.p}{2.2}}$. For $9^{1/3}$, he wrote $\frac{1}{3}9^p$. For $(2\frac{1}{2})^{1/4}$, he wrote $\boxed{\frac{1.p\ 1}{4.2.2}}$. For $4^{3/2}$, he wrote $\boxed{1^p\ \frac{1}{2}}$ 4, also $\boxed{\frac{p.1}{1.2}}$ 4.
1484 Chuquet	7^1	7^2	7^3	For 7, Chuquet wrote 7^0. For $12x^{-1}$, he wrote $12.^{1 \cdot \bar{m}}$.
1545 Cardano	7. pos.	7. quad.	7. cub.	For $7x^4$, Cardano wrote 7 quadr. quad.
1559 Buteo	7ρ	$7 \diamond$	$7 \square$	
1572 Bombelli	$\overset{1}{\underset{7}{\smile}}$	$\overset{2}{\underset{7}{\smile}}$	$\overset{3}{\underset{7}{\smile}}$	
1585 Stevin	$7①$	$7②$	$7③$	Stevin suggested for $x^{\frac{3}{2}}$ the notation ③ when he said, "3/2 in a circle would be the symbol for the square root of ③ [i.e., x^3]"; but he never used this notation.
1590 Viète	7N	7Q	7C	Viète used vowels for unknowns and consonants for constants (except that N, Q, C had already been reserved for powers). He used the first style for polynomial equations in one unknown with numerical coefficients. Both styles are from later editions of his work; ear-
Also	B in A 7 for 7 BA	B in A q 7 for 7 BA²	B in A cu 7 for 7 BA³	

				lier he wrote "B in A quadratum 7" for "B in A q7" [7BA²]. In the second style, Viète wrote "B in A qq 7" for 7BA⁴ and "B in A q cu 7" for 7BA⁵.
1610 Cataldi	$7\,\rlap{\;/}{\rlap{\,/}{+}}$	$7\,\rlap{/}{2}$	$7\,\rlap{/}{3}$	For 7, Cataldi wrote 7 ϕ.
1619 Bürgi	$\overset{\text{i}}{7}$	$\overset{\text{ii}}{7}$	$\overset{\text{iii}}{7}$	For $7x^4$, Bürgi wrote $\overset{\text{iv}}{7}$.
1631 Harriot	$7a$	$7aa$	$7aaa$	
1634 Herigone	$7a$	$7a2$	$7a3$	
1637 Descartes	$7x$	$7xx$	$7x^3$	For $7x^4$, Descartes wrote $7x^4$.
1656 Wallis	$7a$	$7aa$	$7a^3$	For $7x^4$, Wallis wrote $7a^4$. In 1676 Wallis spoke of $$\frac{1}{\sqrt{1}}, \frac{1}{\sqrt{2}}, \frac{1}{\sqrt{3}}, \text{ etc.}$$ as having index $-\tfrac{1}{2}$. However, he never used this notation.
1676 Newton*	$7x$	$7xx$	$7x^3$	For $7x^4$, Newton used $7x^4$. He also used the following notations: $$a^{\frac{1}{2}}, a^{\frac{3}{2}}, a^{\frac{5}{3}} \text{ etc.}; a^{-1}, a^{-2}, a^{-3}, \text{ etc.};$$ $$\overline{P+PQ}\Big|^{\frac{m}{n}} = P^{\frac{m}{n}} + \frac{m}{n}AQ + \cdots$$ where $A = P^{\frac{m}{n}}$; $$\overline{x^{\sqrt{2}} + x^{\sqrt{7}}}\Big|^{\sqrt[3]{\frac{2}{3}}}.$$

---

* In a letter (June 13, 1676) to Henry Oldenburg, secretary of the Royal Society of London, Newton said: "Since algebraists write a^2, a^3, a^4, etc. for aa, aaa, $aaaa$, etc., so I write $a^{1/2}$, $a^{3/2}$, $a^{5/3}$, for $\sqrt{a}$, $\sqrt{a^3}$, $\sqrt{c}\,a^5$; and I write a^{-1}, a^{-2}, a^{-3}, etc. for $\dfrac{1}{a}$, $\dfrac{1}{aa}$, $\dfrac{1}{aaa}$, etc." /Cajori (d): I, 355./

For Further Reading

Boyer (d)   ——— (d): I, 335–60
Cajori (a)   Sanford (d): 155–58

Capsule 90   Waldeck E. Mainville, Jr.

RULE OF FALSE POSITION

THE rule of false position is a method of solving equations by assigning a value to the unknown; if, on checking, the given conditions are not satisfied, this value is altered by a simple proportion. For example, to solve $x + x/4 = 30$, assume any convenient value for x, say $x = 4$. Then $x + x/4 = 5$, instead of 30. Since 5 must be multiplied by 6 to give the desired 30, the correct answer must be $4 \cdot 6$ or 24.

This method was used by the early Egyptians (*c.* 1800 B.C.); many problems appearing on Egyptian papyri seem to have been solved by false position. Diophantus, in his text *Arithmetica*, uses a similar procedure to solve simultaneous equations.

The Hindu Bakhshali manuscript (*c.* A.D. 600?) contains some problems solved by false position. The earliest Arabic arithmetic of al-Khowarizmi explained the rule of false position.

The Italian mathematician Fibonacci (Leonardo of Pisa, *c.* 1200) issued a tract dealing with algebraic problems, all solved by false position. The arithmetic of Johann Widmann, published in Leipzig in 1498, is the earliest book in which the symbols $+$ and $-$ have been found. They occurred in connection with problems solved by false position to indicate excess and deficiency. The first edition of *Summa de arithmetica, geometrica, proportioni et proportionalita* (1494) by the Italian friar Luca Pacioli discussed and applied the rule of false position. In England Robert Recorde included the rule of false position in his arithmetic, *The Ground of Artes* (1542).

For Further Reading

Midonick: 91–105   D. E. Smith (a): II, 437–41
Sanford (d): 155–58

VI

The History of TRIGONOMETRY

an overview by

EDWARD S. KENNEDY

Before involving ourselves with details, it will be useful to point out certain general themes that recur throughout this narrative, to indicate the main directions the account will take, and to issue a warning concerning the limitations inherent in any essay of this sort.

Trigonometry, perhaps more than any other branch of mathematics, developed as the result of a continual and fertile interplay of supply and demand: the supply of applicable mathematical theories and techniques available at any given time and the demands of a single applied science, astronomy. So intimate was the relation that not until the thirteenth century was it useful to regard the two subjects as separate entities.

The same sort of varied interaction that went on between theory and application also took place continually within the body of the theoretical material itself—interaction between numerical analysis and geometry. Algebraic considerations, in the sense of discrete operations performed on classes of objects, played an early and essential role, although the symbolism frequently thought of as the hallmark of algebra was not introduced until the sixteenth century. Thus the history of trigonometry exhibits within itself the embryonic growth of three classical divisions of mathematics: algebra, analysis, and geometry.

The beginnings of this development are lost in prehistory. They can be thought of as the first numerical sequences correlating shadow lengths with the time of day. Between these and Hipparchus' cumber-

333

some but complete technique for the numerical solution of plane figures lies an unbridged gap. This "trigonometry" was based on a single function, the chord of an arbitrary circular arc. Menelaus' theorem, involving either plane or spherical complete quadrilaterals, made possible the extension of this discipline to the sphere. However, other methods of passing from plane to sphere certainly competed with the Menelaus theorem and probably preceded it.

These things originated in the general region of the eastern Mediterranean, were recorded by people writing in Greek, and were well established by the second century of our era. The centroid of activity then shifted to India (where the chord function was transformed into varieties of the sine), and thence it moved part of the way back. In the region stretching from Syria to Central Asia, and in the period from the ninth century through the fifteenth, the new sine function and the old shadow functions (tangent, cotangent, secant, etc.) were elaborately tabulated as sexagesimals. With this development the first real trigonometry emerged, in the sense that only then did the object of study become the spherical or plane *triangle*, its sides and angles.

Subsequently, as the locus of activity in astronomy moved to Europe, so also did the new trigonometry. The same type of work that had occupied Oriental scientists—namely, the computation of tables and the discovery of functional relations between parts of triangles—was continued in the West. A new and fundamental European contribution, however, was the replacement of verbal rules by appropriate symbols. But the invention of the infinitesimal calculus, following hard after, foreshadowed the speedy end of trigonometry as an independent and growing branch of mathematics; for with the discovery and exploitation of the complex domain the whole mass of theory was subsumed into analysis. By the end of the eighteenth century Leonhard Euler and others had exhibited all the theorems of trigonometry as corollaries of complex function theory. (As a school subject, however, especially useful for surveyors and navigators, trigonometry still keeps its separate identity.)

The ever-accelerating development of trigonometry provides a ready illustration of the fact that knowledge tends to accumulate at a rate proportional to the quantity that is already at hand; broadly speaking, its growth is exponential with respect to time.

Facile generalizations of this sort are valid and useful, but it is necessary to hedge them about by stating various inadequacies that are inherent in historical exposition. The historian constructs a tale. He spins a yarn of which, in our case, the earlier lengths are uncertain conjectures pieced out with an occasional fact. For the later lengths there is much more factual material; indeed, that very abundance constitutes a difficulty, since most of the facts cannot be utilized yet all are in some way relevant. The resulting account gives an impression of simplicity, continuity, and rationality which is in many ways illusory. The strand of discourse is necessarily one-dimensional, whereas at any given time the complex of facts has many dimensions. We confine our account to the leaders in the field of working with triangles; their predecessors and rank-and-file contemporaries operated on a more primitive level, but they created the background without which these leaders could not have existed. And knowledge of the subject did not grow steadily. It progressed, instead, by a series of discontinuous jumps. Important advances made at one time and place sometimes spread only slowly—sometimes not at all, disappearing only to be rediscovered later.

Three categories of people confronted or confront situations that are in many ways closely analogous. One category is that of the original inventor of a theorem or technique; he applies to the solution of his problems the mathematical tools he has inherited but slowly or quickly intuits more powerful ways of answering old questions or of posing and answering new ones. A second is that of the historian who seeks, hampered by his own hindsight, to retrace via texts and artifacts the thought processes of the inventor. The third is that of the student, for whom a given problem is as new as ever it was to the man who first contemplated it. People in all three categories share in varying degrees an inability to grasp the full implications of their own accomplishments.

Both the historian and the student can gain understanding from the example of the inventor. It is this increase in understanding which justifies presenting the sections that follow.

Shadow Tables and the Calculus of Chords

The accompanying tables show four examples of simple seqeuences of number-couples. They are legitimate ancestors of the tangent and cotangent functions. Each of the four schemes is a primitive application of the fact that the shadow of a vertical stick (a "gnomon")— or of a person, for that matter—is long in the early morning, shortens to a minimum at noon, then becomes longer and longer as the afternoon wears on. Each scheme associates the termination of a particular daylight hour or other measure of time with a particular length of the gnomon's shadow (or of the shadow's projection on the east-west line).

335

TABLE VI-1 EGYPTIAN		TABLE VI-2 IRANIAN (?)		TABLE VI-3 INDIAN		TABLE VI-4 GREEK	
End of Hour	Shadow	End of Hour	$S_t - S_n$	End of Muhurta	$S_t - S_n$	End of Hour	Shadow During an Equinoctial Month
–	–	1	72	1	96	1	25
2	30	2	36	2	60	2	15
3	18	3	24	3	12	3	11
4	9	4	$14\frac{2}{5}$(?)	4	6	4	8
5	3	5	12	5	5	5	6
				6	3		
				7	2		
Noon	0	6[Noon]	0	Noon[$7\frac{1}{2}$]	0	Noon	5

The dates and places of origin of all four tables are uncertain. The first is recorded on an inscription from the thirteenth century B.C. at Abydos in Upper Egypt /NEUGEBAUER and PARKER: I, 116; LEACH: I, 110–27/. The second is reported by al-Biruni (*c.* A.D. 1000), but doubtless it was ancient in his time /(b): Treatise 2, 158/. In this table and in Table VI-3 the second columns of numbers give, in "digits" of a 12-digit gnomon, the difference between the gnomon shadow at the time being considered (S_t) and the shadow at noon (S_n). The scheme in Table VI-3, cited by al-Biruni, is found also in a Buddhist Sanskrit text from the first or second century A.D.[1] (There are fifteen *muhurta* from sunrise to sunset.) Table VI-4 appears on papyri and Byzantine adaptations but probably stems from Greece of the fifth century B.C. /NEUGEBAUER (c)/. Related material from China of about that same time has been found, and some from Mesopotamia that is five hundred or a thousand years older /NEEDHAM: III, 20–284; WEIDNER/.

All these schemes are in many respects naïve, and it would be fruitless to seek anything very deep in the sequences of the integers shown. The lengths of the "hours" they mark are unequal, and the shadows

---

[1] The source is an anti-caste Buddhist tract called the *Sardulakarnavadana,* and the information is from Professor David Pingree. (See also /PINGREE/.)

vary with the geographical latitude of the gnomon and with the season of the year. (Table VI-4, in other, more complete forms, indeed attempts to compensate for variation during the seasons by prescribing that one unit be added to, or subtracted from, each tabular entry for each of the three months preceding or following the vernal equinoctial month.)

Nevertheless, they bear witness that at least three thousand years ago man was employing implicitly the notion of a function. For each scheme, corresponding to a given time there is a unique shadow length. In the case of Table VI-4 the function has two independent variables, time of day and season of the year.

Between these primitive landmarks and the next recognizable stage in the road leading to trigonometry there lies a great gap—if not in point of time, certainly in subject matter and degree of sophistication.

The basic tool came to be the chord function, still tabulated in engineering handbooks, the precursor of the sine rather than the tangent. For serious computation a place-value system for representing numbers was requisite, but since the second millenium B.C. this had been available in the sexagesimal system developed in Mesopotamia (now Iraq).

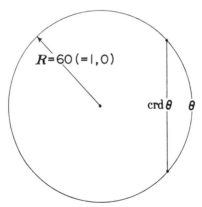

FIGURE VI-1

At the end of Book I, Chapter 11, of the *Almagest* of Claudius Ptolemy (see [91]) there is a table of the function crd θ (defined as shown in Fig. VI-1), calculated to three significant sexagesimal places for the domain $\theta = 1/2°, 1°, 1\,1/2°, \ldots, 180°$ and having a column of tabular differences for interpolation. In Book I, 9, Ptolemy explains how this table was computed. He finds crd $72° = 70;32,3$ the side of a regular pentagon. (See [1] for sexagesimal notation.) To find crd $(\alpha - \beta)$ in terms of crd α and crd β he shows that (in our notation)

$$\text{crd } (\alpha - \beta) \text{ crd } 180° = \text{crd } \alpha \text{ crd } (180° - \beta) - \text{crd } \beta \text{ crd } (180° - \alpha).$$

An expression relating chords of supplementary arcs is

$$\text{crd}^2 \, \theta + \text{crd}^2 \, (180° - \theta) = \text{crd}^2 \, 180° = (2,0)^2,$$

by application of the Pythagorean theorem (see Fig. VI-2).

Hence he computes crd $12° = $ crd $(72° - 60°)$. Now crd $\alpha/2$ in terms of crd α is

$$\text{crd}^2 \, \frac{\alpha}{2} = \tfrac{1}{2} \, \text{crd } 180° \, (\text{crd } 180° - \text{crd } (180° - \alpha)).$$

Using this, he finds successively crd $6°$, crd $3°$, crd $(1\,1/2)°$, and crd $(3/4)°$. By a careful interpolation between the last two he obtains crd $1° = 1;2,50$, whence crd $(1/2)° = 0;31,25$. Now with the aid of an expression for crd $(\theta + 1/2°)$ the whole table can be filled in.

To solve any rectilinear figure with the aid of the table of chords, break it up into suitable right triangles and solve each as follows: Suppose that, as in the right triangle ABC in Figure VI-2, a and A are given. Consider the similar triangle with hypotenuse $c' = 2,0$. We have

$$\frac{c}{2,0} = \frac{a}{\text{crd } 2A} \, , \quad \text{so} \quad c = a\left(\frac{2,0}{\text{crd } 2A}\right) ,$$

and

$$\frac{b}{\text{crd } (180° - 2A)} = \frac{a}{\text{crd } 2A} \, , \quad \text{so} \quad b = a\left[\frac{\text{crd } (180° - 2A)}{\text{crd } 2A}\right].$$

If two legs are given, the Pythagorean theorem may be applied to find the hypotenuse, then the table of chords to find the angles. The *Almagest* abounds in worked-out examples of this technique /NEUGEBAUER (a): 208–14/.

Note that in developing the theory frequent recourse is had to the geometry and geometrical algebra that had been assembled in the *Elements* of Euclid. The general impression is that Ptolemy is giving a systematic exposition of a body of doctrine well known in his time. This would have included also the work of Hipparchus, who in about 150 B.C. (three centuries before Ptolemy) wrote a treatise on chords which has disappeared. So, except for tantalizing hints obtainable from two Old Babylonian cuneiform tablets, there is no way of determining how the trigonometry of chords came into being.

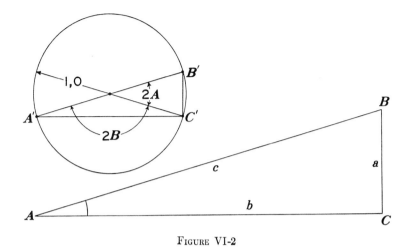

Figure VI-2

One tablet calculates the length of a chord in terms of its versed sine (see "Appearance of the Sine Function," which follows) and the diameter of the circle /NEUGEBAUER (b)/. However, neither the subtended arc nor the angle is involved.

The other tablet is mutilated. The part that remains, however, contains a rectangular array of integers, arranged in four columns and fifteen rows. One column is the set $n = 1, 2, 3, \cdots, 15$. The next two columns we call c_n and b_n. For a given n, each c_n and b_n make up two elements of a "Pythagorean triple" (i.e., there exists a third integer a_n such that $c_n{}^2 = b_n{}^2 + a_n{}^2$). The last column consists of the numbers $(c_n/a_n)^2$. What is most astonishing, however, is the fact that if one considers the sequence of right triangles whose sides satisfy the successive triples, corresponding acute angles β_n run, very nearly, through the set $\beta_1 = 45°$, $\beta_2 = 44°$, $\beta_3 = 43°$, $\cdots$, $\beta_{15} = 31°$. An explanation of the remarkable manner in which the ancient number theorist must have operated in order to construct the table has been given by Neugebauer and Sachs /38–41/. There is not the slightest indication of an awareness of angle or arc magnitudes on the part of the theorist. Nevertheless, and for whatever purpose it was done, ranges of ratios of triangle sides had been calculated a millenium and a half before the time of Hipparchus /PRICE/.

STEREOGRAPHIC PROJECTION AND THE ANALEMMA

For us it is easy to think of the night sky as the upper half of an enormous sphere, rotating once daily from east to west and carrying with it the images of all the celestial bodies. This idea of the celestial sphere has been of fundamental utility in astronomy. Just who originated it, or when, or how, we cannot say; and in the absence of evidence, speculation about it would be as pointless as speculation about the origin of the 360-degree division of the circle [**94**]. About all that can be said with certainty is that by the time of Hipparchus and the emergence of the plane trigonometry of chords the notion of the celestial sphere was well established.

In order to exploit the idea properly, it was necessary to develop a technique for calculating an unknown magnitude on the sphere in terms of known magnitudes. A direct solution would be a calculus dealing with the spherical entities themselves—arcs, angles, and surfaces. Such developments of course took place, and we describe them in the sections following. An alternative proceeding would be somehow to transform the spherical objects into plane entities, then to attack the latter by means of the calculus of chords. Two such methods, both of which anteceded spherical trigonometry as such, we now proceed to sketch.

Consider a sphere S, the plane S' of one of its great circles, and O, one of the circle's poles. Map any point P of the sphere onto the plane S' by drawing the line OP until it intersects S' at P'. This establishes a one-to-one correspondence between all the points of the plane and all the points of the sphere except O. Known as "stereographic projection," this mapping was probably invented by Hipparchus. It has many beautiful properties, some of which (e.g., that circles map into circles) were known to the ancients and utilized by them in analogue computers like the astrolabe /HARTNER/. Apparently they did not know that the mapping is conformal, but the property of conformality is exploited in modern mathematics.

For present purposes the essential fact is that any configuration on the celestial sphere composed of circles maps stereographically into a plane figure composed of straight lines or circles. The latter can be solved by methods described in "Shadow Tables and the Calculus of Chords," above, and the results reinterpreted in terms of the sphere. The notion is simple, but the actual solution of problems by this method is long and complicated. For examples the reader is referred to Ptolemy's *Planisphaerium*. Theorems in spherical trigonometry are still derived by using stereographic projection /DONNAY/.

Another way of evading work on the surface of the sphere as such is to use the fact that although a spherical polygon is not a plane figure, each of its sides is. One seeks therefore to force into a single plane, either by rotations or by orthographic projection, all the arcs of small or great circles involved in a given situation. This having been done, each arc appears in its true size, undistorted, and it can be measured or calculated at will. These descriptive geometric methods went by the name of "the analemma"; they appeared in classical times, proliferated throughout the Middle Ages, and are still useful today /Neugebauer (a): 215; Kennedy (a)/. To illustrate their elegance and simplicity, we introduce an astronomical problem and solve it by an analemma. The same problem will serve later to illustrate other methods.

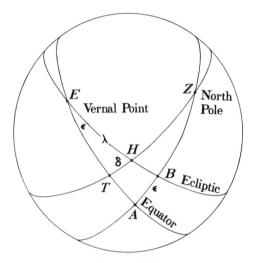

Figure VI-3

Figure VI-3 shows two fundamental circles on the celestial sphere, the equator and the ecliptic, and the constant angle between them, ϵ. For an arbitrary point H on the ecliptic we seek its declination, δ (its distance to the equator), as a function of λ, its celestial longitude.

The analemma appears in Figure VI-4. Project the configuration on the plane of circle ABZ. The equator and ecliptic then project as straight lines intersecting at an angle ϵ. To construct the projection of H, rotate the plane of the ecliptic down into the working plane through a right angle. Then the ecliptic will take the position of circle $E'ZB$ with the vernal point E moving to E'. Lay off the given longitude as arc $E'H'$. The end point of the arc, H', is the folded-down position of H. Now rotate the ecliptic plane back into its original position. As this is done the projection of H moves from H' parallel to EE'

341

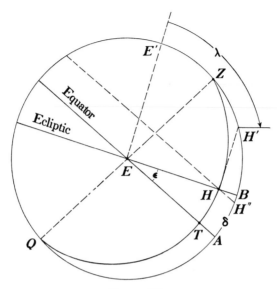

Figure VI-4

FIGURE VI-4

until it merges into EB at H. The projection of circle THZ is the ellipse with center at E and vertex at Z, passing through H. But there is no need actually to construct the ellipse, since all that is desired is the arc TH in true size and shape. To display this arc we rotate its plane about EZ down into the working plane. The projection of T moves to A, and that of H moves parallel to EA, meeting the circle at H''. The required δ is the arc AH''. The reader will find it easy to verify from Figure VI-4 that $\delta = $ arc sin (sin ϵ sin λ), $EA = 1$; and the same result can be expressed in terms of the chord function. Additional illustrations of the technique can be found in the *Analemma* of Ptolemy, a work less well known than the *Almagest*.

APPLICATION OF MENELAUS' THEOREM ON THE SPHERE

A technique for solving spherical figures by working on the surface of the sphere itself was formulated by Menelaus of Alexandria (*c.* A.D. 100), but since his *Sphaerica* exists only in a heavily edited Arabic version, we refer again to the *Almagest* (I, 13), the source from which our examples flow.

The basic theorem has plane and spherical cases. The spherical case is what we need, and we employ the plane case to prove it. The latter is usually formulated nowadays by saying that whenever a triangle

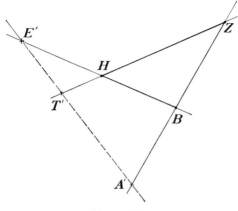

FIGURE VI-5

(see *HZB* in Fig. VI-5) is cut by a transversal (*E'T'A'*) each of the three sides is divided, internally or externally, into two segments, and that of these six segments the product of three having no end point in common (*ZA'* · *T'H* · *E'B*) equals the product of the other three (*A'B* · *ZT'* · *HE'*).

The ancients thought of it rather in terms of a relation between three ratios, namely,

$$(1) \qquad \frac{Z A'}{A' B} = \frac{Z T'}{T' H} \cdot \frac{H E'}{E' B}.$$

Moreover, they stated the theorem as involving four intersecting lines, a complete quadrilateral. For any such configuration any one of the four lines can be thought of as being the transversal and the other three as forming the triangle. Hence four expressions like (1) above are associated with each Menelaus configuration. The plane case can be proved by, say, drawing a line from *B* parallel to *HZ* and working with the pairs of similar triangles thus formed /JOHNSON:47/.

To pass over to the spherical case, we need the following lemma: *The ratio between the chords of twice the arc-segments into which a radius divides an arc (internally or externally) equals the ratio between the corresponding segments of the arc's chord.* As applied to the arc *AG* Figure VI-6, where cases of both internal and external division are displayed, the lemma asserts that

$$(2) \qquad \frac{\text{crd } (2\widehat{A B})}{\text{crd } (2\widehat{B G})} = \frac{A E}{E G}.$$

343

Proof is immediate by use of the shaded similar triangles shown in Figure VI-6.

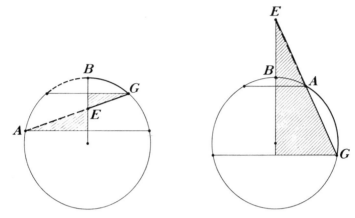

FIGURE VI-6

The spherical case of the main theorem simply substitutes a spherical complete quadrilateral for the plane one and the chord of twice the analogous arc for each of the six plane segments. That is, in Figure VI-7,

$$(3) \qquad \frac{\text{crd } 2\widehat{ZA}}{\text{crd } 2\widehat{AB}} = \frac{\text{crd } 2\widehat{ZT}}{\text{crd } 2\widehat{TH}} \cdot \frac{\text{crd } 2\widehat{HE}}{\text{crd } 2\widehat{EB}}.$$

Proof is immediate after a plane is passed through the three points Z, H, and B. The intersections of the planes of all the great circles with this plane result in a plane Menelaus configuration in which expression (1) holds; then an application of the lemma to each of the sides of triangle HZB effects the passage from (1) to (3).

An illustration of how the theorem is applied is found in *Almagest* I, 14. It is the astronomical problem already posed and solved above. Referring back to Figure VI-3, we note that expression (3) becomes

$$\frac{\text{crd } 2(90°)}{\text{crd } 2\epsilon} = \frac{\text{crd } 2(90°)}{\text{crd } 2\delta} \cdot \frac{\text{crd } 2\lambda}{\text{crd } 2(90°)} ,$$

since Z is the pole of TA, and E the pole of AB.

This is equivalent to

$$(4) \qquad \text{crd } 2\delta = \frac{\text{crd } 2\lambda \cdot \text{crd } 2\epsilon}{120} ,$$

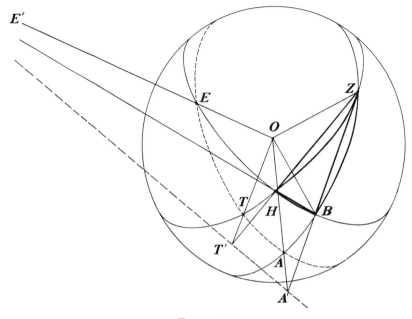

FIGURE VI-7

and since λ and ϵ are both given, δ can be calculated with the aid of the table of chords.

The *Almagest* teems with examples of this sort; indeed, it can be said that for some centuries spherical trigonometry *was* the Menelaus theorem. It is a powerful, although clumsy, tool for the solution of spherical astronomical problems. One must find or construct a complete quadrilateral on which, of the four possibilities, an expression of type (3) exists such that one of the six arcs is the unknown whose magnitude is required and the others are known. When all the original arcs intersect obliquely, the application is complicated indeed. It is stretching a point even to call it trigonometry, since the spherical triangle as such plays no role at all. Neither does the spherical angle enter directly—only arcs are involved.

APPEARANCE OF THE SINE FUNCTION

In the applications of the chord function given above (plane as well as spherical), it is necessary to double the arc before using it as an argument in the table of chords. It would be handier to have a table in which the original arc is itself the independent variable. Eventually someone thought of calculating and using half the chord of double an arc. Once this was done, the sine function had been born [**95**].

The earliest sine tables turn up in India, where they doubtless originated. Their unknown inventors, however, were in possession of Babylonian and Greek mathematical ideas transmitted as by-products of a flourishing Roman trade with South India via the Red Sea and the Indian Ocean. /WHEELER; PINGREE/.

The *Surya Siddhanta* is a compendium of astronomy made up of cryptic rules in Sanskrit verse, with little explanation and no proofs /BURGESS/. It was composed in the fourth or fifth century A.D., but the extant version has been revised so frequently that it is difficult to say which sections are in their original form. We quote several verses translated from Book II:

> 15. The eighth part of the minutes of a zodiacal sign is called the first sine [$S_1 = 30°/8 = 1800'/8 = 225'$]; that, increased by the remainder left after subtracting it from the quotient arising by dividing it by itself, is the second sine [$S_2 = S_1 + (S_1 - S_1/S_1)$].
> 16. Thus, dividing [the sum of] the tabular sines in succession by the first and adding to them in each case [to each successive sine, S_n] what is left after subtracting the quotients from the first

$$(5) \quad \left[\begin{array}{l} S_3 = S_2 + \left(S_1 - \dfrac{S_1 + S_2}{S_1}\right), \\ \cdots\cdots\cdots\cdots\cdots\cdots\cdots\cdots\cdots\cdots\cdots\cdots\cdots\cdots\cdots\cdots\cdots\cdots\cdots \\ S_{n+1} = S_n + \left(S_1 - \dfrac{S_1 + S_2 + S_3 + \cdots + S_n}{S_1}\right), \\ n = 1, 2, \cdots, 23, \end{array}\right]$$

the result is twenty-four tabular sines in order [as shown in Table VI-5].

The next verses give the numbers partially tabulated in the column headed by S_n. These numbers were written out in strings of words which conform to the poetic scheme of the entire work. The text then continues:

> 22. Subtracting these, in reversed order, from the half-diameter gives the tabular versed sines [V_n in the table].

(The reader should note that this table as such does not appear in the Sanskrit original. The *Surya Siddhanta* gives only the numbers recorded in columns S_n and V_n.)

TABLE VI-5

n	$225'n$	S_n	ΔS_n	$\Delta_2 S_n$	V_n
1	$3\frac{3}{4}°$	225'	224	− 1	7'
2	$7\frac{1}{2}°$	449'	222	− 2	29'
3	$11\frac{1}{4}°$	671'	219	− 3	66'
4	15°	890'	215	− 4	117'
5	$18\frac{3}{4}°$	1105'	210	− 5	182'
.	.	.	.	.	.
.	.	.	.	.	.
.	.	.	.	.	.
8	30°	1719'	191	− 8	460'
.	.	.	.	.	.
.	.	.	.	.	.
22	$82\frac{1}{2}°$	3409'	22	−15	2989'
23	$86\frac{1}{4}°$	3431'	7		3213'
24	90°	3438'			3438'

If plotted, the sequence S_n looks like a quarter of a sine wave; in particular, it rises to half its amplitude in traversing a third of the way to its maximum (i.e., $S_8 = S_{24}/2$). It, like almost all the other primitive sine functions, is defined in terms of a circle whose radius, R, is not unity. We distinguish all these from the modern sine function by use of an initial capital S, explicitly displaying the parameter R, when useful, as a subscript. In general, then,

$$\text{Sin}_R \ \theta \equiv R \sin \theta$$

(and analogously for the other trigonometric functions, when they are introduced).

Here in particular, since always $\text{Sin } 90° = R$,

$$S_n = \text{Sin}_{3438'} \ (n \cdot 225') = 57;18 \sin (n \cdot 3;45°).$$

The peculiar-seeming R was so chosen that the resulting function would have the desirable property

$$\text{Sin } \theta \to \theta \quad \text{as} \quad \theta \to 0,$$

which is analogous to the situation with our radians.

That this holds can be demonstrated by regarding the degree not as an angular unit but as a measure of arc length. Then the circumference of any circle is $360° \times 60 = 21,600'$, and, measured in the same units, its radius is $21,600'/2\pi \approx 3,438'$, to the nearest integer.

This also supplies the clue for explaining why 225' is the tabular increment for the independent variable. Commencing with the easily

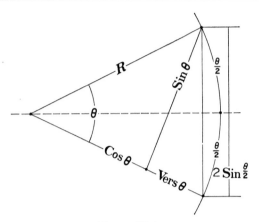

FIGURE VI-8

calculable Sin 30°, apply successively the "half-angle algorithm" expressed in modern symbols as

$$(6) \qquad\qquad \text{Sin } \frac{\theta}{2} = \tfrac{1}{2} \sqrt{\text{Sin}^2 \theta + \text{Vers}^2 \theta},$$

where Vers $\theta \equiv R - \text{Cos } \theta \equiv R(1 - \cos \theta) \equiv R \text{ vers } \theta$, is the "versed
sine," a function still used by surveyors, the last column (V_n) in the
table above (see also Fig. VI-8). This gives the Sines of 15°, then $7\tfrac{1}{2}°$,
and next Sin $3\tfrac{3}{4}° = \text{Sin } 225' \approx 225'$, again to the nearest integer. For
this order of precision there was no point in carrying the process further.

The recursion relation (5) for building up the table from its initial
value alone, with no appeal to geometry, was probably worked out
after the table had been computed. Equation (5) can be written as

$$\Delta S_n = S_1 - \frac{1}{S_1} \sum_1^n S_n,$$

and application of the difference operator ($\Delta_2 U_n = \Delta U_{n+1} - \Delta U_n$)
yields

$$(7) \qquad\qquad \Delta_2 S_n = -\frac{1}{S_1} S_{n+1}.$$

If we take the trigonometric definition of S_n and form its differences
we obtain, putting $\theta = 3;45°$,

$$\Delta S_n = R(\sin \theta(n + 1) - \sin \theta n) = 2R \sin \frac{\theta}{2} \cos (\theta(n + \tfrac{1}{2}))$$

$$\Delta_2 S_n = 2R \sin \frac{\theta}{2} \{ \cos (\theta(n + \tfrac{3}{2})) - \cos (\theta(n + \tfrac{1}{2})) \}$$

348

$$= -4R \sin^2 \frac{\theta}{2} \sin(\theta(n + 1))$$

$$= -\left(4 \sin^2 \frac{\theta}{2}\right) S_{n+1}$$

$$\Delta_2 S_n \approx -\frac{1}{233} S_{n+1},$$

which is close to (7). Evidently someone noticed the linear relation between the function and its second differences (analogous to the differential equation $D^2 \sin x = -\sin x$) and formulated it as (5).

The *Surya Siddhanta* value of R was not completely standard in Indian astronomy. Brahmagupta (c. A.D. 628), in the *Brahmasphuta Siddhanta* used 3,270′, perhaps based on the approximation $\pi \approx 360/109$. In another work, the *Uttara Khandakhadyaka*, he puts R equal to 150′, probably because then the sine of the inclination of the ecliptic (ϵ in Fig. VI-3) is very nearly unity. Other R's were in use, adopted for reasons unknown to us /Biruni (a): 132–49/. Nor was the tabular increment of 3;45° the only one used. An interval of 15°, obtained by a single application of (6), is sometimes found. Under these circumstances linear interpolation becomes quite inaccurate, and the *Uttara Khandakhadyaka* /Brahmagupta: 141/ gives the following interpolation rule for approximating, say, $S(\theta)$ in terms of three tabular values $S_{-1} = S(\theta_{-1})$, $S_0 = S(\theta_0)$, $S_1 = S(\theta_1)$, and the increment

$$d = \Delta\theta = 15° = 900′:$$

> Multiply the residual arc left after division by 900′ $[(\theta - \theta_0)/d]$ by half the difference between the tabular difference passed over and that to be passed over and divide by 900′ $[(\Delta S_0 - \Delta S_{-1})/2d]$. By the result increase or decrease, as the case may be [add it algebraically], half the sum of the same two tabular differences $[(\Delta S_0 + \Delta S_{-1})/2]$. The result, whether less than or greater than the tabular difference to be passed over $[\Delta S_0]$, is the true [modified] tabular difference to be passed over [regard this as a new ΔS_0 upon which to apply linear interpolation].

Put symbolically, the modified tabular difference is

$$\frac{\Delta S_0 + \Delta S_{-1}}{2} + (\theta - \theta_0) \frac{\Delta_2 S_{-1}}{2d},$$

and it is to replace the actual tabular difference, ΔS_0, in the linear interpolation expression

$$S \approx S_0 + \frac{\theta - \theta_0}{d} \Delta S_0.$$

349

That is,

$$S \approx S_0 + \frac{\theta - \theta_0}{d} \left[\frac{\Delta S_{-1} + \Delta S_0}{2} + (\theta - \theta_0) \frac{\Delta_2 S_{-1}}{2d} \right],$$

which is Gauss's forward formula for second differences /MILNE-THOMSON: 63/.

Thus the astronomers of classical India not only originated the sine function but also entered, however intuitively, into subjects later to be called "difference equations" and "interpolation theory."

FROM COMPLETE QUADRILATERAL TO SPHERICAL TRIANGLE

Beginning with the ninth century, the number of people working in trigonometry increased markedly. Astronomers all, they lived in and traveled widely over a region reaching from India to Spain: the Iranian plateau, Iraq, Syria, Egypt, North Africa, and Spain. Of extraordinarily varied ethnic background—Persian, Arabic, Turkish, and so on—almost all shared a common faith, Islam, and a common language of science, Arabic. For the most part they, like poets and historians, were supported by temporal rulers whom many served as court astrologers. A picture (probably typical) of the life of a scientist at such a court has come down to us in a letter written from Samarquand by Jamshid al-Kashi to his father /KENNEDY (b)/. Al-Kashi, a man with no false modesty, regales his correspondent with tales of how he has, in the presence of the sultan, confounded all his mathematical competitors.

For the most part, the manuscripts left by these people have not been looked at in modern times, although even the treatises that have been studied make up a formidable mass. Indian scientific books were the first to receive the attention of Moslem scholarship; some were translated into doggerel Arabic verse in imitation of the Sanskrit *slokas* /BIRUNI (b): 142/.

Later the available Greek works were translated. The sine function was quickly adopted in preference to the chord. In fact, the etymology of the word "sine" indicates the wide variation in background of those who dealt with the function it designates. The Indians called the function *ardhajya*, Sanskrit for "half chord." This was shortened to *jya* and transliterated into three Arabic characters, *jyb*. This can be read as *jayb*, Arabic for "pocket" or "gulf." It was so read by Europeans, who translated it into Latin *sinus*, whence English "sine."

The "rule of four quantities" marked a stage in the transition from a calculus dealing with arcs of a spherical quadrilateral to spherical trigonometry proper, involving the sides and *angles* of a spherical triangle. This theorem states that *in a pair of right spherical triangles having an acute angle* (A *and* A′) *in common or equal, the following relation holds:*

(9) $$\sin a / \sin a' = \sin c / \sin c'.$$

It is immaterial whether modern or medieval sines are written, since only ratios are involved. The theorem is transitional because, although it utilizes triangles, angles are not dealt with.

The reader will find a proof of (9) by means of Menelaus' theorem to be straightforward if (in Fig. VI-9) ACC' is taken as the transversal cutting the sides of triangle PBB'. P is, of course, the pole of the great circle ACC', hence PC and PC' are quadrants. The Menelaus equation (3) can be stated in terms of sines by use of the identity

$$\text{Sin } \theta \equiv \tfrac{1}{2} \text{ Crd } 2\theta.$$

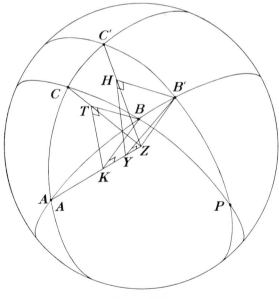

Figure VI-9 also illustrates a proof of the same theorem worked out by Abu'l-Wefa (*c.* 970), an able and prolific mathematician who was born in eastern Iran but worked in Baghdad. This proof furnishes an example of the common technique of proving relations on the surface of the sphere by constructing plane rectilinear objects inside the sphere. Take the radius of the sphere as R. The right (spherical) angles being at C and C', draw BT and $B'H$ perpendicular to the plane of ACC'. Draw also BK and $B'Y$ perpendicular to AZ. Then, the plane right triangles KTB and YHB' being similar, it follows that

$$\frac{KB(= \text{Sin } AB)}{YB'(= \text{Sin } AB')} = \frac{BT(= \text{Sin } BC)}{B'H(= \text{Sin } B'C')},$$

which implies the theorem.

The same configuration was used to prove also the related expression,

$$\text{Sin } b/\text{Sin } b' = \text{Tan } a/\text{Tan } a',$$

the tangent function being well known by the time of Abu'l-Wefa (see the following section).

At about the same time, three individuals independently discovered the sine law for the general spherical triangle:

$$\text{Sin } a/\text{Sin } A = \text{Sin } b/\text{Sin } B = \text{Sin } c/\text{Sin } C,$$

a theorem exemplary of the new trigonometry in that it explicitly employs functions of angles.

Figure VI-10 illustrates Abu'l-Wefa's proof. This is in contrast to the previous theorem; he now operates entirely on the surface of the sphere. Only great circles are employed. ABC is the given triangle; DTE and ZHE have B and C, respectively, as poles. Then EHZ, EAY, and ETD all intersect CB at right angles, arcs TB and HC are quadrants, TD is the measure of angle B, and HZ measures C. Hence application of (9) to the two pairs of right triangles having common angles at B and C, respectively, gives

$$\frac{\text{Sin } BT}{\text{Sin } BA} = \frac{\text{Sin } DT}{\text{Sin } AY},$$

and

$$\frac{\text{Sin } CH}{\text{Sin } CA} = \frac{\text{Sin } HZ}{\text{Sin } AY}.$$

Elimination of the quadrant arcs and the common AY between these yields

$$\text{Sin } CA/\text{Sin } B = \text{Sin } BA/\text{Sin } C,$$

which is the theorem.

352

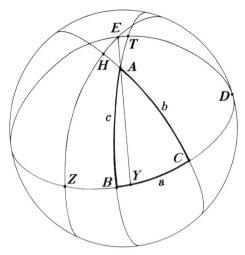

FIGURE VI-10

An application is the astronomical problem already solved by an analemma and the Menelaus' theorem. Applying the sine theorem to the right triangle ETH in Figure VI-3, we immediately obtain

$$\text{Sin } \delta = (\text{Sin } \epsilon \cdot \text{Sin } \lambda)/R,$$

since Sin $90° = R$.

MORE FUNCTIONS AND TABLES

The subject matter of the previous section is primarily geometrical. Its development was accompanied by an accumulation of numerical and computational materials and techniques. By the ninth century, instead of the primitive schemes displayed in Tables VI-1-4, tables of the "(horizontally) extended shadow" were common [96]. The tables gave as a function of the sun's altitude the shadow lengths cast by it on a horizontal plane. (Lengths were measured in units of a standard vertical gnomon.) These are tables of

$$\text{Cot } \theta \equiv R \cot \theta,$$

usually calculated for $\theta = 1°, 2°, 3°, \ldots, 90°$.

Al-Biruni, a great scientist who lived in Central Asia in the eleventh century, wrote an exhaustive treatise on shadow lore, and we can do no better than to take over part of what he says /(b)/. Among Orientals, he asserts, it was customary to use a gnomon of a handspan in length, this being the size of many common objects such as a tent peg, ruler, awl, or knife (the knife of a virtuous man, he is careful to state—that of a malefactor is longer). A span is three handbreadths of four digits (fingerbreadths) each. These were convenient units, available to all; hence for these people R was equal to 12.

Or a man might use himself as a gnomon and step off his own shadow with his feet, taken to be a seventh of his height ($R = 7$). Others noted that a plumb line from the top of the cranium would bisect the feet, hence they preferred $R = 6\frac{1}{2}$. Tables of all three varieties of cotangent function were common in the Middle Ages.

Less widespread were tables of the "reversed shadow," that cast by a horizontal gnomon on a plane normal to it, the gnomon lying in the plane of the altitude circle. These are tables of

$$\text{Tan } \theta \equiv R \tan \theta,$$

where again the argument was originally the solar altitude.

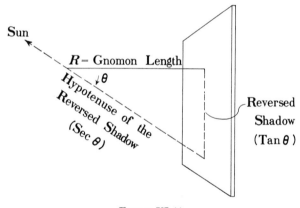

FIGURE VI-11

Rarely tabulated, but explicitly defined and applied in Sanskrit as well as Arabic works, were relations called the "hypotenuse of the shadow." For the reversed shadow this function is illustrated in Figure VI-11. We are justified in applying the modern term "secant" to this function, since it is

$$R/\cos \theta = R \sec \theta = \text{Sec } \theta.$$

In like manner, the "hypotenuse of the extended shadow" is

$$R/\sin \theta = R \csc \theta = \text{Csc } \theta.$$

354

In many rules and discussions the vernacular equivalent of our phrase "sine of the complement of θ" occurs; that is,

$$\text{Sin } (90° - \theta) = R \sin (90° - \theta) = R \cos \theta = \text{Cos } \theta.$$

So, certainly by the end of the ninth century A.D. and probably well before then, all six of the common modern trigonometric functions were well established, and the identities connecting them were in full application [97].

A few people had recognized and corrected the inconvenience of the parameter $R \neq 1$. Since computations in the sexagesimal system were customary, this inconvenience could be minimized by putting $R = 60 = 1,0$, in imitation of the Ptolemaic chord function. The astronomer Habash ($c.$ 830), for instance, tabulated in his astronomical handbook the function $(1,0) \cdot \tan \theta$. Then multiplication or division by R becomes a mere matter (as we would put it) of shifting the sexagesimal point one place to the right or left. The ancients called it "elevating" or "depressing" the number by one place. At least two people, Abu'l-Wefa, and al-Biruni, took the final step of putting $R = 1$, but this measure was of so little interest to their contemporaries and successors that it was ignored until recently.

Precision of tabular material steadily increased. The astronomical handbook of al-Khazini ($c.$ 1120) has a table of Sin θ to three significant sexagesimal places, with columns of first- and second-order differences. But the handbook of Ulugh Beg ($c.$ 1440) has a sine table calculated, for each minute of arc, to five sexagesimal places; the precision of this magnificent example of computational mathematics is one to 60^5. The basis of the table is a determination of Sin $1°$ in terms of a value for Sin $3°$ previously arrived at by methods essentially the same as those Ptolemy used in calculating the table of chords; see the section entitled "Shadow Tables and the Calculus of Chords," above. Our letter-writing acquaintance al-Kashi, the man behind Ulugh's table, then proceeded to solve the "trisection cubic"

$$x = \frac{x^3 + (15,0) \cdot \text{Sin } 3°}{(45,0)}$$

by an iteration process. The root is the desired Sin $1°$. Such iteration methods are at least as old as Ptolemy, and in fact the "Kepler equation,"

$$x = y + k \sin y,$$

was solved a good six centuries or more before the time of Johann

Kepler (1571–1630), probably in India, and by the same iterative method /AABOE (a); KENNEDY and TRANSUE/.

Only the best of the Moslem mathematicians computed in pure sexagesimals. Most of them would convert all given quantities into decimal integer expressions in terms of the lowest order of sexagesimal fractions present (e.g., 2;11,52 = 2,11,52 seconds = 7,912 seconds), proceed with the computation, and then convert the result back into sexagesimals.

The decimal place-value system had been introduced into the Middle East by the eighth century or thereabouts, and from there it was carried to Europe. When, with Regiomontanus (1436–1476), the study of trigonometry began to be activated in Europe, decimals began to replace sexagesimals in trigonometric tables. Regiomontanus first equated R to 60×10^5, the base of the earlier system exerting a residual effect, but he later adopted the value of 10^7. (Decimal fractions had been fully expounded by al-Kashi some decades before this time. They were not widely used, however, until after the publication of *La Disme* by Simon Stevin in 1585.)

Each increase in precision demanded an increase in the integer R. Rheticus (1514–1576) commenced a table of tangents and secants for each ten seconds of arc, taking $R = 10^{15}$. The work was completed by a pupil.

INTRODUCTION OF SYMBOLS

We are so accustomed to expressing relations by means of compact symbols that it is tempting to identify the symbols with mathematics itself. In fact, however, throughout all of antiquity and the Middle Ages, even for a long time after the center of activity had shifted to Europe, the subject matter was written out in ordinary language. The major task of the historian seeking to understand a piece of ancient mathematics is not so much to put it into his own mother tongue as to express it, if possible, in modern symbols and thus to disclose its abstract form. Any development toward a more compact way of displaying complex relationships is a fundamental advance.

Such an advance in the field of trigonometry was made by the French mathematician François Viète (1540–1603; he was also known as Vieta). As an example of his notation we give the "trisection cubic" already encountered. As displayed by him the equation is

Xq in E3 — Ec aequatur Xq in B,

where "B" and "E" have the meanings indicated in Figure VI-11, "q" and "c" stand for "square" and "cube," respectively, the minus sign is as we use it, "aequatur" means "equals," and "in" stands for "multiplication" /BRAUNMÜHL: I, 167/. In modern notation the expression is

$$(x^2 \cdot 3e) - e^3 = x^2 \cdot b,$$

and it is not hard to show that it is equivalent to the identity

$$3 \sin \theta - 4 \sin^3 \theta = \sin 3\theta.$$

Viète worked out equations for the cosines of multiple angles: θ, 2θ, 3θ, . . . , 10θ. His expressions for $2 \cos 7\theta$ and $2 \cos 8\theta$ are

$$1QQC - 7QC + 14C - 7N$$

and

$$1QCC - 8CC + 20QQ - 16Q + 2,$$

respectively, where N stands for $2 \cos \theta$. We would write the same polynomials as

$$n^7 - 7n^5 + 14n^3 - 7n$$

and

$$n^8 - 8n^6 + 20n^4 - 16n^2 + 2,$$

respectively.

The duality subsisting between corresponding sides and angles of pairs of polar spherical triangles was also noted and applied by Viète /SISTER ZELLER: 84, 106/. Thus, to find the angles of a triangle when the three sides are given he applies the formula

$$\frac{\sin a \sin b}{\cos c \mp \cos a \cos b} = \frac{1}{\cos C},$$

previously proved by Regiomontanus in a form that uses the versed sine function encountered by us earlier:

$$\frac{\text{vers } C}{\text{vers } c - \text{vers } (a - b)} = \frac{1}{\sin a \sin b}.$$

Viète wrote down the dual theorem that gives the sides of a spherical triangle in terms of its angles:

$$\frac{\sin A \sin B}{\cos A \cos B \mp \cos C} = \frac{1}{\cos c}.$$

Trigonometric Functions in the Complex Domain[2]

It remains for us to sketch how the breakthrough in analysis, which was occasioned by the invention of the infinitesimal calculus, engulfed old-fashioned trigonometry in a flood of new theory.

Trigonometry began as a tool for the mensuration of geometric figures; today we think of it as a set of relations between complex numbers, with no necessary appeal to arcs or angles.

The transitional process started with the exhibition of infinite-series representations for the trigonometric functions. By the end of the seventeenth century Isaac Newton was in possession of the series

$$\sin x = x - \frac{x^3}{3!} + \frac{x^5}{5!} - \frac{x^7}{7!} + \cdots$$

and

$$\cos x = 1 - \frac{x^2}{2!} + \frac{x^4}{4!} - \frac{x^6}{6!} + \cdots .$$

Since he also knew that (as we would write it)

$$e^x = 1 + x + \frac{x^2}{2!} + \frac{x^3}{3!} + \cdots ,$$

he was already close to a relation between the trigonometric and the exponential functions.

Imaginary numbers had been encountered, long before this, as roots of polynomial equations. In 1545 Girolamo Cardano, with some bewilderment, had found the roots of a certain equation to be $5 + \sqrt{-15}$ and $5 - \sqrt{-15}$. In 1702 Johann Bernoulli adduced a formal connection between imaginaries, the arc tangent (inverse of a trigonometric function), and the logarithm (inverse of an exponential function). He accomplished this by differentiating the expression

$$z = \frac{(t-1)b}{t+1} \sqrt{-1},$$

to obtain the differential equation

$$\frac{dz}{b^2 + z^2} = -\left(\frac{dt}{2bt\sqrt{-1}} \right).$$

[2] The material in this section is from M. Cantor /II, 508, and III, 73, 362, 382/.

Finally, in 1740, Leonhard Euler showed that

$$e^z = \lim_{n \to \infty} \left(1 + \frac{z}{n}\right)^n$$

and that

$$\sin z = \frac{e^{iz} - e^{-iz}}{2i}, \qquad \cos z = \frac{e^{iz} + e^{-iz}}{2},$$

where $i = \sqrt{-1}$. These may be regarded as definitions of the trigonometric functions, whereupon all identities connecting the functions can be derived from the new definitions.

Capsule 91   Larry Mossburg

THE *ALMAGEST* OF PTOLEMY

AT THE beginning of the Christian era one of the outstanding men in mathematics was the astronomer Claudius Ptolemy, who wrote *Syntaxis mathematica* ("Mathematical Collection"). To distinguish this work from less important ones, commentators referred to it as *magiste,* "greatest." Subsequently, Arab commentators prefixed the Arab definite article *al,* and the work became known as *Al-magest* ("the greatest"), then as "the *Almagest.*" Most of the *Almagest* is based on the work of the early Greek astronomer Hipparchus (*c.* 180 to *c.* 125 B.C.). However, it exerted a great influence because of its clear and elegant style, and it was considered to be the standard work on astronomy until Nicolaus Copernicus (1473–1543) and Johann Kepler (1571–1630) introduced their heliocentric theory of the solar system. Ptolemy's original contributions in the *Almagest* consist of a theory of motion of the five planets, for which Hipparchus and others had merely collected observational data.

For the mathematician the *Almagest* is of interest because of the trigonometric identities Ptolemy devised to help him in compiling his table of chords (which is roughly equivalent to a sine table). The circumference of the circle was divided into 360 parts (now called degrees); the diameter was divided into 120 divisions; then each of these divisions was divided into 60 parts called *partes minutae primae* ("first small parts"), and each of the first small parts was subdivided into 60 parts called *partes minutae secundae* ("second small parts"). Hence our use of the terms "minutes" and "seconds." The reason for sixtieths was the influence of the sexagesimal system of the Babylonians, whose work in astronomy was known to the Greeks. Ptolemy's table of chords gave chord lengths for arcs from $\frac{1}{2}°$ to $180°$ in steps of $\frac{1}{2}°$. Thus, the chord length for a $90°$ arc would be

$$\text{crd } 90° = R\sqrt{2} = 84^{\text{p}}51'10''$$

expressed sexagesimally and taking $R = 60$ (Fig. [91]–1).

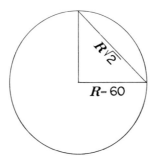

FIGURE [91]-1.

Ptolemy's ingenious use of identities for computing chords is based on the so-called Ptolemy's theorem of which he gives a proof; but the theorem was known before Ptolemy. It asserts

$$AB \cdot CD + BC \cdot AD = AC \cdot BD.$$

Rewriting, we have

$$AD \cdot BC = AC \cdot BD - AB \cdot CD$$

or, since $R = 60$,

(1) $120 \text{ crd } (\gamma - \beta) = \text{crd } \gamma \cdot \text{crd } (180° - \beta) - \text{crd } \beta \cdot \text{crd } (180° - \gamma),$

which is essentially the same as our familiar

(2) $\sin (U - V) = \sin U \cos V - \sin V \cos U;$

and, indeed, we can obtain (2) immediately from (1) by letting crd $\theta = 2R \sin \theta/2$, $\gamma = 2U$, and $\beta = 2V$. (See Fig. [91]-2, in which arc $AB = \beta$, arc $AC = \gamma$, and arc $BC = \gamma - \beta$, and Fig. [91]-3; $\sin \theta/2 = (\frac{1}{2} \text{ crd } \theta)/R$.)

360

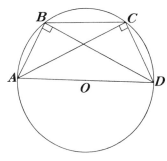

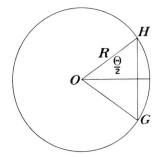

FIGURE [91]-2.   FIGURE [91]-3.

Thus, knowing the chords of two arcs, Ptolemy had a handy device for finding the chord of a third arc. He also used the equivalent of the half-angle formula,

$$\sin^2 \frac{\theta}{2} = \tfrac{1}{2}(1 - \cos \theta).$$

The *Almagest* is written in thirteen books, the contents of which are briefly listed here. Book I includes preliminaries to the Ptolemaic system, with general explanations of different heavenly bodies in relation to the earth as center, and propositions in spherical geometry, as well as the table of chords and their method of calculation. Book II contains more material on subjects covered in Book I. Book III contains information on the length of the year and the motion of the sun on the eccentric and epicycle hypotheses. In Book IV the lengths of the months and a theory of the moon are covered. Book V contains the construction of the astrolabe and more material on theory of the moon. Book VI contains information on conjunction and oppositions of sun and moon, solar and lunar eclipses and their periods. Books VII and VIII catalogue 1,028 fixed stars; the remaining books are devoted to the movements of five planets.

Perhaps the most important contribution of the *Almagest* was its promotion of the idea that a quantitative, mathematical description of natural phenomena is both desirable and possible.

For Further Reading

BRENDAN
AABOE (b): 101–27
HEATH (c): 402–11

HOOPER: 146–56
PTOLEMY (a)

ANGLE

EUCLID says, "A *plane angle* is the inclination to one another of two lines in a plane which meet one another and do not lie in a straight line." This definition was a departure from the earlier Greek idea that an angle was a deflection or a breaking of lines. To allow for a straight angle, Euclid immediately defines as "rectilineal" that angle whose lines are in the same straight line.

Greek philosophers debated whether the angle concept should be considered a quantity, a quality, or a relation—categories originated by Aristotle. Proclus, quoting his teacher Syrianus, says that the concept lies with a combination of these three, for "it needs the *quantity* involved in magnitude, thereby becoming susceptible of equality, inequality, and the like; it needs the *quality* given it by its form; and lastly, the *relation* subsisting between the lines or the planes bounding it."

The German H. Schotten /EUCLID: I, 179/ in 1893 summarized the definitions of angle by placing them in three categories: the difference in direction between two lines; the measure of rotation needed to bring one side from its original position to the position of the other, staying meanwhile in the plane of the angle; and, finally, that portion of the plane contained "between" the two straight lines defining the angle. In more recent years the definition of angle as a set of points has been frequently used.

Pierre Herigone, in a French work in 1634, was apparently the first to use a symbol for angle. He used both $<$ and $\angle$, although $<$ had already been used to mean "less than." The symbol $\angle$ survived, with some variants. In England both 7 and $\wedge$ appeared (*c.* 1750). During the nineteenth century in Europe such forms were used as $\widehat{ab}$ to designate the angle between a and b and $\widehat{ABC}$ to designate angle ABC. The arc on the angle symbol ($\sphericalangle$) first appeared in Germany in the latter half of the nineteenth century. In 1923 the National Committee on Mathematical Requirements, sponsored by the Mathematical Association of America, recommended $\angle$ as the standard symbol for angle in the United States.

362

For Further Reading

CAJORI (d): I, 403–6   D. E. SMITH (a): II, 277
EUCLID: I, 176–81

Capsule 93   *Donald L. Reinking*

RIGHT ANGLES

THE ancients measured the height of an object by placing a stick at right angles to the horizon and then comparing the lengths of the shadows cast by the object and the stick. These early geometers knew that a right angle was one of the basic ideas of geometry.

Euclid, too, knew this. In his *Elements* he gives the following definition:

> *When a straight line set on a straight line makes adjacent angles equal to one another, each of the angles is right, and the straight line standing on the other is called perpendicular to that on which it stands.*

Euclid further realized that it was essential to assert his fourth postulate,

> Postulate 4: *All right angles are equal to one another.*

Thus the right angle served as a fixed standard for measuring other angles. Accordingly, the terms "acute angle" and "obtuse angle" were then defined as being respectively less than and greater than a right angle.

Of greater interest and subtlety, however, is the relation of Postulate 4 to the famous "fifth parallel postulate" of Euclid.

> Postulate 5: *If a straight line falling on two straight lines makes the interior angles on the same side less than two right angles, the two straight lines, if produced indefinitely, meet on that side on which are the angles less than the two right angles.*

It is clear that Postulate 5 would be no criterion at all unless "two right angles" always represented the same fixed magnitude.

After Euclid, there were those who believed that Postulate 4 should be proved. Among these were Proclus (*c.* 460) and Saccheri (1733), both of whom attempted proofs—proofs that were defective because they assumed equivalent statements of Postulate 4 /EUCLID: I, 200/.

Hilbert (1899) approached Postulate 4, and other postulates, more carefully—and in a larger context. After stating six axioms of congruence, he proved several theorems on congruence of triangles and angles and then proved that all right angles are equal.

The symbol ∟ for right angle was used as early as 1698 by Samuel Reyher. Using a vertical line | for equality and ∧ for angle, he writes ∧*A*|∟ for "angle *A* is a right angle." The commonly used "rt. ∠" appeared in America around 1880 in the widely used Wentworth geometry textbook.

For Further Reading

CAJORI (d): I, 406–8   HILBERT: 12–24
EUCLID: I, 181–82, 200–201

Capsule 94   Phillip S. Jones

ANGULAR MEASURE

WHY are there 360° in a complete revolution? There are no reasons for this except historical ones. Even these reasons must eventually give way to hypotheses and these to a mere "because" as is always the case when we continue to ask "why?" and "where?" at each step of an investigation backward into either time or logic. Let's review the story of 360° briefly.

The ancient Babylonians, having settled down (4000–3000 B.C.) to drain marshes, cultivate fields, build cities, and exchange goods, found an interest in astronomy for its own sake, for its relation to religious concepts, and for its connections with the calendar, the seasons, and planting time. They also developed a number system based on 60, using the place-value idea for fractions as well as for whole numbers. (The idea of a decimal point and positions to its right representing tens, hundreds, etc., did not enter our Hindu-Arabic number system until about A.D. 1585—over four thousand years later!) Why the Babylonians chose 60 no one knows, though there are many interesting theories [1]. It might even be that the use of 60 followed from the easy subdivision of a circle into six equal parts using its radius as a chord. Perhaps the original source of the 60 was as $\frac{1}{6}$ of 360. The idea of 360 parts in a circle might have sprung from a slightly erroneous estimation of 360 days in a year. However, it seems likely that the sexagesimal number system preceded the division of the circle into 360 parts; certainly it preceded the subdivision of each part into 60 subparts. In any event, whether 60 or 360 came first, the Babylonians studied astronomy and used a sexagesimal number system in which fractions were written with denominators of 60 and 60 × 60 using a place-value notion much as we write decimal fractions.

Thus when the Greek civilization through trade and conquest partially absorbed the Babylonian culture, it took its sexagesimal fractions along with its astronomy. Hypsicles (*c.* 180 B.C.) was the first Greek astronomer to divide the circle of the zodiac into 360 parts, following the Chaldeans, who had divided it into 12 signs and each sign into 30 (and sometimes 60) parts. Neither Hypsicles nor the Chaldeans used this division for other circles. This generalization is apparently due to the astronomer Hipparchus (*c.* 150 B.C.).

Ptolemy (*c.* 125 A.D.), the famous Greek astronomer and geographer, made a general use of sexagesimal fractions in computation of all sorts, not merely in measuring angles. He did this, he said, to avoid the use of "fractions," thus indicating that the complete idea of a fraction as we teach it in our elementary schools was not clear to this Greek genius, but that he did appreciate the efficiency of the place-value notion in computations involving them. He used a decimal non-place-value arithmetic for integers, however, and did not use sexagesimal fractions for measuring time. This latter idea was introduced by his commentator, Theon of Alexandria (*c.* 390 A.D.).

The Babylonian sexagesimal fractions used in the Arabic translations of the Greek of Ptolemy were called by translators "first small parts" for sixtieths, "second small parts" for sixtieths of sixtieths, and so forth. The first European translations were into Latin, which was the international language of scholarship. In Latin these phrases became *partes minutae primae* and *partes minutae secundae,* from which in turn we have our words "minutes" and "seconds."

These words now in daily use epitomize, then, a story reaching back to prehistoric times which gives the real reasons for our use of 360 parts in dividing a circle. Note, though, that these reasons exist as a part of history but not as a part of logic or thought. The choice of units and of methods of subdivision is always arbitrary, but there may be thoughtfully developed "logical" reasons for selecting as a matter of convenience and simplification a particular unit or subdivision. Such conscious choice did not enter into the setting up of our degree-(hour)-minute-second subdivisions which are, then, the most popular and the least logical systems of angular measure. All other systems and units have been thoughtfully designed and deliberately named: radians, mils, grades, gones, cirs, decimally or centesimally divided degrees, and millicycles.

Radians contrast with degrees in their origin. This contrast is especially sharp if one looks at a strictly factual account as found in the standard works on the history of mathematics. Apparently mathematician Thomas Muir and physicist James T. Thomson independently considered the need for a new angular unit. Later they met and happened to discuss the need and a name for the unit. They settled upon "radian," after having been consulted by Alexander J. Ellis, as a compound of "radial angle," though Thomson may have been motivated by an analogy with "median." They had debated other possibilities, such as "radial" rather than "radian." The first appearance of this term in print was in an examination paper set by Thomson in 1873. Its introductions to the general reading public seem to have been in Ellis' *Algebra Identified with Geometry* (London: 1874) where he discussed the exponential form for complex numbers, and in the 1879 edition of William Thomson and P. G. Tait's *Treatise on Natural Philosophy.* However, Todhunter's *Plane Trigonometry* (1891) used the term "radial," while in this country Professors Oliver, Wait, and Jones of Cornell University in their duplicated manuscript, "Notes on Trigonometry" (1880), discuss the "expressing" of angles in terms of "π." In their printed second edition of 1884 they still do not use the term "radian" but refer to "π-measure," "circular," or "arcual measure." None of these authors explains why he adopted the unit, but their use of it makes it clear that their reasons lay in the resulting simplification of certain mathematical and physical formulas (especially the derivatives and integrals of trigonometric functions, and the expressions for velocities and accelerations in curvilinear motion).

366

Applications involving the radian measure of angles as exact quantities rather than as approximations are to be found not only in pure mathematics and theoretical mechanics, as suggested earlier, but also in such concrete situations as in optics, where the expression $x - \sin x$ has been tabulated, and in the design of gears, especially involute and cycloidal gears.

The mil, too, was deliberately selected and named to serve a particular purpose. Its history seems to be perhaps more recent and certainly more obscure than that of the radian.

Soldiers are often taught that a mil is an angle that subtends one yard at a thousand yards. This is not true. However, this statement is *nearly* true, and this approximation is the key to the chief, and perhaps sole, use of the mil as well as to the reason for its definition.

A few people distinguish between a "true mil" and the mil officially adopted by the United States Army. For such persons a "true mil" is the angle subtended at the center of a circle by an arc equal in length to .001 of the radius. This "one thousandth" accounts for the name of the unit, *mil*, which latter, then, always recalls its relationship to the radian. Further, referring to our earlier discussion of radians, it is easy to see that the sines and tangents of small angles are nearly one one-thousandth of their numerical measure in mils. Thus, in the diagram of an isosceles triangle in Figure [94]–1,

(1) $\frac{1}{2}S = R \tan \frac{1}{2}A.$

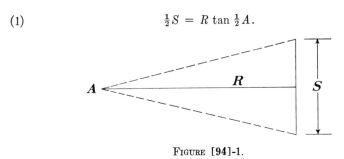

FIGURE [94]-1.

If A were small and measured in radians, we would have $\tan \frac{1}{2}A \approx \frac{1}{2}A$, and since the mil measure of A (or $\frac{1}{2}A$) is 1,000 times its radian measure, we have

(2) $\tan \frac{1}{2} A \approx \frac{1}{2} \left(\dfrac{A}{1,000} \right),$

if A is measured in mils. Substituting this in (1) we have $S \approx R \cdot (A/1,000)$. From this it follows that if $A = 1$ and $R = 1,000$, $S = 1$, or a "true mil" is approximately "the angle subtended by a yard at a thousand yards."

There would be 6,283.18 + "true mils" in a revolution, an awkward (and irrational—even transcendental) number. This would not be a convenient number for teaching, for use, or for graduating sighting and aiming instruments. Hence, the United States Army arbitrarily defined a mil to be 1/6,400 of a circle or revolution. The error thus introduced is slight. The Army's chief use of the mil is of course in directing artillery fire, determining the original range, and making corrections. Forward observers, often using binoculars with built-in scales for measuring in mils the angles subtended by objects of known or estimated size, have easy arithmetic rules for estimating distances (ranges) and angles.

The mil was first introduced by the Swiss army's artillery in 1864. It was adopted by the French in 1879 and by the United States Army in 1900.

This capsule is adapted from an article appearing in the October 1953 issue of the *Mathematics Teacher*.

Capsule 95   Roger D. Lowe and Cynthia Schenck

SINE AND COSINE

THE line segment joining the two end points of an arc of a circle was studied by certain of the pre-Christian Greeks. This chord (from the Latin *chorda*, "bowstring," which in turn comes from the Greek *chorde*, meaning the intestine of an animal and hence the string of a bow) may also be associated with the central angle that cuts off the chord. While the chord of an arc is not a sine, half the chord divided by the radius of the circle is the sine of half the arc (or half the central angle corresponding to the whole arc). Symbolically (Fig. [95]-1), if R is the radius of the circle, and α is the arc AB (or the corresponding central angle AOB) subtended by the chord, c,

(1)
$$\sin \frac{\alpha}{2} = \frac{c/2}{R} = \frac{1}{2R} \operatorname{crd} \alpha.$$

Note, in the figure, that $c = \overline{AB} = \operatorname{crd} \overset{\frown}{AB} = \operatorname{crd} \alpha.$

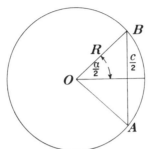

FIGURE [95]-1.

Theon of Alexandria (A.D. 390) in a commentary on earlier works wrote that Hipparchus (140 B.C.) had written twelve books on the computation of chords, including tables of chords. (A table of chords gave the length of chord for a given arc of a circle of radius R, where R was often taken as 60 units. Thus, an arc of 90° would have a chord length of $R\sqrt{2} = 84^+$; an arc of 60° would have a chord of $R = 60$; etc. See Fig. [95]-2.)

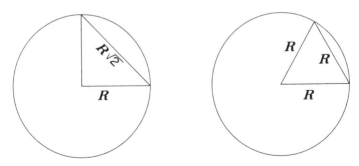

FIGURE [95]-2.

These twelve books are now lost, but Hipparchus says in a surviving commentary that he proved his results "by lines"—that is, by using properties of line segments inscribed in a circle. One such property is the so-called Ptolemy's theorem for a quadrilateral inscribed in a circle (Fig. [95]-3):

(2) $$AB \cdot CD + BC \cdot AD = AC \cdot BD.$$

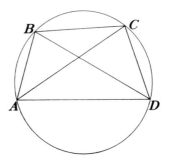

Figure [95]-3

Ptolemy (Claudius Ptolemaeus, *c.* A.D. 150) probably copied much material from Hipparchus, but his *Almagest* survives [91] and the name of Ptolemy is therefore closely associated with trigonometry. Ptolemy's *Almagest* gives a table of chords for arcs from $\frac{1}{2}°$ to 180° in $\frac{1}{2}°$ steps. Looking at formula (1) above, we see that this is equivalent to a table of sines from $\frac{1}{4}°$ to 90° in $\frac{1}{4}°$ steps. He thought of the circle as being divided into 360 degrees and the radius into 60 parts. Using Babylonian sexagesimal notation, he further divided each part into 60 subparts, and each of the latter into sixtieths, and so on. Thus he found that the chord of a 90° arc was 84 parts, 51 sixtieths, and 10 thirty-six hundredths. This was written 84ᵖ51′10″.

By around A.D. 500 the Hindu mathematician Aryabhata was computing half-chords. Somewhat later, tables of both the sine and the versed sine (that is, $1 - \cos \theta$) were computed by Hindu mathematicians. The sine was called *jya*, one of several spellings of the Hindu word which means half-chord. Later the Arabs transliterated this to *jyb*, which was later incorrectly read as *jayb* (Arabic for "pocket," "gulf," or "bosom") by the translator Gherardo of Cremona (*c.* 1150), who, translating from Arabic to Latin, used the Latin equivalent *sinus* which we now use in the form "sine."

Ulugh Beg (fifteenth century, Persian) found sin 1° by an approximate solution of a third-degree equation. Using a radius of 60 parts he found sin 1° to be 1ᵖ2′49″43‴11ⁱᵛ, where 43‴ means 43 sixtieths of a sixtieth of a sixtieth of a part.

Georg von Peurbach (1460) used a radius of 600,000 parts in computing sines, and Regiomontanus a decade later used a radius of 6,000,000 and later 10,000,000 parts. Rheticus (1550) repeated this accuracy and became the first European to discard the arc and use the trigonometric functions as ratios of sides of right triangles. The sine, and other functions, thus could be thought of as pure numbers rather than as lengths. By 1613 Pitiscus had published tables of sines to fifteen decimal places.

The term "cosine," for the sine of the complement of the angle, is due to Edmund Gunter (1620), who suggested combining the terms "complement" and "sine" into *co.sinus*, which was soon modified to *cosinus* and anglicized to "cosine."

Albert Girard (1626) wrote the sine of angle A as an uppercase italicized A, and the cosine of angle A as a lowercase italicized a. Though the abbreviation "sin" appears in a drawing by Gunter in 1626, it was not published in a book until 1634.

For Further Reading

CAJORI (d): II, 142–79   D. E. SMITH (a): II, 614–19
HOOPER: 118–35

Capsule 96   *Ruth Anne Miller*

TANGENT AND COTANGENT

WHILE the sine and cosine concepts had their beginnings in the context of astronomy, the tangent and cotangent emerged from the humbler requirements of practical measurements of heights and distances.

An example of computing the equivalent of the cotangent of an angle is found in Problem 56 of the Rhind papyrus (*c.* 1650 B.C.) [30], which gives the dimensions of a square pyramid and asks for the "seqt"—the number obtained when the horizontal "run" is divided by the vertical "rise" of the face of the pyramid. (See Fig. [96]-1, in

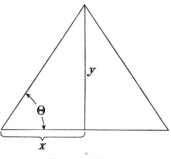

FIGURE [96]-1.

which $y = 250$ cubits, $x = 180$ cubits $= 180 \cdot 7$ "hands," and seqt $= 7x/y = 7 \cot \theta$.)

To the Egyptian the seqt was an indication of the slope of the face of the pyramid, and it corresponds to what we now call the cotangent, except for the Egyptian custom of measuring the run in "hands" (approximately 4 inches) and the rise in cubits (which the Egyptian took as equal to 7 hands). For a pyramid with square base of side 360 cubits and height 250 cubits, the scribe obtains the result (expressed in modern notation):

$$\text{seqt} = \frac{180 \cdot 7 \text{ hands}}{250 \text{ cubits}} = 5\frac{1}{25} \text{ hands per cubit.}$$

This seqt value of 5.04 is, of course, exactly 7 times the corresponding cotangent value of 0.72.

Preceding the tangent and cotangent functions were the ideas associated with shadows cast by the vertical stick or gnomon of sundials which were used in Egypt as early as 1500 B.C. In Figure [96]-2 the

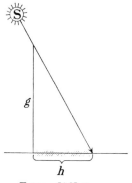

FIGURE [96]-2.

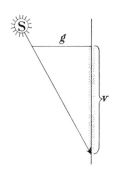

FIGURE [96]-3.

gnomon is represented by g, and h is the horizontal shadow, or *umbra recta*, of the upright gnomon. The basic idea was that a higher elevation of the sun gave a shorter shadow (essentially the cotangent concept). Indeed, "shadow tables" compiled in Egypt as early as 1200 B.C. gave decreasing shadow lengths for consecutive (increasing) hours of the morning.

Later versions of the sundial appeared on the walls of buildings so that the gnomon g was horizontal, and the shadow was vertical. (In Fig. [96]-3, v is the vertical shadow, or *umbra versa*, of the turned gnomon g.) Now higher elevations of the sun corresponded to longer shadows v (essentially the tangent concept).

By the ninth century A.D. the Arabs had compiled tables for vertical shadows v, as well as for horizontal shadows h. At first the gnomon's length was somewhat arbitrary; but by A.D. 830 the Arab mathematician Habash took $g = 60$ (the sexagesimal base) and compiled a table for v. In terms of the decimal system (base ten), his shadow table gave values for

$$v = 10 \tan \theta$$

(as we would now write it)—that is, his table was a "tangent" table except that the "decimal" (sexagesimal) point was shifted one place to the right. About fifty years later the gnomon was taken equal to one unit by the Arab mathematicians Abu'l-Wefa and al-Biruni.

Although the terms tangent and cotangent were not coined until later, the European mathematician Rheticus in 1551 explicitly defined each of these two functions as a ratio. Thomas Fincke (1583) contributed the name "tangent," perhaps because the vertical shadow length v lies along the tangent to the circle of radius g (Fig. [96]-4).

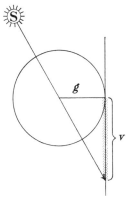

FIGURE [96]-4.

In 1620 Edmund Gunter gave the Latin equivalent of "cotangent of A" for "complement tangent of A," which stood for "tangent of the complement of A." In 1674 Jonas Moore gave the abbreviation "cot" for cotangent.

For Further Reading

CAJORI (d): II, 142–79
HOOPER: 157–63

D. E. SMITH (a): II, 620–22

Capsule 97   Eleanor Hayes

TRIGONOMETRIC IDENTITIES

TRIGONOMETRIC concepts originally involved measurement of chords and arcs. The identities as we know them came along later but were certainly used by Ptolemy (A.D. 150) and most likely by Hipparchus (140 B.C.) upon whose work Ptolemy built, in particular, his table of chords. Ptolemy made quite explicit use of the equivalent of the identities

$$\sin (\alpha \pm \beta) = \sin \alpha \cos \beta \pm \cos \alpha \sin \beta$$

and

$$\sin^2 \frac{\theta}{2} = \tfrac{1}{2}(1 - \cos \theta).$$

When the Hindus introduced the half-chord or sine concept, the Pythagorean identities became more obvious, and we find Varahamihira (*c.* A.D. 505) giving the verbal equivalent of

$$\sin^2 \theta + \cos^2 \theta = 1.$$

Hipparchus and Ptolemy had also used the equivalent of this identity.

The equivalent of the cosine law which we now write in the form

$$c^2 = a^2 + b^2 - 2ab \cos C,$$

is found, in equivalent form, in Propositions 12 and 13 of Book II of Euclid.

The sine law, which we write now as

$$\frac{a}{\sin A} = \frac{b}{\sin B} = \frac{c}{\sin C},$$

was known in an equivalent form by Ptolemy, but it was not set forth clearly until al-Biruni (973–1048) proved that the sides of a triangle have the same ratio as the sines of the opposite angles. The corresponding law of sines for spherical triangles was given by Regiomontanus (1464).

Abu'l-Wefa (940–998) gave the equivalent of

$$\sin 2x = 2 \sin x \cdot \cos x,$$

and also gave the equivalents of the following identities:

$$\tan x{:}1 = \sin x{:}\cos x,$$

$$\tan x{:}1 = 1{:}\cot x,$$

$$\sec x = \sqrt{1 + \tan^2 x}.$$

He also expressed sin x in terms of sine and cosine of $x/2$.

Nasir eddin (1201–1274) wrote *Treatise on Quadrilaterals*, the first book in which trigonometry is treated separately from astronomy. He covered both plane and spherical trigonometry.

François Viète (1540–1603) very systematically applied algebra to trigonometry, thus getting very close to analytic trigonometry. He rediscovered most of the elementary identities and obtained general formulas equivalent to expressions for sin nx and cos nx in terms of sin x and cos x. These had been known only in special cases before. Viète gave the cosine law essentially in the form

$$\frac{2ab}{a^2 + b^2 - c^2} = \frac{1}{\sin (90° - C)}.$$

For Further Reading

D. E. Smith (a): II, 628–31

VII

The History of THE CALCULUS

an overview by

CARL B. BOYER

The infinitives "to calculate," "to compute" (or the Middle English equivalent *counten,* from which we get "to count"), and "to reckon" all have similar meanings related to the carrying out of numerical processes. Of these three phrases, the last two are closely associated, etymologically and in current meaning, with mental processes. The first, on the other hand, carries from its origin a connotation of nondeliberative manipulation, for "to calculate" once meant "to reckon by means of pebbles." The word "calculus" is the diminutive of the Latin *calx,* meaning "stone." In medicine the literal meaning is still quite apparent in the phrase "a calculous person," referring to one suffering from kidney stones.

It is one of the ironies of history, then, that the phrase *"the* calculus" should have become firmly attached to a branch of mathematics which calls for subtlety and sophistication of thought in the highest degree. The inappropriateness of the term is clear from the fact that mastery of *the calculus* would be out of the question for anyone who found it necessary to fall back on pebbles for computational purposes.

CONCEPTIONS IN ANTIQUITY

In the more formal sense, the calculus was fashioned in the seventeenth century of our era; but the questions from which it arose had been asked for more than seventeen centuries before our era began.

Egyptian hieratic papyri and Babylonian cuneiform tablets include problems in rectilinear and curvilinear mensuration that are appropriate to the domain of the calculus; but pre-Hellenic treatment of these problems fell short of full-fledged mathematical stature in two serious respects: (1) there was no clear-cut distinction between results that were exact and those that were approximations only, and (2) relationships through deductive logic were not explicitly brought out. The Rhind papyrus, copied by the scribe Ahmes (or Ahmose) about 1650 B.C., shows that the Egyptians correctly found the volume of the square pyramid to be 1/3 the volume of the rectangular prism having the same base and altitude. No demonstration of the correctness of this comparison was given, and in our century it has been shown that a rigorous proof is impossible without infinitesimal considerations—that is, without the calculus. If the comparison of simple rectilinear configurations calls for such sophistication, one should not expect too much of pre-Hellenic mensuration of curvilinear figures. Ahmes, for example, took the area of a circle as equal to that of a square whose side is 8/9 the diameter of the circle. This is not a bad approximation, for it is tantamount to assuming for pi a value of about 3.16. Nevertheless, although counting the number of correct figures in a decimal approximation for pi is good sport, it is not a reliable measure of the mathematical level of a civilization. It would have been a greater achievement had the Egyptians been able to show that their formula for the area of the circle was *not* exact, while that for the volume of the pyramid *was* exact.

Mathematics in the Mesopotamian Valley was on a higher level than that along the Nile; yet Babylonian views were open to the two criticisms noted above with respect to the Egyptian. As early as the seventeenth century B.C. the Babylonians applied their admirably flexible algebra to a wide variety of practical problems, including the mensuration of configurations. Knowing the Pythagorean theorem, they found the diagonal of a unit square correctly to the equivalent of half a dozen decimal places. The area of a circle they generally took to be 3 times the square on the radius, but on at least one occasion they used for pi the better approximation 3 1/8. Not even the Babylonians, however, had criteria for determining whether they were dealing with exact results or with approximations only.

Perhaps the closest approach of the Babylonians to the calculus lay in an iterative algorithm they devised for finding the square root of any (rational) number. Let a be the number of which the square root is desired, and let a_1 be a first approximation to the square and less than the root. Then a/a_1 ($= b_1$) will be an approximation by excess. Moreover, if a_2 and b_2 are the arithmetic and harmonic means, respectively, of a_1 and b_1, then a_2 and b_2 will be better approximations by excess and defect than a_1 and b_1. By continuing this process, in which a_i and b_i are the arithmetic and harmonic means, respectively, of a_{i-1} and b_{i-1}, one has an infinite process that will take one as close as he pleases to the desired square root. Had the Babylonians possessed any way of knowing or showing that the process is nonterminating, they might now be hailed as the originators of the subject of infinite sequences, a basic part of modern calculus. However, Babylonian skill in algebra was not matched by concern for logic; hence credit for adumbrations of the calculus must go to the ancient people for whom the logical approach to a subject constituted a veritable passion.

The First Mathematicians

It is universally admitted that the Greeks were the first mathematicians—first, that is, in the significant sense that they initiated the development of mathematics from first principles. Hippias (*c.* 425 B.C.) or someone else of about his period showed that, in terms of whole numbers, no exact numerical comparison of the diagonal with a side was possible for a square or a regular pentagon or a cube or a regular hexagon—in fact, for very many familiar geometrical figures. It was a shock to the Greek mathematical community to learn that there are such things as incommensurable line segments and that the occurrence of such situations is appallingly commonplace—that is, that concepts akin to the calculus arise in the most elementary of situations. The dialogues of Plato show that mathematicians of the time were deeply disturbed by this discovery.

The Greek discovery of incommensurability confronted mathematicians directly with an infinite process. Whenever the Euclidean algorithm for finding the greatest common divisor of two integers is applied in arithmetic, the process comes to an end in a finite number of steps, for the set of positive integers has a smallest element, the number 1. If, on the other hand, the analogous scheme is applied in geometric garb to finding the greatest common measure of two incommensurable line segments, the process will go on forever; there is no such thing as a smallest line segment—at least not according to the orthodox Greek view, nor according to conventional modern concepts. The prospect of an infinite process disturbed ancient mathematicians, for here they were confronted with a crisis. They were unable to answer the subtle paradoxes Zeno of Elea proposed at about the same time that the devastating discovery of incommensurables [18] was made. Aristotle and other Greek philosophers sought to answer the paradoxes of Zeno, but the replies were so unconvincing that mathematicians of the time concluded that it was best to shun infinite processes altogether.

This view might seem to preclude any Greek equivalent of the calculus; Eudoxus nevertheless suggested an approach that appeared to mathematicians to be irrefragable and to serve essentially the same purpose as an infinite process. He started with an axiom, often known as "the lemma of Archimedes," which appears in Euclid's *Elements* as Definition 4 in Book V:

> *Magnitudes are said to have a ratio to one another which are capable, when multiplied, of exceeding one another.*

This "definition," which really is an assumption, presumably had been used by Eudoxus in much the same way that it is handled by Euclid in Book X, 1 (and still later by Archimedes) to prove the basic procedure in the "method of exhaustion," the Greek geometrical equivalent of the calculus:

> *Two unequal magnitudes being set out, if from the greater there be subtracted a magnitude greater than its half, and from that which is left a magnitude greater than its half, and if this process be repeated continually, there will be left some magnitude which will be less than the lesser magnitude set out.*

This statement can be generalized by replacing "greater than its half" by "greater than or equal to its half (or its third or any other proper fraction)."

Here, in an awkward geometrical form, is one of the earliest theorems on limits; for the crux of the matter is that if A is the greater of the two given (positive) magnitudes a and A, and *if* $\mu_n = A/2^n$, then

$$\lim_{n \to \infty} u_n = 0 < a.$$

It should be noted that whereas the modern notation has recourse to the symbol for infinity, the ancient language carefully avoids any overt reference to an infinite process. Nevertheless, the two formulations are not far apart in meaning. To show that

$$\lim_{n \to \infty} u_n = 0,$$

one has to demonstrate that, given a positive number ϵ, however small

379

(the equivalent of the lesser magnitude a in Euclid's proposition), one can find an integer N (the equivalent of Euclid's phrase "if this process be repeated continually") such that for $n > N$ we have $u_n < \epsilon$.

Euclid's calculus (derived, presumably, from Eudoxus) may have been less effective than that of Newton and Leibniz two thousand years later; but in terms of basic ideas it was not far removed from the limit concept used crudely by Newton and refined in the nineteenth century. The beginning student who so glibly parrots the remark that the integral of x^3 is $x^4/4 + C$ may have more facility in the techniques of the calculus than Euclid ever achieved, but the chances are that his formulation of the limit concept will fall short of the rigorous logical standards observed by the author of the *Elements*. The purpose of Euclid in the "exhaustion lemma" (Book X, 1) was to prepare the ground for the earliest truly rigorous comparison of curvilinear and rectilinear figures that has come down to us—the proof in Book XII, 2, that the areas of circles have the same proportional relation as do the areas of squares on their diameters.

Up to this point we have stressed the similarity between the ancient method of exhaustion and the rigorous modern formulation of the calculus, but there are also essential differences.

The ancient and the modern are in sharpest contrast with respect to motive. The method of exhaustion provided an impeccable demonstration of a theorem whose truth had been arrived at more informally. The substance of *Elements*, Book XII, 2, for instance, had been assumed by the pre-Hellenic civilizations in the Nile and the Tigris-Euphrates valleys well over a millenium before Euclid's day. The value of the modern calculus, however, lies not so much in its power of rigorous demonstration as in its marvelous efficacy as a device for making new quantitative discoveries.

THE METHOD OF ARCHIMEDES

The ancient Greeks did have an alternative approach to integration, one that served as an effective aid to invention. This device was described by Archimedes, in a letter to Eratosthenes, simply as "a certain method by which it will be possible for you to get a start to enable you to investigate some of the problems in mathematics by means of mechanics." The "certain *method,*" which Archimedes correctly saw would enable his contemporaries and successors to make new discoveries, consisted of a scheme for balancing against each other the "elements" of geometrical figures. A line segment, for example, is to be regarded as made up of points; a plane surface area is thought of as consisting of indefinitely many parallel line segments; and a solid figure is looked upon as a totality of parallel plane elements. Without necessarily subscribing to the validity of such a view in mathematics, Archimedes found this approach heuristically very fruitful. The very first theorem he discovered through a balancing of elements was the celebrated result that the area of a segment of a parabola is 4/3 the area of a triangle with the same base and equal height. This he found by balancing lines making up a triangle against lines making up the parabolic segment. (See [**109**] for proof.)

The applicability of the method will be illustrated here by giving the second proposition of the *Method*, that *the volume of a sphere is 4 times the volume of a cone with base equal to a great circle of the sphere and height equal to the radius.* (Democritus earlier had known that the volume of a cone is 1/3 the volume of a cylinder having the same height and radius.) Archimedes discovered this theorem through an ingenious balancing of circular sections of a sphere and a cone against circular elements of a cylinder, as shown in Figure VII-1.

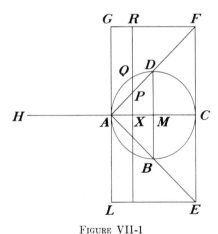

FIGURE VII-1

Let HC be regarded as the bar of a balance with the midpoint A as fulcrum. Let $ABCD$ be a great circle of the sphere and let AEF and $LEFG$ be sections of a right circular cone and a right circular cylinder, respectively, each having axis AC and diameter of the base EF. Then if a plane perpendicular to AC is passed through any point, X, on AC, this plane will cut the cone, the sphere, and the cylinder in circles C_1, C_2, and C_3, respectively. Let the radii of the three circles be $Y_1 = XP$, $Y_2 = XQ$, and $Y_3 = XR$. Now Archimedes found that if one regards the circles as having weights proportional to their areas, then upon placing at H the circles C_1 and C_2, these would exactly balance the circle C_3 where it is in the diagram—that is, with center at X. (This fact can easily be verified through elementary analytic geometry.) Hence, if one were to hang the cone and sphere at H, they would balance the cylinder if it were hung at M, the midpoint of AC. Therefore the sphere and cone together are equal to ½ the cylinder; or, inasmuch as the cylinder is 3 times the cone, the sphere is equal to ½ the cone AEF or 4 times the cone ABD—that is, in our language, the volume of the sphere is $4\pi r^3/3$. One might be inclined to conclude here that the theorem has been proved; yet Archimedes regarded this result as but a plausible conclusion to be verified ultimately by the rigorous method of exhaustion.

The *Method* of Archimedes could have been of greater significance in the development of the calculus if printing had been an invention of ancient times rather than the Renaissance. Copies of the *Method* evidently were never numerous, and for almost two millenia the work remained essentially unknown. Probably no development in the history of mathematics in our century rivals for romance the rediscovery of the Archimedean *Method* [109].

It is of importance to note that there are two aspects of the ancient beginnings in the integral calculus. One of these, stemming from Eudoxus, is represented by the rigorous method of exhaustion (illustrated in the proof of the theorem from *Elements*, Book XII, 2); the other, arising out of atomistic views associated with Democritus, is related to the method of Archimedes. The former, not far removed from nineteenth-century concepts, was a faultless means of establishing the validity of a theorem. The latter, whether making use of indivisibles or of Archimedes' elements of lower dimensionality (more closely resembling the seventeenth-century stage of the calculus), was a device that led to the discovery of plausible conclusions. Archimedes exploited both aspects most successfully indeed. His "mechanical" method led to theorems on areas, volumes, and centers of gravity which had eluded all of his predecessors; but he did not stop there. He went on to demonstrate these theorems in the traditionally rigorous manner through the method of exhaustion.

No one in the ancient world rivaled Archimedes, either in discovery or in demonstration, in the handling of problems related to the calculus. Nevertheless, the most general ancient theorem in the calculus was due not to Archimedes but to Greek mathematicians who lived probably half a dozen centuries later.

In the *Mathematical Collection* of Pappus (*c.* A.D. 320) we find a proposition awkwardly expressed but equivalent to the statement that *the volume generated by the rotation of a plane figure about an axis not cutting the figure is equal to the product of the area of the plane figure and the distance that the center of gravity of the plane figure covers in the revolution.* Pappus was fully aware of the power of this general theorem and of its analogue concerning areas of surfaces of revolution. It includes, he saw, "any number of theorems of all sorts about curves, surfaces, and solids, all of which are proved at once by one demonstration." Unfortunately, Pappus does not tell us how to prove the theorem, and we do not know whether it was either discovered or proved by Pappus himself.

MEDIEVAL CONTRIBUTIONS

Pappus was the last of the outstanding ancient mathematicians; following him, the level of mathematics in the Western world sank steadily for almost a thousand years. The Roman civilization was generally inhospitable to mathematics. Latin Europe in the twelfth and thirteenth centuries became receptive to classical learning, transmitted through Greek, Arabic, Hebrew, Syriac, and other languages, but the level of medieval European mathematics remained far below that of the ancient Greek world.

However, a certain ingenuous originality resulted in a fourteenth-century advance in a direction that had been shunned in antiquity. Archimedean mathematics, like Archimedean physics, had been essentially static; the study of dynamic change had been regarded as appropriate for qualitative philosophical discussion rather than for quantitative scientific formulation. But in the fourteenth century Scholastic argumentation in the universities at Oxford and Padua gave a mathematical twist to the Aristotelian views of change and variation. Scholars began to raise such questions as the following: If an object moves with varying speed, how far will it move in a given time? If the temperature of a body varies from one part to another, how much heat is there in the entire body? One recognizes such queries as precisely those that are handled by the calculus; but medieval scholars had inherited from antiquity no mathematical analysis of variables. The result was that the churchmen of the English, French, and Italian universities developed a primitive integral calculus of their own.

One of the leaders in this movement was Nicole Oresme (*c.* 1323–1382), bishop of Lisieux. In studying, for example, the distance covered by an object moving with variable velocity, Oresme associated the instants of time within the interval with the points on a horizontal line segment (called a "line of longitudes"), and at each of these points he erected (in a plane) a vertical line segment ("latitude"), the length of which represented the speed of the object at the corresponding time. Upon connecting the extremities of these perpendiculars or latitudes, he obtained a representation of the functional variation in velocity with time—one of the earliest instances in the history of mathematics of what now would be called "the graph of a function." It was then clear to him that the area under his graph would represent the distance covered, for it is the sum of all the increments in distance corresponding to the instantaneous velocities. Here, of course, one runs up against all the philosophical and logical difficulties that led to the paradoxes of Zeno and caused the careful Greek mathematicians to avoid the study of variations as such.

Oresme classified his types of variation according to whether they were uniform (that is, constant) or difform (not constant), and he subdivided the latter type into uniformly difform or difformly difform variations. Only in the cases of the uniform and the uniformly difform rates of change was he able to carry out the implied integration; it was an easy matter for him to find the area of a rectangle or a triangle or a trapezoid. For example, he stated that if a body moves with uniformly difform speed (i.e., with uniform rate of change of velocity), starting from rest at A in Figure VII-2, then the graph will be a straight line forming a right triangle with the base line AB (the line of longitudes) and the final ordinate (or latitude) BC. Inasmuch as the area of this triangle is the base times ½ the altitude, Oresme correctly concluded that the distance covered by the object in this case is the same as that covered by another body moving for the same length of time with a uniform speed equal to DE, the speed of the first object at the half-time mark.

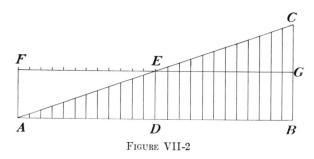

FIGURE VII-2

His argument was that the representation of an object moving during time AB with speed DE will be the rectangle $FGBA$, and the distance covered in this case will be given by the area of this rectangle, which is equal to the area of the triangle ABC. What Oresme was saying is equivalent, of course, to the statement

$$S = \int_0^T Kt\, dt = \frac{KT^2}{2},$$

a result expressed unequivocally (but on the basis of the same type of reasoning) by Galileo Galilei two and a half centuries later.

Both Oresme and Galileo were making use of the ancient notion of geometrical elements; that is, they thought of the distance triangle and the rectangle as made up of indefinitely many vertical straight lines. Such a view is similar to that used by Archimedes in his *Method;* and although this treatise was not extant at the time, the idea that an area could be composed of infinitely many geometrical lines was shared by medieval scholars with both ancient and modern writers.

Of the other aspect of ancient integral calculus, the rigorous method of exhaustion, there was little recognition during the medieval period; and even during the renaissance in mathematics the logical precision of antiquity had few emulators. What geometers of the dawning modern age were looking for was not so much an argument as a method—not an argument to establish beyond doubt a conclusion that seemed to be plausible, but a method of discovery which would lead to new results.

It is important to note, in connection with medieval studies in the latitude of forms, that there was no equivalent of the concept of differentiation. The beginning calculus student of today, looking at the representations of Oresme, probably would first think of the steepness of the velocity-time curve as a measure of the acceleration, the rate of change of velocity with respect to time. Only later is he likely to think of the area under the graph as a measure of distance; today the concept of the derivative is usually presented first in calculus courses, with the notion of the integral coming later. Those textbooks that reverse the roles and place the integral before the derivative in a sense have history on their side, inasmuch as integration preceded differentiation by about two thousand years; but it is true that some of the work of Archimedes in connection with the tangent to the spiral comes close to the differential calculus, and manuscript copies of the graphical representations of Oresme contain strong hints of the differential triangle. It should have been obvious in the latitude of forms that the representation of a rapidly increasing quantity called for a rapidly increasing latitude with respect to longitude; yet no systematic terminology or method of handling such a concept was developed either in antiquity or during the medieval period.

THE NEW ROLE OF INDIVISIBLES

The renaissance in mathematics consisted of a complicated interplay of medieval traditions with newer and older ideas. In particular, the revival of wide interest in the works of Archimedes led, in the seventeenth century, to a search for shortcuts that might simplify the integral calculus. Here the seductive concept of the indivisible in geometry played an important role.

Galileo had been in touch with this concept through medieval contributions to dynamics, and Johann Kepler (1571–1630), in his popular treatise on the gauging of wine casks [107], adopted streamlined Archimedean methods. Overshadowing the integrations of Galileo and Kepler, however, was a treatise written by Galileo's disciple Bonaventura Cavalieri, in 1635, entitled *Geometria indivisibilibus continuorum nova quadam ratione promota*. In this book the indivisible, or fixed infinitesimal, was applied so successfully to problems in the mensuration of areas and volumes that the fundamental postulate, usually bearing the name "Cavalieri's theorem," has remained enshrined in elementary textbooks to this day: *If two solids (or plane regions) have equal altitudes, and if sections parallel to the bases and at equal distances from them are always in a given ratio, then the volumes (or areas) of the solids (or regions) are also in this ratio.* This principle permitted Cavalieri to pass from a strict correspondence of indivisibles in a given ratio to the conclusion that the totalities of these indivisibles (i.e., the figures of higher dimensionalty) were also in this ratio. The idea behind this was not really new in 1635, for it is essentially related to the mechanical method of Archimedes and to the graphical integrations of Oresme and Galileo. Kepler had used the idea when he found the area of the ellipse,

$$\frac{x^2}{a^2} + \frac{y^2}{b^2} = 1,$$

to be πab. Inasmuch as the ordinates in this ellipse are each in the ratio b/a to the corresponding ordinates in the circle $x^2 + y^2 = a^2$, the area of the ellipse must be in the ratio b/a to πa^2 (the area of the circle).

Cavalieri ingeniously applied the idea of indivisible to a wide variety of new problems. The "higher parabolas" of Pierre de Fermat,

$a^{n-1}y = x^n$, were introduced into mathematics at just about the time that Cavalieri was developing his geometry of indivisibles; and in his later works Cavalieri extended Oresme's integration of kt to include the integral of kt^n, stating the equivalent of the well-known formula

$$\int_0^T kt^n \, dt = \frac{kT^{n+1}}{n+1}.$$

In this last respect Cavalieri seems to have been first in publication, but the result had been known independently to others, including (in addition to Fermat) Gilles Persone de Roberval and Évangelista Torricelli. The method of indivisibles, in other words, was not the property of Cavalieri. It was being used widely by men who were *au courant* with mathematical thoughts of the day.

Cavalieri, like his contemporaries, regarded his method of indivisibles as part of geometry; but, even as he was writing, an analytical revolution was sweeping through Europe.

DESCARTES, WALLIS, AND FERMAT

The seeds of the revolution had been sown in the Renaissance period when Regiomontanus (Johann Müller), François Viète (also known as Francis or Franciscus Vieta), and others saw that the increasingly symbolic algebra of their day—which, oddly enough, had been derived from the thoroughly rhetorical algebra of the Arabs—served as an effective handmaiden to geometry. The logical outcome of such ideas was the analytic geometry of Fermat and René Descartes. Descartes's *La géométrie*, published in 1637, only two years after Cavalieri's *Geometria indivisibilibus*, inevitably changed the course of infinitesimal analysis.

It was not so much the use of coordinates that made the work of Descartes (and of Fermat) so important; coordinates had been used effectively in antiquity, especially in the geometry of Apollonius, and again in a more primitive form in the latitude of forms of Oresme. Descartes saw as the objective of his work the cooperation of algebra and geometry to the end that mathematics might have the best aspects of both branches. In the end, however, it turned out that geometry lost popularity in the partnership. Pure geometry was so overshadowed that it made little progress during the next century and a half, during which time infinitesimal analysis went through a process of arithmetization that amounted almost to a revolution.

The altered view is seen promptly and clearly in the *Arithmetica infinitorum of* John Wallis, published in 1655, a score of years after

Cavalieri's classic. In this widely read treatise the author included his own characteristic approach to the integral of kx^n, a proof prompted by Cavalieri's geometry of indivisibles but proceeding by a stark arithmetization of the calculus. To show, for example, that

$$\int_0^1 x^2 \, dx = \frac{1}{3} \, ,$$

Wallis formed the ratio of the sum of the squares of equally spaced ordinates under the curve $y = x^2$ to the sum of the squares of the corresponding ordinates under $y = 1$. If one considers only the ordinates at $x = 0$ and $x = 1$, the ratio is

$$\frac{0^2 + 1^2}{1^2 + 1^2} = \frac{1}{2} = \frac{1}{3} + \frac{1}{6}.$$

If one subdivides the interval between $x = 0$ and $x = 1$ into two equal parts and thinks of each part as a unit interval, the ratio of the squares of the ordinates becomes

$$\frac{0^2 + 1^2 + 2^2}{2^2 + 2^2 + 2^2} = \frac{5}{12} = \frac{1}{3} + \frac{1}{12}.$$

If one subdivides the interval from $x = 0$ to $x = 1$ into three equal subintervals and regards the subintervals as unit segments, the ratio of the squares of the ordinates is

$$\frac{0^2 + 1^2 + 2^2 + 3^2}{3^2 + 3^2 + 3^2 + 3^2} = \frac{7}{18} = \frac{1}{3} + \frac{1}{18}.$$

It became apparent to Wallis, through induction, that

$$\frac{0^2 + 1^2 + 2^2 + 3^2 + \cdots + n^2}{n^2 + n^2 + n^2 + n^2 + \cdots + n^2} = \frac{1}{3} + \frac{1}{6n^3}$$

and that as n increases indefinitely this ratio comes closer and closer to $\frac{1}{3}$, so that for $n = \infty$ the ratio will be $\frac{1}{3}$.

Fermat rightly claimed that Wallis' method of induction was logically inadequate; but the arithmetizing tendencies of Wallis were steps in the right direction nevertheless. Fermat himself was heading in the same direction, both in his invention of analytic geometry and in his contributions to the calculus.

His method of integrating x^n was the most elegant of those available at the time, and it came closer to the modern Riemann integral than any other before the nineteenth century. To find

$$\int_0^T x^2 \, dx,$$

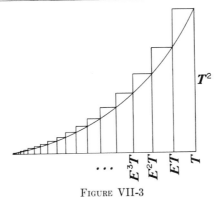

T^2

E^3T E^2T ET T

FIGURE VII-3

for instance, Fermat erected ordinates to the curve $y = x^2$ at the points in Figure VII-3 whose abscissas are, respectively, T, ET, E^2T, E^3T, etc., where $E < 1$. With these ordinates as altitudes, he formed a sequence of rectangles approximating the area under the curve, as shown. The sum of these is given by the infinite geometrical progression

$$T^3(1 - E)(1 + E^3 + E^6 + E^9 + \cdots)$$
$$= T^3(1 - E)\left(\frac{1}{1 - E^3}\right) = \frac{T^3}{1 + E + E^2}.$$

Upon letting E approach more and more closely to 1, the widths of the rectangles tend toward 0, and the sum of the areas of the rectangles tends toward the area under the curve—$T^3/3$. By similar reasoning Fermat showed that, for all rational values of n except $n = -1$,

$$\int_0^T x^n \, dx = \frac{T^{n+1}}{n + 1}.$$

This work by Fermat, unfortunately unpublished at the time, brought to a climax the methods of integration initiated by Eudoxus two millenia before.

But Fermat, the greatest of all amateurs in mathematics, was responsible for an even more significant contribution to the development of the calculus. He was literally the inventor of the process that we now call "differentiation." During the very years in which he and Descartes were inventing analytic geometry, Fermat had discovered an amazingly simple method for finding the maxima and minima of a polynomial curve. In terms of the Cartesian notation (which, however, Fermat rejected in favor of the older notation of Viète), the method is as follows: Let the ordinate of a curve be given by $2x^3 - 5x^2 + 4x - 7$,

389

for example. Then, at a neighboring point with abscissa $x + E$, the ordinate will be $2(x + E)^3 - 5(x + E)^2 + 4(x + E) - 7$. Now, Fermat argued, at a maximum or minimum point the change in the ordinate is virtually imperceptible, provided that the change E in the abscissa is small; hence he set the above two values of neighboring ordinates equal to each other. The result, upon transposing terms, is

$$(6x^2 - 10x + 4)E + (6x - 5)E^2 + 2E^3 = 0.$$

Upon dividing by E and letting E be 0 in the remaining terms (inasmuch as the value of E must be indefinitely small for the rule to apply), one obtains the equation $6x^2 - 10x + 4 = 0$. Solving this equation leads to the values 1 and $\frac{2}{3}$ as the abscissas of the critical points on the curve. This procedure of Fermat's is, in essence, that used today in the calculus when one finds the derivative and sets it equal to 0. Certainly the use of Δx or h where Fermat wrote E is immaterial; and while it is admitted that Fermat did not explain precisely why he set E equal to 0, neither did his successors for a couple of centuries. There is therefore every reason to acknowledge, with Pierre Simon Laplace, that Fermat was the inventor of the differential calculus, as Eudoxus has been recognized as the inventor of the integral calculus.

Did Fermat, who was well aware of the rules for differentiating and integrating, notice the relation between these? He apparently knew full well that in the first case one multiplied the coefficient by the exponent and lowered the exponent by one unit, whereas in the latter case one increased the exponent by one unit and divided the coefficient by the new exponent. Strangely, he seems to have seen nothing significant in this striking inverse relationship, nor did his contemporaries, such as Évangelista Torricelli, James Gregory, or Isaac Barrow.

Barrow, who was a teacher of Isaac Newton's, published a rule of tangents much like Fermat's method of maxima and minima, in which the interplay of coefficients and exponents again was clear. Given a curve such as

$$3x^2y - 5xy^3 - 7x + 6y = 0,$$

he first subtracted the members of this equation from the corresponding members of the equation obtained by replacing each x by $x + e$ and replacing each y by $y + a$. In the resulting equation he then discarded all those terms above the first degree in a and e. From the equation thus obtained he formed the ratio

$$\frac{a}{e} = \frac{-6xy + 5y^3 + 7}{3x^2 - 15xy^2 + 6},$$

which he knew to be the slope of the tangent to the curve at the point (x, y). Here one easily recognizes the familiar rule now expressed as $dy/dx = -(f_x/f_y)$. Barrow published in his *Lectiones Opticae et Geometricae* (1670) not only the above rule of tangents (which he apparently had discussed with Newton) but also rules for reducing inverse tangent problems (antiderivatives) to quadratures (areas).

Fermat's method for maxima and minima and Barrow's tangent rule were by no means the only devices and formulas invented in connection with these topics. René de Sluse and Johann Hudde were among others who noted the play on coefficients and exponents in finding tangents and extrema of polynomials, and the inverse nature of tangent and quadrature problems may have been known to Torricelli and Gregory. These results were finding applications in the science of that time—in Fermat's principle of least time in the refraction of light, and in the dynamics of Christiaan Huygens.

The concept of acceleration had been known to Oresme, but not until the seventeenth century did it come to play a significant role in dynamics. Where classical Greek mathematics had been essentially static in its language and concepts, the mathematics of the "Age of Genius" had become more oriented toward an analysis of variability. The logarithms of John Napier (1550–1617) had been defined in terms of two varying line segments, one of which increased arithmetically with respect to time while the other decreased geometrically. The astronomy of Nicolaus Copernicus, built upon uniform circular motion, required for its study nothing beyond trigonometry; but the new astronomy of Kepler, in which the planets move nonuniformly in elliptical orbits, called for higher mathematics. By the end of the second third of the seventeenth century all the rules needed to handle such problems in areas and rates of change, in maxima and tangents, were available.

The time was now ripe to build the infinitesimal analysis into the subject we know as *the calculus*. No specific new invention was needed; the techniques were at hand. What was wanting was a sense of the universality of the rules.

NEWTON AND LEIBNIZ

This awareness was achieved first by Isaac Newton, in 1665–66, and again independently by Gottfried Wilhelm von Leibniz in 1673–76.

To the predecessors of these two intellectual giants the various rules had appeared as clever devices of limited usefulness. Most of the rules, for instance, were applicable only to problems involving polynomials or to those that by some obvious transformation could be converted into polynomial form. Descartes in 1637 had wished to limit geometry to the study of algebraic curves; but the explosive interest in the later seventeenth century in the cycloid, the catenary, and other curves expressed by trigonometric and logarithmic functions had bestowed mathematical respectability on a vastly increased array of problems. The rules of Cavalieri, Fermat, Wallis, and Barrow involving the raising and lowering of exponents did not seem to apply to the transcendental functions; what the mathematics of the decade from 1666 to 1676 needed so badly was a general algorithm that would apply indifferently to all functions, rational or irrational, algebraic or transcendental. The answer, supplied by Newton and Leibniz, was a new and general infinite analysis.

The key to the new analysis lay in the discovery by Newton and Leibniz of the great utility of infinite-series expansions. So important did Newton regard this aspect of his work that he insisted that it was an essential part of his new analysis, and this is the reason the binomial theorem is so frequently ascribed to him.

The theorem had long been known for integral powers, including even the rule of succession for the terms in the expansion; but Newton was the first one to apply the theorem to the case of fractional exponents, in which the expansion is nonterminating. Through an infinite binomial expansion Newton was in a position to find quadratures that had baffled his predecessors—for example, the integral of $(x - x^2)^{1/2}$, which Wallis had sought in vain. The method could be applied also to find the integrals (Newton called them "fluents") of transcendental functions. It had long been known that the area under the hyperbola, $y = 1/(1 + x)$, had the familiar property of logarithms that, as the abscissa increases geometrically, the area under the curve increases arithmetically. Newton saw that it was possible to write $\ln(1 + x)$ as an infinite series. One simply expanded $1/(1 + x)$ as

$$1 - x + x^2 - x^3 + \cdots ,$$

using either long division or the binomial expansion of $(1 + x)^{-1}$; then to each term in the series one applied the familiar rule of quadratures: raise the exponent by one and divide by the new exponent. The result is

$$\ln(1 + x) = x - \frac{x^2}{2} + \frac{x^3}{3} - \frac{x^4}{4} + \cdots ,$$

usually known as the series of Mercator because it was first published in 1668 by Nicolaus Mercator, although Newton had known it earlier.

Leibniz likewise soon realized the potentials in infinite series, and the equality

$$\frac{\pi}{4} = 1 - \frac{1}{3} + \frac{1}{5} - \frac{1}{7} + \cdots$$

is generally known as Leibniz' series. However, this is but a special case of the series

$$\arctan x = x - \frac{x^3}{3} + \frac{x^5}{5} - \frac{x^7}{7} + \cdots,$$

named for James Gregory, who first published it in 1668. Newton seems to have developed series for trigonometric functions at about the same time.

Newton's first work describing his calculus, *De analysi per aequationes numero terminorum infinitas*, was written in 1669 but was not published until 1711 because of the author's aversion to controversy. In it Newton described, as follows, his extended use of the word "analysis":

> And whatever the common analysis performs by means of equations of a finite number of terms . . . this new method can always perform the same by means of infinite equations; so that I have not made any question of giving this the name of *Analysis* likewise. For the reasonings in it are no less certain than in the other, nor the equations less exact. . . . To conclude, we may justly reckon that to belong to the *Analytic Art* by the help of which the areas and lengths, etc. of curves may be exactly and geometrically determined.

In short, Newton here is arguing that mathematical algorithms dealing with infinite processes are just as respectable as those applied to ordinary algebra; if one must have a single or first inventor for the calculus in this sense, the choice must be Newton. His contribution was not so much a rule for differentiation or integration, or even the disclosure of these as inverse operations; it was the recognition that these constituted parts of a new analysis—the application of infinite processes in the general study of functions of whatever type.

With Leibniz, too, the essential element in his later but independent invention of the calculus was the recognition by 1676 that he was building a new and universal analysis. In his first published papers, in the *Acta eruditorum* for 1684 and 1686, Leibniz described his new method as not impeded by irrational or transcendental functions.

Therefore, inasmuch as algebraists formerly assumed letters or general numbers for the quantities sought, in such transcendental problems I have assumed general or indefinite equations [i.e., infinite series expansion] for the lines [functions] sought . . . and the analytic calculus is extended in this way to those lines which hitherto have been excluded for no greater cause than that they were believed unsuited to it.

Just as Descartes realized that his geometry marked a new stage in the development of the subject, so Newton and Leibniz were aware that their discoveries had forged a new analysis going beyond ordinary algebra. Throughout the eighteenth century the distinction Newton and Leibniz emphasized was retained in the phrases "higher analysis" and "sublime analysis," to distinguish the infinite procedures from the rules of ordinary algebra.

Distinctive banners are an effective aid to crusaders. Leibniz, in particular, gave much thought to the question of appropriate notations. The measure of his success in this matter is the survival to this day of his language and symbols. Newton was more hesitant in this respect in his published works (although he experimented extensively in manuscript), and little of the Newtonian formulation has remained. A manuscript of Newton's from October 1666 gives a full description of his method "to resolve problems by motion," that is, to find such things as areas, tangents, and centers of gravity of "crooked lines" by considering the way in which the coordinates vary with time /NEWTON: 15–64/. Here Newton used p and q as symbols for the rates of change ("velocities") of x and y, respectively, but later he substituted the "prickt letters" $\dot{x}$ and $\dot{y}$ for p and q, and the word "fluxions" for "velocities." Hence his analysis came to be known as the "method of fluxions." In the 1666 form of the calculus, the slope of the curve

$$x^3 - abx + a^3 - cyy = 0$$

was found by Newton as follows: Replace x by $x + po$ and y by $y + qo$. From the equation thus obtained subtract the original equation, divide by o, and omit all terms still containing o (inasmuch as these are "infinitely little"). The result is

$$3pxx - abp - 2cqy = 0,$$

from which the ratio q/p (which we now call the "slope") is easily obtained as $2cy/(3xx - ab)$. In his *De quadratura*, written in 1676 (but not published until 1704), Newton avoided the phrase "infinitely little." He first determined the ratio of the fluxions q/p (or $\dot{y}/\dot{x}$) and then let o vanish, or be "evanescent," finding thus what he called the "prime and ultimate ratio" (what we would now call the "limit of the ratio") of the fluxions. This approach was close to the modern point of view; but the form of the calculus which Newton published in the *Principia* in 1687 reverted to the crude first ideas of "infinitely little" increments or "moments." In this form his calculus so closely resembled that of Leibniz that one may pardon contemporaries of the men for confusing the two points of view.

Whereas Newton wrote several substantial accounts of his methods in the calculus, he published these belatedly. In contrast, Leibniz wrote little but published early. He limited himself to a few short articles that appeared in *Acta eruditorum* shortly after the journal was established. His first article appeared in 1684, three years before Newton's brief (two-page) hint in the *Principia*. In Leibniz' first paper the simple rules of differentiation were crudely justified in language that is now reminiscent of Newton's infinitely little quantities. To find the differential of the product xy, Leibniz replaced x by $x + dx$ and y by $y + dy$ (where dx and dy were the differentials, or infinitely small differences, of x and y). The difference

$$(x + dx)(y + dy) - xy$$

then represented the infinitely small difference in the product corresponding to the infinitely small differences in x and y. Inasmuch as the term $dx \cdot dy$ was regarded as incomparably small in proportion to the other terms in the difference, Leibniz disregarded this and wrote

$$d(xy) = x \, dy + y \, dx.$$

From his formula for the differential of a product Leibniz easily derived the differential of integral powers of a variable such as x through an induction beginning with $x \cdot x$, and analogous formulas for differentials of reciprocals and quotients readily followed. If in Leibniz' product formula one were to divide both sides by dx, one would obtain the modern formula for the derivative, with respect to x, of the product xy, where y is a function of x; but Leibniz did not think in terms of dependent and independent variables or of the derivative of one with respect to the other. Consequently there was a lack of precision in this reasoning which was criticized by his contemporaries even while they admired the power of the new method.

Language, Logic, Symbolism

Newton and Leibniz were not the first ones to use methods equivalent to differentiation or integration or to notice the inverse relation between these, nor were they first in the use of infinite series. The contribution of these men, as we have said before, lay in gathering together devices of limited applicability and developing, from them, methods of universal scope.

Realization of the import of this situation led each man to develop a language, a logic, and a symbolism for the new subject. Neither one was able to present a convincing logical foundation, although in this respect Newton certainly came closer than Leibniz. Newton's best attempt appeared in the *Principia*, where he described his idea of "prime and ultimate ratios." Making use somewhat anachronistically of modern notations, we may paraphrase Newton's formulation by describing $\dot{y}$ as the ratio of the "evanescent" quantities Δy and Δt. In this connection Newton cautioned: "By the ultimate ratio of evanescent quantities is to be understood the ratio of quantities, not before they vanish, nor afterwards, but with which they vanish." Here he came extraordinarily close to the limit concept; but Newton did not fully achieve this concept, for he kept insisting on describing as a ratio of velocities what we now see is nothing but a single real number. The precise definition of real number, however, was an achievement of the late nineteenth century.

The calculus of Leibniz was, from the logical point of view, distinctly inferior to that of Newton, for it never transcended the view of dy/dx as a quotient of infinitely small changes or differences in y and x. Just what was meant by an infinitely small change could not be made clear by either Leibniz or his disciples, nor could they justify the discarding of quantities that were infinitely small in relation to others. From the logical point of view the calculus of Leibniz was a failure, but heuristically it was a resounding success.

Part of this success resulted from a view of the integral which was as imprecise as that of the differential quotient. Newton's fluents, analogues of the modern "indefinite integral," were what we would now call antiderivatives with respect to time. Newton at first used a small square as his symbol of integration (presumably because he realized that it determined an area); later he used a short vertical line over a letter to indicate the fluent of the quantity, writing $\overset{\shortmid}{y}$ where we would use $\int y\,dt$. Leibniz emphasized the summation aspect of the "integral"; he also aided in gaining acceptance of the word itself. However, whereas we think in terms of a limit of a characteristic sum of finite magnitudes, Leibniz thought of an actual sum of infinitely small quantities or infinitesimals. This accounts for the fact that Leibniz used as his symbol of integration (now also our own) an old-style elongated form of the letter S, first letter of *summa* ("sum").

Newton had written up several accounts of his methods some years before Leibniz had hit upon his own formulations; yet most of their contemporaries learned the calculus through differentials rather than fluxions. When Newton had been on the point of publishing *De analysi* in 1672, controversies over his optical discoveries had caused him to delay publication further. Hence even the two-page hint of his analytical methods in the *Principia* in 1687 had been anticipated by the two papers of Leibniz in 1684 and 1686 in the *Acta eruditorum*. The first of these, in only half a dozen pages, announced the differential calculus, with rules (without proof) and applications; the second described the integral calculus. Although the papers were marred by misprints and poor exposition, their great significance became apparent to two Swiss mathematicians, Jakob Bernoulli and his brother Johann.

The Bernoulli brothers (also known as Jacques and Jean) became ardent disciples of Leibniz, and the younger brother composed textbooks in the differential and integral calculus. Among Johann's followers was the wealthy Marquis de L'Hospital, and presumably it was on account of a financial arrangement the marquis had made with his young tutor that publication of Johann Bernoulli's textbooks was side-tracked to make way for the appearance in 1696 of the first real textbook in the calculus, L'Hospital's *Analyse des infiniment petits* [111]. From this time on the spread of Leibnizian methods was rapid indeed.

A large measure of the success of the differential calculus on the Continent is to be ascribed to the fact that Leibniz had the faculty of inspiring eager disciples, whereas Newton's aloofness precluded his establishment of a school of followers. Then, too, the work of Leibniz was more firmly wedded to the arithmetization of analysis, whereas that of Newton often was couched in the language of synthetic geometry. Again, Leibnizian conceptions in the calculus, while further removed from sound logic than those of Newton, were more suggestive of the ways in which the algorithms of the calculus could be applied to geometrical and physical problems. The history of mathematics seems here, as earlier, to show that a premature insistence on logical precision, at the expense of imaginative and plausible reasoning, can have an adverse effect of the development of the subject. This we shall see in a later discussion of the *Analyst* controversy.

If Newton had followed through on his intention to publish his *De analysi* in about 1672, there would have been no problem about priority, for at that time Leibniz had not discovered the new analysis; but dread of criticism led Newton to delay publication. The result was a controversy about priority that was perhaps more bitter than any other the field of mathematics has known. The acrimony of the dispute was caused not so much by personal animosities as by national pride, for by 1700 factional supporters had come to realize that what was at stake was credit for a new and universal branch of mathematics—not merely a new method of tangents or a new algorithm. Newton and Leibniz had seen that series such as

$$x - \frac{x^3}{3} + \frac{x^5}{5} - \frac{x^7}{7} + \cdots ,$$

$$x - \frac{x^2}{2} + \frac{x^3}{3} - \frac{x^4}{4} + \cdots ,$$

and

$$x - \frac{x^2}{2} + \frac{x^3}{4} - \frac{x^4}{8} + \cdots$$

could be used for transcendental functions just as well as for algebraic ones—for tan x and $\ln(1 + x)$ as well as for the rational algebraic function $2x/(2 + x)$. It is natural, therefore, that it should be to one of the inventors of the new infinite analysis that the concept of function would be generally attributed, and this distinction came to Leibniz. Newton was aware of the general idea, and in fact the calculus of Newton brought out more clearly than that of Leibniz the distinction between dependent and independent variables; but Leibniz was the inspiration of the eighteenth century because of the pedagogical quality of his work, as distinct from the logical quality.

Mathematicians in Continental Europe now eagerly and uncritically exploited possibilities inherent in infinite power series. The center of this movement was the Swiss mathematician Leonhard Euler, the intellectual descendant of Leibniz through the teaching of Johann Bernoulli. Euler very appropriately has become known as "analysis incarnate" for his work in the study of infinite processes. In the opening pages of his *Introductio in analysin infinitorum* (1748), Euler defined a function of a variable as "any analytic expression whatsoever composed of that variable quantity and numbers or constant quantities." He did not immediately make clear what an "analytic expression" is; but later he stated, concerning a variable z, that "there will be no doubt but that every function of this variable can be transformed into an infinite expression of the form $Az^\alpha + Bz^\beta + Cz^\gamma + Dz^\delta$, etc." Here Euler did not quite reach the modern view that such a series *defines* a function, but his treatise did make clear the key role that infinite series play in the study of functions.

THE SEARCH FOR RIGOR

While Euler and the Continental mathematicians were eagerly piling discovery upon discovery in the new analysis, with little concern for logical foundations, British scholars were endeavoring (with little success) to defend the method of fluxions from the devastating criticisms of the philosopher and divine George Berkeley, published in 1734 in the *Analyst*. Defenders of fluxions often had argued that the increments did not vanish but were "evanescent," and it was in this connection that Berkeley penned these celebrated lines /STRUIK (e): 338/:

> And what are these fluxions? The velocities of evanescent increments. And what are these same evanescent increments? They are neither finite quantities, nor quantities infinitely small, nor yet nothing. May we not call them the ghosts of departed quantities?

The arguments in the *Analyst* were well taken, but the effect of the work on British mathematics was unfortunate. The inability of Berkeley's opponents to give satisfactory answers evidently dampened British ardor for the new analysis, and the level of achievement in research fell below that on the Continent.

In part the lower level of British analysis may be attributed also to a continuation of synthetic and geometrical emphasis and to the handicap, easily exaggerated, involved in the adoption of the fluxional notations rather than the felicitous symbols of the differential and integral calculus. At all events, while the British were fussing about the meaning of the word "evanescent," Euler in particular was making of the new analysis the subject we know today—the study of infinite processes.

Continental mathematicians, also, had occasional qualms concerning the validity of the methods in the differential and integral calculus. In particular, Bernard Nieuwentijt objected to differentials of higher order, and Michel Rolle at first suspected the calculus of leading to paralogisms; but as British critics became more vocal, those on the Continent lapsed into silence. In the hands of Johann Bernoulli and Leonhard Euler the algorithmic procedures were so productive of new results that mathematicians were loath to question them.

However, in 1772 one of the keenest mathematicians of the century, Joseph Louis Lagrange, tried his hand at an alternative explanation, one which in a way harked back to the emphasis of Newton and Leibniz on infinite series. Most mathematicians felt comfortable, for example, in expanding (by long division or otherwise) such an expression as $1/(1 + x)$ in an infinite series—in this case the series

$$1 - x + x^2 - x^3 + x^4 - \cdots .$$

Lagrange suggested that the coefficient of the $(n + 1)$th term in this expansion, when divided by $n!$, be taken as the definition of the nth derivative of the function at $x = 0$. (For values of x other than $x = 0$ the general Taylor expansion serves the same purpose.) Twenty-five years later (1797) Lagrange made this idea the foundation of his classic treatise *Théorie des fonctions analytiques*, from which stems, in large part, the modern theory of functions of a real variable. Lagrange's approach met with little response and later was seen to be inadequate, but it was his use of the phrase *fonction dérivée* which resulted in the modern word "derivative," and it is essentially his notation that we follow when we write $f'(x)$ or $f^n(x)$.

The École Polytechnique, where Lagrange taught, produced the leading French mathematicians of the nineteenth century. Among these none was more prolific or influential than Augustin Louis Cauchy, to whom the modern college-level presentation of the concepts of the calculus is mainly due. Cauchy, in his *Cours d'analyse de l'École Polytechnique* of 1821, introduced an improved definition of limit: "When the successive values attributed to a variable approach indefinitely a fixed value so as to end by differing from it by as little as one wishes, this last is called the limit of all the others."

In the light of contemporary mathematics this definition is far from acceptable, for it does not make clear the differing roles of dependent and independent variables; but it was adequate for Cauchy's contemporaries. The author went on to define the derivative of a function $f(x)$ as the limit of the ratio

$$\frac{\Delta y}{\Delta x} = \frac{f(x + \Delta x) - f(x)}{\Delta x}$$

as Δx approaches zero. Whereas in the work of Leibniz the differential had been the basic notion, now it was the derivative that was fundamental—the differential dy being simply $f'(x)\Delta x$. Moreover, whereas in the work of Leibniz and, *a fortiori*, in that of Newton, the integral of $f(x)$ had been thought of largely as an antiderivative—that is, another function of which $f(x)$ is the derivative—Cauchy emphasized that the integral of $f(x)$ is defined independently of the derivative, being a limit of a characteristic sum. It is from the Cauchy point of

400

view that the broad modern generalizations of the integral have developed.

It would not be accurate to leave the impression that Cauchy was alone in the early nineteenth-century development of the foundations of the calculus. At that time in Bohemia there was a priest, philosopher, and mathematician, Bernhard Bolzano, who was seeking to arithmetize analysis; and his definition of the derivative was the same as Cauchy's. Moreover, Bolzano and Cauchy independently reached the modern definition of continuity as a function. They defined a function $f(x)$ as continuous in an interval if for any value of x in this interval the difference $f(x + \Delta x) - f(x)$ becomes and remains less than any given quantity for Δx sufficiently small.

The notions of the calculus as presented by Cauchy and Bolzano come very close to the form presented today in a first course, but certain phrases in their exposition are lacking in precision. Just what are "successive values"? What does it mean to "approach indefinitely" or to "become and remain"? And what is a "sufficiently small" quantity? With the increasing arithmetization of mathematics as the nineteenth century progressed such phrases gave way, in the lectures of Karl Weierstrass, to the elegance and precision of the "epsilon-delta" definition of a limit, one in which the vague idea of "approaching" is abandoned for purely numerical language. The function $f(x)$ is said to have a limit L for the value $x = a$ if, given any positive number ϵ, there exists a positive number δ such that

$$|f(x) - L| < \epsilon$$

for any x for which $0 < |x - a| < \delta$.

It is to be noted here that it is the function that has a limit, not the vague something called a "variable." The general idea of a quantity depending on (or varying with) another quantity can be found in the fourteenth-century work of Oresme, and Leibniz had used the word "function" in somewhat the modern sense. Behind the work of Newton there lay very definitely the notion of quantities varying with time and hence depending upon each other. Euler's influential *Introductio* had placed the emphasis on an "analytic expression" in the definition of a function, although in another connection he had suggested that any curve drawn freehand determines a functional relationship.

With the tendencies to arithmetize during the nineteenth century, the notion of correspondence became dominant in analysis; accordingly, P. G. Lejeune Dirichlet called y a function of x if for every value of x there corresponded one or more values of y. (Conventional usage today would omit the words "or more.") The arithmetization of the function became firmly established in what is sometimes called the "Weierstrassian static theory of the variable." An independent variable x, in other words, does not "vary" in the ordinary meaning of the term; it is simply a set of numbers—often, in analysis, the set of all real numbers—today called the "domain." A function $f(x)$ of the independent variable x is then simply another set of numbers, now known as the "range," such that to every value of x there corresponds just one value of $f(x)$.

The closing years of the nineteenth century saw the development of the theory of sets by Georg Cantor, and this approach to the foundations of analysis has since become characteristic of the twentieth century. Consequently the idea of function is now generally defined in the severely unambiguous language of sets: "Given two sets of elements, denoted by A and B respectively, we say that B is a function of A—or A is mapped into B, $A \rightarrow B$—if for every element of A there is a corresponding element of B and if no two distinct elements of B correspond to the same element of A."

In order to sharpen the concept still further, the functional relationship between A and B is frequently defined as "the set of ordered pairs (a, b), where a is an element of the set A and b is an element of the set B, such that if $(r, p) = (r, q)$ then $p = q$."

With refinements in the notion of function there were bound to be changes in the calculus. The notion of the integral, for example, has undergone such modification that one no longer speaks of "the" integral. There are many types of integrals, each formulated to cover the ever wider types of functions which have come under consideration.

Were Eudoxus to reappear in the twentieth century, he might have difficulty in recognizing these descendants of the method of exhaustion; but in at least one respect he would feel completely at home in the mathematics of today. The drive for precision of thought from which the ancient integral calculus arose is matched today by a comparable insistence on rigor in analysis. Eudoxus would thus share the feeling of pride suggested in the continuing use of the phrase "the calculus," which sets the subject apart from the ordinary calculations that all too often are mistaken by the uninitiated as the preoccupation of mathematicians.

ARCHIMEDES AND
HIS ANTICIPATIONS OF CALCULUS

ARCHIMEDES of Syracuse (287–212 B.C.) is by common consent the greatest mathematician of antiquity. In the quantity and difficulty of his problems, in the originality of his methods, and in the rigor of his proofs, he towers above all the others. He was interested in both pure and applied mathematics and originated two branches of physics (statics and hydrodynamics). He was renowned for his mechanical inventions, some of which were put to use in the defense of Syracuse against the attacking Roman army under Marcellus. According to legend, Archimedes was studying a geometrical diagram in the sand when he was killed by a Roman soldier during the capture of the city.

In his work on areas and volumes he further developed the method of exhaustion, whereby the desired quantity is approximated by the partial sums of a series or by terms of a sequence. He obtained approximations to the area of a circle by comparing it with inscribed and circumscribed regular polygons. Using polygons of ninety-six sides, he showed that the area of a circle bears to the square of its radius a ratio which lies between 3 10/71 and 3 10/70, a remarkably good estimate of π. He found the area of a sphere to be four times the area of one of its great circles—a result that enabled him to compare spheres and cylinders. This work was commemorated on his tomb by the engraved device of a sphere inscribed in a cylinder. He also found volumes of spheres and segments cut by planes from various quadric surfaces.

The secret of his discoveries came partially to light as recently as 1906, when a treatise of his was rediscovered. In this work, addressed to his friend Eratosthenes, he explained how he had arrived at some of his results (actually coming close to performing integrations in many important cases), after which he would find proofs. This is well illustrated in the problem of squaring the parabola, as shown in Figure [98]-1.

Let s be the region bounded by a parabola p and a chord AB, with M the midpoint of AB. Let t be the tangent to p at A. From B and M

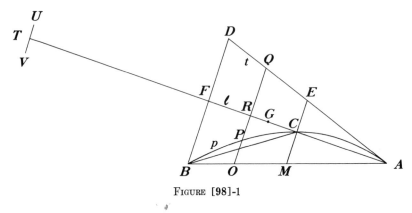

FIGURE [98]-1

let lines parallel to the axis of p meet t in D and E, respectively, and let ME meet p in C, which is called the vertex of s. By a previously known theorem, C is the midpoint of ME. Let ℓ be the line AC, intersecting BD in F.

Archimedes now compares the parabolic segment s with the triangle ABD. Let O be any point on AB. Let a line through O parallel to the axis of p meet p, t, and ℓ in the points P, Q, and R, respectively. By another known theorem,

$$OP : OQ = OB : AB = RF : AF.$$

He now takes an ingenious step: he considers the line ℓ as a lever, with F as fulcrum, and takes a point T on ℓ such that F is the midpoint of AT. At T he "hangs" a segment UV, congruent to OP. Then from the above equation

$$UV : OQ = RF : AF = RF : TF, \quad \text{or} \quad UV \cdot TF = OQ \cdot RF.$$

Thus the segment UV, suspended at its midpoint T, is in equilibrium with the segment OQ, suspended at its midpoint R. Archimedes now thinks of the triangle ABD as the union of all such segments OQ, parallel to the axis. Each segment has a corresponding segment OP congruent to a segment UV, which "hangs" at T. Thus he imagines the triangle as being balanced by the parabolic segment s, thought of as suspended at T. Moreover, as is previously known, the triangle can be considered as suspended at its centroid, which is the point G on ℓ, such that $FG = 1/3\ FA = 1/3\ FT$. Therefore, s and triangle ABD have areas which are in the ratio $1:3$. Finally, triangle ABC is four times as large as triangle ABC, and we have Archimedes' discovery: the area of a parabolic segment is $4/3$ that of a triangle having the same base and vertex.

404

To demonstrate this result, which depends so much on a brilliant intuition, Archimedes uses the method of exhaustion [**109**]. He inscribes in the segment a triangle having the same base and vertex. Next, in each of the segments remaining he inscribes a triangle in the same manner and continues to inscribe triangles in the parabolic segments remaining at each stage. Then he proves that for each triangle the two triangles constructed on its sides have a total area that is 1/4 that of the given triangle. He thus "exhausts" the parabolic segment by taking out successively these inscribed triangles. The total area can be approximated by a sum of areas, which by proper grouping leads to a geometrical progression, each term of which is 1/4 the previous term. The sum of an infinite such progression is 4/3 of the first term. Very carefully Archimedes shows that the area of the parabolic segment cannot exceed 4/3 of the area of the first inscribed triangle and likewise that it cannot fall short of it. Thus he reaches his desired conclusion and by avoiding the pitfalls of infinitesimals and limit operations attains a level of rigor which compares favorably with anything done up to the eighteenth century.

For Further Reading

ARCHIMEDES   MESCHKOWSKI (b): 13–23
BOYER (f): 48–60   D. E. SMITH (a): II, 679–84
HEATH (c): 277–342

Capsule 99   Ralph C. Huffer

SIMON STEVIN

THERE was a gap of eighteen hundred years between the time of Archimedes and that of the next major contributor to the knowledge of hydrostatics and the statics of solids: Simon Stevin (1548–1620). Stevin was developing the principles of statics while Galileo Galilei (1564–1642) was working on dynamics. Stevin and Galileo laid the foundations of applied mechanics.

One problem that was to bring Stevin fame (it led him to invent a method still found in calculus books) was that of finding the total force of water on a dam. Up to about 1586 the only way to attack this problem was Eudoxus' method of exhaustion—a tedious and difficult process. In an effort to find a simpler solution, Stevin made use of the fact, then newly discovered, that under the surface of a liquid the pressure is the same in all directions.

Although he did not use the modern notation that will be employed here, Stevin's solution was equivalent to the following. He considered the dam to be a square, 1 foot on an edge, with one edge in the surface of the water. He conceived of the square as divided into horizontal strips. (At first he treated the case of only 4 strips; later he discussed the general case in which the number was n.)

He thought of each strip as revolved $90°$ about its upper edge until it lay horizontal, subject to the weight of water lying directly above it and supported by it. The uppermost strip would then lie on the surface and would support no water; the second strip would be at a depth of $1/n$ feet below the surface and would support a volume of water of dimensions $1 \times 1/n \times 1/n$; the third, $1 \times 1/n \times 2/n$; and so on until the nth, $1 \times 1/n \times (n-1)/n$. The total weight of water supported (the force on the dam) would thus become w (the density of the water) multiplied by the total volume:

$$\frac{0}{n^2} + \frac{1}{n^2} + \frac{2}{n^2} + \cdots + \frac{n-1}{n^2} = \frac{1}{n^2}(0 + 1 + 2 + \cdots + n - 1)$$

$$= \frac{1}{n^2} \cdot \frac{n}{2} \cdot (n-1)$$

$$= \frac{1}{2} - \frac{1}{2n}.$$

He then repeated the procedure but rotated each strip about its lower edge. This yielded $\frac{1}{2} + 1/2n$.

Since the first rotation had moved each strip into a position higher than its original place and the second rotation into a lower position, the true force on the square would be intermediate to the two results. The machinery was thus in order to consider what would happen as the number of strips was allowed to increase, just as elementary calculus books do today.

From our vantage point we now recognize in this solution the essential properties of a definite integral. It is interesting to observe that the date associated with Stevin's solution of the problem (1586) is exactly one hundred years before the date on which Isaac Newton presented his *Principia* to the Royal Society.

For Further Reading

BOYER (f): 99–104   SANFORD (d): 314–15

Capsule 100   Ralph C. Huffer

JOHANN KEPLER

THE fame of Johann Kepler (1571–1630) rests solidly on the three laws that describe the motion of the planets around the sun. Using the data from Tycho Brahe's lifelong observations of the planets, Kepler discovered these laws empirically after years of tedious computations and after several ingenious but incorrect guesses.

Although calculus was not well developed until the time of Isaac Newton, Kepler had his own rough-and-ready version in which the fine points of rigor were intentionally passed over in order to get at applications, which were his main interest. Kepler based his intuitive calculus on a "principle of continuity" by which limiting cases were covered by the general definitions. For example, he thought of the area of a circle as the area of an inscribed polygon made up of an infinite number of isosceles triangles with vertices at the center of the circle, altitudes equal to the radius, and bases consisting of infinitesimal chords of the circle (Fig. [**100**]-1). By this technique he found the area of the circle to be ½ the product of the radius by the circumference. In modern notation:

$$\text{Area} = \tfrac{1}{2}rc_1 + \tfrac{1}{2}rc_2 + \tfrac{1}{2}rc_3 + \cdots$$
$$= \tfrac{1}{2}r(c_1 + c_2 + c_3 + \cdots)$$
$$= \tfrac{1}{2}r(\text{circumference}).$$

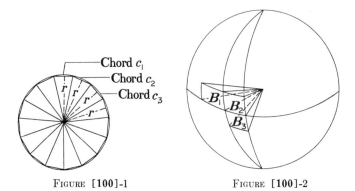

FIGURE [100]-1   FIGURE [100]-2

He similarly thought of a sphere as composed of pyramids with vertices at the center of the sphere and infinitesimal bases near the surface, thereby getting the volume of a sphere as $\frac{1}{3}$ its radius multiplied by its surface area (Fig. [100]-2). In modern notation:

$$\text{Volume} = \tfrac{1}{3}rB_1 + \tfrac{1}{3}rB_2 + \tfrac{1}{3}rB_3 + \cdots$$

$$= \tfrac{1}{3}r(B_1 + B_2 + B_3 + \cdots)$$

$$= \tfrac{1}{3}r(\text{surface area}).$$

Kepler also applied his "principle of continuity" to elementary geometry, treating parallel lines as two lines whose point of intersection has receded to infinity. He recognized the parabola as the limiting case of both the ellipse and the hyperbola in which one of the foci has receded to infinity—a fairly natural idea when these curves are thought of, in the manner of the early Greeks, as sections of a cone.

By rotating segments of conic sections about an axis in their plane, he succeeded in finding volumes for many solids. Apparently he was led into this work through an argument over the correct way to measure the volumes of wine barrels [107]. His solution involved approximations and his principle of continuity. The method turned out to be widely applicable. In particular, he used it in working out his second law (that the vector from sun to planet sweeps over equal areas in equal time intervals) in order to find the area between two focal radii of an ellipse.

For Further Reading

BOYER (f): 106–12   NEWMAN: I, 125–28
——— (g): 354–58   TURNBULL: 73–78
HOFMANN (b): 119–20

BONAVENTURA CAVALIERI

We are all familiar with the fact that Euclid wrote a systematic account of geometry which has dominated the teaching of geometry ever since. What is not so well understood is that Archimedes, within the century following Euclid, anticipated many of the results and methods of the calculus. Archimedes' work was largely neglected until the rediscovery of the calculus in the seventeenth century. Bonaventura Cavalieri (1598–1647) was one of the earliest mathematicians to revive these ideas. His notion of "indivisibles," borrowed from the Greeks, proved inadequate for the later purposes of the calculus. Nevertheless, Cavalieri was able to obtain areas bounded by such curves as $y = x^m$, where m is a positive integer.

In his book *Geometria indivisibilibus continuorum*, which appeared in 1635, Cavalieri asserted that a line was made up of an infinite number of points, each without magnitude; a surface of an infinite number of lines, each without breadth; and a volume of an infinite number of surfaces, each without thickness. These definitions were not actually used by Cavalieri in his investigations; but then, neither did Euclid, except to a small degree, use the array of definitions with which he prefaced his work. Cavalieri used the term "indivisibles" for the components into which he imagined that a line, a surface, or a solid were divided. In actual practice he divided a surface or a volume into a finite number of portions, *n*, and then permitted *n* to approach infinity. In this, his ideas anticipate those used much later.

Cavalieri's theorem, or principle, which is found in his *Geometria* of 1635, may be stated as follows:

> *Two solids of equal altitude have the same volume, if plane cross sections at equal height have the same area.*

We can gain an idea of the significance of the theorem if we conceive of two stacks of cards, *A* and *B*, placed on a table. Let us say that the cards in stack *A* are all circular and that their area is some function $f(h)$ of the height above the table. Let us say also that the cards in

409

stack B are square and that their area is given by the same function $f(h)$ of the height. Then Cavalieri's principle asserts that the two stacks have the same volume. Note that the two stacks may be quite irregular in profile; the theorem is unaffected by this. Note, further, that the shape of the cards is irrelevant; they need not be either circles or squares.

For Further Reading

BOYER (g): 361–64   STRUIK (e): 209–19
D. E. SMITH (c): II, 605–9

Capsule 102   Paul T. Mielke

PIERRE DE FERMAT

PIERRE de Fermat, born in August, 1601, at Beaumont-de-Lomagne, France, was the son of Dominique de Fermat, a leather merchant, and Claire de Long, daughter of a family of parliamentary jurists. He was educated in his native town and at the University of Toulouse in the south of France. His professional life of thirty-four years was spent in the service of France. On May 14, 1631, he was appointed Commissioner of Requests at Toulouse, and in 1648 he was promoted to King's Councilor in the local parliament of Toulouse, at which post he continued for seventeen years until his death. He died January 12, 1665, in the city of Castres, two days after conducting a case there.

Little is known of Fermat's personal life except that he lived temperately and quietly, devoting his spare time, of which there appears to have been ample, to the study of mathematics. On June 1, 1631, he was married to Louise de Long, his mother's cousin. Of this marriage there were five children, three sons and two daughters.

Though mathematics was only an avocation for him, Fermat was one of the truly great mathematicians of all time. E. T. Bell calls him "the prince of amateurs." Arthur Rosenthal states flatly /75/ that he was "the greatest mathematician of the first part of the seventeenth

410

century, not only in general but particularly in the domain of calculus." This latter statement may seem strange, since Newton and Leibniz are usually credited with having invented the calculus, and their works on the subject were not published until at least twenty years after Fermat's death; yet Fermat, as early as 1629, had developed the method that is now standard in the calculus for solving the problem of maximizing or minimizing a function. Fermat's method is as follows: Suppose one wishes to find the maximum or minimum value of an expression $f(A)$, to use the modern functional notation but denoting the unknown by A, as did Fermat (who followed Viète in the practice of denoting unknowns or variables by vowels and known quantities or constants by consonants). Now replace the unknown A by $A + E$. (Fermat thought of E as being an unknown whose value was small in comparison to that of A.) Then assume these two quantities, $f(A)$ and $f(A + E)$, to be approximately equal. In fact, *set* them equal (it did not matter to Fermat that this is not true—the end justified the means!) and cancel all possible terms from both sides of the equation. Now divide both sides by E and drop all terms still containing E. In modern symbols, and using the fact that the equation $f(A + E) = f(A)$ is equivalent to $f(A + E) - f(A) = 0$, the above exercise may be abbreviated as

$$\left[\frac{f(A + E) - f(A)}{E} \right]_{E=0} = 0.$$

The student of calculus recognizes this as the familiar necessary condition that a differentiable function have a maximum or a minimum on an open interval, namely, that the derivative of the function be zero.

As a first example of the technique, Fermat chose the problem of expressing a known number as a sum of two numbers whose product was to be a maximum. Let B be the known number and A the unknown. Then let $f(A) = A(B - A) = AB - A^2$. A maximum for $f(A)$ is sought. Fermat's recipe gives

$$f(A + E) = (A + E)[B - (A + E)]$$

$$= AB - A^2 + EB - 2AE - E^2$$

$$\frac{f(A + E) - f(A)}{E} = B - 2A - E$$

$$\left[\frac{f(A + E) - f(A)}{E} \right]_{E=0} = B - 2A.$$

411

Setting this equal to 0 yields the result that A must be chosen to be $\frac{1}{2}B$. Thus

$$f(\tfrac{1}{2}B) = \tfrac{1}{2}B(\tfrac{1}{2}B) = \tfrac{1}{4}B^2$$

is the maximum value for $f(A)$. Note that this procedure does not really prove that this value of A maximizes $f(A)$. Something more is required, as the student of elementary calculus knows, to establish that $\frac{1}{4}B^2$ is actually the maximum in this case. The method merely singles out the so-called stationary values of $f(A)$, namely, those values of A for which the derivative of $f(A)$ is zero. Now it is clear from his work that Fermat knew that the method failed to distinguish between maxima and minima, but since he confined his attention only to what would now be termed "nice," or "nonpathological" problems, he apparently failed to discover that stationary values are not necessarily either maxima or minima. As an example of this, consider the expression

$$f(A) = 3A^5 - 5A^3.$$

Fermat's method yields three stationary values, namely, $f(-1)$, $f(0)$, and $f(+1)$. The first is a (relative) maximum and the third a (relative) minimum, but $f(0)$ is neither a maximum nor a minimum for $f(A)$. Finally, Fermat did not see that his condition is not even necessary for a maximum or a minimum to occur on a closed interval. As an example of this, consider the following problem: Let a straight wire of length L be cut into two pieces. Wrap one piece in the shape of a square and the other in the shape of a circle. Now determine where the cut should have been made in order to maximize the total area of the two figures. There is no solution to this problem, but if one allows for the possibility of making no cut at all and wrapping the uncut wire in the shape of a circle, this is the correct answer, yet the area function will not have zero derivative at its maximum value. The maximum is a so-called end-point maximum. A final shortcoming of Fermat's method is exhibited by the expression $f(A) = A^{2/3}$ where A is any real number. This expression has the minimum value zero at $A = 0$, but the derivative of $f(A)$ does not even exist at $A = 0$ and is nowhere zero.

It would be less than fair to fault Fermat for these imperfections in his method. It was evidence of his genius that he understood as much as he did about the methods of calculus some fifty-seven years before it was "invented." He used his method to solve many interesting and useful problems, among these the problem of finding the tangent to a curve at a point. In the course of his investigations he also enunciated an important law of physics which is now known as "Fermat's principle of least time," namely, that a ray of light, in passing from one point to another in space, will follow that path which requires the least time for passage.

For Further Reading

BELL (d): 56–72   [5th ed. 297–99]
BOYER (g): 367–401   ROSENTHAL
EVES (c): 329–31

Capsule 103   Amy C. King

JOHN WALLIS

JOHN Wallis (1616–1703) was one of the most able, remarkable, and original mathematicians of his day. His work in analysis did much to pave the way for the great discoveries of Isaac Newton. It was during the Christmas vacation of his fifteenth year that he picked up his brother's algebra book and was delighted by and intrigued with the unfamiliar symbols which he found. Within a short time he had mastered the book; this experience marked the beginning of a most productive career in mathematics.

His *Tractatus de sectionibus conicis ("Conic Sections")*, published in 1655, contained the first discussion of conics as second-degree curves and helped to make the new (1637) but obscurely written analytic geometry of René Descartes intelligible. Although he realized that his use of Bonaventura Cavalieri's "indivisibles" was at times quite unorthodox, Wallis gave almost free rein to his imagination and intuition, as the following example shows (Fig. [**103**]-1).

He thought of a triangle as composed of an infinite number of "very thin" parallelograms whose areas (from vertex to base of the triangle) form an arithmetic progression with 0 for the first term and

$$(A/\infty) \cdot B$$

for the last term—since the last parallelogram (along the base B of

413

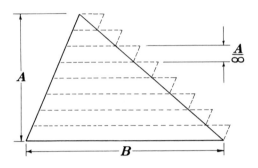

FIGURE [103]-1

the triangle) has altitude A/∞ and base B. The area of the triangle is the sum of the arithmetic progression

$$0 + \cdots + \frac{A}{\infty} \cdot B = \left(\frac{\text{no. of terms}}{2}\right)(\text{first} + \text{last term})$$

$$= \frac{\infty}{2}\left(0 + \frac{A}{\infty} \cdot B\right)$$

$$= \frac{\infty}{2} \cdot \frac{A}{\infty} \cdot B$$

$$= \tfrac{1}{2} A \cdot B.$$

The symbol ∞ for infinity seen above was first used in this work by Wallis.

In his *Arithmetica infinitorum* (1656) Wallis achieved further results in calculus, and his methods were now more arithmetic and less geometric. An interesting example is his expression for $\pi/2$:

$$\frac{2 \cdot 2 \cdot 4 \cdot 4 \cdot 6 \cdot 6 \cdot 8 \cdot \cdots}{1 \cdot 3 \cdot 3 \cdot 5 \cdot 5 \cdot 7 \cdot 7 \cdot \cdots},$$

which he obtained by a sophisticated interpolation, somewhat as follows. He knew how to find the areas represented in modern notation by

$$\int_0^1 (1 - x^2)^0 \, dx, \qquad \int_0^1 (1 - x^2)^1 \, dx, \qquad \int_0^1 (1 - x^2)^2 \, dx,$$

and so on, these areas being 1, $\tfrac{2}{3}$, 8/15, and so on, respectively. Thus the area

$$\int_0^1 (1 - x^2)^{1/2} \, dx$$

in Figure [103]-2 should lie between 1 and $\tfrac{2}{3}$ since the exponent $\tfrac{1}{2}$ lies between 0 and 1.

414

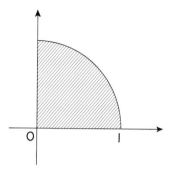

FIGURE [103]-2

But this last integral represents $\frac{1}{4}$ the area of the unit circle; hence $\pi/4$ lies between 1 and $\frac{2}{3}$, and by a complicated process Wallis finally deduced the above expression for $\pi/2$. Wallis did not know how to expand $(1 - x^2)^{\frac{1}{2}}$ into a binomial series, but this example shows that he was one of the first to accept fractional exponents, and he was only a step short of successfully representing complex numbers graphically.

For Further Reading

BOYER (f): 168–74
———— (g): 415–21
KING AND READ (a): 35–38

D. E. SMITH (c): I, 46–54, 217–23
STRUIK (e): 244–53

Capsule 104   *Elinor B. Flagg*

ISAAC BARROW

DURING the early school life of Isaac Barrow (1630–1677) the chances that he would make any significant contribution to the academic world seemed exceedingly small. At his first school, Charterhouse, he was so troublesome and fond of fighting that his father was heard to pray that should God decide to take one of his children, he could best spare Isaac. This was the same Isaac who, years later, was to recognize the genius of another Isaac, his pupil, Isaac Newton, in whose mind he planted the seed from which Newton's calculus grew.

After being removed from Charterhouse, Barrow entered Felstead School, Essex, where he applied himself with such diligence that he was admitted to Trinity College, Cambridge, at the age of fourteen. Four years later (1648) he was graduated, and after another year he became a fellow of Trinity College. In 1663 he became the first occupant of the Lucasian Chair of Mathematics at Cambridge, but only six years later he resigned this chair on his own initiative in favor of his pupil Isaac Newton, whose superior abilities he graciously acknowledged.

When Barrow was writing his *Lectures on Optics and Geometry* (1669) he turned to Newton for help (probably on the material on optics); Newton in turn benefited from those portions of the *Lectures* dealing with new ways of determining areas and tangents to curves. Most significant for the later development of calculus was Barrow's method of determining tangents to curves by use of a "differential triangle," sometimes called "Barrow's triangle."

In order to construct the tangent line t in Figure [**104**]-1 to the curve at the point P, Barrow determined another point T on t as follows. Let Q be a point on the curve; then, since P and Q are neighboring points, $\triangle PTM$ and $\triangle PQR$ are very nearly similar, Barrow said, especially as the little triangle becomes infinitely small, so we can write

(1) $$RP/QR = MP/TM$$

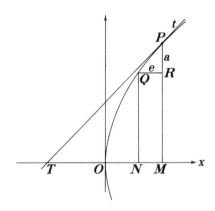

FIGURE [**104**]-1

(approximately). Denoting the coordinates of P and Q by (x, y) and $(x - e, y - a)$, substituting these values into the equation of the given curve, deleting terms involving powers of a and e higher than the first power, we can find the ratio a/e. Since M is a known point, we can now find T (on the x-axis) using the length of segment TM which is given by $y(a/e)$—a direct consequence of (1).

Let us apply Barrow's method to the parabola $y^2 = 4x$ shown in Figure [104]-1. Since Q is on the curve, its coordinates satisfy the equation; hence

$$(y - a)^2 = 4(x - e)$$

$$a^2 - 2ay + y^2 = 4x - 4e.$$

But

$$y^2 = 4x.$$

Hence

$$a^2 - 2ay = -4e.$$

Now if a is sufficiently close to zero (say $a = 0.01$), then we may neglect the term a^2 and obtain

$$-2ay = -4e$$

or

$$\frac{a}{e} = \frac{2}{y}.$$

In particular, if P is the point $(1, 2)$ on $y^2 = 4x$, then TM has length 2 and T is $(-1, 0)$. PT is the required tangent line.

For the later development of calculus the significant aspect of Barrow's method was not the actual construction of the tangent line, but rather the ratio a/e. If we apply elementary calculus to the above example we obtain $dy/dx = 2/y$ for the derivative of $y^2 = 4x$—the same results given by Barrow's a/e. In fact, if Barrow had divided by e and *then* deleted terms, his method would have been almost identical to the steps often followed in applying the modern definition of derivative. This is easily seen by letting $\triangle x = -e$ and $\triangle y = -a$ in Barrow's procedure.

Although Barrow succeeded in finding derivatives (as we say today) for many specific equations, his method clearly required a logical foundation. It was left for Newton and others to lay these foundations (including especially the theory of limits by Cauchy and others) and to create a handy symbolism.

For Further Reading

BALL (a): 309–12   [5th ed. 300–302]
BOYER (f): 181–86   MIDONICK: 106–15
——— (g): 424–26   STRUIK (e): 253–63
EVES (c): 333–35

Capsule 105   Amy C. King

LEIBNIZ

GOTTFRIED Wilhelm von Leibniz (1646–1716) was a versatile genius who applied his energies to mathematics, logic, philosophy, theology, law, economics, linguistics, and history—including a genealogical history of the Brunswick family, in whose employ he spent the last forty years of his life at the court of Hanover. When the Duke of Hanover went to London to become George I, the first German King of England, Leibniz was left behind in neglect, and two years later he died; it is said that only his faithful secretary attended the funeral.

Leibniz' interest in mathematics resulted from a diplomatic mission to Paris in 1672. Here he was fortunate enough to meet Christiaan Huygens, who presented a copy of his work on the oscillation of the pendulum to Leibniz and started the gifted young diplomat (who had already written a little on combinatorial analysis and mechanics) on a career in mathematics. Huygens gladly agreed to instruct Leibniz, who began studying the works of Barrow, Cavalieri, Pascal, Descartes, and others.

Leibniz said that it was while reading Pascal that he suddenly realized that the tangent to (or slope of) a given curve could be found by forming the ratio of the *differences* in the ordinates and abscissas of two neighboring points on the curve as these differences were made smaller and smaller. He further noted that the quadrature of (area under) the curve could be taken as the *sum* of the ordinates or of infinitely many thin rectangles. Most significantly, he observed that these two processes of differencing and summing (i.e., of differentiating and integrating) were inverses of each other.

In a manuscript dated October 29, 1675, Leibniz drew what he called

418

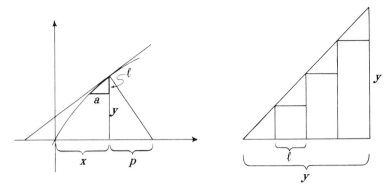

FIGURE [105]-1   FIGURE [105]-2

a "characteristic triangle" (see Fig. [105]-1), an idea previously used by Barrow and Pascal. He proceeded essentially as follows: $\ell/a = p/y$. (Most likely Leibniz thought of these segments as belonging, in some intuitive and approximate sense, to similar triangles.) Hence rect. pa = rect. $y\ell$. Summing these rectangles, he first wrote, in Cavalieri's notation,

(1)   *omn.* pa = *omn.* $y\ell$ (where *omn.* stands for *omnia*, "all");

but since in this same manuscript he said, "It will be useful to write $\cdots \int \ell$ for omn.ℓ, that is, the sum of the ℓ's," we shall use the $\int$ sign a few lines sooner than he did. Equation (1) then becomes

(2) $\int pa = \int y\ell$

where the summation, $\int y\ell$, is from 0 to y and thus gives the area, $\dfrac{y^2}{2}$, of an isosceles right triangle (see Fig. [105]-2). Thus Leibniz obtained the relation

(3) $\int pa = \int y\ell = \dfrac{y^2}{2}$

or

$$\int p\, dx = \int_0^y y\, dy = \frac{y^2}{2}\,,$$

as one might prefer to think of the integration today. Leibniz, of course, was only beginning to develop these ideas. If we assume, as Leibniz tacitly did, that the curve passes through the origin, then we may write $\int p\, dx$ as

$$\int_0^x p\, dx.$$

419

Indeed, from (3) he then deduced such expressions as $\int x = \dfrac{x^2}{2}$ and $\int (x + y) = \int x + \int y$. We notice that he did not quite yet write $\int x$ as $\int x \, dx$, but in this same manuscript he soon came to the matter of "differences," as he always called differentials. At first he wrote dx as $\dfrac{x}{d}$ because he reasoned that if $\int$ (summing segments) *raised* the dimension of segment to area, then d ("differencing," the inverse of summing) should *lower* the dimension.

Ten days later, in a second manuscript, while applying his calculus to various problems, he realized that dx was a better notation, and made the marginal comment, "dx and $\dfrac{x}{d}$ are the same; i.e., the difference between two neighboring x's." Then, in this second manuscript, he speculated, "Let us see whether $dx \, dy$ is the same as $d \, \overline{xy}$, and whether $\dfrac{dx}{dy}$ s the same as $d \, \dfrac{x}{y}$." He correctly concluded that they are not the same, but was apparently unable to say what the correct expressions were until, another ten days later in a third manuscript, he found that $y \, dx = d \, \overline{xy} - x \, dy$, thus obtaining the equivalent of the correct result, $d(xy) = xdy + ydx$. Somewhat later he found also the correct expression for $d(x/y)$ and derived many other rules for differentiating.

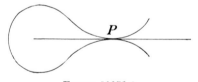

FIGURE [105]-3

Not only did Leibniz have a remarkable knack for notation; he also originated such terms as abscissa, ordinate, coordinate, axis of coordinates, and function. He gave the amusing term "point of osculation" to a point such as P (Fig. [105]-3) where two branches of the curve have a common tangent.

"Leibniz' theorem" enables us easily to write down the nth derivative of a product of two functions. Thus, if $y = uv$ where u and v are functions of x, we can write the fourth derivative, for example, as follows:

(6) $\quad \dfrac{d^4 y}{dx^4} = y^{(4)} = u^{(4)}v^{(0)} + 4u^{(3)}v^{(1)} + 6u^{(2)}v^{(2)} + 4u^{(1)}v^{(3)} + u^{(0)}v^{(4)}$

and, in general,

$$(7) \qquad y^{(n)} = \sum_{k=0}^{n} C_{n,k} u^{(n-k)} v^{(k)}$$

where in (6) the coefficients 1, 4, 6, 4, 1 are simply the binomial co-efficients for the fourth power expansion and where in (7) the $C_{n,k}$ are the general coefficients in the binomial expansion.

Leibniz generalized the binomial theorem into the multinomial theorem that gives an expression for $(a + b + c + \ldots + z)^n$, and he invented determinants in the context of solving linear systems of equations.

While visiting London in 1673, Leibniz was introduced to the method of infinite series, and the following series bear his name (though priority of discovery seems to go to James Gregory):

$$\frac{\pi}{4} = 1 - \frac{1}{3} + \frac{1}{5} - \frac{1}{7} + \cdots ,$$

$$\tan^{-1} x = x - \frac{x^3}{3} + \frac{x^5}{5} - \frac{x^7}{7} + \cdots .$$

The familiar alternating series test used in beginning calculus is due to Leibniz.

For Further Reading

BALL (a): 353–65

BELL (d): 117–30

BOYER (f): 187–223

———— (g): 437–45

KING and READ (a): 33–35

MESCHKOWSKI (b): 47–59

D. E. SMITH (c): I, 173–81, 229–31, 267–70; II, 619–26

STRUIK (e): 270–84

Capsule 106   Robert H. Dykstra

NEWTON'S *DOT*-AGE VERSUS LEIBNIZ' *D*-ISM

ISAAC Newton considered geometrical quantities as generated by continuous motion; thus, for example, a moving point generates a line.

The quantity, x, thus generated he called a "fluent"; its rate of change, which he denoted by $\dot{x}$ (i.e., dx/dt in modern notation), he called the "fluxion" of x. Further, the small amount by which a fluent, x, increases in a small interval of time, o, he called the "moment" of the fluent; and he denoted this moment by $\dot{x}o$ (i.e., $(dx/dt)(dt)$ or dx as we would now write it).

Then Newton commented on the ratio $\dot{y}o/\dot{x}o$ as follows: "The moments of flowing quantities [or fluents] are as the velocities of their flowing or increasing [i.e., $\dot{y}o/\dot{x}o = \dot{y}/\dot{x}$]"—a statement we would now write as $dy/dx = (dy/dt)/(dx/dt)$.

The following example from Newton's *Method of Fluxions* (written in 1671 but not published until 1736) will illustrate his use of fluxions and moments. Except for substituting a simpler example, we quote /D. E. SMITH (a): II, 693–94/:

> If the moment of x be represented by the product of its celerity $\dot{x}$ into an indefinitely small quantity o (that is, $\dot{x}o$), the moment of y will be $\dot{y}o$, since $\dot{x}o$ and $\dot{y}o$ are to each other as $\dot{x}$ and $\dot{y}$. Now since the moments as $\dot{x}o$ and $\dot{y}o$ are the indefinitely little accessions of the flowing quantities, x and y, by which these quantities are increased through the several indefinitely little intervals of time, it follows that these quantities, x and y, after any indefinitely small interval of time, become $x + \dot{x}o$ and $y + \dot{y}o$. And therefore the equation which at all times indifferently expresses the relation of the flowing quantities will as well express the relation between $x + \dot{x}o$ and $y + \dot{y}o$ as between x and y; so that $x + \dot{x}o$ and $y + \dot{y}o$ may be substituted in the same equation for those quantities instead of x and y.
>
> Therefore let any equation
>
> $$xy - a = 0$$
>
> be given, and substitute $x + \dot{x}o$ for x and $y + \dot{y}o$ for y, and there will arise
>
> $$xy + x\dot{y}o + \dot{x}oy + \dot{x}o\dot{y}o - a = 0.$$
>
> Now, by supposition,
>
> $$xy - a = 0,$$
>
> which therefore being expunged and the remaining terms being divided by o, there will remain
>
> $$\dot{y}x + \dot{x}y + \dot{x}\dot{y}o = 0.$$
>
> But whereas o is supposed to be infinitely little that it may represent the moments of quantities, the terms which are multiplied by it will be nothing in respect to the rest. Therefore I reject them and there remains:
>
> $$x\dot{y} + y\dot{x} = 0.$$

Leibniz developed (*c.* 1675) his differential notation and applied it to find the differential of such expressions as xy. His use of differentials (from the Latin *differentia* for "difference") is nicely illustrated by the following excerpt from a letter he wrote to John Wallis in 1699 /D. E. SMITH (a): II, 696–97/:

> It is useful to consider quantities infinitely small such that when their ratio is sought, they may not be considered zero, but which are rejected as often as they occur with quantities incomparably greater. Thus if we have $x + dx$, dx is rejected. But it is different if we seek the difference between $x + dx$ and x, for then the finite quantities disappear. Similarly we cannot have xdx and $dxdx$ standing together. Hence if we are to differentiate xy we write:
>
> $$(x + dx)(y + dy) - xy = xdy + ydx + dxdy.$$
>
> But here $dxdy$ is to be rejected as incomparably less than $xdy + ydx$. Thus in any particular case the error is less than any finite quantity.

For Newton's fluxion $\dot{x}$, Leibniz wrote $\dfrac{dx}{dt}$ (as well as $dx{:}dt$ and dx ad dt).

For what we now call the integral of y, Newton wrote $\overset{\shortmid}{y}$ or $\boxed{y}$ or $[y]$. Leibniz wrote the same integral as $\int y$ and, a little later, as $\int y \, dx$. The $\int$ sign, an old form of the letter S, stood for summations which both Newton and Leibniz knew could be accomplished by antidifferentiation (the method of "inverse tangents," as it was then called).

Although Leibniz realized the importance of the theoretical basis of calculus, he did not give it the careful study found in Newton's work. One of Leibniz' main concerns was good notation, and he succeeded in developing for calculus the notation we use today, which was ignored by the English for half a century because of their loyalty to Newton and their dislike of Leibniz. The turning point came with the organization, following Charles Babbage's entrance into Cambridge University in about 1810, of the Analytical Society. The Society saw the advantages of Leibniz' dx notation over Newton's $\dot{x}$ notation and, as Babbage put it, "advocated the principles of pure *d*-ism as opposed to the *dot*-age of the university."

For Further Reading

BOYER (f): 187–223   SCHRADER (c)
CAJORI (d): II, 180–206   D. E. SMITH (c): II, 613–26

423

GAUGING—VOLUMES OF SOLIDS

Johann Kepler (1571–1630), in addition to working in astronomy and in various fields of mathematics, made an extensive study of volumes of various solids of revolution. On the occasion of his second marriage he had reason to become interested in methods of measuring the volumes of wine casks and, as a result, other volumes of revolution. His work was published in 1615 in *Stereometria doliorum vinorum* ("solid geometry of wine barrels"), which included a study of volumes of solids obtained by rotating segments of conic sections about an axis in their plane. Kepler considered the volume of such a solid as the sum of numerous thin layers, each of which was a cylindrical disk.

Otto Toeplitz /82–83/ refers to Kepler's purchase of the wedding wine. The merchant measured the volume of a barrel (see Fig. [**107**]-1) by inserting a foot rule into taphole S until it reached the lid at D. The merchant used the length $SD = d$ to calculate the price. Kepler saw that a narrow high barrel might have the same measurement d as a wide one and hence the same price, though its volume would be much smaller.

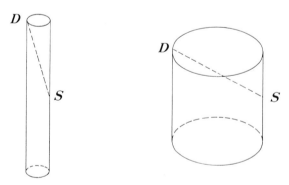

Figure [**107**]-1

Assuming the barrel to be approximately a cylinder, taking the measurement d, the radius r of the base, and the height h, Kepler approximated the volume V as follows:

By the Pythagorean theorem

$$d^2 = \left(\frac{h}{2}\right)^2 + (2r)^2$$

and hence

$$r^2 = \frac{d^2}{4} - \frac{h^2}{16}.$$

The volume is given by

$$V = \pi r^2 h$$

$$V = \pi\left(\frac{d^2}{4} - \frac{h^2}{16}\right)h$$

(5)
$$V = \frac{\pi d^2}{4} h - \frac{\pi}{16} h^3.$$

Now, for fixed d and r, Kepler wanted to know what value of h would give the largest volume V, and he showed that h should satisfy the condition $3h^2 = 4d^2$ in order to have the largest volume.

Today we would maximize V by differentiating (5), setting the result equal to zero, thus obtaining $3h^2 = 4d^2$:

$$\frac{dV}{dh} = \frac{\pi d^2}{4} - \frac{3\pi}{16} h^2 = 0.$$

Hence

(7)
$$4d^2 - 3h^2 = 0 \quad \text{or} \quad \frac{h}{d} = \frac{2}{\sqrt{3}},$$

and

$$h = \frac{2}{\sqrt{3}} d$$

is the required value of h which maximizes V.

Kepler, of course, did not have available the concept of derivative (which, at least in the above form, was a later invention), but he observed that the method of using d to calculate the price of wine was quite legitimate for Austrian barrels because their shape showed remarkable agreement with (7); and even if the barrels' shape deviated somewhat from this proportion, the method of using d could still be utilized because, as he showed, near its maximum a function changes only slowly. (Today we would be more inclined to say, "near" an ordinary maximum the rate of change is "close" to zero.) This intuitive, subformal use of the derivative concept was given a more explicit formulation a few years later (1629) by Fermat, whose method was sufficiently close to the present-day procedure for finding ordinary maxima and minima that some of our textbooks call it "Fermat's method."

As time went on, others studied the problem of finding the volumes of wine barrels. Fairly precise methods were developed and were referred to as "gauging." It became somewhat of an art to measure the volumes of all the differently shaped casks.

A very complete and detailed method was given by Nicholas Pike (*c.* 1790) /LEAVITT: 214–15/:

> Rule.—Take the dimensions of the cask in inches, viz. the diameter at the bung and head, and the length of the cask; subtract the head diameter from the bung diameter and note the difference.
>
> If the staves of the cask be much curved or bulging between the bung and head, multiply the difference between the bung and head diameter by .7; if not quite so much curved, by .65; if they curve yet less by .6; and if they are almost or quite straight, by .55; and add the product to the head diameter; the sum will be a mean diameter, by which the cask is reduced to a cylinder.
>
> Square the mean diameter, thus found, then multiply it by the length; divide the product by 294 for wine, or by 359 for ale or beer, and the quotient will be the content in gallons.

Pike noted that since 231 cubic inches equal one wine gallon, the volume of wine in gallons is obtained by the rule just given because, for a cylinder,

$$V = \pi\left(\frac{d}{2}\right)^2 \ell \text{ cu. in.}$$

$$= \frac{(\pi/4)d^2\ell}{231} \text{ wine gallons}$$

$$= \frac{d^2\ell}{\dfrac{231}{(\pi/4)}} \approx \frac{d^2\ell}{294} \text{ wine gallons,}$$

where d is the mean diameter and ℓ is the length of the cask. For ale or beer the numbers 231 and 294 above would be replaced by 282 and 359, respectively. Pike also suggested that proper allowances be made for the thickness of the wood of the cask.

As an example of Pike's method, we give the following problem and the answers:

What is the content in wine and ale (or beer) gallons of a cask, the staves being much curved, whose bung diameter is 35 inches, head diameter 27 inches, and length 45 inches?

Bung diameter	35
Head diameter	27
Difference	8
Multiplier	.7
Product	5.6
Head diameter	27.0
Mean diameter	32.6
	32.6
	1956
	652
	978
Squared	1,062.76
Length	45
	531380
	425104
	47,824.20

The answers are given below, first for wine gallons and then for ale (or beer) gallons.

$$47,824.20 \div 294 = 162.69^{+}.$$

$$47,824.20 \div 359 = 133.21^{+}.$$

For Further Reading

STRUIK (e): 192–97   TOEPLITZ: 82–83

Capsule 108   Melcher Fobes

FINITE DIFFERENCES

BROOK Taylor (1685–1731) originated the "calculus of finite differences" and published his results in his *Methodus incrementorum directa*

427

et inversa (1715), which also contains Taylor's theorem on series, familiar to all calculus students [**114**]. In this book Taylor drew many analogies between his new finite calculus and the ordinary (infinitesimal) calculus, and by applying his new calculus to the problem of transverse vibration of strings he was the first to succeed in reducing the problem to mechanical principles.

But the earliest real elaboration of finite differences was given by James Stirling (1692–1770) who also, twenty-five years before Maclaurin, gave the so-called Maclaurin expansion, which, of course, is merely a special case of Taylor's theorem. Stirling is probably best known for the so-called Stirling's formula, due to Abraham De Moivre (1667–1754), which approximates $n!$ by

$$\sqrt{2\pi n}\ (n^n)(e^{-n}).$$

In his *Methodus differentialis* (1730) Stirling applied the calcalus of finite differences to summation of series and interpolation, and he introduced the Stirling numbers, still important in the theory. Though he was not one of the great mathematicians, he was one of the interesting and versatile men of the early years of calculus. At the age of twenty-two he was expelled from Oxford University for his political beliefs, and he made his way to Venice. He studied and taught there for ten years before he returned hastily to England because he had discovered some of the closely guarded trade secrets of the glass blowers of Venice and feared assassination. For the rest of his life his colleagues usually referred to him as "the Venetian."

The calculus of finite difference was further developed by many other men after Stirling, including Leonhard Euler, whose Δ notation replaced the dot notation of Taylor, and George Boole, whose textbook *Finite Differences* (1860) provides the motivation for the following illustrative material. The notation $x^{(3)}$ means $x(x-1)(x-2)$, and the parallel column on the right gives the corresponding idea in ordinary calculus.

The finite derivative of $F(x)$ is defined by

$$\Delta F(x) = \lim_{h \to 1} \frac{F(x+h) - F(x)}{h} \qquad\qquad DF(x) = \lim_{h \to 0} \frac{F(x+h) - F(x)}{h}$$

$$= F(x+1) - F(x). \qquad\qquad\qquad\qquad = \frac{dF}{dx}.$$

428

Thus, for example,

$$\Delta x^{(3)} = \Delta x[(x - 1)(x - 2)]$$
$$= 3x(x - 1) \hspace{3cm} Dx^3 = 3x^2.$$
$$= 3x^{(2)}.$$

The inverse operation, the finite antiderivative, is illustrated by

$$\Delta^{-1}x^{(2)} = \frac{x^{(3)}}{3} + C. \hspace{2cm} D^{-1}x^2 = \int x^2\,dx = \frac{x^3}{3} + C.$$

The analogue of the fundamental theorem equates summation, Σ, with Δ^{-1}; thus,

$$\sum_{x=a}^{x=b} x^{(2)} \hspace{3cm} \int_a^b x^2\,dx = [D^{-1}x^2]_a^{b+0} = \left[\frac{x^3}{3}\right]_a^{b+0}.$$
$$= [\Delta^{-1}x^{(2)}]_a^{b+1} = \left[\frac{x^{(3)}}{3}\right]_a^{b+1}.$$

Applying this fundamental theorem for finite derivatives to the summation

$$2 \cdot 1 + 3 \cdot 2 + 4 \cdot 3 + \cdots + 10 \cdot 9 = \sum_{x=2}^{10} x(x - 1),$$

we have

$$\sum_{x=2}^{10} x(x - 1) = \sum_{x=2}^{10} x^{(2)} = [\Delta^{-1}x^{(2)}]_2^{10+1} = \left[\frac{x^{(3)}}{3}\right]_2^{11}$$
$$= \left[\frac{x(x - 1)(x - 2)}{3}\right]_2^{11} = \frac{11 \cdot 10 \cdot 9}{3} - 0$$
$$= 330.$$

It is interesting to note that from the start Brook Taylor drew many analogies between the new finite calculus and the ordinary calculus—as Boole did, also, in his text of 1860.

For Further Reading

Boyer (g): 465–69 $\hspace{3cm}$ Struik (e): 328–33

ARCHIMEDES
AND THE "METHOD OF EXHAUSTION"

THE versatile Archimedes (287–212 B.C.) used his own early version of integral calculus to find areas and volumes, and in some respects it was very similar in spirit to our present calculus. In a letter to Eratosthenes, Archimedes explained his "lever method" for discovering formulas for areas and volumes, but when he published proofs for these formulas he used the "method of exhaustion" in order to conform to the standards of rigor of his time. The reader is familiar with the Greek attack on the area problem for the circle: The increasing areas of the inscribed polygons increase to the area of the circle, while the decreasing areas of the circumscribed polygons decrease to the area of the circle.

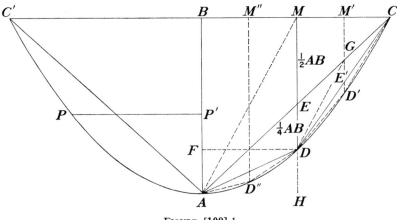

FIGURE [109]-1

Let us look at Archimedes' "exhaustion" proof, in somewhat modified form, that the area of a parabolic segment is 4/3 the area of the inscribed triangle [98]. Suppose Figure [109]-1 represents a portion of a parabola cut off by a chord $C'C$ perpendicular to its axis AB. We assume as the definition of the parabola that it is the set of points P such that AP' is proportional to $(P'P)^2$—that is, in modern notation, that our parabola has the equation $y = kx^2$. Archimedes shows that the area of this portion of the parabola is 4/3 the area of $\triangle C'AC$, or equivalently, that the area bounded by AB, BC, and the parabola is 4/3 that of $\triangle ABC$. To do this, he "exhausts" the parabolic area by first adding to $\triangle ABC$ the triangle ADC, in which D is the point where a parallel to AB through the midpoint M of BC cuts the parabola, and showing that $\triangle ADC = 1/4 \, \triangle ABC$. Next, he constructs parallels to AB through M' and M'', the midpoints of MC and BM, cutting the parabola at D' and D'' and shows that $\triangle AD''D + \triangle DD'C = 1/4 \, \triangle ADC = 1/(4^2) \, \triangle ABC$. This procedure, continued indefinitely, leads to the conclusion that the parabolic area is approximated by

$$(1) \qquad \triangle ABC + \tfrac{1}{4}\triangle ABC + \frac{1}{4^2} \, \triangle ABC + \cdots + \frac{1}{4^n} \, \triangle ABC,$$

which, as n increases, comes closer and closer to $4/3 \, \triangle ABC$.

The proof that $\triangle ADC = \tfrac{1}{4} \, \triangle ABC$ goes as follows, using the notation and construction lines of the figure. From the definition of the parabola, $AF = k(FD)^2$ and $AB = k(BC)^2$. Since $FD = BM = \tfrac{1}{2}BC$, it follows that $AF = HD = \tfrac{1}{4}AB$. By similar triangles, $EM/AB = MC/BC = \tfrac{1}{2}$, so that $EM = \tfrac{1}{2}AB$. Hence,

$$DE = AB - HD - EM = AB - \tfrac{1}{4}AB - \tfrac{1}{2}AB = \tfrac{1}{4}AB.$$

Thus $\triangle ADE$ and $\triangle AEM$ have the same altitude AH and bases $DE = \tfrac{1}{4}AB$ and $EM = \tfrac{1}{2}AB$, respectively. Thus $\triangle ADE = \tfrac{1}{2} \, \triangle AEM$. Similarly $\triangle DEC = \tfrac{1}{2} \, \triangle EMC$; so that, by addition, $\triangle ADC = \tfrac{1}{2} \, \triangle ACM$. Moreover, $\triangle ACM$ and $\triangle AMB$ have equal bases (MC and BM) and equal altitudes (AB), and thus $\triangle ADC = \tfrac{1}{4} \, \triangle ABC$.
Q.E.D.

The student may find it interesting, and a bit more difficult, to prove by similar methods, using the construction lines shown in the figure, that $\triangle DD'C = \tfrac{1}{4} \, \triangle DCE$ and $\triangle AD''D = \tfrac{1}{4} \, \triangle ADE$, so that $\triangle AD''D + \triangle DD'C = \tfrac{1}{4} \, \triangle ADC = 1/(4^2) \, \triangle ABC$, thus completing the second stage of the proof.

It is also a matter of considerable interest that Archimedes gives a second proof of this result by means of inscribed and circumscribed rectangles, almost identical in spirit with our present method of formation of upper and lower Riemann sums. In this case the Riemann sums can be evaluated simply so that the use of the fundamental theorem is not necessary.

For Further Reading

Archimedes: 233–52; supple- ——— (g): 100–103, 142–46
ment, 5–51 Meschkowski (b): 13–23
Boyer (f): 31–37, 51–56 Toeplitz: 11–14

Capsule 110   Marlow Sholander

CONVERGENCE

The following analogy has some merit. Mathematics is a supermarket. Numerous authors have arranged window displays. Convergence is a major department of the store. A capsule on convergence is a representative shopping-bag-size selection. The reader will indulge the author if, after choosing an item or two, he then heads for a handy display (in some standard work on the history of mathematics) and if most of his selections are (oversimplifications like) slices, crumbs, and peanut butter.

As a starter, we might argue that Zeno's paradox (*c.* 450 B.C.) on the impossibility of traversing a race course simplifies to the falsity $1 > \frac{1}{2} + \frac{1}{4} + \frac{1}{8} + \cdots$. However, one of the first men known to grapple with limits and series was Archimedes (*c.* 220 B.C.). In our notation, he found that

$$\lim_{n \to \infty} 4^{-n} = 0$$

and that

$$\sum_{n=1}^{\infty} 4^{-n} = \tfrac{1}{3},$$

and gave the formula

(1) $$\frac{a}{1-r} = a + ar + ar^2 + \cdots .$$

There was now an interval in which these gains were consolidated (a span equal to that from Nero to Castro). Around 1673, Leibniz was led by the contrast between

(2) $$1 - \tfrac{1}{2} + \tfrac{1}{3} - \tfrac{1}{4} + \cdots$$

and

(3) $$1 + \tfrac{1}{2} + \tfrac{1}{3} + \tfrac{1}{4} + \cdots$$

432

to the idea of convergence and divergence. His test for alternating series, the first convergence test, did not appear until 1705. A midway event, Newton's *Principia*, marks the peak of the intuitive period (in which men were guided to discoveries by whispers from nature). Newton had the perspicacity to be uneasy about the rationalia cloaking the calculus. His derivatives, strictly speaking, were ratios rather than limits.

Meanwhile, back in the study, so many pretty formulas were being found that intuitionism was giving way to formalism (in which nature speaks to man through formulas). About 1700, men were debating the message borne by (1) in the special case

(4) $\frac{1}{2} = 1 - 1 + 1 - 1 + \cdots .$

Leibniz, noting the subtotals 1, 0, 1, 0, $\cdots$, thought probability was involved. Euler, under whom formalism crested (*c.* 1750), remained calm in the face of this and companion puzzlers like

(5) $-1 = 1 + 2 + 4 + 8 + \cdots ,$

if the formal binomial theorem is applied to $(1 - 2)^{-1}$. By the time someone observed that

(6) $0 = 0 + 0 + \cdots = (1 - 1) + (1 - 1) + \cdots$

           $= 1 - 1 + 1 - 1 + \cdots$

           $= 1 - (1 - 1) - (1 - 1) - \cdots = 1,$

formalism was outward bound. Euler was almost led to define "sums" of series. D'Alembert said something ought to be done about "limits." Lagrange agreed but adopted inferior remedies.

An era of rigor was ushered in by Gauss (*c.* 1800). It went almost unnoticed until Cauchy (1830) began moving analysis built on sand to a proper foundation. Rigor has been fashionable ever since. We owe Weierstrass (*c.* 1860) grim thanks for the full "epsilon-delta" hairshirt version.

It is noteworthy that it took many years and much evidence to convince people that $a_1 + a_2 + \cdots + a_n$ and $a_1 + a_2 + \cdots + a_n + \cdots$ were of different species. Even Cauchy and Abel were loath to conclude that sums must be denied to series like (3), (4), and (5). It took time to ascertain that grouping or ungrouping, as in (6), might change convergence to divergence. Infinite sums, in other words, might not be associative. Note that our definition matches the natural grouping

(7) $\{[(a_1 + a_2) + a_3] + a_4\} + \cdots .$

Cauchy used series like (3) to demonstrate that commutativity also must be sacrificed. Any sum desired can be obtained by rearranging its terms.

It is unlikely that students in beginning calculus profit by being told that (3), (4), (5), and so forth behave properly in other environments. Abel shunned such series as works of the devil. Cauchy returned to divergent series and by 1840 had laid the groundwork for their theory. Divergent series acquired major status with Poincaré's 1886 investigations. Earlier, in 1882, Hölder was the first of many to invent new definitions, with valid applications, for the sum of a series. By some summability methods, (4) becomes true. Even the worst example, (5), escapes ostracism through valuation theory initiated by Ostrowski in 1918.

Notwithstanding the foregoing, the usual definition of sum is the universally useful one. Rules for handling series stem from it. Not knowing the rules of the game did not bar Fourier and Heaviside from major discoveries. Euler rearranged series, unaware of attendant dangers. Gauss interchanged repeated limits without apologizing. Cauchy, in a similar case, went astray but recovered, inventing uniform convergence in the process. Surefooted instincts and suitable precautions protected them from fatal error.

However, the finest intuitions in mathematical history have floundered in the fogs of infinite processes. Ill befitting to lesser mortals are cavalier sneers at, say, tests for series convergence. A variety of examples that exhibit the perverseness of infinity are available to counter such tendencies. The following deserves wider circulation. You are given an urn A containing an infinity of tags. Alongside are empty urns B and C. You transfer two tags from A to B and then move one tag from B to C. You repeat these steps an infinite number of times. How many tags are now found in B? You can defend (and hence attack) any answer from zero to infinity. The method is well illustrated by the procedure which validates, say, 3. Tags entering B are numbered 1 and 2, 3 and 4, 5 and 6, and so forth. Remove from B tags 2, 4, 6, 7, 8, 9, and so forth. Tags 1, 3, and 5 are left.

For Further Reading

BELL (a): 282–95
BOYER (g): 477, 486–89, 566, 598–617

CAJORI (e): 373–77
STRUIK (a): 124–27 [4th ed. 124–26]
——— (e): 383–91

434

THE ORIGIN OF L'HOSPITAL'S RULE

THE so-called rule of L'Hospital, which states that

$$\lim_{x \to a} \frac{f(x)}{g(x)} = \frac{f'(a)}{g'(a)}$$

when $f(a) = g(a) = 0$, $g'(a) \neq 0$, was published for the first time by the French mathematician G. F. A. de L'Hospital (or de l'Hôpital) in his *Analyse des infiniment petits* (Paris, 1696). The Marquis de L'Hospital was an amateur mathematician who had become deeply interested in the new calculus presented to the learned world by Leibniz in two short papers, one of 1684 and the other of 1686. Not quite convinced that he could master the new and exciting branch of mathematics all by himself, L'Hospital engaged, during some months of 1691/92, the services of the brilliant young Swiss physician and mathematician Johann Bernoulli, first at his Paris home and later at his château in the country. When Bernoulli left for his hometown, Basel, the marquis kept up correspondence with his tutor, at the same time publishing some original contributions of his own findings. When, in 1696, L'Hospital's book appeared, he acknowledged his indebtedness to Leibniz and Bernoulli, but only in general terms:

"I have made free with their discoveries (*je me suis servi sans façon de leur découvertes*), so that whatever they please to claim as their own I frankly return to them."

The question of the actual dependence of L'Hospital on Bernoulli remained unanswered, and acquired in the course of the years somewhat the character of a mystery. Bernoulli, after L'Hospital had sent him a copy of the book, thanked him courteously and praised it. But subsequently, in some private letters written during the lifetime of the marquis, he claimed that much of the content of the *Analyse des infiniment petits* was really his own property. In 1704, after L'Hospital's death, he made a public claim to that section, No. 163, which contains the rule for 0/0. Mathematicians interested in such priority puzzles have been philosophizing about this supposed dependence of L'Hospital on Johann Bernoulli ever since, weighing Bernoulli's acknowledged greatness as a mathematician against his equally acknowledged reputation for nastiness. A generally acceptable conclusion was not reached until recent times.

Considerable clarification came in 1922, when Johann Bernoulli's manuscript on the differential calculus, dating from 1691/92, was at last published. (The corresponding manuscript on the integral calculus was known from Johann Bernoulli's *Opera*, published in 1742 during the lifetime of the author.) Comparison of these notes by Bernoulli and the text of L'Hospital's book revealed that there was a considerable overlapping, so that it seemed that Bernoulli had fathered much of the nobleman's intellectual offspring. But the true situation came to light only in 1955, when Bernoulli's early correspondence was published. It then appeared that in 1694 a deal was actually made between the marquis and his former tutor, by which L'Hospital offered him a yearly allowance of 300 livres (and more later) provided that Bernoulli agreed to three conditions:

1. To work on all mathematical problems sent to him by the marquis
2. To make all the discoveries known to him
3. To abstain from passing on to others a copy of the notes sent to L'Hospital

This settled the priority question. Here is a translation of the section of the letter which contains the unusual proposition sent by L'Hospital in Paris to Johann Bernoulli in Basel, March 17, 1694:

> I shall give you with pleasure a pension of three hundred livres, which will begin on the first of January of the present year, and I shall send two hundred livres for the first half of the year because of the journals that you have sent, and it will be one hundred and fifty livres for the other half of the year, and so in the future. I promise to increase this pension soon, since I know it to be very moderate, and I shall do this as soon as my affairs are a little less confused. . . . I am not so unreasonable as to ask for this all your time, but I shall ask you to give me occasionally some hours of your time to work on what I shall ask you—and also to communicate to me your discoveries, with the request not to mention them to others. I also ask you to send neither to M. Varignon nor to others copies of the notes that you let me have, for it would not please me if they were made public. Send me your answer to all this and believe me,
> *Monsieur tout à vous*
>
> le M. de Lhospital

436

Bernoulli's answer has not been found, but from a letter of July 22, 1694, we know that he had accepted the proposal. It must have been a little windfall for the impecunious young scientist, just married and still looking for a position (which he obtained the following year at the University of Groningen in the Netherlands). How long this interesting relationship lasted we do not know, but Bernoulli's finances improved and those of L'Hospital did not become any better. By 1695 it may have come to an end.

Several letters from Bernoulli to his patron with answers to questions have now been published, and the one dated July 22, 1694, contains the rule for 0/0. The formulation is very much like the one we find in the *Analyse des infiniment petits* and is based on a geometrical consideration. In our words, if

$$y = \frac{f(x)}{g(x)}$$

and both curves $y = f(x)$ and $y = g(x)$ pass through the same point P on the x axis, $OP = a$, so that $f(a) = g(a) = 0$, and if we take an ordinate $x = a + h$, then the figure shows immediately that

$$\frac{f(a + h)}{g(a + h)}$$

is almost equal to the quotient of $hf'(a + h)$ and $hg'(a + h)$ when h is small. In the limit we find, now in Bernoulli's words:

> In order to find the value of the ordinate (*appliquée*) of the given curve
>
> $$\left[y = \frac{f(x)}{g(x)} \right]$$
>
> in this case it is necessary to divide the differential (*la différentielle*) of the numerator of the general fraction by the differential of the denominator.

Bernoulli's examples are almost the same that L'Hospital uses:

(1)
$$y = \frac{\sqrt{2a^3x - x^4} - a\sqrt[3]{a^2x}}{a - \sqrt[4]{ax^3}}$$

for $x = a$. Then

$$y = \left(\frac{16}{9}\right)a.$$

This example is used by both Bernoulli and L'Hospital.

(2)
$$y = \frac{a\sqrt{ax} - xx}{a - \sqrt{ax}}$$

for $x = a$. Then $y = 3a$.

This example of Bernoulli is changed by L'Hospital into

$$y = \frac{aa - ax}{a - \sqrt{ax}}$$

for $x = a$. Then $y = 2a$.

The situation has thus been clarified. When L'Hospital's book appeared, Bernoulli was bound by his promise not to reveal which sections of the book belonged to him. He could express himself only privately. Then, after the death of the marquis, he felt that he need not be so silent any more, and claimed as his own the most striking result of the book—the rule for 0/0. But he could not prove his assertion. At present he stands vindicated.

From this discovery of the origin of L'Hospital's rule we should not conclude that from now on it should be called after Bernoulli. First of all, there are already plenty of rules and theorems called after Bernoulli (due to at least three members of the Bernoulli family—Jakob, Johann, and Daniel). But there is a more weighty consideration. When we begin to change the names of rules and theorems in accordance with the strict laws of priority, we soon come to the dreary conclusion that our science will lose many of its most familiar expressions. Pythagoras' theorem was known to the Babylonians more than a millennium before the sage of Crotona lived. The Cauchy-Riemann equations were known to D'Alembert and Euler. Taylor's theorem would be Gregory's until another claimant pops up—it is not impossible that Indian mathematicians may have come close to it, c. 1500. The Indians also played with Pell's equation long before John Pell studied it—or better, did not study it, since Pell's connection with the equation is rather remote. Fourier's series were used by Euler and Daniel Bernoulli. Pascal's triangle was known to the Chinese mathematician Yang Hui (thirteenth century) and probably is even older; and his contemporary Chhin Chiu-shao worked with Horner's method in the theory of algebraic equations as with an ancient tool. And so on.

The names attached to mathematical discoveries are often the names of persons who made these results better known or understood through their own outstanding work. Any new historical discovery may disturb the delicate balance of nomenclature again. Let the good marquis keep his elegant rule; he paid for it and made it public property. After all, he deserves some fame; his book on the new calculus was not only the first to be published, and contained contributions of his own, but it was good enough to hold its prominent position for half a century and longer. Even after other and better textbooks appeared it continued to be used as a good first introduction into the calculus; we know of an edition as late as 1790. It appeared in an English and a Latin translation, and there also exist commentaries, like that of L'Hospital's friend Varignon. We should have some respect for it.

For Further Reading

COOLIDGE (c): 147–70   STRUIK (e): 312–16

This capsule is adapted from an article appearing in the April 1963 issue of the *Mathematics Teacher*.

Capsule 112   Bernard J. Yozwiak

CALCULUS OF VARIATIONS

AN IMPORTANT problem in differential calculus is one which attempts to find those values of the *independent variables* for which a given function assumes a greatest or a least value, that is, a maximum or a minimum. Analogously, in the calculus of variations, a basic problem is to find one or more unknown *functions* so that a given definite integral involving those functions becomes a maximum or a minimum. The problem may be stated as follows. Consider the definite integral

$$(1) \qquad I = \int_{x_1}^{x_2} f\left(x, y, \frac{dy}{dx}\right) dx.$$

When x_1 and x_2 have definite numerical values, when f is a given function of x, y, and $\frac{dy}{dx}$, and when y is given as a function of x, then the integral has a definite value. We may then make comparisons as to the value of this integral when different choices of y as a function of x are used. The basic problem of the calculus of variations is to find that particular function $y = y(x)$ which renders (1) a maximum or minimum.

This type of problem has concerned mathematicians throughout history, and many such problems were solved by very ingenious methods. The ancient Greeks knew that of all curves of a given perimeter the circle is the one that encloses the greatest area. Many other such problems that were proposed were solved mainly by the intuition of the individual. The systematic development of the calculus of variations began when Johann Bernoulli (1667–1748) reproposed the brachistochrone problem, which Galileo had studied earlier. The brachistochrone (shortest time) problem seeks to find that one of all possible curves joining two points in a vertical plane such that a particle sliding (without friction) down this curve under the influence of gravity will go from the upper to the lower point in a minimum of time. That is, the path of quickest descent was sought. Both Johann and his brother Jakob solved the problem, and their independent solutions were published in 1696. The problem involves minimizing the integral

$$T = \frac{1}{\sqrt{2g}} \int_0^{x_0} \sqrt{\frac{1 + (y')^2}{y}}\, dx$$

and yields the path of a cycloid as its solution.

Jakob Bernoulli's method of solving the brachistochrone problem was more general than Johann's, and the method was powerful enough to solve a large variety of problems, including an isoperimetric problem (one in which the perimeter is held fixed), which he had proposed in reply to his brother's problem. Leonhard Euler, a pupil of Johann Bernoulli, made many contributions to the study of the calculus of variations, including the development of the differential equation,

$$\frac{\partial f}{\partial y} - \frac{d}{dx}\left(\frac{\partial f}{\partial y'}\right) = 0,$$

whose solution yields the minimizing function of (1), provided that the minimum exists.

For Further Reading

BELL (a): 377–81 CAJORI (e): 232, 251, 369–72
BOYER (g): 534 STRUIK (e): 391–413

440

A JUNGLE OF DIFFERENTIAL EQUATIONS

Sown late in the seventeenth century, the seeds of calculus sprang into blossom and quickly proliferated a jungle of mathematical equations to be classified as "differential equations." Pioneers in this jungle, including Isaac Newton, Gottfried Wilhelm von Leibniz, and Jakob and Johann Bernoulli, found themselves pondering such gnarled equation-trees as

$$\log x \frac{d^2 y}{dx^2} + x^2 y \frac{dy}{dx} + y^3 = \sin x,$$

which is only an "ordinary" differential equation. These early investigators hoped to "solve" such equations for relationships between y and x which utilized known elementary functions such as $\sin x$, $\log x$, x^2, and so forth. It was eventually demonstrated that such elementary functions were inadequate /BELL (a): 401/. More than just mathematicians' chopping wood, such equations provided formulations of the laws governing the physical universe; their solutions were indispensable to the growing fields of mechanics and astronomy. For example, a ball thrown downwards from the height of 300 feet with an initial velocity of 10 feet per second would, neglecting air resistance, be describable by the differential equation

$$\frac{d^2 y}{dx^2} = -32.2,$$

where y stands for the height of the ball and x for the time of its travel. This description is based on Newton's second law of mechanics. This differential equation has as its solution (fitting the conditions described above)

$$y = 300 - 10x - 16.1x^2,$$

so that (for instance) after three seconds the height of the ball could be predicted to be $300 - 10(3) - 16.1(3^2)$ or approximately 125 feet.

Newcomers came to work in the jungle; probably every mathematician of the period tried at least to carve his name on some differential equation. But few large tracts could be claimed. By 1800 the jungle had overgrown the fences of one general theory after another, and despite their potential utility, some of the hardier equations had yet to yield a chip. Nevertheless, in 1743 Leonhard Euler was able to discuss completely the general linear homogeneous ordinary differential equation with constant coefficients.

Mathematicians were well into a second century of work on differential equations when Augustin Louis Cauchy emphasized the need for rigor. His exposure of faulty solutions sent many a careless colleague back to reconsider previous claims. For example, the solving of differential equations by infinite series was an important technique, popularized by Euler. But even Euler's reasoning was often dubious. In one instance /Cajori (e): 238/, Euler claimed that

$$\sin x - 2 \sin 2x + 3 \sin 3x - 4 \sin 4x + \cdots = 0.$$

As the story goes, even that Paul Bunyan of mathematics and physics, Pierre Laplace, listened abashed as Cauchy discoursed on the convergence of infinite series and hurried home to check claims he had made in *Mécanique Céleste* /Cajori (e): 374/. Along with better understanding of the problems inherent in the solving of differential equations during this period came better implements and methods with which to attack previously inaccessible equations. Outstanding among these new tools were Émile Picard's method of successive approximations, Laplace's transforms, and the symbolic operators of Oliver Heaviside.

Toward the close of the nineteenth century, Sophus Lie's theory of continuous groups was brought to bear on differential equations, together with the theory of equations. Early in our own century the heavy machineries of abstract algebra, topology, and functional analysis were moved in, and the jungle began to give way to hundred-acre lots. Today's high-speed computers have made complicated numerical approximation and reiteration methods not only feasible but advantageous, and much research in differential equations is being directed toward the refinement of these computer methods. The jungle of differential equations has been greatly cleared.

For Further Reading

Bell (a): 400–419   Cajori (e): 383–92
Boyer (g): 493–96

MACLAURIN AND TAYLOR
AND THEIR SERIES

BROOK Taylor (1685–1731) was an eminent British mathematician whose relatively short lifetime roughly coincided with the latter half of the lifetime of Isaac Newton. He was an unusually talented youth with a wide range of interests, including music and art as well as mathematics and philosophy. A derivation of the series that bears his name was included in his principal mathematical work, *Methodus incrementorum directa et inversa*, which was published in London in 1715. At that time the mathematical world was deeply embroiled in the Newton-Leibniz priority controversy. Taylor was an ardent supporter of Newton's claim, and the book is filled with biased references to Newton and his work, no mention being made of other contributors. The book was primarily devoted to the development of a branch of mathematics known today as the calculus of finite differences [108]. Much of its content was then new, and the book acquired considerable renown in spite of the fact that the author's obscure style, his complicated notation, and other defects made it very difficult to read. Put into modern notation, Taylor's derivation of his series as given in *Methodus incrementorum* was approximately as follows.

Suppose $y(x)$ is a function of x, Δx is an arbitrary real number, and

$$\Delta y(x) = y(x + \Delta x) - y(x)$$

is the change in the value of the function corresponding to a change of Δx in the value of the argument x. We also define higher-order differences of the function:

$$\Delta^2 y(x) = \Delta(\Delta y(x)) = \Delta y(x + \Delta x) - \Delta y(x)$$

$$\Delta^3 y(x) = \Delta(\Delta^2 y(x)) = \Delta^2 y(x + \Delta x) - \Delta^2 y(x)$$

$$\Delta^n y(x) = \Delta(\Delta^{n-1} y(x)) = \Delta^{n-1} y(x + \Delta x) - \Delta^{n-1} y(x).$$

It follows, of course, that if $z(x)$ is any other function of x, then

$$\Delta(y(x) + z(x)) = \Delta y(x) + \Delta z(x).$$

443

Now then, a being an arbitrary number, we have

$$y(a + \Delta x) = y(a) + \Delta y(a)$$

$$y(a + 2\Delta x) = y(a + \Delta x) + \Delta y(a + \Delta x)$$

$$= (y(a) + \Delta y(a)) + \Delta(y(a) + \Delta y(a))$$

$$= y(a) + \Delta y(a) + \Delta y(a) + \Delta^2 y(a)$$

$$= y(a) + 2\Delta y(a) + \Delta^2 y(a)$$

$$y(a + 3\Delta x) = y(a + 2\Delta x) + \Delta y(a + 2\Delta x)$$

$$= (y(a) + 2\Delta y(a) + \Delta^2 y(a)) + \Delta(y(a) + 2\Delta y(a) + \Delta^2 y(a))$$

$$= y(a) + 2\Delta y(a) + \Delta^2 y(a) + \Delta y(a) + 2\Delta^2 y(a) + \Delta^3 y(a)$$

$$= y(a) + 3\Delta y(a) + 3\Delta^2 y(a) + \Delta^3 y(a),$$

and, in general,

$$y(a + n\Delta x) = y(a + (n - 1)\, \Delta x) + \Delta y(a + (n - 1)\, \Delta x)$$

$$= y(a) + n\Delta y(a) + \frac{n(n - 1)}{2!}\, \Delta^2 y(a)$$

$$+ \frac{n(n - 1)(n - 2)}{3!}\, \Delta^3 y(a) + \cdots + \Delta^n y(a).$$

Letting $n\Delta x = v$ then gives

$$y(a + v) = y(a) + \frac{v}{\Delta x}\, \Delta y(a) + \frac{\dfrac{v}{\Delta x}\left(\dfrac{v}{\Delta x} - 1\right)}{2!}\, \Delta^2 y(a)$$

$$+ \frac{\dfrac{v}{\Delta x}\left(\dfrac{v}{\Delta x} - 1\right)\left(\dfrac{v}{\Delta x} - 2\right)}{3!}\, \Delta^3 y(a) + \cdots + \Delta^n y(a)$$

$$= y(a) + v \cdot \frac{\Delta y(a)}{\Delta x} + \frac{v(v - \Delta x)}{2!}\, \frac{\Delta^2 y(a)}{(\Delta x)^2}$$

$$+ \frac{v(v - \Delta x)(v - 2\, \Delta x)}{3!}\, \frac{\Delta^3 y(a)}{(\Delta x)^3} + \cdots + \Delta^n y(a).$$

At this point Taylor argued that if we consider v fixed and let n become infinite, then Δx approaches 0,

$$\frac{\Delta^K y(a)}{(\Delta x)^K} \quad \text{and} \quad \frac{v(v - \Delta x)(v - 2\, \Delta x) \cdots (v - (K - 1)\, \Delta x)}{K!}$$

become

$$\frac{d^K y(a)}{dx^K} \quad \text{and} \quad \frac{v^K}{K!}$$

respectively for each K, and the number of terms becomes infinite so that we get

$$y(a + v) = y(a) + v\,\frac{dy(a)}{dx} + \frac{v^2}{2!}\,\frac{d^2 y(a)}{dx^2} + \frac{v^3}{3!}\,\frac{d^3 y(a)}{dx^3} + \cdots,$$

which, of course, is Taylor's series.

It was this sort of fuzzy argument regarding limits which left not only Taylor but also Newton, Leibniz, and all their contemporaries vulnerable to the attack of George Berkeley, bishop of Cloyne, and led later mathematicians to develop the proofs that students learn today.

Colin Maclaurin (1698–1746) was a brilliant Scottish mathematician who, at the age of nineteen, on the basis of a competitive examination, became professor of mathematics at the University of Aberdeen. He devised a different derivation of the series, which appeared in his book *Treatise on Fluxions*, published in Edinburgh in 1742. Maclaurin's argument was intended to meet the objections raised by Berkeley and, in modern notation, was essentially as follows.

Assume that

$$y(x) = A_0 + A_1 x + A_2 x^2 + A_3 x^3 + \cdots$$

where the coefficients $A_0, A_1, A_2, A_3, \cdots$ are fixed numbers whose values are to be determined. Then

$$\frac{dy(x)}{dx} = A_1 + 2 A_2 x + 3 A_3 x^2 + \cdots,$$

and setting $x = 0$ gives

$$\frac{dy(0)}{dx} = A_1.$$

Similarly,

$$\frac{d^2 y(x)}{dx^2} = 2 A_2 + 3 \cdot 2 A_3 x + 4 \cdot 3 A_4 x^2 + \cdots,$$

so that

$$\frac{d^2 y(0)}{dx^2} = 2 A_2,$$

and it is clear that

$$\frac{d^n y(x)}{dx^n} = n!\,A_n + (n + 1)n\cdot(n - 1) \cdots 2 A_{n+1} x$$
$$+ (n + 2)(n + 1) \cdots 3 A_{n+2} x^2 + \cdots,$$

from which

$$\frac{d^n y(0)}{dx^n} = n! \, A_n.$$

Thus

$$y(x) = y(0) + x\frac{dy(0)}{dx} + \frac{x^2}{2!}\frac{d^2 y(0)}{dx^2} + \frac{x^3}{3!}\frac{d^3 y(0)}{dx^3} + \cdots.$$

It is apparent that in this argument the following questions are ignored: What functions have such series representations? For what values of x does the series represent the function? Does term-by-term differentiation of the series yield the derivative of the given function?

Neither Taylor nor Maclaurin has a valid claim to having discovered the series to which their names are attached, since it appeared in previously published work of Johann Bernoulli with which both of them were probably familiar. However, both men made significant contributions to mathematics and probably deserve the measure of immortality which the association of their names with the series has given them.

For Further Reading

Boyer (g): 422, 462, 469   Struik (e): 328–41

This capsule is adapted from an article appearing in the March 1968 issue of the *Mathematics Teacher*.

Capsule 115   Merrill Shanks

FROM $n!$ TO THE GAMMA FUNCTION

Consider the factorial function given by

$$x! = x(x - 1) \cdots (2)(1).$$

It is defined, and the formula makes sense, only if x is a positive integer. Is it possible there is another function F defined for *all positive real* numbers such that $F(x) = x!$ when x is a positive integer? It is easy to see that if there is one such function, F, then there are infinitely many. For the function given by $F(x)(1 + A \sin 2\pi x)$ is also equal to $x!$ when x is a positive integer, and this is true whatever the constant A may be. It is also easy to construct *one* such function by the simple device of requiring that $F(x) = x!$ when x is a positive integer, n, and by having the function be linear between the successive integers, n and $n + 1$:

(1) $\qquad F(x) = n! + n \cdot n! \, (x - n), \qquad \text{if} \quad n \leq x \leq n + 1.$

Yet, in some sense, the solution given by (1) is "cheating." There should be a function given by a single "nice" formula which is not obtained by patching together infinitely many pieces of functions as is done by formula (1).

John Wallis (1616–1703) was perhaps the first to consider $n!$ for values of n other than the positive integers. His work on

$$\left(\frac{1}{2}\right)! = \frac{\sqrt{\pi}}{2}$$

and his formulas related to the gamma function were of fundamental importance for the further development of the theory.

This problem—extending the domain of the factorial function—attracted many mathematicians in the early years of the eighteenth century. It was solved by Leonhard Euler about 1730 in a paper entitled, when translated from the Latin, *On Transcendental Progressions Whose General Term Cannot Be Expressed Algebraically*. The genius of Euler is illustrated by his finding the "right" formula—in the sense that his function, among the infinitude of possibilities, has all sorts of unexpected applications. His first solution to the problem was the now familiar infinite product representation

(2) $\qquad x! = \lim_{n \to \infty} \frac{n! \, (n + 1)^x}{(x + 1)(x + 2) \cdots (x + n)}.$

The right side of (2) converges if x is any real number greater than zero, and this solves the problem.

Turning his attention to integrals Euler found, after a dubious argument, that

(3) $\qquad x! = \int_0^1 (-\log u)^x \, du,$

if x is a positive integer. The simple proof of (3) by induction is left to the reader. Observe that the improper integral in (3) converges if

$x > -1$, and so the integral defines an extension of the factorial function. A change of variable puts (3) in a familiar form. Let $-\log u = t$. Then the integral becomes

$$\int_0^\infty t^x e^{-t}\, dt.$$

Nowadays we define the gamma function, Γ, by

(4) $$\Gamma(x) = \int_0^\infty t^{x-1} e^{-t}\, dt,$$

a notation due to Adrien Marie Legendre. Then $\Gamma(x + 1) = x!$ if x is a positive integer. The integral for Γ also converges for complex x for which $\mathrm{Re}(x) > 0$. It is easy to show, by integration by parts, that $\Gamma(x + 1) = x\,\Gamma(x)$ and since $\Gamma(1) = 1$, again $\Gamma(x + 1) = x$, if x is a positive integer.

For Further Reading

Davis (a)

Capsule 116   John M. H. Olmsted

PARTIAL DERIVATIVES

In the closing years of the seventeenth century it was already becoming evident, as disclosed in the writings of Isaac Newton and Gottfried Wilhelm von Leibniz, that in both the theory and application of mathematics proper attention must be given to the concept of partial derivatives, wherein differentiation of a function of several variables is performed with respect to one variable at a time. In most of the early researches in which partial derivatives appeared, no special notation was used. Thus the symbol $\dfrac{du}{dx}$ was alternatively interpreted, according to context, to mean the ordinary or total derivative of u with respect to x or the partial derivative of u with respect to x with other independent variables being held fixed, now usually denoted $\dfrac{\partial u}{\partial x}$. The need for a

448

distinct notation for partial derivatives, however, initiated a conflict in symbols that persisted throughout most of the eighteenth and nineteenth centuries. Of the dozens of proposals made, most died out rather quickly. Under consideration were the symbols d, D, δ, ϑ, and ∂, often in combination with subscripts or superscripts or both. For a period toward the end of the eighteenth century it seemed possible that the letters d and D might be exclusively appropriated by those working in finite differences, and the round ∂ was introduced for ordinary derivatives. The round ∂ was used in 1770 by Marquis de Condorcet for partial differentials, and in 1776 by Leonhard Euler in the form $\dfrac{\partial^\lambda}{p} \cdot V$, now written $\dfrac{\partial^\lambda V}{\partial p^\lambda}$. This symbol was first used in the modern combination $\dfrac{\partial v}{\partial x}$ in 1786 by Adrien Marie Legendre, and the letter δ was introduced in an identical role in 1824 by William Rowan Hamilton. However, this use of the "round dee" (∂) was not generally adopted until close to the end of the nineteenth century, and Carl Gustav Jacob Jacobi, who outlined the advantages of the symbol ∂ in 1841, is often incorrectly credited with its invention [**117**].

A substantial proportion of the applications of mathematics, especially to physics and astronomy, rests on partial derivatives and in particular on solutions of partial differential equations. One of the early pioneers in this area was Daniel Bernoulli (1700–1782). In 1747, Jean Le Rond d'Alembert solved the fundamental problem of the vibrating string by formulating it in terms of the differential equation

$$\frac{\partial^2 u}{\partial t^2} = a^2 \frac{\partial^2 u}{\partial x^2}$$

and expressing the solution in the form

$$u = f(x + at) + \phi(x - at).$$

Others whose early work in partial differential equations had far-reaching effects in the history of mathematics and its applications are Euler; Joseph Louis Lagrange (1736–1813); Pierre Simon Laplace (1749–1827), who introduced and studied the equation

$$\frac{\partial^2 u}{\partial x^2} + \frac{\partial^2 u}{\partial y^2} + \frac{\partial^2 u}{\partial z^2} = 0;$$

and Jean Joseph Fourier (1768–1830), who exploited his celebrated series in the study of the heat equation

$$\frac{\partial u}{\partial t} = a^2 \frac{\partial^2 u}{\partial x^2}.$$

Partial differential equations continue to be a subject of active research activity today.

For Further Reading

CAJORI (d): II, 197–242   STRUIK (e): 351–68
——— (e): 384–88

Capsule 117   Philip E. Bedient

MULTIPLE INTEGRALS AND JACOBIANS

THE extension of the definition of the definite integral from functions of a single real variable to functions of two or more variables occupies an important position in most textbooks on calculus. The subject is usually begun by treating functions defined over a closed region R in the plane where the boundary of R is not too complicated and where rectangular coordinates are used as the means for formulating the definition. Once this has been done, methods are introduced for transforming from rectangular coordinates to other coordinate systems by means of equations of the form $x = f(u, v)$ and $y = g(u,v)$. These equations can be interpreted as transforming the region R in the xy-plane into a region Q of the uv-plane. Then under suitable restrictions on the functions f and g, we may use the formula

$$\iint_R \varphi(x, y)\, dx\, dy = \iint_Q \varphi(f(u, v), g(u, v)) \frac{\partial(f, g)}{\partial(u, v)}\, du\, dv.$$

The symbol $\dfrac{\partial(f, g)}{\partial(u, v)}$ is called the "Jacobian" of the transformation and is defined as the functional determinant

$$\frac{\partial(f, g)}{\partial(u, v)} = \begin{vmatrix} \dfrac{\partial f}{\partial u} & \dfrac{\partial f}{\partial v} \\[2mm] \dfrac{\partial g}{\partial u} & \dfrac{\partial g}{\partial v} \end{vmatrix}.$$

It is interesting to note the form in which this theorem appears in S. F. Lacroix's treatise on the calculus. This three-volume study was considered a standard reference work in the early part of the nineteenth century /CAJORI (d): II, 226/. Lacroix presents the theorem on transformation of coordinates without the advantage of the modern notation for determinants and, of course, without reference to Jacobi (Carl Gustav Jacob Jacobi, born in 1804, only ten years earlier than Lacroix). If we write

$$P = \frac{\partial f}{\partial u}, \qquad Q = \frac{\partial f}{\partial v},$$

$$P' = \frac{\partial g}{\partial u}, \qquad Q' = \frac{\partial g}{\partial v},$$

then we can write, as Lacroix did,

$$dx \, dy = (PQ' - P'Q) \, du \, dv.$$

This notation for partial derivatives had been used first by Euler, in 1728. Lacroix gives credit for the formula to a 1769 study by Euler. The corresponding theorem for three variables is also presented by Lacroix and credited by him to a work by Joseph Louis Lagrange, written in 1773.

At the time Lacroix's treatise was printed, the study of determinants was being pursued with great industry by some of the greatest mathematicians of the day. In 1812 Augustin Louis Cauchy had given proofs of most of the basic theorems, but he was handicapped by the lack of a suitable notation. In 1841 the theory of determinants was unified by Jacobi's masterful papers, which included much new material, including his own work with "functional determinants."

In the notation of Jacobi's papers the determinant, which we now call the Jacobian, would have been written

$$\sum \pm \frac{\partial f}{\partial u} \cdot \frac{\partial g}{\partial u},$$

a notation which was not as suggestive as the one we use today. In 1854 W. F. Donkin introduced the notation

$$\frac{\partial(f, \, g)}{\partial(u, \, v)}$$

for the same determinant. The modern notation for a determinant appeared in 1855 in an article by Arthur Cayley. The name "Jacobian" was first used by James Joseph Sylvester and was included in his

"Glossary of Mathematical Terms," in *Philosophical Transactions,* 1853.

Felix Klein, in his beautifully written account of this period /(c): I, 164/, emphasizes the great influence of the textbooks written by George Salmon. These books were written in English and translated into many other languages; they had a profound effect on the adoption of vocabulary and symbols in the mathematical world. In the first edition of Salmon's *Modern Higher Algebra,* the word "Jacobian" is not used in connection with the "functional determinants." However, the second edition of the book (1866) uses the word, crediting it to Sylvester. The Preface to the second edition reads in part /iii/:

> With respect to the use of new words I have tried to steer a middle course. In this part of Algebra combinations of ideas require to be frequently spoken of which were not of important use in the older Algebra. This made it necessary to employ some new words in order to avoid an intolerable amount of circumlocution. But feeling that every strange term makes the science more repulsive to a beginner, I have generally preferred the use of a periphrasis to the introduction of a new word which I was not likely often to have occasion to employ.

Salmon then proceeds to use "Jacobian" and the symbolism with which we are familiar today.

For Further Reading

Bell (a): 425–27   Cajori (d): II, 87–101, 220–42

Capsule 118   Jack Bedient

CALCULUS IN JAPAN

According to a somewhat tenuous tradition, the founder of Japanese calculus, called *yenri,* was probably Seki Kowa (1642–1708), who was born in the same year as Newton. The *yenri* ("circle theory") was a method for determining the length of a circular arc by using infinite series to find the limit of the sum of the 2^n equal, inscribed chords spanning successively bisected arcs.

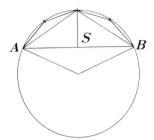

FIGURE [118]-1

To illustrate the method we refer to Figure [118]-1, which shows an example taken from a work published (*c.* 1722) by Takebe, who was a student of Seki. Takebe begins by inscribing equal chords in a circular arc, the number of chords being a power of two. (For $n = 2$ there are $2^n = 2^2 = 4$ chords in the set.) Then, by means of a recursion relation (giving the length of a chord from the set of 2^n chords in terms of the length of a chord from the preceding set of 2^{n-1} chords), he finds the limit of the square of the sum of the half-chords and produces the formula

$$(1) \qquad (\tfrac{1}{2}\,\text{arc})^2 = sd\left[1 + \frac{2^2}{3\cdot 4}\frac{s}{d} + \frac{2^2\cdot 4^2}{3\cdot 4\cdot 5\cdot 6}\left(\frac{s}{d}\right)^2 \right.$$
$$\left. + \frac{2^2\cdot 4^2\cdot 6^2}{3\cdot 4\cdot 5\cdot 6\cdot 7\cdot 8}\left(\frac{s}{d}\right)^3 + \cdots \right],$$

where d is the diameter and s is the *sagitta* ("height") of the given arc AB. In the original publication the series is written, of course, in an older notation. All this is done in such an obscure way, using numerical cases to determine coefficients in many of the relations, that some authorities think Takebe did not originate the above series. Rather, it is considered possible that he might have obtained it by reading the work published (*c.* 1713) by the Chinese writer Mei, who included some series brought to China (*c.* 1700) by the Jesuit missionary Pierre Jartoux. It is known that Jartoux had some correspondence with Leibniz.

Takebe was also interested in approximating the value of π by calculating the perimeters of regular polygons of 2^n sides inscribed in a circle of unit diameter. A curious aspect of his work is illustrated by the following example. The perimeters, a, b, c, of polygons of 2^{15}, 2^{16}, 2^{17} sides, respectively, are found to be

$$a = 3.14159\ 26487\ 76985\ 6708-$$

$$b = 3.14159\ 26523\ 86591\ 3571+$$

$$c = 3.14159\ 26532\ 88992\ 7759-.$$

Then, to get an even better approximation for π, he uses the following formula (which he neither proves nor explains),

$$(2) \qquad p = b.+ \frac{(b-a)(c-b)}{(b-a)-(c-b)},$$

where p is alleged to approximate π better than the three previous values, a, b, c. In this particular case, for the above values of a, b, c, formula (2) does in fact give a closer approximation, namely,

$$p = 3.14159\ 26536-.$$

Takabe claims to have approximated π correctly to forty-one places by repeated applications of (2).

For Further Reading

CAJORI (e): 78–82   D. E. SMITH (a): II, 701–2

MIKAMI: *See index*

Capsule 119   Jack Bedient

INFINITESIMALS IN INDIA

NEAR the beginning of the Christian era the invention of zero to help in the writing of numbers in the decimal scale posed a special challenge to Hindu mathematicians. Their investigations into indeterminate forms made use of analytic methods that, however primitive, might have contributed to mathematical progress in Europe if avenues of communication had existed.

The Hindu mathematician Brahmagupta, who was active in the seventh century, was aware of the pitfalls in attempting "division by zero"; he stated that "positive or negative divided by cipher [zero] is a fraction with that for denominator." This expression was called *kha-cheda* ("the quantity with zero as denominator"). Regarding its meaning or significance Bhaskara (*c.* 1150) remarked /MIDONICK: 135/:

> In this quantity consisting of that which has cipher [zero] for its divisor, there is no alteration, though many may be inserted or extracted; as no change takes place in the infinite and immutable God, at the period of the destruction or creation of worlds, though numerous orders of beings are absorbed or put forth.

A rough transcription of this quotation would be

$$\frac{a}{0} = \frac{a}{0} + k = \infty .$$

Although Bhaskara spoke of quantities with divisor zero, it seems that in some rudimentary way he was thinking of fractions with infinitesimal denominators. Today the above idea might be expressed in the form

$$\lim_{\epsilon \to 0} \frac{a}{\epsilon} = \lim_{\epsilon \to 0} \left(\frac{a}{\epsilon} + k \right) = \infty , \quad \text{where} \quad \epsilon > 0.$$

It must be emphasized that for $\epsilon > 0$ the symbolism

$$\lim_{\epsilon \to 0} \frac{a}{\epsilon} = \infty$$

is defined to mean that by taking a positive ϵ sufficiently close to 0, a/ϵ can be made arbitrarily large. Thus, as $\epsilon \to 0$ through positive values, a/ϵ has no limit but increases without (an upper) bound.

Krsna (*c.* 1550), a commentator on Bhaskara, notes that $a/0 = b/0$. We might prefer to write this in the form

$$\lim_{\epsilon \to 0} \frac{a}{\epsilon} = \lim_{\epsilon \to 0} \frac{b}{\epsilon} , \quad \text{where} \quad \epsilon > 0,$$

subject to the above definition; that is, both a/ϵ and b/ϵ can be made arbitrarily large by taking a positive ϵ sufficiently close to 0.

Brahmagupta remarked that $0/0 = 0$, a statement that is incorrect if interpreted to mean

$$\lim_{\epsilon \to 0} \frac{\epsilon}{\epsilon} = 0;$$

455

this limit is, of course, 1. (It is most improbable that Brahmagupta meant

$$\lim_{\epsilon \to 0} \frac{\epsilon^2}{\epsilon},$$

which is also an "indeterminate form of the type 0/0" and, in this case, has the limit 0.)

Bhaskara stated that $(a \cdot 0)/0 = a$, a result we would accept, again with the same interpretation as above; namely,

$$\lim_{\epsilon \to 0} \frac{a}{\epsilon} \cdot \epsilon = a.$$

Bhaskara also gave some examples of correct and incorrect valuations of the unknown in equations involving limits; as a solution of equation

$$(1) \qquad \frac{x \cdot 0 + \dfrac{x \cdot 0}{2}}{0} = 63$$

he gives $x = 42$, which is correct if (1) is interpreted to mean

$$\lim_{\epsilon \to 0} \left[\frac{x \cdot \epsilon + \dfrac{x \cdot \epsilon}{2}}{\epsilon} \right] = 63;$$

and as an alleged solution of

$$(2) \qquad \left[\left(\frac{x}{0} + x - 9 \right)^2 + \left(\frac{x}{0} + x - 9 \right) \right] \cdot 0 = 90$$

he gives an incorrect value, 9, for x. Example (1) is easily verified. But it is more interesting to discover how the mistake occurred in example (2). As usual, we replace 0 by ϵ, and rewrite (2) in the equivalent form

$$(3) \qquad \lim_{\epsilon \to 0} \left[\frac{x^2}{\epsilon^2} \cdot \epsilon + (2x^2 - 17x) + (x^2 - 17x + 72) \cdot \epsilon \right] = 90,$$

$$\text{where} \quad \epsilon > 0.$$

At this point Bhaskara, who wrote the first term of (3) as $(x^2 \cdot 0)/0^2$, replaced the 0^2 in the denominator by 0, obtaining $(x^2 \cdot 0)/0$. This is equivalent to replacing the first term in (3) by $x^2\epsilon/\epsilon$. The error here, of course, is in assuming that

$$\lim_{\epsilon \to 0} \frac{x^2 \epsilon}{\epsilon^2} = \lim_{\epsilon \to 0} \frac{x^2 \epsilon}{\epsilon}.$$

Thus Bhaskara incorrectly concluded that the limit of the first term was x^2. In this way he incorrectly arrived at the quadratic equation

(4) $$x^2 + (2x^2 - 17x) + 0 = 90,$$

which does, in fact, have a root of 9. (The other root is $-(10/3)$; but the Hindus were often content with finding only one root, even when both roots were positive. We have used modern notation in writing expressions like equation (4); the reader is referred to [79] where equations like (4) are written in Hindu notation of Bhaskara's time.)

For Further Reading

DATTA and SINGH: *See index*   BOYER (g): 241–46

Capsule 120   A. R. Lovaglia

THE DEFINITE INTEGRAL

HISTORICALLY the definite integral evolved from the attempt to define and compute the area of the plane bounded by the graph of the function $y = f(x)$, the x-axis, and the vertical lines $x = a$ and $x = b$, $a < b$ (Fig. [120]-1).

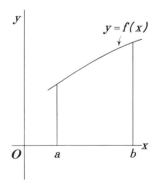

FIGURE [120]-1

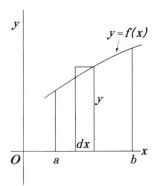

FIGURE [120]-2

457

The region in this figure was thought of as being divided up into "infinitely many infinitesimal rectangles" of width dx and height y (Fig. [120]-2). The area was then obtained by "summing" the areas of these rectangles. This sum is denoted by the integral sign $\int$, which is an elongated S from the Latin *summa*. Thus the area in question can be written

$$\int_a^b f(x)\ dx, \qquad \int_a^b y\ dx,$$

or simply $\int y\ dx$. This notation is in general use today and is read "the integral from a to b of ydx." The integral symbol was first introduced in 1675 by Gottfried Wilhelm von Leibniz in a manuscript in which he wrote $\int \ell$ for *omn. ℓ*, that is, for the sum of all the ℓ's [105]. The word "integral" was first used by Jakob Bernoulli in 1690.

The ordinary summation symbol Σ is the Greek uppercase letter sigma. Thus, in the spirit of Leibniz, one would write

$$\sum y\ dx = \int_a^b y\ dx = \text{area}.$$

The "modern" theory of integration began with Augustin Louis Cauchy, who at the beginning of the nineteenth century developed the integral as a "limit." For a function $y = f(x)$, continuous in the interval $[a, b]$, he formed the sum of products

$$S_n = (x_1 - x_0) \cdot f(x_0) + (x_2 - x_1) \cdot f(x_1) + \cdots + (x_n - x_{n-1}) \cdot f(x_{n-1}),$$

where $a = x_0 < x_1 < x_2 < \cdots < x_{n-1} < x_n = b$ (Fig. [120]-3).

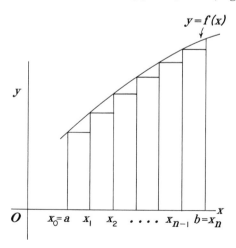

FIGURE [120]-3

If the differences $x_i - x_{i-1}$, where $i = 1, 2, \ldots, n$, decrease to 0, the value of S_n will "finally attain a certain limit" S, which will depend uniquely on the given function $f(x)$ and on the values of a and b. This limit is called the "definite integral."

Although the symbol "$f(x)$" is still used to denote a function, the x in $f(x)$ is entirely superfluous, if not misleading. With the advent of set theory in the latter half of the nineteenth century, a function came to be regarded as a correspondence between two sets, A and B. This correspondence is denoted by a single letter, say f. The function f associates with each x in A a unique element, $f(x)$, in B, $f(x)$ depending on x. The notation "$f(x)$" is read as "the value of f at x." Since the integral is determined completely by the function f and the limits a and b, it is entirely correct and sufficient to denote the integral by $\int_a^b f$. This notation has come into fairly wide usage during the last twenty years.

For Further Reading

BOYER (f): 206, 229

—— (g): 441

CAJORI (d): II, 242–52

NEWMAN: I, 53–58

ROSENTHAL

VIII

Development of
MODERN MATHEMATICS

an overview by

R. L. WILDER

To select points on the time scale of human events and to designate the intervals thereby demarcated as "eras" is always an arbitrary act. Inevitably it is influenced by the point of view, the experience, and the purposes of the selector. In particular, any designation of a specific date as marking the commencement of the "modern era" in mathematics would be affected by the mathematical interests of the designator. Even though one were a "universalist" in mathematics, possessing wide knowledge of all its branches, his judgment in this regard might not find acceptance. Probably one of the last mathematicians who could be termed a universalist was the late John von Neumann, who expressed the view that "the calculus was the first achievement of modern mathematics" /I, 3/—a judgment not many would be likely to accept. Certainly the two centuries following the formulations of the calculus by Newton and Leibniz constituted a period of high productivity in mathematical research. But it was a period characterized by the rough methods of the pioneer and seemingly imbued with the philosophy (attributed to d'Alembert), "Forge ahead, and faith will come to you."

460

Not until the nineteenth century do we find the inception of those features that appear to distinguish modern mathematics from its progenitors of the preceding two centuries. In the works of Cauchy and other early-nineteenth-century mathematicians the attention paid to greater rigor points to a change in the philosophy underlying mathematical research. And not until the latter part of the nineteenth century do we find clearly exhibited those characteristics that appear to distinguish modern mathematics from earlier forms of the science. It would seem reasonable, therefore, to say that by "modern mathematics" is meant the mathematics of, roughly, the past century.

More important than dates, however, are the general characteristics of modern mathematics.

General Characteristics of Mathematics in the Modern Era

A new point of view.—A prominent feature of modern mathematics —one that serves to distinguish it from the mathematics of earlier eras—is a *point of view*, a new conception of the nature of mathematics. From this point of view mathematics is seen, not as a description of an external world of reality, nor merely as a tool for studying such a world, but rather as science in its own right. No longer simply a servant of the natural sciences, mathematics has achieved a status seemingly independent of the natural sciences while still lending aid to the latter. Having moved from its earlier dependence on natural phenomena for the inspiration of new concepts, mathematics now finds most of its stimulus for new ideas from within itself.

The achievement of such a status was not, of course, the result of determination on the part of certain individuals to declare the independence of mathematics from natural phenomena. Rather, it was the result of a natural evolution one can detect as long ago as Babylonian times, when meager attempts at what we would call "number theory" were made by the temple scribes. When sufficiently developed, mathematics breeds a fascination its devotees find irresistible. No doubt during the Sumerian-Babylonian era the needs of a fast-developing society, requiring quicker aids to computation, lay at the root of the development of multiplication tables, tables of reciprocals, and the like. But inevitably certain curious aspects of numbers, when so manipulated and tabulated, evinced themselves and inculcated in the observer a desire to know "why."

The "why" that probably had most to do with development of the

461

new viewpoint in modern mathematics was a question related to the parallel axiom of Euclid's geometry. As is well known, the question whether the addition of this axiom to Euclid's other postulates was necessary stimulated a great deal of research from the third century B.C. to the time of its solution in the non-Euclidean geometries of the nineteenth century. Could one prove this axiom as a theorem deduced from the other axioms (termed "postulates" in the *Elements*)? Today we would call this a question of the "independence" of the parallel axiom in the Euclidean system. Influenced, no doubt, by the prevailing opinion that Euclidean geometry embodied a description of the structure of physical space, the major work on the problem was directed at trying to prove the axiom as a theorem. Probably the classic example is the work done by Saccheri and published in 1733 in his *Euclides ab omni naevo vindicatus* (commonly translated "Euclid Freed of Every Flaw"), where he presumed to show that the denial of the axiom would lead to contradiction, thus providing a proof of the axiom by the method of *reductio ad absurdum*.

It is interesting to conjecture just how much influence such work, particularly Saccheri's, had on the ultimate outcome, soon to follow. There is evidence that Saccheri's faulty reasoning, as well as the two cases to which his denial of the axiom led, stimulated thinking along the lines of counterexamples. Certainly something of the sort was in the air; for early in the nineteenth century Gauss, Bolyai, and Loba-chevsky, working independently and in different countries, concluded that denial of the parallel axiom is consistent with the other axioms of Euclidean geometry. It was not until the latter part of the century, however, that the import of this conclusion was fully realized.

Parallel with these developments were the generalizations of number being invented by the algebraists. Prominent among the proponents of such generalizations were Hamilton and Grassmann, who realized that the axiomatic method could be applied just as effectively to algebra as to geometry. Much as Bolyai and Lobachevsky, in order to found their non-Euclidean geometry, substituted for the parallel axiom a statement contradicting it, so Hamilton finally decided to defy the commutative "law" of algebra for multiplication. Certainly such work contributed to the new point of view, especially when it became apparent that the new types of noncommutative systems had applications in the physical sciences.

By the end of the nineteenth century the arbitrary character of any system of algebra or geometry, whether it was a classical form or one of the newly invented types, was becoming evident to the mathematical world. And when classical geometry and algebra lost their previously assumed property of "necessity," this loss could not help but create the new point of view. Moreover, the effect was ultimately to be observed in all areas of human thought, not only in the world of science but in philosophy and literature as well. Its influence has been an ever-growing phenomenon, something like the advance of a tidal wave. In mathematics the practical result was the flowering of the modern form of the axiomatic method, which has subsequently proved so important for research and is certainly to be regarded as one of the major achievements of the nineteenth century.

It is interesting to note here an opinion expressed in 1908 by another great universalist, Henri Poincaré. Apparently having no conception of the possibilities of the axiomatic method as a research tool, he termed it a means of classification of mathematical theories by which one strove to "enumerate the axioms and postulates, more or less concealed, which form the foundation of the different mathematical theories." And he concluded, "But, when everything has been enumerated, there will be ways of classifying it all. A good librarian always finds work to do, and each new classification will be instructive for the philosopher." /POINCARÉ: 45./ This may, of course, have been an unconsciously motivated protest against the direction in which mathematics was proceeding, since he evidently had equal scorn for the newly introduced theory of sets (see below). If so, he was not alone in thus protesting; others, before and since, have made their feelings known in like fashion. Thus Bell quotes Du Bois-Reymond as predicting in 1882 that the tendencies in mathematics would reduce it to "a mere play with symbols, in which arbitrary meanings would be attached to the signs as if they were the pieces on a chessboard or playing cards" /BELL (a): 182/.

A higher level of abstraction.—A natural concomitant of the new form of the axiomatic method was the greater abstraction that characterizes modern mathematics. An important feature of the axiomatic method in its modern form is its treatment of basic terms as undefined; thus in geometry, instead of trying to give a spurious definition of "point" as in Euclid, one usually attributes to the term no meaning except what may be implied by the axioms. This attitude was, perhaps, a result of (1) the emergence of models in both algebra and geometry in which the basic terms no longer had their usual intuitive meanings— as, for instance, geometries in which the "points" were objects totally unlike the classical conception—and (2) the development of the

463

rigorous mathematical (as opposed to the dictionary) type of definition, which forced the realization that continual definition could lead
only to endless regression or to a vicious circle; one cannot give explicit
definition of every basic concept of an axiomatic system. An important
by-product of this treatment is that one can interpret the undefined
terms in any concrete way he pleases so long as the axioms turn out
to be true in the interpretation; one then has all the theorems of the
theory at his disposal in the new interpretation.

One may protest that abstractness is not peculiar to modern mathematics. Certainly the geometry of Euclid was a grand abstraction from
physical space. And what greater effort of abstracting has ever been
made by the human race than that which was involved in the evolution
of the ordinary numbers of arithmetic—the "counting numbers"? But
the type of abstraction to be found in modern mathematics is of an
even higher order. On the one hand, the abstractions of primitive
number and geometry were only once removed from physical reality;
men used the numbers to count *things* and the geometry to measure
objects. On the other hand, while modern mathematics is still useful
for such purposes, it is more concerned with a conceptual world that
is at least one stage removed from concepts that are derived directly
from sense perceptions; the objects, relations, and operations with
which it is concerned are already themselves abstractions. It is this
kind of abstractness that characterizes modern mathematics.

It should perhaps be emphasized that this is not "Platonism."
Mathematical abstractions are not transcendent "forms" independent
of our world with all its particularity and concreteness. The concepts
with which mathematics deals are part of man's cultural heritage.
There has never yet been a mathematical discovery or invention that
did not have its connections with the mathematical culture existing
at the time. One recalls here Newton's famous statement: "If I have
seen further, . . . it is by standing upon the shoulders of Giants." The
virtual simultaneity of Bolyai's and Lobachevsky's works on non-
Euclidean geometry was no accident, and their similarity was certainly
not due to plagiarism. A strong case can be made for the thesis that
simultaneous discovery or invention is the rule rather than the
exception /MERTON (a, b)/.

If all that the axiomatic method had accomplished were classification, to use Poincaré's word, then it would certainly have died a natural
death. It is interesting to note that as late as 1931 another great
universalist, Hermann Weyl, decried the extent to which axiomatization was infiltrating mathematics; but twenty years later, in a review

of the mathematical advances of the first half of the twentieth century, he stated: "One very conspicuous aspect of twentieth-century mathematics is the enormously increased role which the axiomatic approach plays. Whereas the axiomatic method was formerly used merely for the purpose of elucidating the foundations on which we build, it has now become a tool for concrete mathematical research." /WEYL: 523./ Clearly he had found it impossible to write a summary of this kind without coming to grips with the achievements of the axiomatic method. He followed the remarks just quoted with the example of how neatly the axiomatic method serves to separate the two aspects of the real number system: its operational or algebraic aspect, and its structural or topological aspect. "Modern axiomatics," he said, "simpleminded as it is (in contrast to modern politics), does not like such ambiguous mixtures of peace and war, and therefore, cleanly separated both aspects from each other." (Weyl's article is divided into two parts, one stressing algebraic matter and the other geometric—an organization that was perhaps suggested by this separation.) Consideration of the algebraic aspects of the real numbers resulted in broad generalizations, such as those of rings and fields, which not only contributed to an expansion of algebra per se but gave a penetrating perspective to elementary arithmetic. (The use of commutative, associative, and like properties in modern systems of elementary teaching testifies to the recognition of this.) "Abstract algebra" became a commonly used term.

Introduction of the theory of sets.—Research in the structural (topological) aspect of the real number system gave rise to one of the most fundamental concepts of modern mathematics—a concept which itself forms one of the characterizing features of modern mathematics, namely, the notion of "set." That the notion was about to evolve into full flower during the latter part of the nineteenth century may be seen in the works of Riemann, Dedekind, Weierstrass, and Cantor. Weierstrass gave an example of a function continuous over the real numbers but having no derivative at any point. Both Riemann and Cantor investigated properties of functions defined by Fourier-series expansions which exhibited such curious properties as to be given the epithet "pathological" (although modern set theory and topology have taught us that they are not deserving of such a label). It became clear that the lack of a more penetrating description of the intuitively conceived structure of the real number system was a veritable scandal in mathematics. Only by making a more precise analysis of the structure as a

465

whole—what might be termed a "global" analysis like that exemplified in Dedekind's theory of "cuts," as opposed to studying the properties of individual numbers—could a better approximation be achieved. And thus, out of sheer necessity, the theory of sets was born. As one looks back it seems amazing that so much was accomplished by the analysts of the seventeenth, eighteenth, and early nineteenth centuries without any firm foundation in a more clearly defined real number system. "Forge ahead, and faith will come to you"; the prophecy of the second half of this quotation found realization in the new foundation theories of Dedekind, Méray, and Weierstrass.

However, the real power of the theory of sets, and the circumstances that gave it ultimately its key position in modern mathematics, may be found in the extension of Cantor's researches into the real number continuum, which carried him on into an investigation of the nature of number. It is chiefly to elucidate the nature of number that the teacher uses the set concept in modern systems of teaching arithmetic. For by a consideration of sets and operations with them (union, intersection, etc.) one can arrive at a much better intuitive understanding of the nature of the natural numbers and operations with them. Even pioneering logicians such as Frege and Russell, who (vainly) tried to establish that mathematics is only an extension of logic, seized upon the set concept as the most suitable tool for definition of the cardinal and ordinal numbers.

At this point we may speculate on (1) how much the newly developing point of view regarding the nature of mathematics helped to achieve the new foundation of the real numbers and (2) how much the act of setting up such a foundation contributed, in turn, to the development of the new point of view. Certainly so long as one regarded numbers as having an absolute character, or as having a so-called Platonic existence, one could only hope to *discover* them and their properties through some sort of revelation or some apprehending of reality. To give up the idea that such a revelation or apprehension would ever occur and to set forth to *build up* a suitable concept of number must have taken great courage. At the very least, this course of action must have been stimulated by the newly evolving point of view regarding mathematics. And, once taken, it must have hastened acceptance of the new point of view by the evidence of its own success.

On the algebraic side, the notion of abstract group had been evolving, notably through the work of Cauchy, Galois, and their successors. Besides forming a great unifying force for both algebra and geometry, it helped bring out the power of the axiomatic approach and the abstraction fostered thereby. Klein's well-known *Erlanger Programm,* which not only emphasized the role of groups in geometry but exhibited a mode for classifying the classical geometries, forms one of the best examples of the unifying power of the group notion. The role of the group concept in defining such fundamental concepts as rings and fields is becoming common knowledge to an increasing number of mathematics students early in their studies.

Greater rigor; "self-sufficiency."—Another aspect of modern mathematics, helping to distinguish it from the mathematics of earlier eras, is the greater demand for *rigor* in the demonstrations of proofs for theorems and the validation of proposed theories. Unquestionably the development of the modern axiomatic method contributed to this, since the method made more precise the assumptions underlying a proof and allowed less play to vague intuition (except as an aid to discovery). But the contributions of the nineteenth century to such topics as the theory of limits, the real number system, foundations of geometry, the extensions of the notion of number in algebra, and symbolic logic seemed to make possible a precision that was formerly unattainable. One could now justify one's assertions by appealing to fundamental concepts that were clearly defined and commonly accepted (except by the minority group known as "intuitionists"; see under "Crisis," below). It seemed no longer necessary to appeal to "common sense" or to notions generally accepted but not well defined. And this helped tremendously to foster in the mathematical community of the late nineteenth century a feeling of *self-sufficiency,* for mathematics appeared to be independent of the physical world for its theoretical justification. It had laid down its own foundations.

CRISIS

All seemed right in this "best of all possible [mathematical] worlds." However, the feeling of confidence in the new foundations and mathematical rigor received a blow around the beginning of the twentieth century—a blow that gradually assumed the proportions of a crisis. Usually mathematics has benefited from its crises, a classical case being the crisis in Greek mathematics resulting from the discovery of incommensurable line segments and the paradoxes of Zeno. The discovery of contradictions in set theory (by Burali-Forti and Russell around 1900) precipitated more soul-searching. The new foundations constructed by the nineteenth-century mathematicians seemed to have developed a fissure that raised a question as to how rigorous the newly found rigor really was. It is characteristic of the working mathematician that, like his progenitors, he forges ahead with the "faith" already cited, even though his foundations may not be secure. Nevertheless, although analysts, algebraists, and geometers continued to develop their theories, seemingly oblivious of the situation in the foundations, the crisis began to claim the attention of more and more mathematicians.

The philosophy propounded by Kronecker (1823–1891) was recalled. Kronecker had decried the "infinitistic" methods underlying the work of Weierstrass, Cantor, and the other creators of the new foundations. He maintained that "real" mathematics was only that which could be derived from the natural numbers ("intuitively given"), using only constructive methods. Nonconstructive existence proofs (for example, proving that every algebraic equation has a root by showing that assumption of nonexistence of such a root leads to contradiction) were to him worthless, and by his criteria only the rational numbers (or, more generally, the algebraic numbers), not the general real numbers, were admissible. After the contradictions in set theory were found, the Dutch mathematician L. E. J. Brouwer took up Kronecker's philosophy (c. 1908) and developed it further, gathering about him a small but influential group of cobelievers. Both Poincaré and Weyl tended to this "intuitionism" in their philosophy. However, it soon became evident that to adhere faithfully to this doctrine would result in having to abandon most of the new foundations.

Leaders among those who refused to accept such an extreme "cure" were Russell, Whitehead, and Hilbert. Taking their cue from the late nineteenth-century logicians (especially Frege and Peano), Russell and Whitehead sought a secure foundation in the tautologies of elementary logic, together with implication rules that seemed safe from error. Although their work *Principia mathematica* was never completed, it went far enough to exhibit the necessity for bringing in axioms concerning the infinite and the hierarchy of types. These axioms, if not of questionable validity, at least could not be considered purely "logical" in character.

Hilbert and his students did not begin actively carrying out their program (called "formalism") until about the 1920's, and they were obviously influenced by the findings of Russell and Whitehead as well as the doctrines of Brouwer. They took the attitude that safety can be secured by acting as though mathematics is developed purely formally, the axioms and theorems being exhibited as pure formulas (in much the same way as in *Principia mathematica*) devoid of meaning for the purposes of the investigation, while the methods of deriving theorems (new formulas) are confined to such as are purely finitistic. If a formula having the form "Both F and *not-F* hold" (i.e., $F \cdot \sim F$) is encountered, then one has contradiction; hence the purpose was to show that such a formula could not be derived. That this program, too, could not be carried to completion became evident in the 1930's, following Gödel's famous work of 1931 /NEWMAN: III, 1668–95/.

It is characteristic of science that what may seem like failure often turns out to be the opposite. It is as important to show what cannot be done as to show what can be done. In the sense that their programs could not be carried to successful completion, one may say that the "logistic" school (Russell *et al.*) and the "formalist" school (Hilbert and his students) were failures. But from the standpoint of their influence on later developments, they were quite successful; for they may be said to have given the major impulse for the creation of modern "mathematical logic," now a recognized field of mathematics. The resulting impact on the modern point of view was to be considerable. The crisis had not been solved, but it had not stultified mathematics any more than had the earlier crises; and, as before, the ultimate influence was beneficial.

First Quarter of the Twentieth Century

Modern mathematics reached full momentum soon after the turn of the century. Despite the crisis in logic and the theory of sets, there was a feeling of confidence that the new foundations for analysis and geometry, and the beginnings in abstract algebra, could be safely used to build new theories. New types of integrals, based on measure theories of sets of points, were introduced into analysis; by "measure" one meant an extension of the notion of "length" to other types of sets of points than the linear interval. The topology which in the hands of Riemann, Poincaré, and others had consisted chiefly in studying the global properties of configurations was now extended by Schoenflies, Fréchet, Hausdorff, and others to properties in the small—the set-theoretic topology.

Coincidentally with these developments, applications of the newer theories were made. One of the aspects of the evolution of mathematics which has long puzzled historians and challenged philosophers for an explanation is the manner in which mathematical theories originally invented purely for mathematical purposes have ultimately found application in other sciences. Frequently cited is the invention of the conic sections by the Greeks (presumably for the purpose of duplicating the cube, although Neugebauer /(a): 226/ has conjectured that they originated in the theory of sundials) and their later application in astronomy by Kepler. Carl Boyer has cited other cases /(h)/.

This type of phenomenon has been frequently repeated during the current century. One of the best-known cases, no doubt, is the use made of the non-Euclidean geometry of Riemann in the theory of relativity. Measure theory, mentioned above as a device for the extension of integrability to a wider class of functions, began to find application in probability. Group theory, matrix theory, and the like found application in the new quantum theory of physics. Earlier developments such as these led Whitehead /Macm. ed., 48/ to comment:

> Nothing is more impressive than the fact that as mathematics withdrew increasingly into the upper regions of ever greater extremes of abstract thought, it returned back to earth with a corresponding growth of importance for the analysis of concrete fact. . . . The paradox is now fully established that the utmost abstractions are the true weapons with which to control our thought of concrete fact.

The tendency to greater abstraction has not been confined to mathematics, of course. Physics, in particular, has exhibited the same tendency; and as other sciences become more mature, they also become more abstract. This phenomenon in the natural sciences goes hand in hand with the ever-growing abstraction and generality of the mathematics they use. It is certainly no exaggeration to say that if mathematics had not developed ever-greater abstractions, it would have lagged behind the other sciences. Perhaps no greater tribute has been paid to the new point of view in mathematics and its impact on the sciences than that made by Einstein in his famous 1921 lecture "Geometric und Erfahrung" /NAGEL et al.: 619/:

> The progress entailed by axiomatics consists in the sharp separation of the logical form and the realistic and intuitive contents. . . . The axioms are voluntary creations of the human mind. . . . To this interpretation of geometry I attach great importance because if I had not been acquainted with it, I would never have been able to develop the theory of relativity.

The trends in mathematics in the United States during the first quarter of the twentieth century were profoundly influenced by E. H. Moore and his school at the University of Chicago. One need only

470

recall such names as those of Veblen, Birkhoff, and R. L. Moore, who were products of his school, to appreciate this fact. But the influence of E. H. Moore extended beyond his scholarly interest in the development of mathematics along the new axiomatic lines. In addition, he had an intense interest in mathematical education. This was exhibited in 1902 when he devoted much of his retiring address as president of the American Mathematical Society to a discussion of the role of axiomatics in the teaching of mathematics in the primary and secondary schools /MOORE/. He espoused the historical or genetic method whereby the pupil is led from structures or patterns observable in the physical laboratory to abstract descriptions of them in axiomatic form. Earlier, Moore had been active in the formation of the Mathematical Association of America, an association formed by members of the American Mathematical Society who felt that a separate organization was needed to foster the interests of collegiate mathematics, and he had personally encouraged the formation and growth of the *American Mathematical Monthly*. Soon after this, he was appointed to a committee of the Association to look into the teaching of mathematics in the secondary schools. According to H. E. Slaught, one outcome of this committee's influence was the organization of the National Council of Teachers of Mathematics. (See /SLAUGHT/ for a discussion of Moore's pedagogical interests.)

During this first quarter of the twentieth century the flow of mathematics students from the United States to Europe for their doctoral work was greatly decelerated. This change was due to the growth of graduate departments of mathematics in American universities and, also, to the First World War and its effect on European universities. (The European universities recovered from the war quickly; however, the political situation of the 1930's drove many of the greatest mathematicians of central Europe to seek refuge in the United States.)

SECOND QUARTER
OF THE TWENTIETH CENTURY

Founding of the Institute for Advanced Study.—One of the most influential events in the mathematical world in the second quarter of the twentieth century was the founding of the Institute for Advanced Study in Princeton, New Jersey. In addition to Einstein, such scholars as von Neumann, Weyl, and Gödel—typical of the flood of European notables soon to join their mathematical colleagues in the United States—were "charter members" of its faculty. One of the most im-

471

portant forces implementing cultural evolution is that of diffusion, by means of which concepts pass from one culture to another. And the two events just mentioned (establishment of a center where research mathematicians could congregate and mix ideas over an extended period of time, and the movement of so many of the mathematically elite from foreign centers to the United States) were to furnish a profound stimulus to the production of new mathematics not only in the United States but in the entire world. The establishing of new journals and, ultimately, new abstracting agencies made known to the entire world of mathematics the accelerated tempo of research in the United States, resulting not only in the attraction of visitors to this country but in invitations to American mathematicians to foreign research centers. The impact on mathematics was worldwide.

The axiomatic method as a research tool.—The importance of the axiomatic method as a research tool was now becoming thoroughly apparent. In a report in 1939 on progress in algebra, MacLane termed algebra a study concerned with "the postulational description of certain systems of elements in which some or all of the four rational operations are possible . . . supplemented by an investigation of their 'structure' " /17–18/. Topology was making use of postulational methods to broaden its scope and applications far beyond the limitations formerly implied in its role of a geometry as defined by Klein. Its applications within geometry, especially algebraic geometry, were being matched and surpassed by its applications to algebra and analysis, fields destined to broaden their horizons far beyond their original confines. Even the classical field of number theory found uses for such notions as were obtainable from the abstract formulation of "group" and its generalizations (for example, semigroups).

The Second World War and its aftermath.—The impact of World War II and its aftermath on mathematics, as on other sciences, was incomparably greater than was that of World War I. Moreover, while the effect of the earlier war was perhaps a retardation, that of World War II was quite the opposite. Mathematicians were called upon to leave their ivory towers and to study physical and social structures that presented entirely new problems to them. Probably no other period in the history of mankind can furnish such an illustration of the influence of the cultural environment on the development of science, and in particular mathematics. Being compressed within such a short interval of time, the influence is much more manifest than when operating over long periods (as in the development of Babylonian and Greek mathematics).

Oddly enough, the abrupt change of occupation to which many mathematicians felt themselves called was occasioned not so much by a demand for the special techniques they had learned as by a need for the mental habits and disciplines they had acquired. Although frequently something of a technical nature in their previous training might be necessary for the task in hand, it was more likely that what was needed was their adeptness "in analyzing relations, in distinguishing what is essential from what is superficial in the statement of these relations, and in formulating broad and meaningful problems" /FRY: 76/. The inevitable result was the opening up of new fields.

Naturally the new activities were not entirely divorced from the current state of mathematics; nor, as already observed, were they independent of the previous interests of the individual. In some cases the new fields represented accelerated development of theories whose origins were prewar. But who can doubt, for example, that the rapid development of computer theory and of the automatic computer has been due to the war and its aftermath? The period from the invention of the abacus to the first automatic computer, completed in 1944, is paralleled by that from the beginnings of the study of logic in Greece to the applications of modern mathematical logic to the theory of automata. Yet the essential development of computer and automata theory is almost completely compressed into the past quarter century— chiefly the period since the war. Although the case of an individual cannot furnish a stereotype of a historical trend, a glance at the complete list of published works of the renowned John von Neumann /I: 645–52/ reveals the impact that the war and subsequent years could exert on the work of a creative mathematician.

The new phenomenon of governments pouring money into grants and contracts for research in pure science, magnifying and all but usurping what was formerly a "prerogative" of academic institutions and scholarly foundations, has been one of the most powerful forces for the acceleration of mathematical invention, as well as for the increase in the number of professional mathematicians. New fields such as game theory and information theory received a strong stimulus to rapid development. Linear programming, operations research, and the like were strongly influenced by wartime and postwar demands. There were new developments, also, in numerical analysis, probability and statistics, matrix theory, set theory, Boolean algebra, and mathematical logic, as well as new applications thereof. Industrial as well as military demands began to draw many established mathematicians, and many younger men just starting their careers, away from academic life. Meanwhile the applications of mathematics to the social sciences were growing—applications not only of new statistical methods but of the more abstract parts of algebra and topology. Over a broad range, from the aircraft industry to an extensive array of academic subjects, the demand for mathematicians far exceeded the supply.

Revision of curricula.—It became apparent that two factors accounted for the shortage of mathematicians who could meet the new demands. In addition to the wider applications being found for mathematics, there was the rapid expansion and changing character of mathematics itself. Although graduate mathematics curricula had been continually revised to take advantage of newer and more powerful methods, similar revision had not taken place at the more elementary levels. An examination of the grade school curricula revealed that not only was much time being lost in needless repetition and outworn methods but the content of courses was the same as a century earlier; little advantage had been taken of newer and more elegant methods susceptible of being taught on the elementary level. This situation was perhaps a reflection of the lack of communication between instructors in teacher-training institutions and creative mathematicians. For not all the "new mathematics" requires an acquaintance with the traditional secondary and undergraduate college curriculum in order to be understood. The elementary parts of set theory, for instance, require no previous mathematics and are probably more natural to the uninhibited child than to the nineteenth-century mathematician schooled in traditional doctrine. Moreover, they make possible a far better conception of number, for instance, than did the traditional drill methods.

Attempts by local, regional, and national groups to pump new blood into the elementary curriculum took the form of specific content recommendations and suggestions for writing new textbooks.

Aid from the federal government soon made it possible to train both secondary school teachers and college teachers in the newer approaches to traditional ideas. The burden that this training placed on already overworked teachers presented a problem needing to be worked out with supervisors and school boards, not always sympathetic. Fears were expressed that the proposed curricula would lead to neglect of aspects of traditional mathematics needed in practical applications, especially such as were needed in the natural sciences.

But it seemed to be generally agreed that reform was long past due; only its character caused disagreement.

Despite all the problems encountered, an inevitable change was in the making. Anyone familiar with the history of mathematical curricula should not be surprised at the new developments. It was not many years ago that the undergraduate mathematics program in an American college or university did not go beyond the calculus; and now one finds the calculus being taught in not a few of the private academies and larger high schools of the country. Even with specialization, the amount of mathematics that the professional has to learn today would have been inconceivable less than a century ago. Without improved methods the situation would have been impossible, and belated recognition of this fact resulted in the construction of new secondary school curricula. History will record that great commendation was earned by those secondary school teachers and university mathematicians who found a common bond in their interest in the problems of secondary education and developed amicable relations for the purpose of their solution.

Retrospect and Prospect

In looking back over the past century one is struck by the way in which mathematics has undergone a process of maturation. That such an observation can be made regarding what is perhaps the oldest of sciences, which underwent significant growth even so long ago as pre-Hellenic and Hellenic times, must seem odd. Nevertheless, considering that mathematics has recently spread out in all directions, extending old branches and bringing forth new ones, the statement seems justified. Without losing its influence on other sciences, it has grown from within and more clearly exhibited its own nature and possibilities. Abstraction has proved a unifying force, making it possible for diverse and seemingly unrelated subjects to be seen as parts of one organic structure. The resultant simplification has made it possible for the young mathematician to progress more rapidly than previous generations of mathematicians.

One of the best indications of maturity is an awareness of one's own limitations. Not only has mathematics matured in the sense that it has grown, but mathematicians have matured in the sense that they have become aware of the limitations of their particular field of knowledge. Today the mathematician who is aware of the newer developments in mathematical logic must conclude that mathematics, far from dealing with absolute truth, is just as much a science as physics or chemistry, dependent as they are on natural phenomena and on presuppositions that cannot be justified except in terms of special purposes, coherent systems, and pragmatic results. It has been shown quite conclusively that to hope for a secure foundation, free from contradiction and yet complete enough for all our needs, is no longer realistic. The axiomatic method, for a time considered the answer to all problems about foundations, has revealed its Achilles heel in the logic and set theory that it presumes. Stripped to its essentials, it is seen to be ultimately dependent upon the particular logic and set theory employed. And neither of these is a unique thing.

This is not a situation to be viewed with pessimism. The mature standpoint counsels that the mathematician, like the physicist, should face the reality that has finally been exposed to him after centuries of illusion. His task is now clear, and he will not waste further time and energy pursuing a will-o'-the-wisp Truth that has constantly evaded him. Rather, he will view his creations from the standpoint of their usefulness and their adaptability to circumstances, remaining open-minded about the methods used to achieve them. That certain methods may lead to contradiction when used indiscriminately does not mean that they should be abandoned; such a situation only points to the need for determining the areas in which the methods are safe.

Perhaps it is fortunate that most mathematicians of the modern era have applied themselves to their daily tasks without worrying about the "crisis" in the foundations; for it begins to appear that they were, after all, correct in forging ahead with their creations, possessed of the same faith that inspired the mathematician of d'Alembert's day. Such faith and creativity will undoubtedly continue to mark mathematicians in the years to come. It seems likely that mathematics will not only grow more rigorous because of our present understanding of its nature and limitations but will exhibit the organic unity of its structure as it continues to grow new branches and bear fruit.

Mathematics never exhausts itself. Today we know why this is so. We know that mathematics is not a completeable subject—not even in its oldest forms, such as arithmetic. Always new vistas will open for research. Terrestrial exploration may no longer offer challenge to one possessed of a pioneering spirit, but mathematics will never cease to challenge him who is capable of exploring its mysteries. Never again should anyone heed the pessimists who declare that all worthwhile problems have been solved and mathematics has come to the end of its road.

476

IX

The SCIENCE of PATTERNS:
A Perspective on Contemporary
Mathematics

LYNN ARTHUR STEEN

T hroughout the twentieth century mathematics has grown in extent and diversified in form. Now, near the end of the century, what appeared of greatest importance one hundred years ago is just a small part of the entire mathematical landscape. Two events triggered periods of explosive growth: the Second World War, which forced the development of many new and powerful methods of applied mathematics, and the development of electronic computing, which provided mathematicians with a tool of immense power and potential.

During the first half of the century, mathematical growth was stimulated primarily by the power of abstraction. The axiomatization of mathematics on a foundation of logic and sets at the end of the nineteenth century made possible the grand theories of algebra, analysis, and topology that dominated most of mathematics research and teaching for the first two-thirds of the century.

The formalism of sets and axioms made clear the common logical roots of algebra and analysis, of the discrete and the continuous. Clarity in logic became the major goal of a grand synthesis of mathematics, epitomized in the French Bourbaki volumes and subsumed in many university mathematics courses around the world. What emerged has been called "pure" mathematics, or more recently, "core" mathematics—a paradigm of mathematics that rested the analysis of Newton on the deductive method of logic.

But the scientific paradigm established by Newton involves more than deduction. Newton perceived patterns in the accumulated astronomical data of his time; he abstracted from these patterns certain general principles (whence *Principia*); and he used these principles to deduce patterns both known and unknown in the behavior of planetary bodies. His was a science of patterns—rooted in data, supported by deduction, confirmed by observation.

Classical mathematics has remained rooted in the Newtonian mathematics of analysis, a synthesis of algebra and geometry applied to the study of how things change. But now this core has been supplemented by major developments in other mathematical sciences—in number theory, logic, statistics, operations research, probability, computation, topology, and combinatorics, in addition to algebra, geometry, and analysis.

In each of these subdisciplines, applications parallel theory. Even the most esoteric and abstract parts of mathematics—number theory and logic, for example—are now used routinely in applications (e.g., in computer science and cryptography). Fifty years ago G. H. Hardy could boast of number theory as being the most pure and least useful part of mathematics; today number theory is studied as an essential prerequisite to many applications of coding, including data transmission from remote satellites, protection of financial records, and efficient algorithms for computation.

In 1960, at a time when theoretical physics was still the central jewel in the crown of applied mathematics, Eugene Wigner wrote about the "unreasonable effectiveness" of mathematics in the natural sciences: "The miracle of the appropriateness of the language of mathematics for the formulation of the laws of physics is a wonderful gift which we neither understand nor deserve" (p. 14). Indeed, over the last several decades theoretical physics has continued to adopt (and occasionally invent) increasingly abstract mathematical models as the logical foundation for current theories: Lie groups and gauge theories—exotic expressions of symmetry—have joined fermions and baryons as fundamental tools in the physicist's search for a unified theory of both microscopic and macroscopic forces of nature.

During this same period, striking applications of mathematics have emerged across the entire landscape of natural, behavioral, and social science. Moreover, applications of one part of mathematics to another—of geometry to analysis, of probability to number theory—provide new evidence of the fundamental unity of mathematics. Despite the ubiquity of connections among problems in science and mathematics, the discovery of new links remains surprisingly unpredictable and serendipitous. Whether planned or unplanned, the cross-fertilization

478

between science and mathematics in problems, theories, concepts, and paradigms has rarely been greater than it is now in the last quarter of the twentieth century.

The tremendous growth and power of mathematical applications parallel the phenomenal impact of computing. It is ironic but indisputable that computers were made possible by application of abstract theories of such mathematicians as Boole, Cantor, Turing, and von Neumann, theories that just a few decades ago were widely derided by critics of the "new math" as wild abstractions irrelevant to practical purposes. It is doubly ironic that the computer is now the most powerful force changing the nature of mathematics. Even some mathematicians who never use computers devote their entire research careers to problems generated by the presence of computers. In all fields of mathematics, computers have posed new problems for research, provided new tools to solve old problems, and introduced new research strategies.

Although the public often views computers as a replacement for mathematics, each is in reality a power tool for the other. Indeed, just as computers provide new opportunities for mathematics, so also mathematics makes computers so incredibly effective. Mathematics provides abstract models for natural phenomena as well as algorithms for implementing these models in computer languages. Applications, computers, and mathematics form a closely interrelated system yielding results never before possible and ideas never before imagined.

The Mathematical Sciences

Rapid growth in the nature and applications of mathematics means that the Newtonian core—calculus, analysis, differential equations—is now just one part of a more diverse mathematical landscape. Yet most scientists have explored only this original territory, because that is all that was part of their curriculum in school, college, and graduate school. With the exception of statistics—an old science widely employed across all disciplines that has become largely mathematical during the twentieth century—the narrow Newtonian legacy of analysis is the principal interface between practicing scientists and the broad mathematical foundations of their disciplines. The dramatic changes in the mathematical sciences of the last quarter century are largely invisible to those outside the small community of research mathematicians.

Today's mathematical sciences, like yesterday's Gaul, can be divided into three parts of roughly comparable size: statistical science, core mathematics, and applied mathematics. Each of these three major areas is led (in the United States) by a few thousand active researchers and is studied each year by more than half a million college and university students. Although these areas overlap considerably, each province has an identifiable character that corresponds well with the three stages of the mathematical paradigm established by Newton: data, deduction, and observation.

Statistical science studies problems associated with uncertainty in the collection, analysis, and interpretation of data. Its tools are probability and inference; its territory includes stochastic modeling, statistical inference, decision theory, and experimental design. Statistical science influences policy in agriculture, politics, economics, medicine, law, engineering, and science. Advances in instrumentation and communication (to gather and transmit data) have posed new challenges to statistics, leading to rapid growth in new methods and new applications.

Core mathematics establishes properties of number and space—ideas rooted in antiquity. Its tools are abstraction and deduction; its edifices include functions, equations, operators, and infinite-dimensional spaces. Core mathematics comprises the traditional subjects of number theory, algebra, geometry, analysis, and topology. After a half century of explosive growth in specialized disciplines, core mathematics is enjoying a renaissance; the unexpected but welcome discoveries of deep links among its various components have given it renewed integrity.

Applied mathematics fits mathematical methods to scientific observations and theories. Through applied mathematics, scientific ideas stimulate mathematical innovation and mathematical tools solve scientific problems. Traditional methods of applied mathematics include differential equations, numerical computation, control theory, and dynamical systems; such traditional methods are today entering major new areas of application, such as combustion, turbulence, optimization, physiology, and epidemiology. In addition, new tools from game theory, decision science, and discrete mathematics are being applied in the human sciences, where choices, decisions, and coalitions rather than continuous change are the apt metaphors for description and prediction.

All attempts to divide mathematics into parts are necessarily artificial: statistical science, core mathematics, and applied mathematics represent just one of many possible structures that may help one understand the whole. They do not represent intrinsic divisions in the nature of mathematics so much as differences in style, purpose, and history; they may more aptly describe types of mathematicians than types of mathematics. Others have attempted to portray the nature of modern mathematics in somewhat different terms. What is important about these labels is that they help focus attention on certain characteristics of the mathematical sciences, not that they themselves represent inherent or essential compartments.

A fact that *is* inherent in mathematics and that is essential to understand its role in society is this: today's mathematical sciences are very different from the mathematics of a quarter century ago that most of today's scientists and engineers studied. Computers, applications, and cross-fertilization have combined to transform the mathematical sciences into an extraordinarily diverse and powerful collection of tools for science. Without even asking permission, mathematicians have quite literally rebuilt the foundations of science. The work is not finished, but its new shape is sufficiently visible that all who teach and use mathematics should take the time to explore its new features.

STATISTICAL SCIENCES

Computers and statistics both deal with data: what computers record, transform, and manipulate, statisticians interpret, summarize, and display. This confluence of problem sources and problem solvers has radically transformed (some would say restored) statistical science to a data-intensive discipline. Phenomena described by traditional statistical distributions (normal, Poisson, etc.) represent only a tiny part of the enormous quantity of data captured by computers all over the world.

Spatial Statistics. The increasing use of electronic scanning devices (for example, in tomography, in airborne reconnaissance, in environmental monitoring) has produced an urgent need for sophisticated analysis of data with inherent spatial structure. Image enhancement—the visual clarification of blurred images—is one common application; another is the statistical compression of data to permit efficient storage and subsequent analysis.

Research in spatial statistics employs a wide variety of mathematical, statistical, and computational techniques. Problems of separating signal from noise borrow ideas from engineering; ill-posed scattering problems employ methods of numerical linear algebra; smoothing of data requires statistical techniques of regularization. A further consideration is the inherent geometry of each problem, which in many cases is dynamic and nonlinear.

Bootstrap and Jackknife Statistics. Many applications of statistics (for example, clinical data from innovative medical protocols) involve small data sets from which one would like to infer meaningful (significant) patterns. Bradley Efron at Stanford University has pioneered a method of using limited data to generate more data with the same statistical characteristics (hence the term *bootstrap methods*).

Bootstrap methods use computers to resample the given data repeatedly in order to generate millions of similar possible data sets, which yield accurate approximations to various complex statistics. By comparing the values of these statistics for the given sample with the distribution obtained for all possible samples, one can determine whether the observed values are significant. Jackknife methods are related to bootstrap techniques; they reduce bias in the statistical procedures by repeatedly slicing away part of the data.

CORE MATHEMATICS

The external forces that impinge on mathematics—primarily applications, computers, and cross-fertilization—influence the core ("pure") parts of the subject in profound ways. To illustrate the nature of the changes, we will use two rather different areas of impact—computation and geometry—as widely separated mileposts along a vast continuum of mathematics.

Computation. Core mathematics has changed under the influence of computers as much as the more applied areas of the mathematical sciences, but in different ways. Most noticeable is the shift in research interests to questions motivated by computation. But computers have also changed the way conjectures are invented and tested, the way proofs are discovered, and—in an increasing number of cases—the nature of proof itself.

The archetypal event in computer-assisted mathematics was the 1976 proof of the century-old Four Color Conjecture based on a computer analysis of thousands of reduction patterns; this reduction bridged a gap between mathematical theory and human analysis of cases. At the time, this event shook the very epistemology of mathematics; yet ten years later, in 1986, nearly half the plenary lectures at the International Congress of Mathematicians were devoted to topics linked in some way with computation.

For example, Louis de Branges used computers to help discover a proof of the seventy-year-old Bieberbach Conjecture concerning the size of coefficients in the power-series expansions of certain analytic functions of a complex variable. At a crucial stage in the proof, de Branges had reduced the entire argument to verifying an inequality between two polynomials; this verification was done by computer to a degree sufficient to provide convincing circumstantial evidence of the validity of this line of argument. The final link in the formal argument was supplied by a theoretical proof of this inequality, actually known and proved in the theory of special functions long before de Branges needed it.

Computers are largely responsible for renewed interest in one of the oldest problems in mathematics—how to find the prime factors of an integer. Integer factorization moved from a backwater to high priority in mathematics simply because of its application in computer-based cryptography: a code based on the product of two large prime numbers cannot be broken with current algorithms because there is no known efficient method of recovering the two factors from knowledge of only the product. The problem of factoring has recently been attacked by means of elliptic curves—abstract graphs of cubic polynomials—whose properties lead directly to the fastest algorithm yet discovered for factoring large numbers.

The theory of computational complexity—the analysis of the inherent difficulty in solving problems—is now shedding new light on a very old equation: $ax = b$. The traditional solution, division, has not been of interest to mathematicians since the middle ages, yet now it is a subject of intense research. Only recently has it been proved that the known method of performing division of complex numbers is the best way possible—in the sense that no method can involve fewer real arithmetical operations. The methods employed in this fundamental analysis of arithmetic presage similar analysis for the entire range of computer algorithms, enabling mathematicians to determine the limits of computational feasibility as well as areas ripe for further improvement.

Geometry. If computers typify the modern era of mathematics, geometry epitomizes its classical roots. Historically, geometry—the study of space—has been one of the major pillars of core mathematics. For various reasons, its role in the mathematics curriculum has declined over the past twenty years, so that even those with university degrees in mathematics often have little acquaintance with modern geometry. In sharp contrast to this curricular decline is the renaissance of geometry in research mathematics. In a very real sense, geometry is once again playing a central role on the stage of mathematics, much as it did in the Greek period.

A principal actor in modern geometry is a *manifold*, a term used by geometers to describe surfaces and spaces that behave locally like Euclidean space. Manifolds form the natural locus for solutions to differential equations, and in turn their geometry imposes structure on the analytic nature of these solutions. Thus, manifolds are important not just to geometry but to all areas of classical analysis.

Two of the three 1986 Fields Medals—the "Nobel Prizes" of mathematics—went to Michael Freedman of the University of California at San Diego and Simon Donaldson of Oxford University for work in the geometry of four-dimensional manifolds. Freedman's methods showed that the topological classification of four-dimensional manifolds mimics the algebraic classification of quadratic forms; Donaldson exploited the wave-particle duality of matter to develop an entirely new approach to the study of fundamental problems of geometry. One result of this work was the discovery of exotic four-dimensional spaces similar but not identical to the conventional space-time continuum of our physical world.

Geometry and computers intersect in one of the most lively and attractive interstices of the mathematical sciences: computer graphics. To produce realistic graphics on a computer requires considerable theoretical interaction among geometrical representation, algebraic encoding, and computer algorithms. In return, computer graphics have provided crucial assistance in many mathematics problems: new minimal surfaces have been found with the aid of computer graphics, and the visual displays of iterative maps (in "fractal" pictures) reveal patterns that would never be noticed by analytic means alone. Fractal patterns give an apt description of a large class of physical phenomena ranging from the fracturing of glass to the texture of surfaces.

Geometric computing is beginning to prove very useful also in core areas of mathematics remote from geometry, since supercomputers can calculate and display in visual form the roots of equations and other mathematical objects. This enables mathematicians for the first time to "see" the content of the abstract theorems they prove, and thereby make new conjectures suggested by the eye rather than the mind.

APPLIED MATHEMATICS

Applied mathematics is distinguished from core mathematics not so much by content or method as by objective: in applied mathematics, the value (or significance) of new methods is measured by the degree to which they improve scientific understanding or technological applications.

The roots of the scientific revolution lie in the empirical methods introduced by Galileo to replace the speculative explanations of classical

Greek natural philosophy. Newton introduced theoretical science by showing that mathematical results deduced from basic axioms can explain empirical data. In our time, John von Neumann pioneered the computational paradigm, which uses the results of theoretical science to simulate reality on a modern computer. As a consequence, computational methods now pervade all aspects of applied mathematics.

Necessity is the mother of invention in mathematics as in life. Since the needs of science stimulate the growth of mathematics, as science expands and grows, so does mathematics. This relationship has sparked explosive growth in the nature and range of applied mathematics. Two very different areas illustrate both the diversity and the innovation of present research.

Biological Science. Nothing better illustrates the potential for mathematics in the biological sciences than the many traces of mathematics behind the Nobel Prizes. For example, Allan Cormack won the 1979 Nobel Prize in medicine for his application of the Radon transform, a well-known technique from advanced classical analysis, to the development of tomography and CAT scanners. The 1984 prize in chemistry was awarded to biophysicist Herbert Hauptman, who took his Ph.D. in mathematics and who is president of the Medical Foundation of Buffalo, for fundamental work in Fourier analysis pertaining to X-ray crystallography.

Indeed, recent research in the mathematical sciences suggests dramatically increased potential for fundamental advances in the life sciences using methods that depend heavily on mathematical and computer models. Structural biologists have become genetic engineers, capturing the geometry of complex macromolecules in supercomputers and then simulating interaction with other molecules in their search for biologically active agents. Using these computational methods, biologists can portray on a computer screen the geometry of a cold virus—an intricate polyhedral shape of uncommon beauty—and search its surface for molecular footholds on which to launch their biological assault.

Geneticists are beginning the monumental effort to map the entire human genome—an enterprise requiring expertise in statistics, combinatorics, artificial intelligence, and data management to organize billions of bits of information. Ecologists—the first mathematical biologists—use the extensive theories of population dynamics to predict the behavior and interaction of species. Neurologists now use the theory of graphs to model networks of nerves in the body and the neural tangle in the brain. And physiologists employ contemporary algorithms applied to nineteenth-century equations of fluid dynamics to determine such things as the effects of turbulence in the blood caused by swollen heart valves or clumps of cholesterol.

Knot Theory. Fifty years ago, motivated in part by the need to find mathematical models of quantum mechanics, mathematicians developed the theory of operators. Operator theory flourished as a branch of analysis, pursued for its pure mathematical interest as well as for its continued applicability to quantum physics. In recent years, the investigation of new types of operators has yielded new methods of classification, which in turn have led to the discovery of important relations between operator algebras and the classification of knots—a vexingly difficult problem that had previously defied all attempts at solution.

The key to the classification of knots is a scheme to encode knot patterns in algebraic terms so that algebraic manipulations correspond to physical actions on the knot. This makes it possible, for the first time, to determine whether one knot can be transformed into another or unlinked completely into a straight line. As happens so often in mathematics, the significance of the new structure emerged in recognition of its ubiquity: that the same pattern appeared in several places—in knot groups, in statistical mechanics, and in certain classes of solvable equations—is precisely why the technique takes on special power.

Recently, biologists studying the replication of DNA have teamed up with mathematicians working in knot theory, since DNA in the cell is normally coiled into a tight knot. How DNA can replicate and then pull apart if it is tightly knotted is difficult to imagine—like the magician's trick of effortlessly separating two intertwined rope knots. The scope of operator theory, from motivation and application in quantum mechanics through esoteric research in pure mathematics to knots and the unfolding of DNA, provides an amazing, albeit not atypical, example of the many interconnections among diverse parts of mathematics.

THE SCIENCE OF PATTERNS

The foregoing examples from contemporary mathematical science illustrate metaphors of the mathematical method that originated three hundred years ago in the Newtonian synthesis: data, deduction, and observation. These examples also reveal the effects of the major contemporary forces for change: computers, applications, and cross-fertilization. Hundreds of other examples could have illustrated the same points; those chosen here are neither the deepest nor the most important. They do suggest, however, the variety and scope of today's mathematics.

Mathematics is often defined as the science of space and number—as the discipline rooted in geometry and arithmetic. Although the diversity of modern mathematics has always exceeded this definition, it wasn't until the recent resonance of computers and mathematics that a more apt definition became fully evident.

Mathematics is the science of patterns. The mathematician seeks patterns in number, in space, in science, in computers, and in imagination. Mathematical theories explain the relationships among patterns; applications of mathematics use patterns to "explain" and predict natural phenomena that fit these patterns. Patterns suggest other patterns, often yielding patterns of patterns. In this way mathematics follows its own logic, beginning with patterns from science and deriving new patterns from the initial ones.

To the extent that mathematics is the science of patterns, computers change not so much the nature of the discipline as its scale: computers are to mathematics what telescopes and microscopes are to science. They have increased by a millionfold the portfolio of patterns investigated by mathematical scientists. As this portfolio grows, so do the applications of mathematics and the cross-linkages among the subdisciplines of mathematics.

Because of computers, we see more than ever before that mathematical discovery is like scientific discovery. It begins with the search for pattern in data—perhaps in numbers, but often in geometric or algebraic structures. Theories emerge as patterns of pattern, and significance is measured by the degree to which patterns in one area link to patterns in other areas. Subtle patterns with the greatest explanatory power become the deepest results, forming the foundation for entire subdisciplines.

Texas physicist and Nobel laureate Steven Weinberg, echoing Harvard mathematician Andrew Gleason, has suggested that mathematics' uncanny ability to provide just the right patterns for scientific investigation may be because the patterns investigated by mathematicians are *all* the patterns there are. If patterns are what mathematics is all about, then the "unreasonable effectiveness" of mathematics may not be so unreasonable after all.

READING LIST

Albers, Donald J., and G. L. Alexanderson. *Mathematical People: Profiles and*

Interviews. Cambridge, Mass.: Birkhäuser Boston, 1985.

American Mathematical Society. "Mathematics: The Unifying Thread in Science." *Notices of the American Mathematical Society* 33 (1986): 716–33.

Davis, Philip J., and Reuben Hersh. *The Mathematical Experience.* Cambridge, Mass.: Birkhäuser Boston, 1980.

Gleik, James. *Chaos.* New York: Viking Press, 1987.

Guillen, Michael. *Bridges to Infinity: The Human Side of Mathematics.* Boston: Houghton Mifflin, 1983.

Hofstadter, Douglas R. *Gödel, Escher, Bach: An Eternal Golden Braid.* New York: Vintage Press, 1980.

Jaffe, Arthur. "Ordering the Universe: The Role of Mathematics." In *Renewing U.S. Mathematics: Critical Resource for the Future*, pp. 117–62. Washington, D.C.: National Academy Press, 1984.

Kitcher, Philip. *The Nature of Mathematical Knowledge.* New York: Oxford University Press, 1983.

Kline, Morris. *Mathematics and the Search for Knowledge.* New York: Oxford University Press, 1985.

Lakatos, Imre. *Proofs and Refutations: The Logic of Mathematical Discovery.* Cambridge, England: Cambridge University Press, 1976.

Mac Lane, Saunders. *Mathematics: Form and Function.* New York: Springer-Verlag, 1986.

National Academy of Sciences. *Mathematical Sciences: A Unifying and Dynamic Resource.* Washington, D.C.: The Academy, 1986.

———. *Mathematical Sciences: Some Research Trends.* Washington, D.C.: The Academy, 1988.

Peitgen, H.-O., and P. H. Richter. *The Beauty of Fractals: Images of Complex Dynamical Systems.* New York: Springer-Verlag, 1986.

Rucker, Rudy. *Infinity and the Mind: The Science and Philosophy of the Infinite.* Cambridge, Mass.: Birkhäuser Boston, 1982.

Steen, Lynn Arthur. *Mathematics Today: Twelve Informal Essays.* New York: Springer-Verlag, 1978.

———. *Mathematics Tomorrow.* New York: Springer-Verlag, 1981.

Tymoczko, Thomas. *New Directions in the Philosophy of Mathematics.* Cambridge, Mass.: Birkhäuser Boston, 1986.

———. "The Four Color Problem and Its Philosophical Significance." *Journal of Philosophy* 76 (1979): 57–83.

Wigner, Eugene P. "The Unreasonable Effectiveness of Mathematics in the Natural Sciences." *Communications on Pure and Applied Mathematics* 13 (1960): 1–14.

CONTRIBUTORS OF CAPSULE ARTICLES

(Capsules are identified by number after each author's name.)

Alberti, Furio—46
Alexander, Howard—101
Alexander, John W., Jr.—48
Amundson, Harlen E.—42
Anderson, Lee—51, 53
Barnett, I. A.—11, 12, 13
Baravalle, Hermann von—44
Bedient, Jack—118, 119
Bedient, Philip E.—117
Benner, Carl V.—7, 19
Brannan, Samuel J.—107
Bowsher, Lester E.—61
Byrkit, Donald R.—6
Charp, Sylvia—49
Cummins, Kenneth—66
Deal, Duane E.—57
Dessart, Donald J.—89
Dykstra, Robert H.—106
Ellis, Wade—67
Fey, James—15, 16, 23, 24, 26, 27, 36, 41, 45, 47, 48
Flagg, Elinor B.—104
Fobes, Melcher—108, 109
Foltz, Tom—63
Fuller, Leonard E.—72, 73
Ganis, Sam E.—22
Habegger, Philip—52
Hayes, Eleanor—97
Heck, William—3, 24
Heinke, Clarence H.—37
Hellmich, Eugene W.—76
Hirschy, Harriet D.—30, 58
Hood, Rodney T.—71, 98
Huffer, Ralph C.—99, 100
Jones, Phillip S.—94
Keihn, Meta Darlene—50
Kennedy, Joe—82
King, Amy C.—103, 105
Klaasen, Daniel L.—60, 64
La Mar, Diana—2
Lott, Fred W.—14
Lovaglia, A. R.—120
Lowe, Roger D.—56, 95

Mainville, Waldeck E., Jr.—35, 90
Mann, Nathaniel III—39
Merick, Lloyd C., Jr.—8
Mielke, Paul T.—102
Miller, Leland—25, 36, 43
Miller, Ruth Anne—96
Morgan, T. J.—62
Mortlock, R. S.—29
Mossburg, Larry—91
Olmsted, John M. H.—116
Park, Richard M.—69, 70
Perlis, Sam—75, 77
Porter, James—113
Pratt, Gertrude V.—78
Ragan, Esther—82
Read, Cecil B.—80
Reinking, Donald L.—93
Retz, Merlyn—50, 65
Robold, Alice I.—55
Rossmeissl, John W.—18
Sanchez, George I.—6
Schenck, Cynthia—54, 95
Schrader, Dorothy—21
Selby, Samuel—54
Seybold, Anice—74
Shanks, Merrill—115
Shively, L. S.—88
Sholander, Marlow—110
Shreves, Jerry W.—92
Sister M. Stephanie Sloyen—81
Somers, Donald L.—59
Struik, Dirk J.—111
Tatham, Elaine J.—68
Trimble, Harold C.—9
Vogeli, Barry D.—1, 4, 5
Webber, Frederick A.—18
Western, Donald W.—85
Wetherbee, William B.—17, 20, 32, 38
Wolfe, Dorothy—86, 87
Wollan, G. N.—114
Wrestler, Ferna E.—79
Yozwiak, Bernard J.—40, 112

489

APPENDIX
Resources beyond This Yearbook

CHARLES V. JONES

with the assistance of JANE M. KEISER

IN THE twenty years since the 31st Yearbook was published, many developments in the history of mathematics have made it desirable to extend this bibliographic section. Since 1969 the history of mathematics has become a discipline: it has scholarly journals, research programs and centers, a canon of criticism, national and international professional organizations, conferences and congresses, professionally trained historians of mathematics—in fine, a social structure with all the trappings of a fully developed scholarly discipline. The result has been an exponential growth in literature on the history of mathematics. As the literature mounts, specialization and fragmentation develop, with special journals for special subdisciplines. Locating information, a difficult task two decades ago, is now an overwhelming one. However, new bibliographic aids have appeared that facilitate searching the literature.

This growth in the history of mathematics has not been the work of any one individual. But there is one person whose activities set into motion several programs that have provided much of the disciplinary structure in the history of mathematics and, in no insignificant way, made this updating of the 31st Yearbook necessary. Kenneth Ownsworth May (1915–1977) was the focus of activity in the history of mathematics at the University of Toronto from 1967 to his death. In addition to writing, editing, and starting a graduate program in the history of mathematics, he launched a bibliographic study culminating in 1973 in the basic reference, *Bibliography and Research Manual of the History of Mathematics* (hereafter, May's *Bibliography*). This was followed by his founding in 1974 of the journal *Historia Mathematica*. Before his death in 1977, May also saw the publication of his *Index of the American Mathematical Monthly, Volumes 1 through 80 (1894–1973)*. Other projects and activities, at the time of May's work and since, have made investigation in the history of mathematics much easier than it was before.

When the 31st Yearbook first appeared it was virtually one of a kind, but when we look at it now it is one resource among many. This bibliography contains all citations that appear at the ends of the capsules or that are cited in the essays introducing the main divisions. There was no attempt to include all sources on a given topic; still, this bibliography is a rich resource. Clearly this yearbook with its bibliography was a harbinger of the expanding study of the history of mathematics. Along with the "Historically Speaking" series that ran in the *Mathematics Teacher*, the 32nd Yearbook (*A History of Mathematics Education in the United States and Canada*), and many other NCTM publications with historical themes, it has answered a continuing need among mathematics teachers as well as historians for guidance in searching the literature in the history of mathematics. Although it has been supplemented in a variety of ways, it remains one of the few publications that deals with how to bring history into the mathematics classroom. Some of its capsules are gems of brevity and precision; collectively, they are a mine of information. However, some of the information or interpretations might now be changed. The purpose of this essay on bibliographic sources is to indicate where the more serious reader might look to extend her or his investigation beyond this yearbook.

Bibliography A (p. **499**) is the original bibliography as published in the 31st Yearbook in 1969. Bibliography B (p. **516**) is an annotated list of sources published since 1969. It also lists later editions of some of the works listed in Bibliography A.

BIBLIOGRAPHIES

Since the publication of this yearbook, Kenneth O. May's *Bibliography and Research Manual of the History of Mathematics*, with about 31 000 separate entries arranged under about 3700 topics (by May's own count), has become the standard bibliographic reference in the history of mathematics. The book contains three main parts: "Research Manual," "Bibliography," and "Appendices." The Research Manual contains a succinct discussion of the issues one should be aware of in doing historical research. It also describes a filing system for keeping track of bibliographic, factual, and interpretative information as it is acquired in the process of reading. It is an amalgam of traditional methods, well thought-out for the needs of a researcher today—for example, the library card catalog or the traditional note-taking techniques described in such sources as Barzun and Graff's *The Modern Researcher* (revised 1970, Harcourt, Brace). It has been widely tested, as in the construction of May's *Bibliography*. (Much of the structure of this system may be directly transferred to a computer data-base system; however, some aspects may not. In particular, the note-taking system permits taking notes from library sources that do not circulate and ordinarily would not be near a computer.)

The categories used in the main Bibliography are Biography, Mathematical Topics, Epimathematical Topics, Historical Classification, and Information Retrieval; these are clearly described in the book. "Epimathematical" is a term coined by May and subsumes many topics of interest to teachers. There is one major drawback in the design of May's *Bibliography*: in an effort to get a maximum of entries into a minimum of space, titles of articles from periodicals are not given. Instead, topics dealt with in the article are sometimes given in square brackets. As an example, an article by Phillip S. Jones entitled "Tangible Arithmetic I: Napier's and Genaille's Rods," which appeared in the *Mathematics Teacher*, volume 47, November 1954, pp. 482–87, appears in May's format as

Jones P S 1954 MT 47, 482–487 [rods].

In order to get the title, one would have to know that "MT" stands for *Mathematics Teacher* (an alphabetical list of abbreviations for about 3000 journals is in the Appendices) and then retrieve and inspect the particular issue of the journal. This entry appears under "Napier" in the Biography section; there is no separate listing of "Napier's Rods" (nor of Genaille). Books are given more complete citations, including titles, as the entry for this yearbook illustrates; it is listed under Epimathematical Topics, "Education: Using history in teaching mathematics":

Hallerberg Arthur E + 1969 *Historical Topics for the Mathematics Classroom*, Wash DC (NCTM), 524 p.

Although May's *Bibliography* was published in 1973, systematic coverage of the literature went only to about 1968. Entries appear that were published after 1968 (this yearbook being a case in point), but the coverage becomes more spotty.

Another bibliography, edited by Joseph W. Dauben, appeared in 1985: *The History of Mathematics from Antiquity to the Present: A*

493

Selective Bibliography. Unlike May's *Bibliography*, Dauben supplies fewer citations (less than 2400) but gives complete citation information, usually with an annotation. The entry for this yearbook and a separate entry for P. S. Jones's essay "I" appear as follows in part 6 ("The History of Mathematics: Selected Topics; Mathematics Education; The Use of History in the Teaching of Mathematics"):

> 2165. Hallerberg, A., ed. *Historical Topics for the Mathematics Classroom*. (National Council of Teachers of Mathematics 31st Yearbook.) Washington, D.C.: 1969.
>
> A classic in sourcebooks for the school and college teacher. Contains surveys on number, computation, algebra, geometry, trigonometry, and many short "capsules" of information on a wide variety of topics.
>
> 2166. Jones, P. S. "The History of Mathematics as a Teaching Tool." In Hallerberg, item 2165, 1–17.

Dauben has divided the bibliography into six parts: General Reference Works; Source Materials; General Histories of Mathematics; The History of Mathematics: Chronological Periods; The History of Mathematics: Sub-Disciplines; The History of Mathematics: Selected Topics. Although many individuals contributed to this effort, Dauben has imposed uniformity on the entries, and informative paragraphs introduce some sections and subsections. Another contrast with May's *Bibliography* is the goal: instead of trying to be a window on all literature in the field, Dauben's *Selective Bibliography* attempts to provide sources that give basic historical introductions to topics. Hence, although there is overlap between these two bibliographies, Dauben's *Selective Bibliography* does not continue the work of May.

If these two bibliographies are not sufficient, other options are available, although sometimes they are less convenient. May had intended to continually update his *Bibliography* in the "Abstracts" section of the journal *Historia Mathematica*, which he founded in 1974. Almost every issue (there are four each year) contains abstracts of books and articles with complete entries and often annotations with key words highlighted. Like May's *Bibliography*, this abstracting service tries to cover all literature in the history of mathematics. However, this bibliographic source is inefficient to use: no cumulative list of the many abstracts has been published, and the entries are not organized in any particular way, although more recent issues are alphabetical by author and key terms have always been highlighted.

Less comprehensive coverage of the history of mathematics is given by the annual critical bibliographies of the journal *Isis* and the several volumes of the *Isis Cumulative Bibliography* that consolidate the annual critical bibliographies. This too is a continuing bibliographic service of *Isis*, the journal of the History of Science Society. They are organized by topics or time periods; mathematics appears separately among the topics and is mixed in among other sciences in some time periods. The citations in these bibliographies are complete and sometimes have annotations.

Several specialized bibliographies have historical sources identified. Gaffney and Steen's *Annotated Bibliography of Expository Writing in the Mathematical Sciences* has a section on history and one on biography. The International Study Group on the History and Pedagogy of Mathematics (HPM) includes a bibliography in each issue of its newsletter, and these have been consolidated into cumulative bibliographies. Citations in these include sources useful in using history to teach mathematics and range widely in quality and sophistication. May's *Index of The American Mathematical Monthly* gives the table of contents of each volume of the journal along with an author index and a subject index. For example, the subject index contains the following:

Napier bones (rods) 28-447 33-326.

In the table of contents for volume 28 (1921) is found

UHLER, H. S.—Multiplication of large numbers 447

and for volume 33 (1926)

VANHEE, Louis—Napier's rods in China 326.

The indexes to the *Mathematics Teacher* and to the *Arithmetic Teacher* have the history of mathematics as a broad category and only a few subcategories. Articles are listed with title and author. However, titles are sometimes not very descriptive of the content of articles, so judging appropriateness may require actually seeing the articles. These last three indexes all serve only one journal each; they are not appropriate for searching all the literature.

HISTORIES, SOURCE BOOKS, AND BIOGRAPHIES

Another major resource that has appeared since the 31st Yearbook was published is the *Dictionary of Scientific Biography* (DSB). It is listed in the original bibliography of this yearbook as forthcoming and is now complete. The DSB constitutes a work of the highest quality, but its usefulness goes beyond biography. In much historical research, biographical or not, individuals are identified, making it possible to consult the DSB. Many mathematicians are included in the DSB, and its articles almost always contain descriptions of technical contributions. Each article has a bibliography of both primary and secondary sources, often with references to analyses of technical work.

Among the general histories, Carl Boyer's *A History of Mathematics* has been reissued unchanged in paperback. Boyer drew on a wide learning, providing insight into both the technical details of mathematics and the broader philosophical context of the developments. He includes bibliographic sources both in footnotes and at the end of each chapter. A revised edition of Boyer by Uta Merzbach—Boyer is deceased—contains new material on nineteenth- and twentieth-century mathematics. A recent history, David Burton's *The History of Mathematics: An Introduction*, includes much more information about general history and bibliographic sources that complement those in Boyer. Howard Eves's *An Introduction to the History of Mathematics* continues to be updated, the fifth edition being the current one, with the sixth edition in press. Its level of difficulty is lower than that of Boyer, but Eves covers the late Middle Ages better and excels in providing problems. His chapter bibliographies give standard sources. Morris Kline produced a prodigious general history, *Mathematical Thought from Ancient to Modern Times*, not as a textbook but a book for the general book market. It is highly technical in most of its discussions and has found its place among the standard resources. Kline relied far more on primary sources than the other general histories, and this has produced some contrasting historical interpretations and generated historical discussions. Dirk Struik's *A Concise History of Mathematics* has been reissued in a fourth revised edition. As the title claims, it is a concise treatment and therefore foregoes detailed discussions. However, it discusses social and economic factors that have influenced the development of mathematics.

More specialized histories have also appeared. *The Historical Roots of Elementary Mathematics*, by Bunt, Jones, and Bedient, gives detailed discussions of aspects of history sometimes slighted in the general histories. *The Historical Development of the Calculus*, by C. H. Edwards, Jr.; *From the Calculus to Set Theory, 1630–1910*, edited by Ivor Grattan-Guinness; *Origins of Modern Algebra*, by Lubos Novy; and *A History of Algebra*, by B. L. van der Waerden, restrict their historical treatments to specific topics or time periods and provide more detailed discussions as a consequence. Edna Kramer's *The Nature and Growth of Modern Mathematics* and Ettore Caruccio's *Mathematics and Logic in History and in Contemporary Thought* /Caruccio/ deal with current mathematics. Howard Eves has also produced several smaller works that treat specific topics in separate essays; *Great Moments in Mathematics* in two volumes ("Before 1650" and "After 1650") focuses especially on the human dimension of important mathematical breakthroughs.

Source books in the history of mathematics have also proliferated. Midonick, Newman, and David Eugene Smith /Smith (c)/ appear in the original bibliography along with Struik, whose source book /Struik (e)/ has now been reissued in paperback. Add to these *A Source Book in Greek Science*, by Cohen and Drabkin; *Greek Mathematical Works*, by Ivor Thomas; *A Source Book in Medieval Science*, by Edward Grant; *A Source Book in Classical Analysis*, by Garrett Birkhoff; and *From Frege to Gödel, A Source Book in Mathematical Logic, 1879–1931*, by Jean van Heijenoort. Two more recent source books collect together many of the items in the earlier source books and supplement them with commentary, additional translations or items from other sources, or source material from histories and a variety of other sources: *Classics of Mathematics*, edited by Ronald Calinger; and *The History of Mathematics: A Reader*, edited by Fauvel and Gray. In the spirit of Newman's *The World of Mathematics* is Campbell and Higgins's three-volume *Mathematics: People, Problems, Results;* it complements the Newman anthology by bringing together a variety of readings, many from the twentieth century.

RESEARCH STRATEGY: GOING BEYOND THIS BIBLIOGRAPHY

Even though the resources in the history of mathematics have expanded significantly since this yearbook was first published, researching a given topic may not be easy. There is no "algorithm" for finding the appropriate literature; research remains a mix of plunging in and reading, taking notes, building bibliography, more reading, more bibliography, more notes, writing a portion, and so on. Organizing this process requires an "information storage and retrieval system," such as that described in May's *Bibliography* or some other. It should be able to accommodate this organic process, which continually shifts back and forth among several kinds of activity; it certainly must not hinder this basic character of research. With a filing and note-taking system, there are strategies for starting a research project, the following being but one.

Once a topic is identified, this yearbook should be consulted for appropriate capsules. Next, consult general and special histories of mathematics; however, it may be necessary to guess which citations apply to the topic, since bibliographies at the ends of chapters usually do not indicate this.

Mathematics encyclopedias have not been strong in history. However, if the topic is not too arcane, general encyclopedias can be one of the better places to search. In addition to giving an overview of the topic, good bibliographic sources may be included, and these sources in turn could lead to other sources. Moreover, the better encyclopedias have been successful in getting recognized experts to write articles, including good historians in some cases. But the quality of encyclopedias varies: the *Encyclopedia Britannica* and the *Encyclopedia Americana* are two that have traditionally supplied reliable information, sometimes with a high level of sophistication; the *World Book Encyclopedia* has used recognized historians for its articles on the history of mathematics, although they are written at a generally lower level of difficulty.

Proper names are very helpful with any research topic because biographical material is much more extensively indexed. As the example above shows, references to Napier's rods can only be found in the "Biography" section of May's *Bibliography*. Proper names also permit use of the DSB. The annual critical bibliographies of *Isis*, in addition to being organized by time periods, also have a name index in the back of each issue; proper names make this resource much more efficient to use. The other bibliographies described above are also essential; consulting them should make the search process much easier.

Bibliography is at the heart of the process of communicating knowledge, especially in a careful or scholarly way. Very often it is through bibliography that a reader acquires understanding: the bibliography leads to the sources of information or interpretation where more detailed treatment may be found. Moreover, bibliography permits the reader to replicate the writer's research techniques and conclusions, and replication of results is the hallmark of scholarly work.

Bibliography A

Organizations and periodicals to which frequent reference is made are indicated by abbreviations as follows:

AAAS	American Association for the Advancement of Science
AMM	*American Mathematical Monthly*
AT	*Arithmetic Teacher*
MAA	Mathematical Association of America
MT	*Mathematics Teacher*
NCTM	National Council of Teachers of Mathematics
NMM	*National Mathematics Magazine*
SMSG	School Mathematics Study Group
SSM	*School Science and Mathematics*

AAAS, Cooperative Committee on the Teaching of Science and Mathematics. "Preparation of High School Science Teachers," *Science*, CXXXI (April 8, 1960), 1024–29.

AABOE, ASGER (a). "Al-Kashi's Iteration Method," *Scripta Mathematica*, XX (March-June 1954), 24–29.

——— (b). *Episodes from the Early History of Mathematics.* ("New Mathematical Library.") New York: Random House, L. W. Singer Co., 1964.

AL-KHOWARIZMI. *See* KHOWARIZMI, AL-.

APOLLONIUS OF PERGA (a). "On Conic Sections." Translated by R. CATESBY TALIAFERRO, ("Great Books of the Western World," XI, 595–804.) Chicago: Encyclopaedia Britannica, 1952.

——— (b). *Treatise on Conic Sections*, ed. THOMAS LITTLE HEATH. 1896. New York: Barnes & Noble, n.d.

ARCHIBALD, RAYMOND CLAIRE (a). "Golden Section; A Fibonacci Series," *AMM*, XXV (1918, 232–38.

——— (b). *Outline of the History of Mathematics.* Buffalo, N.Y.: MAA, 1949.

ARCHIMEDES. *Works* [including *The Method of Archimedes*], ed. THOMAS LITTLE HEATH. 1897, 1912. Reprint. New York: Dover Publications, 1953.

BALL, WALTER WILLIAM ROUSE (a). *A Short Account of the History of Mathematics* (4th ed.). New York: Dover Publications, 1960.

——— (b). *String Figures* (3d ed., 1929). Reprinted in *String Figures and Other Monographs* (3d ed.). New York: Chelsea Publishing Co., 1960.

BARAVALLE, HERMANN VON (a). "The Geometry of the Pentagon and the Golden Section," *MT*, XLI (January 1948), 22–31.

―――― (b). "The Number *e*—the Base of Natural Logarithms," *MT*, XXXVIII (December 1945), 350–55.

―――― (c). "The Number π," *MT*, XLV (May 1952), 340–48. (Reprinted in *MT*, LX [May 1967], 479–87.)

BARDIS, PANOS D. "Evolution of Pi: An Essay in Mathematical Progress from the Great Pyramid to ENIAC," *SSM*, LX (January 1960), 73–78.

BARNETT, I. A. (a). "Mathematics as an Art—the Higher Arithmetic," *MT*, LXI (April 1968), 424–31.

―――― (b). *Some Ideas About Number Theory.* Washington, D. C.: NCTM, 1961.

BARZUN, JACQUES MARTIN. *Teacher in America.* New York: Doubleday & Co., 1954.

BAUMGART, JOHN K. "Axioms in Algebra—Where Do They Come From?" *MT*, LIV (March 1961), 155–59.

BECK, ANATOLE; BLEICHER, MICHAEL N.; and CROWE, DONALD W. *Excursions into Mathematics.* New York: Worth Publishers, 1969.

BECKER, JERRY P. "On Solutions of Geometrical Constructions Utilizing the Compasses Alone," *MT*, LVII (October 1964), 398–403.

BEILER, ALBERT H. *Recreations in the Theory of Numbers: The Queen of Mathematics Entertains.* London: Constable & Co., 1964; New York: Dover Publications, 1964.

BELL, ERIC TEMPLE (a). *The Development of Mathematics* (2d ed.). New York: McGraw-Hill Book Co., 1945.

―――― (b). *The Last Problem.* New York: Simon & Schuster, 1961.

―――― (c). *Mathematics, Queen and Servant of Science.* New York: McGraw-Hill Book Co., 1951.

―――― (d). *Men of Mathematics.* New York: Simon & Schuster, 1937.

BERGAMINI, DAVID, et al. *Mathematics.* ("Life Science Library.") New York: Time-Life Books, 1963.

BERNSTEIN, JEREMY. *The Analytical Engine: Computers—Past, Present, and Future.* New York: Random House, 1964.

BIDWELL, JAMES K. "Mayan Arithmetic," *MT*, LX (November 1967), 762–68.

BIRUNI, AL- (a). *Al-Biruni on Transits.* Translated by MOHAMMAD SAFFOURI and ADNAN IFRAM, commentary by E. S. KENNEDY. ("Oriental Series No. 32.") Beirut: University of Beirut, 1959.

―――― (b). *Rasa'il'u'l-Biruni.* Hyderabad, Deccan: Osmania Oriental Publications Bureau, 1948.

BLY, F. VAN DER. "Combinatorial Aspects of the Hexagrams in the Chinese Book of Changes," *Scripta Mathematica*, May 1967, pp. 37–49.

BOCHNER, SALOMON. *The Role of Mathematics in the Rise of Science.* Princeton, N. J.: Princeton University Press, 1966.

BONOLA, ROBERTO. *Non-Euclidean Geometry.* Translated by H. S. CARSLAW. 1911. New York: Dover Publications, 1955.

500

BOOLE, GEORGE. *An Investigation into the Laws of Thought.* 1854. New York: Dover Publications, 1953. Published also as Vol. II of *Logical Works.* La Salle, Ill.: Open Court Publishing Co., n.d.

BOURBAKI, NICOLAS. *Éléments d'histoire des mathématiques.* Paris: Hermann, 1960.

BOWDEN, BERTRAM VIVIAN, ed. *Faster than Thought: A Symposium on Digital Computing Machines.* New York: Pitman Publishing Corp., 1953.

BOYER, CARL B. (a). "Cardan and the Pascal Triangle," *AMM*, LVII (June-July 1950), 387–90.

—— (b). *The Concepts of the Calculus.* New York: Hafner Publishing Co., 1949. *See* (f).

—— (c). "The First Calculus Textbooks," *MT*, XXXIX (April 1946), 159–67.

—— (d). "Fractional Indices, Exponents, and Powers," *NMM*, XVIII (November 1943), 81–86.

—— (e). *History of Analytic Geometry.* New York: Scripta Mathematica, 1956.

—— (f). *The History of the Calculus and Its Conceptual Development.* New York: Dover Publishing Co., 1959. (Reprint of *The Concepts of the Calculus,* 1949.) *See* (b).

—— (g). *A History of Mathematics.* New York: John Wiley & Sons, 1968.

—— (h). "Mathematical Inutility and the Advance of Science," *Science,* CXXX (July 3, 1959), 22–25.

—— (i). "Myth, the Muse, and Mathesis," *MT*, LVII (April 1964), 242–53.

—— (j). "Note on Egyptian Numeration," *MT*, LII (February 1959), 127–29.

—— (k). "Viète's Use of Decimal Fractions," *MT*, LV (February 1962), 123–27.

—— (l). "Zero: The Symbol, the Concept, the Number," *NMM*, XVIII (May 1944), 323–30.

BRADFIELD, DONALD L. "The Majesty of Numbers," *MT*, LX (October 1967), 588–92.

BRAHMAGUPTA. *Uttera Khandakhadyaka.* Translated by P. C. SENGUPTA. Calcutta: University of Calcutta, 1934.

BRAUNMUHL, ANTON VON. *Vorlesungen über Geschichte der Trigonometrie.* 2 vols. Leipzig: B. G. Teubner, 1900–1903.

BRENDAN, BROTHER T. "How Ptolemy Constructed Trigonometry Tables," *MT*, LVIII (February 1965), 141–49.

BUNT, LUCAS N. H. "The Training of a Mathematics Teacher in the Netherlands," *AMM*, LXX (June-July 1963), 660–64.

BURGESS, E., tr. *Surya Siddhanta.* Calcutta: University of Calcutta, 1935. (Reprinted from *Journal of the American Oriental Society,* VI [1860], 141–498.)

CAJORI, FLORIAN (a). "The Evolution of Our Exponential Notation," *SSM,* XXIII (June 1923), 573–81.

—— (b). *A History of Elementary Mathematics.* 1917. New York: Macmillan Co., 1950.

—— (c). *A History of the Logarithmic Slide Rule.* 1909. Reprinted in W. W. Ball *et al., String Figures and Other Monographs* (3d ed.). New York: Chelsea Publishing Co., 1959.

—— (d). *A History of Mathematical Notations* (2d ed.). 2 vols. La Salle, Ill.: Open Court Publishing Co., 1951, 1952.

—— (e). *A History of Mathematics* (2d ed., rev. and enl.). 1919. New York: Macmillan Co., 1926.

CANTOR, GEORG. *Contributions to the Founding of the Theory of Transfinite Numbers.* Translated by PHILIP E. B. JOURDAIN. ("Open Court Series of Classics of Science and Philosophy," No. 1.) La Salle, Ill.:)pen Court Publishing Co., 1915; New York: Dover Publications, n.d.

CANTOR, MORITZ. *Vorlesungen über Geschichte der Mathematik.* 4 vols. 1900–1908. New York: Johnson Reprint Co., n.d.

CARDANO, GIROLAMO. *The Great Art or the Rules of Algebra (Ars Magna).* Translated and edited by T. RICHARD WITMER. Cambridge, Mass.: M.I.T. Press, 1968.

CARNAHAN, W. H. (a). "Geometric Solution of Quadratic Equations," *SSM,* XLVII (October 1947), 687–92.

—— (b). "History of Algebra," *SSM,* XLVI (January, February 1946), 7–12, 125–30.

CARRUCCIO, ETTORE. *Mathematics and Logic in History and in Contemporary Thought.* Chicago: Aldine Publishing Co., 1964.

CARSLAW, HORATIO SCOTT. *The Elements of Non-Euclidean Plane Geometry and Trigonometry.* 1916. Reprinted in W. W. Ball *et al., String Figures and Other Monographs* (3d ed.). New York: Chelsea Publishing Co., 1959.

CHACE, A. B., *et al.,* eds. *The Rhind Mathematical Papyrus.* 2 vols. Buffalo, N.Y.: MAA, 1927, 1929.

CHENEY, WILLIAM FITCH, JR. "Can We Outdo Mascheroni?" *MT,* XLVI (March 1953), 152–56.

CONANT, LEVI LEONARD. *The Number Concept, Its Origin and Development.* New York: Macmillan Co., 1931.

COOLIDGE, JULIAN LOWELL (a). *A History of the Conic Sections and Quartic Surfaces.* Toronto: Oxford University Press, 1945; New York: Dover Publication, n.d.

—— (b). *A History of Geometrical Methods.* 1940. New York: Dover Publications, 1963.

—— (c). *The Mathematics of Great Amateurs.* 1949. New York: Dover Publications, 1963.

—— (d). "The Number *e*," *AMM,* LVII (November 1950), 591–602.

COURANT, RICHARD, and ROBBINS, HERBERT E. *What Is Mathematics?* New York: Oxford University Press, 1941.

COURT, NATHAN ALTSHILLER. "Mascheroni Constructions," *MT,* LI (May 1958), 370–72.

COXETER, H. S. M. "The Problem of Apollonius," *AMM*, LXXV (January 1968), 5–15.

Cumulative Index: The Mathematics Teacher, 1908–1965. Washington, D. C.: NCTM, 1967.

DANTZIG, TOBIAS (a). *The Bequest of the Greeks.* New York: Charles Scribner's Sons, 1955.

———— (b). *Number, the Language of Science* (4th ed., rev. and enl.). New York: Macmillan Co., 1954; Doubleday & Co. ("Anchor Books"), 1956.

DATTA, B., and SINGH, A. N. *History of Hindu Mathematics.* Bombay: Asia Publishing House, 1962.

DAVIS, HAROLD THAYER, *et al. Tables of the Mathematical Functions.* 3 vols. San Antonio, Tex.: Trinity University Press. Vols. I and II, 1933, rev. 1963; Vol. III, *Arithmetical Tables,* 1962.

DAVIS, PHILIP J. (a). "Leonhard Euler's Integral: A Historic Profile of the Gamma Function," *AMM*, LXVI (December 1959), 849–69.

———— (b). *The Lore of Large Numbers.* ("New Mathematical Library.") New York: Random House, L. W. Singer Co., 1961.

DE SANTILLANA, GIORGIO. *The Origins of Scientific Thought.* Chicago: University of Chicago Press, 1961; New York: New American Library of World Literature, n.d.

DESCARTES, RENÉ (a). *Discourse on Method.* 1637. Translated by JOHN VEITCH. ("Open Court Classics.") La Salle, Ill.: Open Court Publishing Co., 1960.

Translated by ARTHUR WOLLASTON. Baltimore, Md.: Penguin Books, 1960; New York: Simon & Schuster, Washington Square Press, n.d., and E. P. Dutton & Co., n.d.

———— (b). *Geometry.* 1637. Translated by DAVID EUGENE SMITH and MARCIA L. LATHAM. La Salle, Ill.: Open Court Publishing Co., 1912; New York: Dover Publications, 1954.

DIANA, LIND MAE. "The Peruvian Quipu," *MT*, LX (October 1967), 623–28.

DICKSON, LEONARD EUGENE (a). *Elementary Theory of Equations.* New York: John Wiley & Sons, 1914.

———— (b). *First Course in the Theory of Equations.* New York: John Wiley & Sons, 1922.

———— (c). *History of the Theory of Numbers.* 3 vols. 1919. New York: Chelsea Publishing Co., 1952.

Dictionary of Scientific Biography. New York: Charles Scribner's Sons, forthcoming.

DIOPHANTUS. *Diophante d'Alexandrie.* Translated, with introduction and notes, by PAUL VER EECKE. Brussels: Desclée de Brouwer, 1926.

DODD, W. A. "A New Qualification for Mathematics Teachers in the United Kingdom." *MT,* LVI (May 1963), 311–13.

DODGSON, CHARLES LUDWIDGE. *Euclid and His Modern Rivals.* London, 1879.

DONNAY, JOSEPH D. H. *Spherical Trigonometry After the Cesàro Method.* New York: Interscience Publishers, 1945.

DÖRRIE, HEINRICH. *100 Great Problems of Elementary Mathematics: Their History and Solution*. 1958. Translated by DAVID ANTIN. New York: Dover Publications, 1965.

DUNNINGTON, G. WALDO. *Carl Friedrich Gauss: Titan of Science*. New York: Hafner Publishing Co., 1955.

EELLS, WALTER CROSBY. "One Hundred Eminent Mathematicians," *MT*, LV (November 1962), 582–88.

Encyclopédie des Sciences Mathématiques. Paris: Hermann, 1904.

ESPENSHADE, PAMELA H. "A Text on Trigonometry by Levi ben Gerson (1288–1344)," *MT*, LX (October 1967), 628–37.

EUCLID. *The Thirteen Books of Euclid's Elements*. Translated by THOMAS LITTLE HEATH. 3 vols. 1908. New York: Dover Publications, 1956.

EVANS, GEORGE W. "Some of Euclid's Algebra," *MT*, XX (March 1927), 127–41. (Reprinted in *MT*, LXI [April 1968], 405–14.)

EVES, HOWARD (a). "A Comment on Professor Charles L. Smith's Paper 'On the Origin of ">" and "<","'" *MT*, LVII (November 1964), 481.

—— (b). "A History-of-Mathematics Time Strip," *MT*, LIV (October 1961), 452–54.

—— (c). *An Introduction to the History of Mathematics* (rev. ed.). New York: Holt, Rinehart & Winston, 1964.
  [A fifth edition appeared in 1983. It was not practicable to insert new page numbers in every reference to this book. However, 5th edition page numbers replace 3rd edition page numbers, which were used wherever the book was listed under "For Further Reading" at the end of each capsule.]

—— (d). "The Names 'Ellipse,' 'Parabola,' and 'Hyperbola,'" *MT*, LIII (April 1960), 280–81.

—— (e). "Omar Khayyam's Solution of Cubic Equations," *MT*, LI (April 1958), 285–86.

—— (f). "The Prime Numbers," *MT*, LI (March 1958), 201–3.

—— (g). *A Survey of Geometry*. 2 vols. Boston: Allyn & Bacon, 1963, 1964.

—— (h). *In Mathematical Circles*. Boston: Prindle, Weber & Schmidt, forthcoming.

FELDMAN, RICHARD W., JR. (a). "The Cardano-Tartaglia Dispute," *MT*, LIV (March 1961), 160–63.

—— (b). "History of Elementary Matrix Theory, I–VI," *MT*, LV (October–December 1962), 482–84, 589–90, 657–59; LVI (January–March 1963), 37–38, 101–2, 163–64.

FINK, KARL. *A Brief History of Mathematics*. Translated by WOOSTER WOODRUFF BERNAN and DAVID EUGENE SMITH. ("Open Court Mathematical Series.") La Salle, Ill.: Open Court Publishing Co., 1900.

FISCHER, IRENE (a). "How Far Is It from Here to There?" *MT*, LVIII (February 1965), 123–30.

—— (b). "The Shape and Size of the Earth," *MT*, LX (May 1967) 508–16.

FISHER, R. A. "The Sieve of Eratosthenes," *Mathematical Gazette*, XIV (1929), 546–66.

FREEMAN, H. A. *A History of Mathematics*. New York: Macmillan Co., 1961.

FREITAG, HERTA T., and FREITAG, ARTHUR H. (a). *The Number Story*. Washington, D. C.: NCTM, 1960.

———— (b). "Using the History of Mathematics in Teaching on the Secondary School Level," *MT*, L (March 1957), 220–24.

FREUDENTHAL, HANS. "The Main Trends in the Foundations of Geometry in the Nineteenth Century," in *Logic, Methodology, and Philosophy of Science*, ed. ERNEST NAGEL, PATRICK SUPPES, and ALFRED TARSKI. Stanford, Calif.: Stanford University Press, 1962.

FRY, THORTON C. "Mathematics as a Profession Today in Industry," *AMM*, LXIII (February 1956), 71–80.

FUNKHOUSER, H. GRAY. "A Short Account of the History of Symmetric Functions of Roots of Equations," *AMM*, XXXVII (August-September 1930), 357–65.

GANDZ, S. "The Origin of the Term 'Algebra,' " *AMM*, XXXIII (November 1926), 437–40.

GARDNER, MARTIN (a). "Mathematical Games," *Scientific American*, CCI (August 1959), 128–34.

———— (b). *The Scientific American Book of Mathematical Puzzles and Diversions*. New York: Simon & Schuster, 1959; paper, 1965.

———— (c). *The 2nd Scientific American Book of Mathematical Puzzles and Diversions*. New York: Simon & Schuster, 1961; paper, 1965.

———— (d). *Martin Gardner's New Mathematical Diversions*. New York: Simon & Schuster, 1966.

GILLINGS, R. J. (a). "Problems 1 to 6 of the Rhind Mathematical Papyrus," *MT*, LV (January 1962), 61–69.

———— (b). "The Remarkable Mental Arithmetic of the Egyptian Scribes, I and II," *MT*, LIX (April, May 1966), 372–81, 476–84.

———— (c). " 'Think-of-a-Number' Problems 28 and 29 of the Rhind Mathematical Papyrus (B.M. 10057–8)," *MT*, LIV (February 1961), 97–100.

———— (d). "The Volume of a Truncated Pyramid in Ancient Egyptian Papyri," *MT*, LVII (December 1964), 552–55.

GRAESSER, R. F. "Archytas' Duplication of the Cube," *MT*, XLIX (May 1956), 393–95.

GREENE, ELNICE E. "Loan Words in Mathematics," *MT*, LV (October 1962), 484–89.

GREITZER, SAMUEL L. "Credit Where Credit Is Due!" *MT*, LX (February 1967), 155–57.

Growth of Mathematical Ideas, Grades K–12. Twenty-fourth Yearbook of the NCTM. Washington, D. C.: NCTM, 1959.

GROZA, VIVIAN SHAW. *A Survey of Mathematics—Elementary Concepts and Their Historical Development*. New York: Holt, Rinehart & Winston, 1968.

505

GUGGENBUHL, LAURA (a). "Mathematics in Ancient Egypt: A Checklist (1930–65)," *MT*, LVIII (November 1965), 630–34.

―――― (b). "The New York Fragments of the Rhind Mathematical Papyrus," *MT*, LVII (October 1964), 406–10.

HADAMARD, JACQUES SALOMON. *Essay on the Psychology of Invention in the Mathematical Field*. 1945. New York: Dover Publications, 1954.

HAGA, ENOCH J. "History of Digital Computing Devices," *SSM*, LXII (March 1962), 197–205.

HALLERBERG, ARTHUR E. (a). "The Geometry of the Fixed-Compass," *MT*, LII (April 1959), 230–44.

―――― (b). "Georg Mohr and *Euclidis Curiosi*," *MT*, LIII (February 1960), 127–32.

HARDGROVE, CLARENCE ETHEL, and MILLER, HERBERT F. *Mathematics Library—Elementary and Junior High School* (2d ed.). Washington, D. C.: NCTM, 1968.

HART, PHILIP J. "Pythagorean Numbers," *MT*, XLVII (January 1954), 16–21.

HARTNER, W. "Astrolabe," in *Encyclopedia of Islam*, Vol. I, ed. B. LEWIS and JOSEPH SCHACT. New York: Humanities Press, 1960.

HEATH, THOMAS LITTLE (a). *Diophantus of Alexandria: A Study in the History of Greek Algebra*. 1910. New York: Dover Publications, 1964.

―――― (b). *A History of Greek Mathematics*. 2 vols. London and New York: Oxford University Press, 1921. *See* (c).

―――― (c). *A Manual of Greek Mathematics* (an abridged form of *A History of Greek Mathematics*). New York: Oxford University Press, 1931; Dover Publications, 1963.

HILBERT, DAVID. *The Foundations of Geometry*. Translated by E. J. TOWNSEND. ("Open Court Classics.") La Salle, Ill.: Open Court Publishing Co., 1950.

HLAVATY, JULIUS H. "Mascheroni Constructions," *MT*, L (November 1957), 482–87.

HOBSON, ERNEST WILLIAM. *Squaring the Circle: A History of the Problem*. 1913. Reprinted in *Squaring the Circle and Other Monographs*. New York: Chelsea Publishing Co., n.d.

HOFMANN, JOSEPH EHRENFRIED (a). *Classical Mathematics*. Translated by HENRIETTA O. MIDONICK. New York: Philosophical Library, 1959; London: Vision Press, 1960.

―――― (b). *The History of Mathematics*. Translated by FRANK GAYNOR and HENRIETTA O. MIDONICK. New York: Philosophical Library, 1957.

HOGBEN, LANCELOT. *Mathematics in the Making: Introduction to Techniques*. New York: Doubleday & Co., 1960.

HOOPER, ALFRED. *Makers of Mathematics*. New York: Random House, Vintage Books, 1948.

Insights into Modern Mathematics. Twenty-third Yearbook of the NCTM. Washington, D. C.: NCTM, 1927.

JAMES, ROBERT C., and JAMES, GLEN. *Mathematics Dictionary* (3d ed.). Princeton, N. J.: D. Van Nostrand Co., 1968.

JOHNSON, ROGER ARTHUR. *Advanced Euclidean Geometry*, ed. JOHN WESLEY YOUNG. 1929. Reprint. New York: Dover Publications, 1960; Toronto. McClelland & Stewart, 1960.

JONES, PHILLIP S. (a). "Angular Measure—Enough of Its History to Improve Its Teaching," *MT*, XLVI (October 1953), 419–26.

———— (b). "The Binary System," *MT*, XLVI (December 1953), 575–77.

———— (c). "Complex Numbers: An Example of Recurring Themes in the Development of Mathematics—I–III," *MT*, XLVII (February, April, May 1954), 106–14, 257–63, 340–45.

———— (d). "The History of Mathematics as a Teaching Tool," *MT*, L (January 1957), 59–64.

———— (e). "Irrationals or Incommensurables, I–V," *MT*, XLIX (February–April, October, November 1956), 123–27, 187–91, 282–85, 469–71, 541–43.

———— (f). "The Pythagorean Theorem," *MT*, XLIII (April 1950), 162–63.

———— (g). "Recent Discoveries in Babylonian Mathematics, I—Zero, Pi, and Polygons," *MT*, L (February 1957), 162–65.

———— (h). "Tangible Arithmetic, I—Napier's and Genaille's Rods," *MT*, XLVII (November 1954), 482–87.

———— (i). "Tangible Arithmetic, IV—Finger Reckoning and Other Devices," *MT*, XLVIII (March 1955), 153–57.

———— (j). "Word Origins," *MT*, XLVII (March 1954), 195–96.

KARPINSKI, LOUIS CHARLES. *The History of Arithmetic*. Chicago: Rand Mc-Nally & Co., 1925. New York: Russell & Russell Publishers, 1965.

KEMENY, JOHN G. "Report to the International Congress of Mathematicians," *MT*, LVI (February 1963), 66–78.

KENNEDY, E. S. (a). "Biruni's Graphical Determination of Local Meridian," *Scripta Mathematica*, XXIV (1959), 251–55.

———— (b). "A Letter of Jamshid al-Kashi to His Father—Scientific Research and Personalities at a Fifteenth Century Court," *Orientalia*, XXIX (1960), 191–213.

KENNEDY, E. S., and TRANSUE, W. R. "A Medieval Iterative Algorism," *AMM*, LXIII (February 1956), 80–83.

KHOWARIZMI, AL-. *Robert of Chester's Latin Translation of the Algebra of al-Khowarizmi*, ed. LOUIS CHARLES KARPINSKI. New York: Macmillan Co., 1915.

KING, AMY C., and READ, CECIL B. (a). *Pathways to Probability*. ("Holt Library of Science.") New York: Holt, Rinehart & Winston, 1963.

———— (b). "A Study, Through Biographies, of the History of the Elementary Concepts of Probability," *SSM*, LXII (March 1962), 165–76.

KING, CHARLES. "Leonardo Fibonacci," *Fibonacci Quarterly*, I (December 1963), 15–17.

KINSELLA, JOHN, and BRADLEY, A. DAY. "The Mayan Calendar," *MT*, XXVII (November 1934), 340–43.

BIBLIOGRAPHY A

KLEIN, FELIX (a). *Elementary Mathematics from an Advanced Viewpoint.* Translation from 3d German ed. by E. R. HEDRICK and C. A. NOBLE. 2 vols. I, *Arithmetic, Algebra, Analysis;* II, *Geometry.* 1932, 1939. New York: Dover Publications, n.d.

———— (b). *Famous Problems of Elementary Geometry.* Translated by WOOS-TER WOODRUFF BEMAN and DAVID EUGENE SMITH. 1930. Reprint. New York: Dover Publications, 1956.

———— (c). *Vorlesungen über die Entwicklung der Mathematik im 19. Jahr-hundert.* 2 vols. New York: Chelsea Publishing Co., 1956.

KLEIN, JACOB. *Greek Mathematical Thought and the Origin of Algebra.* Trans-lated by EVA BRANN. Cambridge, Mass.: M.I.T. Press, 1968.

KLINE, MORRIS (a). *Mathematics, a Cultural Approach.* Reading, Mass.: Addison-Wesley Publishing Co., 1962.

———— (b). *Mathematics in Western Culture.* New York: Oxford University Press, 1953.

————, ed. (c). *Mathematics in the Modern World: Readings from "Scientific American."* San Francisco: W. F. Freeman & Co., 1968.

KOJIMA, TAKASHI. *Advanced Abacus.* Rutland, Vt.: Charles E. Tuttle Co., 1963.

KOLMOGOROV, ANDREI NIKOLOEVICH. *Foundations of the Theory of Probability* (2d ed.). New York: Chelsea Publishing Co., 1956.

KOSTOVSKII, ALEKSANDR NIKITICH. *Geometrical Constructions Using Com-passes Only.* Translated by HALINA MOSS. New York: Random House, 1961; Oxford: Pergamon Press, 1961.

KRAMER, EDNA E. *The Main Stream of Mathematics.* Greenwich, Conn.: Fawcett Publications, 1961.

KREITZ, HELEN MARIE, and FLOURNOY, FRANCES. "A Bibliography of Histor-ical Materials for Use in Arithmetic in the Intermediate Grades," *AT*, VII (October 1960), 287–92.

LARRIVEE, JULES A. "A History of Computers—I and II," *MT*, LI (October, November 1958), 469–73, 541–44.

LARSEN, HAROLD D. "The Witch of Agnesi," *SSM*, XLVI (January 1946), 57–62.

LEACH, E. R. "Primitive Time Reckoning," in *From Early Times to the Fall of Ancient Empires. (A History of Technology,* ed. CHARLES SINGER *et al.,* Vol. I.) London and New York: Oxford University Press, 1954.

LEAVITT, DUDLEY. *Pike's System of Arithmetic Abridged.* Concord, N.H.: J. B. Moore, 1832.

LOOMIS, ELISHA SCOTT. *The Pythagorean Proposition.* 1927, 1940. Reprint ("Classics in Mathematics Education," Vol. I). Washington, D. C.: NCTM. 1968.

MAA (a). Commission on the Training and Utilization of Advanced Students of Mathematics. "Report on the Training of Teachers of Mathematics," *AMM*, XLII (May 1935).

———— (b). Committee on the Undergraduate Program in Mathematics. "Recommendations for the Training of Teachers of Mathematics," *MT*, LIII (December 1960), 632–38, 643.

MacLane, Saunders. "Some Recent Advances in Algebra," *AMM*, XLVI (January 1939), 3–19.

Maish, A. M. "Letter to the Editor," *Mathematics Magazine*, XXIX (September-October 1955), 58.

Manheim, Jerome H. *The Genesis of Point Set Topology.* New York: Pergamon Press, 1964.

Marks, Robert W., ed. *The Growth of Mathematics.* New York: Bantam Books, 1964.

The Mathematical Sciences: A Collection of Essays. Committee on Support of Research in the Mathematical Sciences, ed. Cambridge, Mass.: M.I.T. Press, 1969.

May, Kenneth O. (a) "Mathematics and Art," *MT*, LX (October 1967), 568–72.

———— (b). "The Origin of the Four-Color Conjecture," *MT*, LX (May 1967), 516–19.

Menninger, Karl. *Number Words and Number Symbols.* Cambridge, Mass.: M.I.T. Press, 1969.

Merton, Robert K. (a). "Priorities in Scientific Discovery: A Chapter in the Sociology of Science," *American Sociological Review*, XII (1957), 635–59.

———— (b). "Singletons and Multiples in Scientific Discovery: A Chapter in the Sociology of Science," *Proceedings of the American Philosophical Society*, CV (1961), 470–86.

Meschkowski, Herbert (a). *Evolution of Mathematical Thought.* Translated by J. H. Gayl. San Francisco: Holden-Day, 1965.

———— (b). *Ways of Thought of Great Mathematicians.* Translated by J. Dyer-Bennet. San Francisco: Holden-Day, 1964.

Midonick, Henrietta O., ed. *The Treasury of Mathematics.* New York: Philosophical Library, 1965.

Mikami, Yoshio. *The Development of Mathematics in China and Japan.* 1913. New York: Chelsea Publishing Co., n.d.

Miller, George Abram (a). *Collected Works.* 5 vols. Urbana, Ill.: University of Illinois Press, 1935–58.

———— (b). "The Development of the Function Concept," *SSM*, XXVIII (1928), 506–16.

———— (c). "Historical Note on the Solution of Equations," *SSM*, XXIV (May 1924), 509–10.

Miller, G. H. "The Evolution of Group Theory," *MT*, LVII (January 1964), 26–30.

Milne-Thomson, Louis M. *The Calculus of Finite Differences.* London: Macmillan & Co., 1933; New York: St. Martin's Press, 1951.

Moore, Eliakim Hastings. "On the Foundations of Mathematics," *MT*, LX (April 1967), 360–74. (A reprinting of the 1902 address, first published in *Science*, 1903, and later included in the First Yearbook of the NCTM, 1926.)

BIBLIOGRAPHY A

MORITZ, ROBERT EDOUARD (a). "On Certain Proofs of the Fundamental Theorem of Algebra," *AMM*, X (1903), 159.

—— (b). *On Mathematics and Mathematicians* (originally entitled *Memorabilia Mathematica*). 1914. New York: Dover Publications, 1958.

MORRISON, PHILLIP, and MORRISON, EMILY. "The Strange Life of Charles Babbage," *Scientific American*, CLXXXVI (April 1952), 66–73.

MUIR, JANE. *Of Men and Numbers*. New York: Dodd, Mead & Co., 1961; Dell Publishing Co., 1962.

MURRAY, FRANCIS JOSEPH. *Mathematical Machines*, 2 vols. New York: Columbia University Press, 1961.

NAGEL, ERNEST; SUPPES, PATRICK; and TARSKI, ALFRED, eds. *Logic, Methodology, and Philosophy of Science*. Stanford, Calif.: Stanford University Press, 1962.

NEEDHAM, JOSEPH. *Mathematics and the Sciences of the Heavens and the Earth*. (*Science and Civilization in China*, Vol. III.) New York: Cambridge University Press, 1959.

NEUGEBAUER, OTTO (a). *The Exact Sciences in Antiquity* (2d ed.). Providence, R. I.: Brown University Press, 1957; New York: Harper & Bros., Harper Torchbooks ("Science Library"), 1962.

—— (b). *Quellen und Studien zur Geschichte der Mathematik, Astronomie, und Physik*. Division A, III, 180; Division B, I, 90. Berlin.

—— (c). "Über griechische Wetterseichen und Schattentafeln," Österrichische Akademie der Philologische-historische Klasse, Sitzungsberichte, 240, Abhandlung (1962), II, 29–44.

NEUGEBAUER, OTTO, and PARKER, RICHARD A., eds. *The Early Decans*. (*Egyptian Astronomical Texts*, Vol. I.) London, 1960. Providence, R. I.: Brown University Press, 1961.

NEUGEBAUER, OTTO, and SACHS, A. J., eds. *Mathematical Cuneiform Texts*. New Haven, Conn.: American Oriental Society and the American Schools of Oriental Research, 1945.

NEWMAN, JAMES ROY, ed. *The World of Mathematics*. 4 vols. New York: Simon & Schuster, 1956; paper, 1962.

NEWSOM, CARROLL V. *Mathematical Discourses: The Heart of Mathematical Science*. Englewood Cliffs, N. J.: Prentice-Hall, 1964.

NEWTON, ISAAC. *Unpublished Scientific Papers of Isaac Newton: A Selection from the Portsmouth Collection*, ed. ALFRED RUPERT HALL and MARIE BOAS HALL. New York: Cambridge University Press, 1962.

NICOMACHUS OF GERASA. *Introduction to Arithmetic*. Translated by MARTIN LUTHER D'OOGE, with studies in Greek arithmetic by FRANK EGLESTON ROBBINS and LOUIS CHARLES KARPINSKI. 1926. ("Great Books of the Western World," Vol. XI.) Chicago: Encyclopaedia Britannica, 1952.

NIVEN, IVAN (a). *Irrational Numbers*. New York: John Wiley & Sons, 1956.

—— (b). *Numbers: Rational and Irrational*. New York: Random House, L. W. Singer Co., 1961.

OMAR KHAYYAM. *The Algebra of Omar Khayyam*, ed. DAOUD S. KASIR. New York: Columbia Teachers College, 1931.

"On the Mathematics Curriculum of the High School," *MT*, LV (March 1962), 191–95.

ORE, OYSTEIN (a). *Cardano, the Gambling Scholar.* 1953. New York: Dover Publications, n.d.

—— (b). *Niels Henrik Abel, Mathematician Extraordinary.* Minneapolis: University of Minnesota Press, 1957.

—— (c). *Number Theory and Its History.* New York: McGraw-Hill Book Co., 1948.

—— (d). "Pascal and the Invention of Probability Theory," *AMM*, LXVII (May 1960), 409–19.

PHILLIPS, J. P. "Brachistochrone, Tautochrone, Cycloid—Apple of Discord," *MT*, LX (May 1967), 506–8.

PINGREE, DAVID. "Astronomy and Astrology in India and Iran," *Isis*, LIV (1963), 229–46.

POINCARÉ, HENRI. *Science and Method.* Translated by FRANCIS MAITLAND. 1914. New York: Dover Publications, n.d.

PRICE, D. J. "The Babylonian 'Pythagorean Triangle' Tablet," *Centaurus*, X (1964), 1–13.

PTOLEMY, CLAUDIUS (a). *The Almagest.* Translated by F. E. ROBBINS. In *Tetrabiblos*, bound with *Aegyptiaca*, etc., by MANETHO. ("Loeb Classical Library Series," No. 350.) Cambridge, Mass.: Harvard University Press, 1941.
Translated by R. CATESBY TALIAFERRO. ("Great Books of the Western World," Vol. XVI.) Chicago: Encyclopaedia Britannica, 1952.

—— (b). "Das Planisphaerium des Claudius Ptolemaeus," translated by J. DRECKER, *Isis*, IX (1927), 255–78.

RAAB, JOSEPH A. (a). "A Generalization of the Connection Between the Fibonacci Sequence and Pascal's Triangle," *Fibonacci Quarterly*, I, No. 3 (1963), 21–31.

—— (b). "The Golden Rectangle and Fibonacci Sequence, as Related to the Pascal Triangle," *MT*, LV (November 1962), 538–43.

RANSOM, WILLIAM R. " 'One Over . . . ,' " *MT*, LIV (February 1961), 100–101.

RAPPORT, SAMUEL, and WRIGHT, HELEN, eds. *Mathematics.* ("The New York University Library of Science.") New York: New York University Press, 1963; Simon & Schuster, Washington Square Press, 1964.

READ, CECIL B. (a). "Archimedes and His Sandreckoner," *SSM*, LXI (February 1961), 81–84.

—— (b). "Debatable or Erroneous Statements Relating to the History of Mathematics," *MT*, LXI (January 1968), 75–79.

—— (c). "Historical Oddities Relating to the Number π," *SSM*, LX (May 1960), 348–50.

—— (d). "The History of Mathematics: A Bibliography of Articles in English Appearing in Seven Periodicals," *SSM*, LXVI (February 1966), 147–79.

—— (e). "Varying Usage of the Concept of Function," *SSM*, LXIII (December 1963), 726.

REGIOMONTANUS [JOHANN MÜLLER]. *Regiomontanus on Triangles.* Translated by FATHER BARNABAS HUGHES. Madison, Wis.: University of Wisconsin Press, 1967.

RICHESON, A. W. "The Number System of the Mayas," *AMM*, XL (November 1933), 542–46.

ROBINSON, L. V. "Pascal's Triangle and Negative Exponents," *AMM*, LIV (November 1947), 540–41.

ROGERS, JAMES T. *The Pantheon Story of Mathematics for Young People.* New York: Pantheon Books, 1966.

ROLF, HOWARD L. "Friendly Numbers," *MT*, LX (February 1967), 157–60.

ROSENTHAL, ARTHUR. "The History of the Calculus," *AMM*, LVIII (February 1951), 75–86.

RUDNICK, JESSE A. "Numeration Systems and Their Classroom Roles," *AT*, XV (February 1968), 138–47.

SALMON, GEORGE. *Lectures* [formerly *Lessons*] *Introductory to the Modern Higher Algebra* (5th ed.). New York: Chelsea Publishing Co., 1964.

SALYERS, GARY D. "The Number System of the Mayas," *Mathematics Magazine*, XXVIII (September–October 1954), 44–48.

SANCHEZ, GEORGE I. *Arithmetic in Maya.* Private publication: 2201 Scenic Dr., Austin, Tex., 1961.

SANFORD, VERA (a). "Algorisms: Computing with Hindu-Arabic Numerals," *MT*, XLIV (February 1951), 135–37.

——— (b). "The Problem of the Lion in the Well," *MT*, XLIV (March 1951), 196–97.

——— (c). "Robert Recorde's *Whetstone of Witte*, 1557," *MT*, L (April 1957), 258–66.

——— (d). *A Short History of Mathematics.* Boston: Houghton Mifflin Co., 1930.

SARTON, GEORGE (a). *An Introduction to the History of Science.* 3 vols. Baltimore, Md.: Carnegie Institution of Washington, 1927, 1948.

——— (b). *The Study of the History of Mathematics.* 1936. New York: Dover Publications, 1957.

SCHAAF, WILLIAM L. (a). *The High School Mathematics Library* (3d ed.). Washington, D. C.: NCTM, 1968.

——— (b). "How Modern Is Modern Mathematics?" *MT*, LVII (February 1964), 89–97.

——— (c). "Logarithms and Exponentials," *MT*, XLV (May 1952), 361–63.

——— (d). *Recreational Mathematics* (3d ed.). Washington, D. C.: NCTM, 1963.

———, ed. (e). *Our Mathematical Heritage.* New York: Macmillan Co., Collier Books, 1963.

SCHRADER, DOROTHY V. (a). "The Arithmetic of the Medieval Universities," *MT*, LX (March 1967), 264–78.

——— (b). "*De arithmetica,* Book I, of Boethius," *MT*, LXI (October 1968), 615–28.

—— (c). "The Newton-Leibniz Controversy Concerning the Discovery of the Calculus," *MT*, LV (May 1962), 385–96.

SCHUMAKER, JOHN A. "Trends in the Education of Secondary School Mathematics Teachers," *MT*, LIV (October 1961), 413–22.

SCOTT, J. F. *A History of Mathematics from Antiquity to the Beginning of the Nineteenth Century.* London: Taylor & Francis, 1958.

SEIDENBERG, A. (a). "The Ritual Origin of Counting," *Archive for History of Exact Sciences*, II (1962), 1–40.

—— (b). "The Ritual Origin of Geometry," *Archive for History of Exact Sciences*, I (1961/62), 488–527.

Selected Topics in the Teaching of Mathematics. Third Yearbook of the NCTM. New York: Bureau of Publications, Teachers College, Columbia University, 1928.

SHANKS, DANIEL. *Solved and Unsolved Problems in Number Theory.* Washington, D. C.: Spartan Books, 1962.

SHELTON, JULIA B. "A History-of-Mathematics Chart," *MT*, LII (November 1959), 563–67.

SIERPINSKI, W. *Elementary Theory of Numbers.* Warsaw, Poland, 1964.

SISTER M. STEPHANIE. "Venn Diagrams," *MT*, LVI (February 1963), 98–101.

SISTER MARY CLAUDIA ZELLER. *The Development of Trigonometry from Regiomontanus to Piticus.* Ph.D. dissertation, 1944, University of Michigan. Joliet, Ill.: Sisters of St. Francis of Mary Immaculata, 1946.

SISTER MARY OF MERCY FITZPATRICK. "Saccheri, Forerunner of Non-Euclidean Geometry," *MT*, LVII (May 1964), 323–32.

SLAUGHT, H. E. "Eliakim Hastings Moore," *AMM*, XL (1933), 191–95.

SMAIL, LLOYD L. "Some Geometrical Applications of Complex Numbers," *AMM*, XXXVI (1929), 505.

SMELTZER, DONALD. *Man and Number.* New York: Emerson Books, 1958; Macmillan, Collier Books, n.d.

SMITH, CHARLES L. "On the Origin of '>' and '<,'" *MT*, LVII (November 1964), 479–81.

SMITH, DAVID EUGENE (a). *History of Mathematics.* 2 vols. 1923, 1925. New York: Dover Publications, 1958.

—— (b). *Number Stories of Long Ago.* Washington, D. C.: NCTM, 1919. Reprint ("Classics in Mathematics Education," Vol. II). Washington, D.C.: NCTM, 1969.

—— (c). *A Source Book in Mathematics.* 1929. Reprint (2 vols.). New York: Dover Publications, 1959.

SMITH, DAVID EUGENE, and GINSBURG, JEKUTHIEL. *Numbers and Numerals.* Washington, D. C.: NCTM, 1937.

SMITH, DAVID EUGENE, and KARPINSKI, LOUIS CHARLES. *The Hindu-Arabic Numerals.* Boston: Ginn & Co., 1911.

SMITH, DAVID EUGENE, and MIKAMI, YOSHIO. *A History of Japanese Mathematics.* La Salle, Ill.: Open Court Publishing Co., 1914.

SMSG. "Reprint Series," ed. WILLIAM SCHAAF. Pasadena, Calif.: A. C. Vroman, 1966, 1967, 1969.

I.   *Structure of Algebra*
II. *Prime Numbers and Perfect Numbers.*
III. *What Is Contemporary Mathematics?*
IV. *Mascheroni Constructions.*
V.   *Space, Intuition, and Geometry.*
VI. *Nature and History of π.*
VII. *Computation of π.*
VIII. *Mathematics and Music.*
IX. *The Golden Measure.*
X.   *Geometric Constructions.*
XI. *Memorable Personalities in Mathematics: Nineteenth Century.*
XII. *Memorable Personalities in Mathematics: Twentieth Century.*
XIII. *Finite Geometry.*
XIV. *Infinity.*
XV. *Geometry, Measurement, and Experience.*

STRUIK, DIRK J. (a). *A Concise History of Mathematics* (3d rev. ed.). New York: Dover Publications, 1967.

—— (b). "On Ancient Chinese Mathematics," *MT*, LVI (October 1963), 424–32.

—— (c). "The Origin of L'Hôpital's Rule," *MT*, LVI (April 1963), 257–60.

—— (d). "Simon Stevin and the Decimal Fractions," *MT*, LII (October 1959), 474–78.

—— (e). *A Source Book in Mathematics, 1200–1800.* Cambridge, Mass.: Harvard University Press, 1969.

SULLIVAN, JOHN WILLIAM NAVIN. *The History of Mathematics in Europe.* London: Oxford University Press, 1925.

*Surya Siddhanta. See /*BURGESS/.

Teaching Mathematics in Secondary Schools. (Ministry of Education Pamphlet No. 36.) Her Majesty's Stationery Office, 1958.

TELLER, JAMES D. "A Calendar of the Birthdays of Mathematicians," *MT*, XXXV (December 1942), 369–71.

TIETZE, HEINRICH. *Famous Problems of Mathematics.* Translation of 2d German ed. (1959) by BEATRICE KEVITT HOFSTADTER and HORACE KOMM. Baltimore, Md.: Graylock Press, 1965.

TODHUNTER, ISAAC. *A History of the Mathematical Theory of Probability.* 1865. New York: Chelsea Publishing Co., 1965.

TOEPLITZ, OTTO. *The Calculus: A Genetic Approach,* ed. GOTTFRIED KÖTHE. Translated by LUISE LANGE. ("Phoenix Science Series.") Chicago: University of Chicago Press, 1963.

TURNBULL, HERBERT WESTREN. *The Great Mathematicians.* 1929. New York: New York University Press, 1961; Simon & Schuster, 1962. (Reprinted also in/NEWMAN. I, 75–168/.)

VAN DER WAERDEN, B. L. *Science Awakening*. Translated by ARNOLD DRESDEN. New York: Oxford University Press, 1961; ("Science Editions") John Wiley & Sons, 1963.

VARBERG, DALE E. "The Development of Modern Statistics—I and II," *MT*, LVI (April, May 1963), 252–57, 344–48.

VOGEL, KURT. *Vorgiechische Mathematik*. Hannover: Hermann Shroedel Verlag, 1958.

VON NEWMANN, JOHN. *Collected Works*, ed. A. H. TAUB. 6 vols. New York: Pergamon Press, 1961.

VOROBYOV, N. N. *The Fibonacci Numbers*. Waltham, Mass.: Blaisdell Publishing Co., 1962; Boston: D. C. Heath & Co., 1963.

WATSON, GLEN, and CALHOUN, EDWARD C. "Brief History of Computers," *SSM*, LX (February 1960), 87–94.

WEIDNER, E. F. "Ein babylonisches Kompendium der Himmelskunde," *American Journal of Semitic Languages and Literatures*, XL (1923/24), 186–208.

WEYL, HERMANN. "A Half-Century of Mathematics," *AMM*, LVIII (October 1951), 523–53.

WHEELER, ROBERT ERIC MORTIMER. *Rome Beyond the Imperial Frontiers*. London: G. Bell & Sons, 1954; Toronto: Clarke, Irwin & Co., 1954.

WHITEHEAD, ALFRED NORTH. *Science and the Modern World*. New York: Macmillan Co., 1925; paper, Macmillan Co., Free Press, 1967.

WHITEHEAD, ALFRED NORTH, and RUSSELL, BERTRAND (a). *Principia mathematica* (2d ed.). 3 vols. London and New York: Cambridge University Press, 1925–27.

——— (b). *Principia mathematica: to *56* (an abbreviated text of Vol. I of *Principia mathematica*). London and New York: Cambridge University Press, 1962.

WIENER, NORBERT. *Cybernetics: Or Control and Communication in the Animal and the Machine* (2d ed.). Cambridge, Mass.: M.I.T. Press, 1961.

WILDER, RAYMOND L. *Evolution of Mathematical Concepts*. New York: John Wiley & Sons, 1968.

WILLERDING, MARGARET F. "Infinity and Its Presentation at the High School Level," *SSM*, LXIII (June 1963), 463–74.

WILLIAMS, TREVOR I., ed. *A Biographical Dictionary of Scientists*. New York: John Wiley & Sons, Wiley-Interscience, 1969.

WOLFE, HAROLD E. *Introduction to Non-Euclidean Geometry*. New York: Holt, Rinehart & Winston, 1945.

WOLFF, PETER. *Breakthroughs in Mathematics*. New York: New American Library of World Literature, 1963.

World Who's Who in Science, 1700 B.C.–1968 A.D. Chicago: A. N. Marquis Co., forthcoming.

WRENCH, J. W., JR. "The Evolution of Extended Decimal Approximations to π," *MT*, LIII (December 1960), 644–50.

YATES, ROBERT C. *The Trisection Problem.* Ann Arbor, Mich.: Edwards Bros., 1947.

YOUNG, JACOB WILLIAM ALBERT, ed. *Monographs on Topics of Modern Mathematics Relevant to the Elementary Field.* 1911. New York: Dover Publications, 1955.

YOUNG, JOHN WESLEY. *Lectures on Fundamental Concepts of Algebra and Geometry.* New York: Macmillan Co., 1911.

Bibliography B

Sources referred to in this appendix are listed here if they are not listed in Bibliography A or if a later edition has appeared.

Arcavi, Abraham. "History of Mathematics and Mathematics Education: A Suggested Bibliography." *Zentralblatt für Didaktik der Mathematik* 85:1 (1985): 26–29.

Bidwell, James, and Robert G. Clason, eds. *Readings in the History of Mathematics Education.* Washington, D.C.: National Council of Teachers of Mathematics, 1970.
A source book in mathematics education that complements the 32nd Yearbook. Cf. Jones, below.

Birkhoff, Garrett, ed. *A Source Book in Classical Analysis.* Cambridge: Harvard University Press, 1973.

Boyer, Carl B. *A History of Mathematics.* Princeton, N.J.: Princeton University Press, 1985.
This is an unchanged paperback edition of the original 1968 book issued by John Wiley & Sons.

Boyer, Carl B., and Uta Merzbach. *A History of Mathematics.* 2d ed. New York: John Wiley & Sons, 1989.
Merzbach's revision of the 1968 edition (see previous entry) with substantial changes in citations, bibliography, and nineteenth- and twentieth-century histories. The first twenty-two chapters are unchanged.

Bunt, Lucas N. H., Phillip S. Jones, and Jack D. Bedient. *The Historical Roots of Elementary Mathematics.* New York: Dover Publications, 1988.
Babylonian, Egyptian, and Greek sources of concepts in contemporary mathematics. A reissue of the 1976 edition.

Burton, David M. *The History of Mathematics: An Introduction.* Boston: Allyn & Bacon, 1985.

Calinger, Ronald, ed. *Classics of Mathematics*. Oak Park, Ill.: Moore Publishing Co., 1982.
A source book drawing together in one book readings from several earlier source books. (Current distributor: Interstate Printers & Publishers, Inc., 19 N. Jackson St., Danville, IL 61834.)

Campbell, Douglas M., and John C. Higgins, eds. *Mathematics: People, Problems, Results*. 3 vols. Belmont, Calif.: Wadsworth International, 1984.
An anthology of essays on mathematics very similar in spirit and appeal to Newman's *The World of Mathematics*.

Cohen, Morris R., and I. E. Drabkin. *A Source Book in Greek Science*. Cambridge: Harvard University Press, 1948.
Contains sections on "Mathematics," "Astronomy," "Mathematical Geography," and "Physics" among other sciences.

Cumulative Index: The Arithmetic Teacher, 1954–1973. Vols. 1–20. Reston, Va.: National Council of Teachers of Mathematics, 1974.

Cumulative Index: The Mathematics Teacher, 1908–1965. Vols. 1–58. Reston, Va.: National Council of Teachers of Mathematics, 1967.

Cumulative Index: The Mathematics Teacher, 1966–1975. Vols. 59–68. Reston, Va.: National Council of Teachers of Mathematics, 1976.

Dauben, Joseph W. *The History of Mathematics from Antiquity to the Present: A Selective Bibliography*. New York: Garland Press, 1985.

Dictionary of Scientific Biography. 16 vols. C. C. Gillispie, Editor-in-Chief. New York: Charles Scribner's Sons, 1970–1980.

Edwards, C. H., Jr. *The Historical Development of the Calculus*. New York: Springer-Verlag, 1979.
Beginning with incipient calculus concepts in Babylonian and Egyptian mathematics and continuing into this century, discusses technical details in modern notation. Exercises included.

Eves, Howard. *An Introduction to the History of Mathematics*. 5th ed. Philadelphia: Saunders College Publishing, 1983.
A general history particularly noteworthy for the number of student exercises included.

————. *Great Moments in Mathematics (Before 1650)*. Dolciani Mathematical Expositions, vol. 5. Washington, D.C.: Mathematical Association of America, 1980.
Twenty "lectures" each treating a single event in a form accessible to a general college audience with no mathematical prerequisites, designed as mathematics appreciation.

————. *Great Moments in Mathematics (After 1650)*. Dolciani Mathematical

517

Expositions, vol. 7. Washington, D.C.: Mathematical Association of America, 1983.

Twenty more "lectures" following the previous citation.

———. *In Mathematical Circles: A Selection of Mathematical Stories and Anecdotes*. Boston: Prindle, Weber & Schmidt, 1969.

———. *Mathematical Circles Revisited: A Second Collection of Mathematical Stories and Anecdotes*. Boston: Prindle, Weber & Schmidt, 1971.

———. *Mathematical Circles Squared: A Third Collection of Mathematical Stories and Anecdotes*. Boston: Prindle, Weber & Schmidt, 1972.

———. *Mathematical Circles Adieu: A Fourth Collection of Mathematical Stories and Anecdotes*. Boston: Prindle, Weber & Schmidt, 1977.

———. *Return to Mathematical Circles: A Fifth Collection of Mathematical Stories and Anecdotes*. Boston: PWS-Kent Publishing Co., 1988.

Fauvel, John, and Jeremy Gray, eds. *The History of Mathematics: A Reader*. London: Macmillan Press, 1987.

An anthology of readings used as a text in the Open University in Great Britain; contains both primary and secondary sources.

Gaffney, Matthew P., and Lynn Arthur Steen. *Annotated Bibliography of Expository Writing in the Mathematical Sciences*. With the assistance of Paul J. Campbell. Washington, D.C.: Mathematical Association of America, 1976.

Grant, Edward, ed. *A Source Book in Medieval Science*. Cambridge: Harvard University Press, 1974.

Several selections in mathematics, as well as other exact sciences such as astronomy, physics, and optics.

Grattan-Guinness, I., ed. *From the Calculus to Set Theory, 1630–1910*. London: Gerald Duckworth & Co., 1980.

Six chapters, each written by a different historian; included are H. J. M. Bos, R. Bunn, Joseph W. Dauben, Thomas Hawkins, I. Grattan-Guinness, and Kristi Møller Pedersen.

Historia Mathematica. International Journal of History of Mathematics. New York: Academic Press, 1974–.

Research journal published four times a year. Each issue contains bibliography in abstracts section.

Howson, A. G. *A History of Mathematics Education in England*. New York: Cambridge University Press, 1982.

Traces history through biographies of nine people, including Robert Recorde, Samuel Pepys, and Augustus De Morgan.

HPM. International Study Group on the Relations between History and Pedag-

ogy of Mathematics. *Newsletter*. Muncie, Ind: Ball State University (Department of Mathematical Sciences).
Newsletter for teachers at all levels interested in using history in teaching.

Isis. International Review Devoted to the History of Science and Its Cultural Influences. Official Journal of the History of Science Society. Critical bibliography (one issue per year). Philadelphia: University of Pennsylvania (Department of History and Sociology of Science).
Research journal in five numbers a year, one of which is devoted entirely to annotated bibliography in all of science including mathematics.

Isis Cumulative Bibliography. Formed from *Isis* Critical Bibliographies 1–90, 1913–65. Edited by Magda Whitrow. London: Mansell, 1971–1982.
A merging of the critical bibliographies in the previous citation for the years indicated.

Isis Cumulative Bibliography. Formed from *Isis* Critical Bibliographies 91–100, 1965–74. Edited by John Neu. London: Mansell, 1980–1985.
A continuation of the previous citation.

Jones, Phillip S., ed. *A History of Mathematics Education in the United States and Canada*. 32nd Yearbook of the National Council of Teachers of Mathematics. Washington, D.C.: National Council of Teachers of Mathematics, 1970.
History based on reports and government documents, among other sources. Cf. Bidwell and Clason, above.

Kline, Morris. *Mathematical Thought from Ancient to Modern Times*. New York: Oxford University Press, 1972.
A thorough and exhaustive general history of over 1200 pages, requiring an undergraduate preparation in mathematics.

Kramer, Edna E. *The Nature and Growth of Modern Mathematics*. New York: Hawthorn Books, 1970.
A personal interpretation of the historical roots of contemporary concepts; very few sources cited.

May, Kenneth O. *Bibliography and Research Manual of the History of Mathematics*. Toronto: University of Toronto Press, 1973.

———. *Index of The American Mathematical Monthly*. Vols. 1–80 (1894–1973). Washington, D.C.: Mathematical Association of America, 1977.

Nový, Luboš. *Origins of Modern Algebra*. Leiden: Noordhoff International Publishing, 1973.

Phillips, Esther R., ed. *Studies in the History of Mathematics*. Studies in Mathematics, vol. 26. Washington, D.C.: Mathematical Association of America, 1987.
A collection of essays written by historians of mathematics covering upper-undergraduate and graduate topics.

Struik, Dirk J. *A Concise History of Mathematics.* 4th rev. ed. New York: Dover Publications, 1987.

Basically extends the history in the third revised edition (1967) into the first half of the twentieth century.

Struik, Dirk J., ed. *A Source Book in Mathematics, 1200–1800.* Princeton, N.J.: Princeton University Press, 1986.

An unaltered reissue of the 1969 edition from Harvard University Press.

Thomas, Ivor, ed. and trans. *Greek Mathematical Works.* Vol. I: Thales to Euclid; vol. II: Aristrachus to Pappus of Alexandria. (Loeb Classical Library) Cambridge: Harvard University Press, 1939, 1941.

Translations facing texts in Greek with some interpretative notes. (Translator now uses the name Ivor Bulmer-Thomas.)

Van der Waerden, B. L. *A History of Algebra: From al-Khwarizmi to Emmy Noether.* New York: Springer-Verlag, 1985.

A special history written by one of the participants, and author of *Science Awakening* /Van der Waerden/.

Van Heijenoort, Jean, ed. *From Frege to Gödel: A Source Book in Mathematical Logic, 1879–1931.* Cambridge: Harvard University Press, 1967.

Index

The capsule symbol used throughout this book (a boldface numeral within brackets) appears again in the index. When an entry is the subject of an entire capsule, the symbol for that capsule precedes the listing of pages.

Pronunciations are given for certain words, usually personal names, where such guidance might prove helpful. The symbols used are the familiar diacritical marks used in *Webster's Biographical Dictionary* and in most older dictionaries (such as *Webster's New International Dictionary*, 2d ed., and *Webster's New Collegiate Dictionary*, 6th ed.).

Abacists (ăb′ a·sĭsts), 119
Abacus (ăb′ à·kŭs), [28], 28, 46, 94, 117–20, 130, 242
  as counting table, 91, 117–20
  Chinese (*suan phan*), 91–92, 119
  contest with, 92–93
  etymology of, 91, 117
  Japanese (soroban), 91–92, 119
  Roman, 91–92, 118
Abel (ä′bĕl), Niels Henrik, *1802–1829*, 115, 246, 251, 253, 254, 255, 256, 266, 278, 311, 433, 434
Abstract algebra, 234, 284–87
Abstract spaces, 188–89
Abu Kamil, *c. 850—c. 930*, 307–8
Abu'l-Wefa (à·bool′ wĕ·fä′), *940–998*, 103, 177, 195, 351, 352, 355, 373, 375
Abundant numbers, [12], 59–61
Acta eruditorum, 229, 393, 397
Actuarial mathematics, 247
Adams-Bashford method, 115
Additive principle, 6, 23, 27, 37, 38, 41, 42
Adelard (ăd′ ĕ · lärd) of Bath, *c. 1120*, 48, 119
Agnesi (ä·nyâ′ zê), Maria Gaetana, *1718–1799*, 211
  witch of, [56], 210–11
"Aha" problems, 126
Ahmes (ä′ mĕs), A'h-mosè or Ahmose (ä′ mōs), *c. 1650* B.C., 123
Ahmes papyrus. *See* Rhind papyrus

Aiken, Howard H., *1900–*, 112, 162
"al-" names. *See following element*
Alcuin (ăl′ kwĭn) of York, *735–804*, 121
Alembert (à′läɴbâr′), Jean Le Rond lē rôɴ′) d', *1717–1783*, 214, 318, 433, 438, 449, 460
Aleph null, 85
Alexander, James Waddell, *1888–*, 187
Alexandria, 73
Algebra. *See also* Equations
  abstract, 234, 284–87
  Arabic, [80], 76–77, 241–42, 260–61, 305–8, 332
  associative division, 297
  Babylonian, 235–36, 237, 240, 276, 312, 377–78
  Boolean, [74], 257–58, 284–87
  Egyptian, 237, 332
  etymology of, 233
  European, [81], 242–45, 309–11
  fundamental theorem of, [84], 186, 278, 311, 316–18
  geometric, 172, 236, 237–40, 242, 276, 298–301, 308, 338
  Greek, [78], 236, 237–40, 261–62, 298–300, 332
  Hindu, [79], 241–42, 261–62, 301–5
  laws of, 248–51, 256–57, 279–80, 285–86, 296–97
  linear, 283
  of logic, 257

matrix, 257–59, 281–83
  in medieval Europe, 309
  modern, 234, 284–87
  names for unknown, 242
  noncommutative, 256–57, 259, 279–80, 296–97
  quaternionic, 280
  renaissance of, 242–45, 277–78, 309–10
  rhetorical, 234, 235–36, 237, 242, 260, 303, 305, 387
  stages in, 234
  symbolic, 234, 243–44, 387
  symbolism in, 181, 241, 243–44, 245, 291, 310
  syncopated, 234, 240–41, 243, 260–62, 277, 303
L'Àlgebra (Bombelli), 327–28
Algebra (Wallis), 213, 292
Algebraic numbers, [26], 83–84
Algebraic symbolism, 260–63
Algorism. *See* Algorithm
Algorismus vulgaris (Sacrobosco), 48
Algorists, 119
Algorithm (or algorism)
  etymology of, 48, 234
  Euclidean, [17], 69, 267, 378
  square root, 4–5
Al-jabr (al-Khowarizmi), 233, 242, 260, 306
Almagest (Ptolemy), [91], 50, 90, 96, 101, 242, 337, 338, 342, 344, 359–61, 370
  contents of, 361
  etymology of, 359
"al-" names. *See following element*
Alternating series test, 421, 433
American Mathematical Monthly, 471
Amicable (friendly) numbers, [11], 58–59
Ampère (aɴ'pâr'), André, *1775–1836*, 160
Analemma, 339–42
Analemma (Ptolemy), 342
Analyse des infiniment petits (L'Hospital), 397, 435
Analyst, 397, 399
Analytic geometry, 146, 179, 180–82, 190, 247
Anaxagoras (ăn'ăk·săg'ô·răs), *c. 440* B.C., 202
Angle, [92], 362
  definitions of, 362
  degrees in, 365–66
  measure of, 363–64, 364–68

mil, 367–68
  radians, 366–67
  right, [93], 363–64
  symbol for, 362, 364,
  trisection of, [52], 173, 193, 199–201
Angular measure, [94], 364–68
Annuities, [88], 325–26
Anthologia Palatina, 240
Anthoniszoon, Adriaen, *c. 1543–1607*, 109
Antiphon (ăn'tĭ·fŏn), *c. 430* B.C., 202
Apices, 118
Apollodorus (à·pŏl'ô·dō'răs), *c. 100*, 215
Apollonius (ăp'ŏ·lō'nĭ·ŭs) of Perga (pûr' gà), *c. 225* B.C., 174, 175, 178, 180, 182, 222, 223, 239, 387
  circle of, 175
  Conic Sections, 175, 178, 222, 239
Application of areas, 222
Arabic (modern) numeration system, 29, 49
Archibald, R. C., *1875–1955*, 107
Archimedes (är'kĭ·mē'dēz), *c. 287–212* B.C., [98], 4, 73, 82, 89–90, 108, 132, 150, 174, 182, 199, 202, 380–82, 385, 386, 403–5, 430–31, 432
  anticipation of the calculus, [98], 403–5, 409
  area of parabolic segment, 403–5
  determination of pi, 403
  lemma of, 379
  The Measurement of the Circle, 93, 100, 132, 150
  Metaphysics, 51
  Method, 381–82, 385
  method of, [109], 380–82, 385, 403–4, 430–31
  On the Sphere and Cylinder, 174
  postulate of, 174
  The Sand-Reckoner, 89–90
  spiral of, 202–4
Archytas (är·kī'tăs) of Tarentum (tà·rĕn'tŭm), *c. 400* B.C., 198, 222, 276
Area of circle, 126, 149–50, 169, 202, 377, 386, 403, 407, 430, 440
Argand, Jean Robert, *1768–1822*, 15, 246, 292, 293
Aristophanes (ăr'ĭs·tŏf'à·nēz), *c. 400* B.C.,
Aristotle (ăr'ĭs·tŏt'l), *384–322* B.C., 11, 51, 64, 70, 73, 212, 362
Arithmetic (theory of numbers), 32, 36
  Calandri's, 134

etymology of, 233
Treviso, 94, 97, 134
Arithmetica (Diophantus), 57, 79, 240, 261, 332
Arithmetica infinitorum (Wallis), 387–88, 414
Arithmetica integra (Stifel), 140
Arithmetica universalis (Newton), 319–20, 322
Arithmetic mean, 377
Arithmetic triangle. *See* Pascal's triangle
Arithmetische und geometrische Progress-Tabulen (Bürgi), 106, 143
Arithmetization of analysis, 401–2
Ars conjectandi (Bernoulli), 159, 266
Ars magna (Cardano), 212, 244, 277, 278, 310
Artis analyticae praxis (Harriot), 319
Art of Numbring By Speaking-Rods (Leybourn), 141
Aryabhata (är′yà·bŭt′à), *c. 475—c. 550*, 69, 150, 301, 370
Asoka (à·sō′kà), *c. 250* B.C., 47
ASSC, 112, 162–63
Astrolabe, 340
Astronomy, 175, 177, 333, 339–42, 470
Arabic, 350
astronomical instruments, 194
Babylonian, 37, 360, 364–65
concepts of, 38
Greek, 12, 359–61, 365
Hindu, 140, 346–50
Ausdehnungslehre (Grassmann), 279, 280
Automata theory, 473
Automatic Sequence Controlled Calculator (ASCC), 112, 162–63
Axiomatic method, 462–67, 476
as a research tool, 465, 472
Axiomatics, 2–3, 173, 189–90, 191
interpretation of, 189
role of, in teaching, 471
Axiomatic structure, 9
Axioms, 9, 15–16
Euclid's 9, 16
independence of, 462

Babbage (băb′ij), Charles, *1792–1871*, 112–13, 161, 423
analytic engine, 112, 162
difference engine, 112, 161–62
Babbage, H. P., son of Charles, 112, 162

Babylonian algebra, 377–78
Babylonian astronomy, 37–38
Babylonian numeration system, [11], 4, 6, 28, 36–38, 47, 49, 89, 130–32, 135, 360, 364–65, 370
Bakhshali manuscript, 50, 136, 332
Baldwin, Frank S., *1838–1925*, 112
Ball, W. W. Rouse, *1850–1925*, 160, 214
Bank, 120
Barlow, Peter, *1776–1862*
New Mathematical Tables, 101, 108
Barrow, Isaac, *1630–1677*, [104], 153, 390–91, 392, 415–17, 418, 419
differential triangle, 416–17
Lectiones Opticae et Geometricae, 391, 416
Battani, al- (ăl′băt·tä′nē), *850–929*, 103
Bede, the Venerable, *c. 673–735*, 88–89, 122
De temporum ratione, 89
Beg, Ulugh, *1393–1449*, 355, 370
Bell, Eric Temple, *1883–1960*, 127, 170
Beltrami (bal·trä′mê), Eugenio, *1835–1900*, 185
ben Ezra, Abraham, *1097–1167*, 136
Benedetti, Giovanni B., *1530–1590*, 195
Berkeley, George, *1685–1753*, 399, 445
Bernoulli (*Ger.* běr·nool′ê; *Fr.* běr′noo′yē′), Daniel, *1700–1782*, 438, 449
Bernoulli, Jakob or Jacques, *1654–1705*, 181, 229, 266, 397, 440, 441, 458
Ars conjectandi, 159, 266
Bernoulli, Johann or Jean, *1667–1748*, 145, 227, 228, 312, 358, 397, 398, 399, 435–39, 440, 441, 446
Bertrand, Joseph L. F., *1822–1900*, 63
Bessel, F. W., *1784–1846*, 246
Beyer, Johann Hartmann, *1563–1625*, 138
Bhaskara (bäs′kà·rà), *1114—c. 1185*, 30, 68, 218, 241, 304, 455, 456–57
Bidder, George Parker, *1863–1953*, 160
Binary numeration system, 6–9, 89, 93, 130
Binary quadratic form, 323
Binet (bê′nê′), Jacques P. M., *1786–1856*, 78
Binomial coefficients, 156–57. *See also* Pascal's triangle
Binomial theorem, [67], 114, 264–66, 294, 310, 392, 421, 433
Birkhoff George D., *1884–1944*, 471
Biruni, al- (ăl′bē·rōō′nê), *973–1048*, 336,

353, 355, 373, 375
Blater, J., *c. 1888*, 78, 105
Boethius (bō·ē′thĭ·ŭs), *c. 475–524*, 95, 119, 219
Institutis arithmetica, 95
Bolyai (bō′lyoi), Farkas, *1775–1856*, 209
Bolyai, Janos, *1802–1860*, 10, 184, 185, 209, 462, 464
Bolzano (bōl·tsä′nō), Bernhard, *1781–1848*, 401
Bombelli, Rafael, *c. 1526—c. 1573*, 243, 263, 267, 291, 292, 310, 327–28, 330
L'Àlgebra, 327–28
Boole, George, *1815–1864*, 115, 257, 284, 428, 429
Finite Differences, 428
Laws of Thought, 258, 284
Mathematical Analysis of Logic, 284
Boolean algebra, [74], 257–58, 284–87
Brachistochrone, 228, 440
Brahe (brä′ĕ), Tycho, *1546–1601*, 407
Brahmagupta (brŭ′má·goop′tà), *c. 628*, 68, 69, 151, 176, 241, 243, 261, 262, 301, 349, 455–56
Brahmasphuta Siddhanta, 349
Uttara Khandakhadyaka, 349
Brahmasphuta Siddhanta (Brahmagupta), 349
Brianchon (brē′äɴ′shôɴ′), Charles J. *c. 1783–1864*, 179, 230
Bride's Chair, 218
Briggs, Henry, *1561–1630*, 98, 106, 145, 247
Brossin, George. *See* Méré, George Brossin, chevalier de
Brouncker or Brounker (brŭng′ kĕr), William, *c. 1620–1684*, 151, 269
Brouwer, L. E. J., *1882–1966*, 111, 187, 468
Brouwer's number, 111
Buffon (bü′fôɴ′), Georges Louis Leclerc, comte de, *1707–1788*, 153
Buffon's needle problem, 153
Bugilai tribes, 19–21
Burali-Forti, Cesare, *1861–1931*, 85, 467
Bürgi (bür′gĕ), Jobst or Joost, *1552–1632*, 15, 106, 138, 143, 329, 331
Arithmetische und geometrische Progress-Tabulen, 106, 143
Burroughs, William S., *1857–1899*, 113
Buteo (bü′tä′ō′), Jean, *1492—c. 1565*, 263, 329, 330
Buxton, Jedediah, *1705–1774*, 160

Cajori, Florian, *1859–1930*, 214
Calandri, Filippo, *c. 1491*, 134
Calculating machines, 111–14. *See also* Computers, digital
Leibniz', 112, 161
Pascal's, 112, 161
Calculating prodigies, [48], 159–60
Calculating rods, 281
Calculation. *See also* Computation
etymology of, 87
Calculator, desk, 92–93
Calculus, 173, 182, 275
conception in antiquity, 376–78
differential, 418
etymology of, 376
of finite differences, 115
integral, 170, 174, 405–6, 419
in Japan, [118], 452–54
medieval contributions to, 383–86
notations in, 394–95
priority controversy in, 397–98, 443
search for rigor in, 399–402
tensor, 183, 280
of variations, [112], 439–40
Calendar, Mayan, 89
Campanus (kăm·pā′nŭs), Johannes, *c. 1260*, 177
Canon foecundus (Reinhold), 103
Cantor, Georg, *1845–1918*, 36, 72, 84, 85–86, 313, 402, 465, 468
Cardinal numbers, 21
Cardan, Jerome. *See* Cardano
Cardano (kär·dä′nô), Girolamo, or Jerome Cardan (kär′dăn; -d′n), *1501–1576*, 157, 158, 195, 212, 243, 244, 245, 263, 277, 291, 309, 310, 317, 319, 329, 330, 358
Cardan's formula, 277
Carnot (kàr·nō′), Lazare, *1753–1823*, 179
Carroll, Lewis. *See* Dodgson, Charles Lutwidge
Casting out nines, [38], 140
Cataldi, Pietro, *c. 1548–1626*, 269, 329, 331
Catenary, 392
Cauchy (kō′shē′), Augustin Louis, *1789–1857*, 115, 246, 247, 253, 255, 256, 282, 400, 401, 433, 434, 442, 451, 458, 461, 466
Course d'analyse, 400
Cauchy-Riemann equations, 438
Cavalieri (kä′vä·lyâ′rĕ), Bonaventura,

1598–1647, [**101**], 183, 386, 387, 388, 392, 409–10, 413, 418, 420

Geometria indivisibilibus, 386, 409

method of indivisibles, 386–87, 388, 409, 413

principle of, 386–87, 409

Cayley, Arthur, *1821–1895*, 160, 182, 251, 255, 258, 259, 282, 283, 311, 451

Celestial sphere, 340

Ceulen (kû′lĕn), Ludolph van, *1540–1610*, 109, 151

Chace, Arnold B., *1845–1932*

  The Rhind Mathematical Papyrus, 124

Characteristic, 106

Chasles (shäl), Michel, *1793–1880*, 179

Chhin or Chin Chiu-shao, *c. 1201– c. 1261*, 274, 438

Chiarino, Giorgio, *c. 1481*, 147

Chinese–Japanese numeration systems [**5**], 43–44, 48

Chinese rods, 128–29

Chiu Chang Suan Shu, 225

Chords, 96, 334, 368, 374

  etymology of, 368

  tables of, 12, 98, 101, 337–38, 359–60, 369–70

  trigonometric, 96, 335–39, 343–45

Chou Pei Suan Ching, 216, 225

Chuquet (shü′kĕ′), Nicolas, *c. 1484*, 148, 291, 329, 330

  Le triparty en la science des nombres, 329

Chu Shih-chieh, *c. 1303*, 156

  Precious Mirror of the Four Elements, 156

Cipher. *See* Zero, name

Ciphered numeration systems, 40

Circle

  area of, 126, 149–50, 169, 202, 377, 386, 403, 407, 430, 440

  circumference of, 168–69

  definition of cycloid, 227

  division into 360 parts, 37, 359–60, 364–66, 370

  Feuerbach, 230

  inscribed regular polygons, 193, 195, 202

  length of arc, 453

  nine-point, [**65**], 230–31

  squaring of, [**53**], 84, 149, 150, 152–53, 173, 193, 201–4 (*see also* π [pi])

Circle areas

  ratio of, 150, 380

Circles of Proportion (Oughtred), 271

Circumference of circle, 168–69

Cissoid of Diocles, 100

Clairaut (klĕ′rō′), Alexis Claude, *1713–1765*, 181

Clavis mathematicae (Oughtred), 139, 271

Clavius (klä′vĕ·oos), Christopher, *1537–1612*, 212

Clebsch (klāpsh), Alfred, *1833–1872*, 259

Colburn, Zerah, *1804–1839*, 160

Collapsible compass (Euclidean) 192–93

Colson, John, *c. 1736*, 211

Commandino, Federigo, *1509–1575*, 178

Commensurability, 70

Commentary on Euclid, Book I (Proclus), 171, 178

Compass. *See also* Constructions

  collapsible (Euclidean), 192–93

  compass-only constructions, 193–94

  modern, 193

  rusty (fixed-), 177, 194–96

  and straightedge, [**50**], 65, 150–52, 175, 192–96, 197–98, 199–201, 204

Complete quadrilateral, 350–51

Complex numbers, [**76**], 3, 15, 16, 35, 36, 83–84, 244, 245–47, 256, 259, 278, 279, 290–94, 311, 358–59

Composite numbers, [**15**], 35, 62–63, 64–66

Computation, 26, 32, 87–117, 124–25, 130–35

  by calculators and computers, [**49**], 161–64

  of cube roots, 100–101

  of *e*, 110

  finger, 122

  methods of, 77

  modern analytical methods, 114–15

  of pi, 108–10

  by rods, 128

  social implications of, 115–17

  of square roots, 98–101

  of trigonometric functions, 101–4

Computers, 287

  analogue, 113–14

  digital, 111–13

  electronic digital, 8

Computers, digital, [**49**], 161–64. *See also* Calculating machines

EDSAC, 164

ENIAC, 163–64

Harvard Mark I–IV, 162–63
Computer theory, 473
Computing machines, social implications of, 115–17
Comrie, Leslie J., *1893–1950*, 107
  An Index of Mathematical Tables, 107–8
Concept of continuity, 175, 186
Conchoid, 193
  of Nicomedes, 100, 200
Condorcet (kôn′dôr′sĕ′), A. N. de Caritat, marquis de, *1743–1794*, 449
Congruence (number theory), [75], 288–90
Conic Sections (Apollonius), 175, 178, 222, 239
Conic sections, 180, 193, 198, 201, 226, 239, 242, 276, 283, 308, 318, 408, 470
  etymology of, [61], 222–23
Constructions
  compass-only, 193–94
  compass and straightedge, [50], 65, 150–52, 175, 192–96, 197–98, 199–201, 204
  Euclidean, 192–93
  fixed-compass and straightedge, 194–96
  impossible, 193, 197–98, 199–201, 204
  insertion principle of, 200
  postulates, 192
  with limited means, 193–96
Continued fractions, [68], 100, 109–10, 155, 267–70, 304
Continuity, 401
  concept of, 175, 186
Convergence, [110], 432–34
Coordinates, 387
  polar, [64], 181, 204, 229
Copernicus, Nicolaus, *1473–1543*, 103, 359, 391
  De revolutionibus orbium coelestium, 103
Cosine, [95], 358, 359, 368–71
  etymology of, 371
Cosine law, 375
Cotangent, [96], 169, 335, 354, 371–74
  etymology of, 374
Course d'analyse (Cauchy), 400
Cramer, Gabriel, *1704–1752*, 282
Crelle (krĕl′ĕ), August Leopold, *1780–1855*, 95, 255
  Journal, 255, 266, 278
  Rechentafeln, 95
Crelle's Journal, 255, 266, 278

Cremona (kra·mōnä), Luigi, *1830–1903*, 179
Cube roots, 265
Cubic equations, 177, 198, 236, 242, 244, 247, 276–78, 308, 309–10, 317, 321
Cuneiform numerals, 37
Curves, higher plane, 175
Cybernetics (Wiener), 116
Cycloid, [63], 227–28, 392, 440

d'Alembert, Jean Le Rond. *See* Alembert, Jean Le Rond d'
De aequationum recognitione et emendatione (Viète), 245
De analysi (Newton), 393, 397
Decimal fractions, [36], 6, 49, 97–98, 106, 136–37, 137–38, 356
Decimal point, 96, 98, 137, 138, 365
Decimal system. *See* Numeration systems, Hindu-Arabic
Dedekind (dā′dĕ·kĭnt), Richard, *1831–1916*, 36, 72, 85, 259, 465, 466
Dedekind cut, 466
De divina proportione (Pacioli), 78, 207
de Fermat, Pierre. *See* Fermat, Pierre de
Deficient numbers, [12], 59–61
Definite integral, [120], 457–59
  notation, 458–59
Definition, mathematical type of, 463–64
Definitions, 11
Degree measure, 359–60
Degrees of Mortality (Halley), 326
Delamain, Richard, *c. 1630*, 273
  Grammelogia, 273
del Ferro, Scipione. *See* Ferro, Scipione del
Delian problem. *See* Duplication of the Cube
de Méré, chevalier. *See* Méré, George Brossin, chevalier de
Democritus (dē·mŏk′rĭ·tŭs), *c. 410* B.C., 381, 382
De Moivre (dē·mwä′vr′), Abraham, *1667–1754*, 159, 247, 294, 428
  Doctrine of Chances, 159
de Montmort, Pierre Rémond. *See* Montmort, Pierre Rémond de
De Morgan, Augustus, *1806–1871*, 97, 286
De quadratura (Newton), 394–95
De revolutionibus orbium coelestium

(Copernicus), 103
Derivative, 400, 417
Desargues (dä'zàrg'), Gérard, 1593–1661, 179
Descartes (dä·kärt'; Fr.dä'kàrt'), René, 1596–1650, 3, 61, 146, 179, 180, 181, 186, 198, 201, 212, 227, 243, 246, 247, 263, 291, 310, 312, 317, 318, 327, 331, 387, 392, 394, 413, 418
Discours de la méthode, 146, 181
La géométrie, 181, 310, 317, 318, 319, 387
rule of signs, [85], 310, 317, 318–20
De temporum ratione (Bede), 89
Determinants, [73], 258, 259, 281–83, 421, 450–51
Die Coss (Rudolff), 148, 244
Difference equations, 348–50
Differential calculus, 390, 397
Differential equations, [113], 441–42, 449–50
Differential (characteristic) triangle, 385, 416–17, 418–19
Differentials, 395, 397, 420, 423
Differentiation, 389–90, 395
Digit, 88
Dinostratus (dī·nŏs'trà·tŭs), c. 350 B.C., 150
Dinocles (dĭ'ô klēz), c. 180 B.C., 198
Diophantus (dĭ'ô·fän'tŭs) of Alexandria, c. 75 or c. 250, 57, 68, 96, 236, 240, 243, 251, 260, 261, 291, 332
age problem, 240
Arithmetica, 57, 79, 240, 261, 332
dating of, 260
Dirichlet (dē'rê·klä'), P. G. Lejeune, 1805–1859, 63, 80, 269, 313, 401
Discours de la méthode (Descartes), 146, 181
Discriminant, [87], 283, 323–24
La Disme (Stevin), 97, 138
Distributive property, 124, 250–51, 285–87
Division, [34], 93–95, 130–35
Babylonian method of, 131–32
Egyptian method of, 125, 130
galley method of, 94–95, 134–35
Hindu method of, 134
modern form of, 95
scratch method of, 94, 134
d'Ocagne, Maurice. See Ocagne, Maurice d'
Doctrine of Chances (De Moivre), 159
Dodgson, Charles Lutwidge [pseud.

Lewis Carroll], 1832–1898, 78, 220
Donkin, William Fishburn, 1814–1869, 451
Du Bois-Reymond (dü bwä'rä'môɴ'; dü bwà'-), Paul, 1831–1889, 463
Duplation and mediation, 93, 130–31
Duplication of the cube, [51], 173, 193, 197–98, 222, 276, 470
Eratosthenes, 73
origin of problem for, 197
Dürer (dü'rēr), Albrecht, 1471–1528, 81, 195
e (base of natural logarithms), [45], 154–55, 266
computation of, 110
continued fraction for, 110
series expansion for, 110
symbol, 155
transcendence of, 84

Earth, measurement of, 73–74
Eckert, John Presper, Jr., 1919–, 163
Eddington, A. S., 1882–1944, 90
EDSAC, 164
Egyptian numeration system, [2], 23, 38–40, 136, 237
demotic, 39, 40
hieratic, 23, 39, 40, 136
hieroglyphic, 23, 38–40, 136
Einstein, Albert, 1879–1955, 280, 470
Electronic Delayed Storage Automatic Computer, 164
Electronic Numerical Integrator and Calculator, 110, 163–64
Elements (Euclid), 35, 57, 59, 62, 68, 69, 71, 150, 173, 174, 177, 180, 226, 238, 242
angle definition, 362
construction postulates, 192
converse of parallel postulate, 183–84, 208
cosine law, 375
distributive property, 250–51
equilateral triangle, 193
Euclidean algorithm, 69
geometric algebra, 237–39, 298–301, 338
identities, 4
infinitude of primes, 35, 62
isosceles triangle, 219
"lemma of Archimedes," 379
mean and extreme ratio, 204–5
method of exhaustion, 71, 379–80
parallel postulate, 183–84, 207–10, 363

perfect numbers, 59
Pythagorean theorem, 66, 216–18
ratio of circle areas, 150, 380
regular polyhedra, 206, 220–21
right angle, [93], 363–64
s.a.s. 217
theory of proportions, 71
Eléments de géométrie (Legendre), 184
Elements of Geometry (Playfair), 184
Ellipse, 386. *See also* Conic sections
  etymology of, 175, 222–24
Ellis, Alexander J., *1814–1890*, 366
ENIAC, 110, 163–64
Enumeration, 19–20, 25–26
Epsilon-delta, 401, 433
Equations. *See also* Algebra
  approximation to roots of, 274, 310
  biquadratic (quartic), 236, 245, 248,
    253, 278, 310, 317, 318
  cubic, 177, 198, 236, 242, 244, 247, 276–
    78, 308, 309–10, 317, 321
  geometric solution, 177, 222 (*see also*
    Algebra, geometric)
  how written [**66**], 260–63
  indeterminate, 69, 240, 241, 303–5, 308
  Pell, 241, 304, 438
  polynomial, 83, 288–89
  quadratic, 29, 145, 222, 241, 246, 261–
    63, 270, 291–92, 298, 301–3, 306–8,
    317, 323–24
  quartic, 236, 245, 248, 253, 278, 310,
    317, 318
  quintic, 254–55, 278–79, 311
  simultaneous, 281–83
  solution of higher-degree, [**71**], 253–
    55, 276–79
  solution of, by rod numerals, 129
  theory of, 245–47, 275
Equilateral triangle, 193
Eratosthenes (ĕr′ *à*·tŏs′ thĕ·nēz), *c. 230*
  B.C., [**19**], 64, 72–74, 197, 403
  measurement of earth, 73–74
  sieve of, 64, 72–73
Erlanger Programm, 187–88, 191, 259,
  467
Escott, E. B., *1879*, 65
Ethnography, 18–21
Euclid, *c. 300* B.C., 4, 10, 57, 59, 61,
  62–63, 64, 150, 173, 174, 177, 180,
  236, 237, 242, 362, 363
  Elements. See *Elements* (Euclid)
Euclidean algorithm. *See* Algorithm,
  Euclidean
Euclidean compass (collapsible), 192–93

Euclides ab omni naevo vindicatus
  (Saccheri), 462
Euclides Danicus (Mohr), 194, 195
Euclidis Curiosi (Mohr), 195
Eudemian Summary (Proclus), 171, 172
Eudemus (ū·dē′mŭs), *c. 335* B.C., 171
Eudoxus (u·dŏk′ sŭs), *408–355* B.C.,
  29, 35–36, 71, 204, 379, 380, 382, 389,
  390, 402
Euler (oi′lēr), Leonhard, *1707–1783*, 58,
  60, 62, 65, 80, 81, 110, 145, 153, 154–
  55, 159, 181, 186, 229, 230, 246, 247,
  252, 253, 270, 278, 286, 291, 312, 318,
  322, 323, 334, 359, 398, 399, 401, 428,
  433, 434, 438, 440, 442, 447, 449, 451
  Introductio in analysis infinitorum,
    154, 398, 401
Euler circles, 286
Euler line, 230
Euler ϕ-function, 252
Euler's number, 155
European peasant multiplication, 122
Eutocius of Ascalon, *c. 560*, 93, 132,
  197
Even numbers, 33, 52, 64
Eves, Howard, 259
Exchequer, 120
Exhaustion. *See* Method of exhaustion
Exponential notation, [**89**], 327–31
Exponents
  complex, 266
  negative, 266
  rational, 265
Ezra, Abraham ben. *See* ben Ezra,
  Abraham

Factorial n, [**115**], 446–48
False position, method of. *See* Method
  of false position
False position, rule of, [**90**], 237, 308,
  332
  double, 308
False roots, 319
Famous problems of antiquity. *See*
  Angle, trisection of; Circle, squaring of; Duplication of the cube
Ferguson, D. F., 110
Fermat (fĕr′mà′), Pierre de, *1601–1665*,
  [**102**], 57, 64, 79–80, 158, 180, 181,
  210, 227, 247, 252, 310, 386, 387,
  388–89, 390, 392, 410–13, 425–26
Fermat's "last theorem," [**23**], 79–80
  method of maxima and minima, 389–
  90, 411–12, 425–26

Fermat numbers, 64–65
Ferrari, Ludovico, *1522–1565*, 195, 244, 278, 310
Ferro, Scipione del, *1465–1526*, 244, 277, 309
Feuerbach circle, 230
Feuerbach (foi′ĕr·bäк), Karl, *1800–1834*, 230
Fibonacci (fē′bô·nät′chĕ) [Leonardo of Pisa], *1180–1250*, [22], 48, 77–79, 94, 140, 207, 242, 274, 309, 325, 332
Liber abaci, 48, 77, 94, 242, 309, 325
Fibonacci (sequence) numbers, [22], 77–79, 157, 207, 309
Field, 249, 259
Figurate numbers, [10], 16, 33–34, 51–52, 53–58, 66–67
Fincke, Thomas, *1561–1656*, 373
Finger numbers, 88–89, 120–23
Finger reckoning, [29], 88–89, 120–23
Finite differences, [108], 427–29
notation, 428
Finite Differences (Boole), 428
Fior, Antonio Maria, *c. 1535*, 277
Fletcher, A., *1903–*, 107
Fluents, 392, 422–23
Fluxions, 3, 394–95, 397, 399, 422–23
Force of water on dam, 406
Formalist school, 468–69
The Foundations of Geometry (Hilbert), 180–90
Foundations of the Theory of Probability (Kolmogorov), 159
Four-color problem, 186
Four-dimensional geometry, [57], 211–14
Fourier (foo′ryā′), Jean Joseph, *1768–1830*, 313, 434, 449
Fourier's series, 438
Fourth-degree equation. *See* Equations, quartic
Fractional exponents, 415
Fractions, [35], 135–37, 95–98
Arabic, 136
Babylonian, 135
common, 136–37
continued, [68], 100, 109–10, 155, 276–70, 304
decimal, [36], 6, 49, 97–98, 106, 136–37, 137–38, 356
etymology of, 137
Greek, 96, 136
Hindu, 97, 136
Roman, 90, 136

sexagesimal, 37, 96–97, 131–32, 135, 365
symbolism for, 43, 95–98
unit, 40, 43, 50, 95–96, 124–25, 136
Fréchet, Maurice, b. *1878*, 187, 188–89, 191, 469
Frege (frā′gà), Gottlob, *1848–1925*, 258, 466, 468
Fuller, Thomas, *1710–1790*, 160
Function, [82], 312–13
concept of, 337, 398
definition of, 401–2
exponential, 154
graph of, 384
Function notation, 459
Functions, trigonometric, 12–13, 169, 177, 355 (*see also specific functions*)
Fundamental theorem of algebra, [84], 186, 278, 311, 316–18
Fundamental theorem of arithmetic, 35

Galileo (gä′lê·lâ′ô) Galilei (gä′lê lâ′ê), *1564–1642*, 52, 85, 158, 227, 385, 386, 405
Galois (gà′lwà′), Évariste, *1811–1832*, 246, 251, 255, 256, 279, 311, 466
Galois's theory, 255
Game theory, 473
Gamma function, [115], 446–48
notation, 448
Ganita-Sara-Sangraha (Mahavira), 30
Gardiner, William, *1742*
Tables of Logarithms, 145
Gauging, [107], 386, 424–27
Gauss (gous), Carl or Karl Friedrich, *1777–1855*, 3, 15, 65, 159, 183, 184, 185, 186, 209, 246, 247, 253, 278, 288, 291, 292, 310, 311, 316, 320, 324, 350, 433, 434, 462
Gelfond, Aleksander O., *1906–*, 84
Gematria, [20], 34, 74–75, 76, 140
number of the beast, 75
Genaille, Henri, c. *1885*, 142
Genaille's rods, 141–42
General trattato (Tartaglia), 326
Genetic method, 3, 5, 471
Geometria del compasso (Mascheroni), 193, 194
Geometria indivisibilibus (Cavalieri), 386, 409
Geometric algebra. *See* Algebra, geometric

Geometry, 146, 179, 180–82, 190, 247.
    See also Constructions, geometrical
  of abstract spaces, 188–89
  Arabic, 176–77
  Babylonian, 168–69
  in China, [62], 225–26
  definition of, 187–88, 190–92
  demonstrative (systematic), 170–74, 190
  descriptive, 179
  differential, 182–83
  Egyptian, 167, 169–70
  elliptic, 185
  Erlanger Programm, 187–88, 191, 259, 467
  etymology of, 167
  four-dimensional, [57], 211–14
  Greek, 170–76
  Hindu, 176–77
  hyperbolic, 185
  in medieval Europe, 177–78
  modern, 174–75, 190–92
  n-dimensional, 182, 187, 279–80
  non-Euclidean, [55], 10–11, 183–85, 190, 207–10, 224, 462, 464, 470
  origins of, 165–68
  parabolic, 185
  projective, 178–80, 188
  scientific, 166–68, 190
  solid, 174
  solid analytic, 181
  subconscious, 165–66, 190
Gerbert (zhĕr′bâr′) [Pope Sylvester II], *940–1003,* 118
Gergonne (zhĕr′ gôn′), Joseph Diaz or Diez, *1771–1859,* 179, 180
Ghaligai, F., *1521,* 263
Gherardo (jĕr·âr′dō) or Gerard of Cremona, *c. 1114–1187,* 370
Gibbs, Josiah Willard, *1839–1903,* 258, 280
Girard, Albert, *c. 1590—c. 1633,* 245–46, 291, 317, 322, 371
  Invention nouvelle en l'algèbre, 322
Gnomon (nō′mon)
  number form, 52, 67
  sundial, 73, 225, 335–37, 353–54, 372–73
Gödel, Kurt, *1906–,* 469, 471
Goldbach, Christian, *1690–1764,* 65
Golden rectangle, 206
Golden section (mean, ratio), [54], 78, 204–7

φ (phi), τ (tau), 205
Goldstine, Herman H., *1913–,* 163
Golenischev papyrus. *See* Moscow papyrus
Googol, 83
Grammelogia (Delemain), 273
Grandi (grän′dê), Guido, *1671–1742,* 210
Graphical solution
  of algebraic equations, 145–46
  of spherical triangles, 145
Grassmann (gräs′män), Hermann, *1809–1877,* 182, 257, 258, 259, 279, 280, 282, 462
  Ausdehnungslehre, 279, 280
Graunt (gränt), John, *1620–1674*
  Natural and Political Observations, 326
Greatest common divisor (G.C.D.), 69, 267, 378
Greek numeration system, [4], 42–43
  alphabetic (Ionic), 27–29, 33, 42–43, 47, 50, 75, 90, 132–33, 262, 298
  Attic (Herodianic), 42, 82, 298
Greenwood, Isaac, *1729,* 140
Gregory, David, *1627–1720,* 153
Gregory, James, *1638–1675,* 109, 114, 115, 154, 247, 390, 393
The Ground of Artes (Recorde), 119, 332
Groups, 246–53, 311, 466–67, 472
  characteristic, 279
  concept of, 259
  cyclic, 251, 253
  definition of, 248–49
  finite Abelian, 256
  permutation, 247–48, 253, 254–56
  of transformations, 188, 191
Group theory, 187–88, 247–56, 311, 470
Grundlagen der Geometrie [*The Foundations of Geometry*] (Hilbert), 180–90
Gua de Malves, Jean Paul de, *c. 1712–1786,* 320
Gunter, Edmund, *1581–1626,* 111, 273, 371, 374

Habash, *830,* 355, 373
Hadamard (à′dà′màr′), Jacques, *1865–1963,* 65
Half-angle formula, 348, 361
Halley, Edmund, *1656–1742,* 326
  Degrees of Mortality, 326

Hamilton, William Rowan, *1805–1865*, 3, 246, 256, 257, 280, 282, 292, 295, 449, 462
Lectures on Quaternions, 257, 280
Hardy, G. H., *1877–1947*, 14
Harmonic mean, 377
Harpenodoptai ("rope stretchers"), 68, 215
Harriot, Thomas, *1560–1621*, 243, 263, 308, 310, 319, 331
Artis analyticae praxis, 319
Harvard Mark I–VI Calculators, 162–63
Hassar, al-, *c. 1200*, 139, 140
Hausdorff, Felix, *1868–1942*, 469
Heath, Thomas Little, *1861–1940*, 94
Heaviside (hĕv'ĭ sīd), Oliver, *1850–1925*, 434, 442
Hebrew numeration system, 26, 27, 75
Heisenberg, August, 259
Hemisphere, area of, 127
Herigone, Pierre, *1634*, 331, 362
Hermann, Jacob, *1678–1733*, 229
Hermite (ĕr'mēt'), Charles, *1822–1901*, 84, 155
Hero. *See* Heron or Hero of Alexandria
Herodotus (hê·rŏd'ô·tŭs), *c. 450* B.C., 120, 167, 206
Heron or Hero of Alexandria, *c. 75*, 4, 99, 139, 175, 291
Stereometrica, 291
"Heron's method," 99–101
Hesiod (hē'sĭ·ŏd; hĕs'ĭ-), *8th cent.* B.C., 88
Hilbert, David, *1862–1943*, 84, 190, 191, 364, 468, 469
Grundlagen der Geometrie [*The Foundations of Geometry*], 180–90
Hindu numeration system, 28–29, 47, 48
Hindu Reckoning (ibn-Lebban), 133
Hindu-Arabic numeration system, [7], 6, 28, 46–49, 76–77, 94–95, 118, 133–34, 242, 243, 260, 309, 365
Hipparchus (hĭ·pär'kŭs) of Nicaea nĭ·sē'à), *c. 180—c. 125* B.C., 101, 145, 333, 338, 340, 359, 365, 369, 374
Hippasus, *c. 400* B.C., 71
Hippias (hĭp'ĭ ăs) of Elis (ē'lĭs), *c. 425* B.C., 150, 201, 202, 378
Hippocrates (hĭ·pŏk'rà·tēz) of Chios (kī'ŏs), *c. 440* B.C., 150, 172, 198, 202
Hippolytos or Hippolytus (hĭ·pŏl'ĭ· tŭs), *c. 225*, 140

Hisab al-jabr w'al muqabalah. See Al-jabr (al-Khowarizmi).
Hobbes (hŏbz), Thomas, *1588–1679*, 119, 153
Hohenberg, Herwart von, *1610*, 95
Hölder (hûl'dēr), Otto, *1859–1937*, 255, 434
Hollerith, Herman, *1880*, 113
Horner, William George, *1786–1837*, 275
Horner's method, [70], 274–75, 438
Horologium oscillatorium (Huygens), 228
Hubble, Edwin, *1889–1953*, 110
Hudde, Johann, *1633–1704*, 69, 246, 391
Hurwitz, Alexander H., 62
Huygens (hoi'gĕns), Christiaan or Christian, *1629–1695*, 158, 183, 228, 269, 391, 418
Horologium oscillatorium, 228
Hypatia (hī·pā'shĭ·à; -shà), d. *415*, 99
Hyperbola, 392. *See also* Conic sections etymology of, 175, 222–24
Hypercomplex numbers, 257–58, 259, 279–80
Hypsicles, *c. 180* B.C., 57, 365

Iamblichus (ĭ·ăm'blĭ·kŭs) of Chalcis (kăl'sĭs), *c. 325*, 58, 60
ibn-Lebban, Kushyar, *c. 971—c. 1029*, 133
Hindu Reckoning, 133
ibn-Qorra, Tabit, *826–901*, 218
Identities, 4
Imaginary numbers, 3, 15, 16, 35, 36, 83–84, 244, 245–47, 256, 259, 278, 279, 290–94, 311, 358–59
The Immortality of the Soul (More), 213
Impossible constructions (geometrical), 201, 204
Incommensurability, [18], 36, 70–72, 172, 239, 378–79, 467
Indeterminate equations, 69, 240, 241, 303–5, 308
Indeterminate forms, 435–38, 454–56
An Index of Mathematical Tables (Comrie), 107–8
Indian. *See* Hindu references
Indivisibles, 382, 386–87
Infinite series, 109, 114–15, 392–93, 398, 400. *See also* Series
for *e*, 155
for *π*, 152
paradoxes of, 433

Infinite set, 85
of primes, [14], 62–63
of transfinite cardinals, 85
Infinitesimals in India, [119], 454–57
Information theory, 473
Insertion principle, 200
Institute for Advanced Study, Princeton, 471–72
Institutis arithmetica (Boethius), 95
Integral calculus, 382, 390, 397, 402
Integration, 380, 385, 388–90, 396, 400
Irrational numbers, [18], 70–72
Interdependence in mathematics, 15–16
Interest, [88], 325–26
Interpolation theory, 349–50
Intersection of sets, 285–87
Intuitionist school, 467, 468
Introductio arithmetica (Nichomachus of Gerasa), 53, 60, 94
Introductio in analysis infinitorum (Euler), 154, 398, 401
Invariants, 283, 311, 324
Invention nouvelle en l'algèbre (Girard), 322
Inverse tangent problem, 391
Inversion, 194
An Investigation of the Laws of Thought (Boole), 258, 284
Irrational numbers, [18], 70–72, 84, 152, 155, 239
Isosceles triangle, 219

Jacobi (yä·kō′bê), Carl or Karl Gustav Jacob, *1804–1851,* 449, 451
Jacobians, [117], 450–52
notation for, 451
Jacquard, Joseph Marie, *1752–1834*
Jacquard loom, 162
Jartoux, Pierre, *1670–1720,* 453
Joachim, Georg. *See* Rheticus
John of Halifax or Holywood. *See* Sacrobosco
John of Seville, *c. 1140,* 137
Jones, William, *1675–1749,* 145, 153
Jordan, (zhôr′dän′), Camille, *1838–1922,* 255
Juvenal (joo′vĕ·n′l), *c. 60—c. 140,* 121

Kant (känt), Immanuel, *1724–1804,* 213
Karkhi, al-, *c. 1020,* 308

Kashi, Jamshid al-, *c. 1430,* 138, 350, 355, 356
Kasner, Edward, *1878–1955,* 83
Keill (kēl), John, *1671–1721,* 213
Kemeny, John, 3
Kepler, Johann or Johannes, *1571–1630,* [100], 77–78, 207, 359, 386, 391, 407–8, 424–25, 470
planetary laws of motion, 391, 407–8
Stereometria doliorum vinorum, 424
Khayyam, Omar. *See* Omar Khayyam
Khazini, al-, *c. 1120,* 355
Khowarizmi, Mohammed ibn-Musa al- (ăl·ĸoo·wä′rĭz·mē′), *c. 825,* [21], 28, 48, 76–77, 140, 233, 236, 242, 260, 261, 306, 332
Al-jabr, 233, 242, 260, 306
Liber algorismi (see *Liber algoritmi de numero Indorum*)
Liber algoritmi de numero Indorum, 48, 234, 242, 260
Kirchhoff (kĭrĸ′hôf), Gustav Robert, *1824–1887,* 186
Klein, Felix, *1849–1925,* 185, 187, 191, 259, 452, 467, 472
Knotted cords, 129
Kochansky, Adamas, *1631–1700,* 81, 151
Kolmogorov, Andrei Nikoloevich, *1903–*
Foundations of the Theory of Probability, 159
Königsberg bridge problem, 186
Korean number rods, [32], 128–29
Kowa, Seki, *1642–1708,* 281, 452
Kronecker (krō′nĕk′ĕr), Leopold, *1823–1891,* 69, 256, 468
Krsna, *c. 1550,* 455
Kulik, J. P., *1773–1863,* 65
Kummer (koom′ĕr), Ernst, *1810–1893,* 80
Kushyar ibn-Lebban. *See* ibn-Lebban, Kushyar

Lacroix (là′krwä′), Sylvestre François, *1765–1843,* 451
La géométrie (Descartes), 181, 310, 317, 318, 319, 387
Lagrange là′gränzh′), Joseph Louis, *1736–1813,* 115, 214, 247, 253, 254, 256, 270, 313, 318, 322, 323, 399–400, 433, 449, 451
Théorie des fonctions analytiques, 400
Lal ("subtract"), 37
Lalanne (là′làn′), Léon, *1811–1892,* 146

Lallemand (làl'mäɴ'), Charles, 146
Lambert (läm'bẽrt), Johann Heinrich, *1728-1777*, 152, 184, 208, 270
Lamé (là'mä'), Gabriel, *1795-1870*, 69, 80
Laplace (là'pläs'), Pierre Simon, *1749-1827*, 115, 159, 247, 282, 390, 442, 449
Large numbers, [**25**], 82–83, 89–90
Laws of Thought (Boole), 258, 284
Lectiones Opticae et Geometricae (Barrow), 391, 416
Lectures on Optics and Geometry (Barrow), 391, 416
Lectures on Quaternions (Hamilton), 257, 280
Lefschetz, Solomon, b. *1884*, 186–87
Legendre (lẽ·zhän'dr'), Adrien Marie, *1752-1833*, 80, 184, 208, 247, 257, 270, 448–49
    Eléments de géométrie, 184
    Théorie des Nombres, 257
Lehmer (lā'mẽr), Derrick Henry, *1905-*, 110
Leibniz (lĭp'nĭts) or Leibnitz, Gottfried Wilhelm von, *1646-1716*, [**105**], 7, 8, 97, 112, 139, 152, 154, 161, 182, 186, 246–47, 258, 281, 312, 320, 380, 391–98, 418–23, 432–33, 441, 448, 453, 458
    introduction of notation, 419–20, 423
Lemma of Archimedes, 379
Lemoine (lẽ·mwàn'), Émile, *1840-1912*, 196
Leonardo of Pisa. *See* Fibonacci
Leonardo da Vinci. *See* Vinci, Leonardo da
Leybourn (lē'bẽrn; -bûrn), William, *1626—c. 1700*
    Art of Numbring By Speaking-Rods, 141
L'Hospital (lô'pē'tàl') or L'Hôpital (lô'), G. F. A. de [Marquis de Saint-Mesme], *1661-1704*, [**111**], 281, 397, 435–39
Liber abaci (Fibonacci), 48, 77, 94, 242, 309, 325
Liber algorismi. See Liber algoritmi
Liber algoritmi de numero Indorum (al-Khowarizmi), 48, 234, 242, 260
Lie (lē), Sophus, *1842-1899*, 187, 442
Limit concepts, 379–80, 396
Linear programming, 473
Lindemann (lĭn'dẽ·män), Ferdinand, *1852-1939*, 84, 152, 204

Liouville (lyoo'vēl'), Joseph, *1809-1882*, 84, 155
Listing, J. B., *1808-1882*, 186
    Vorstudien zur Topologie, 186
Lobachevsky or Lobachevski (lō'bà·chẽf'ski), Nikolai Ivanovich, *1793-1856*, 10, 184, 185, 209, 462, 464
Logarithms, [**40**], 15, 104–6, 138, 142–45, 154–55, 247, 271–73, 391, 392, 358
    common (or Briggsian), 106, 145
    computation of, 105–6
    definition of, 105–6, 143, 145
    etymology of, 106
    Naperian, 106, 142–45, 154
    natural, 65, 144–45, 154–55
    tables of, 98, 105–6, 143–45, 154, 248
Logic, 11
Logistic, 32
Logistic school, 469
Logistica speciosa (Viète), 245
Loomis, Elisha Scott, 216
    The Pythagorean Proposition, 216
Loubère, Antoine de La, *1600-1644*, 81
Lucas, Édouard Anatole, *1842-1891*, 62, 78, 142
Ludolphian number, 109, 151
Lunes, 202
    of Hippocrates, 150

Machin, John, *1680-1751*, 109, 152, 153
Maclaurin, Colin, *1698-1746*, 322, 445–46
    Treatise on Fluxions, 445–46
Maclaurin series, [**114**], 443–46
Magic squares, [**24**], 80–82
Mahavira, *c. 850*, 30, 291, 301, 303
    Ganita-Sara-Sangraha, 30
Mamun, al- (äl·mä·moon'), *786-833*, 76
Mantissa, 106
Mascheroni (mäs'kä·rō'nê), Lorenzo, *1750-1800*, 193, 194
    Geometria del compasso, 193, 194
Mascheroni constructions, 193–94
Massau, J., *1884*, 146
Mathematical Analysis of Logic (Boole), 284
Mathematical Collection (Pappus), 175, 383
Mathematical induction, [**83**], 79, 156, 313–16
Mathematical logic, 469, 473, 476
Mathematical Tables and Aids to Computation (ed. Archibald), 107

Mathematics
  as an art, 9–10
  as a language, 18
Mathematics of Computation, 107
Matrices, [73], 281–83
Matrix algebra, 281–83
Matrix theory, 257–59, 470
Mauchley, John W., *1907–*, 163
Maurolico, Francesco, *1494–1575*, 314
  Arithmetic, 314
  Tabula beneficia, 103
Maxima and minima, 389–90, 411–12, 425–26, 440
Maxwell, James Clerk, *1831–1879*, 3, 187
Mayan calendar, 46
Mayan numeration system, [6], 45–46, 50, 89
Mean
  arithmetic, 51
  harmonic, 51
Mean and extreme ratio, 78, 204–5
The Measurement of the Circle (Archimedes), 93, 100, 132, 150
Measure theory, 470
Mei Wen-ting, *1633–1721*, 453
Menaechmus, *350 B.C.*, 180, 198, 222
Menaechmian triads, 222
Menelaus, *c. 100*, 175, 342
  Sphaerica, 342
Menelaus' theorem, 334, 342–45, 531
Méray, H. C. R., *1835–1911*, 466
Mercator (mûr·kā′tēr; *Ger.* měr·kä′-tôr), Nicolaus, *1620–1687*, 114, 392
Méré, George Brossin, chevalier de, *1610–1685*, 158
Mersenne (měr′sěn′), Marin, *1588–1648*, 61
Mersenne numbers, [13], 61–62
Metaphysics (Aristotle), 51
Method (Archimedes), 381–82, 385
Method of exhaustion, [109], 379–80, 382, 385, 402, 403, 405, 406, 430–31
Method of false position, 126
Method of Fluxions (Newton), 229, 422
Method of indivisibles, 183
Methodus differentialis (Stirling), 428
Methodus incrementorum directa et inversa (Taylor), 427–28, 445
Miller, George Abram, *1863–1951*, 253
Miller, J. C. P., *1906–*, 107
Milne method, 115
Mils, 367–68
Minos, myth, 197

Minutes, 360, 365–66
Mirifici logarithmorum canonis constructio (Napier), 105
Mirifici logarithmorum canonis descriptio (Napier), 105, 143
Möbius (mû′bĕ·oos), August Ferdinand, *1790–1868*, 186, 187
Möbius strip, 187
Model of postulate set, 463
Models, mathematical and physical, 8–9, 11
Modern algebra, 234, 284–87
Modern Higher Algebra (Salmon), 452
Modern mathematics, 17
  characteristics of, 461–67
  curriculum revision, 474–75
Modulus, 288–90
Mohist Canon, 225, 226
Mohr, Georg, *1640–1697*, 194, 195
  Euclides Danicus, 194, 195
  Euclidis Curiosi, 195
Mohr-Mascheroni constructions, 193–94
Monge (môNzh), Gaspard, *1746–1818*, 179, 183, 259
Montmort, Pierre-Rémond de, *1678–1719*, 159
Moore, Eliakim Hastings, *1862–1932*, 260, 471
Moore, Jonas, *1617–1679*, 374
Moore, R. L., b. *1882*, 471
More, Henry, *1614–1687*, 213
  The Immortality of the Soul, 213
Moschopoulos, Manuel, *c. 1300*, 81
Moscow papyrus, [31], 127, 169–70, 237
Muir (mûr), Thomas, *1844–1934*, 366
Müller, Johann. *See* Regiomontanus
Multiple integrals, [117], 450–52
Multiplication, 93–95, 130–34
  algorithm, 134
  Babylonian, 93, 130–32
  duplation and mediation method of, 93, 130–31
  European peasant method of, 122
  Egyptian, 93, 124, 130
  gelosia method of, 94, 134, 141
  Genaille's rods, 141–42
  grating method of, 94, 134, 141
  Greek, 132
  Hindu, 133–34
  Napier's rods, [39], 111, 141–42, 161
  Russian, 93, 130–31
  scratch method of, 133–34
  tables, 93–95, 130–31

Multiplication table, principal value of, 131
Multiplicative principles, 6, 21, 39, 41, 43, 44
Multiply-perfect numbers, 61
Murray, F. J., 116
Musical scale, 51

Napier (nā'pǐ·ĕr; nȧ·pēr'), John, *1550–1617*, 15, 75, 98, 104, 105, 106, 111, 138, 141–45, 154, 247, 391
  logarithms, [**40**], 106, 142–45, 154
  Mirifici logarithmorum canonis constructio, 105
  Mirifici logarithmorum canonis descriptio, 105, 143
  Rabdologia, 98, 111, 138, 141
Napier's rods, [**39**], 111, 141–42, 161
Nasir eddin, *1201–1274*, 177, 208, 375
  Treatise on Quadrilaterals, 375
National Committee on Mathematical Requirements, 362
National Council of Teachers of Mathematics, 471
Natural and Political Observations (Graunt), 326
Negative numbers, 3, 129
Neugebauer, Otto, *1899–*, 68, 215
Neusis, 200
New Logarithms (Speidell), 154
New Mathematical Tables (Barlow), 101, 108
Newton, Isaac, *1642–1727*, [**106**], 3, 100, 111, 114, 115, 154, 182, 229, 234, 245, 247, 265, 275, 310, 319–20, 322, 331, 358, 380, 390, 391–98, 416, 421–23, 433, 442, 448, 464
  Arithmetica universalis, 319–20, 322
  De analysi, 393, 397
  De quadratura, 394–95
  fluxions and fluents, 422–23
  Method of Fluxions, 229, 422
  Principia, 396, 397, 395
Newton-Raphson method, 100
Newton's method, 265
Nicobar Islands, inhabitants of, 21
Nicomachus (nǐ·kŏm'ȧ·kŭs; nǐ-) of Gerasa, *c. 100*, 53, 54, 56, 57, 60, 94
  Introductio arithmetica, 53, 60, 94
Nicomedes (nǐk'ô·mē'dēz), *c. 240* B.C., 100, 198, 200
Nieuwentijt, Bernard, *1654–1718*, 399
Nine Chapters, 274
Nine-point circle, [**65**], 230–31

Nines, casting out, [**38**], 140
Nomography, [**41**], 115, 145–46
Noncommutative systems, 462
Non-Euclidean geometry, [**55**], 10–11, 183–85, 190, 207–10, 224, 462, 464, 470
  consistency proofs, 185
Nonpositional numeration systems, 23, 27–28, 31
Null set, 284
Number
  beliefs [**9**], 9, 29, 33, 51–52, 58, 64 (*see also* Gematria)
  concept, 11–12, 19–21, 25
  systems, 26, 29 (*see also* Numeration systems)
  theory, 32, 247
  words, 7, 20–21
Number of the beast, 34
Number representations
  Chinese rods, 128–29
  finger numbers, 88–89, 120–23
  knotted cords, 129
  Korean rods, [**32**], 128–29
Numbers
  abundant, 12, 59–61
  algebraic, [**26**], 83–84
  amicable, [**11**], 58–59
  cardinal, 21
  complex, [**76**], 3, 15, 16, 35, 36, 83–84, 244, 245–47, 256, 259, 278, 279, 290–94, 311, 358–59
  composite, [**15**], 35, 62–63, 64–66
  deficient, [**12**], 59–61
  even, 33, 52, 64
  Fermat, 64–65
  Fibonacci (sequence), [**22**], 77–79, 157, 207, 309
  figurate, [**10**], 16, 33–34, 51, 52, 53–58, 66–67
  finger, 88–89, 120–23
  hypercomplex, 257–58, 259, 279–80
  imaginary, 3, 15, 16, 35, 36, 83–84, 244, 245–47, 256, 259, 278, 279, 290–94, 311, 358–59
  irrational, [**18**], 70–72, 84, 152, 155, 239
  large, [**25**], 82–83, 89–90
  Mersenne prime, [**13**], 61–62
  multiply-perfect, 61
  negative, 3, 129
  oblong, 56
  odd, 16, 33, 52, 54, 60, 73
  pentagonal, 55–57

perfect, [**12**], 51, 53, 59–61, 61–62
prime, [**14, 15**], 34–35, 59–60, 61–62,
   62–63, 64–66
properties of, 26, 29, 32
rational, 35, 84
real, 83
relatively prime, 68, 69
square, 33–34, 52, 53–57, 66, 67
theory of, 32, 247
transcendental, [**26**], 83–84, 152, 155
transfinite, [**27**], 85–86
triangular, 34, 51, 53–57
Numeration, 7, 18–21, 25, 31
Numeration systems, 6–9, 26, 29
  additive principle in, 6, 23, 27, 37, 38,
   41, 42
  Arabic (modern), 29, 49
  Babylonian, [**1**], 4, 6, 28, 36–38, 47,
   49, 89, 130–32, 135, 360, 364–65, 370
  binary, 6–9, 89, 93, 130
  Chinese-Japanese, [**5**], 43–44, 48
  Chinese rod, 44
  ciphered, 40
  decimal (*see* Numeration systems,
   Hindu-Arabic)
  degree measure, 90
  development of, 88–91
  early, 7, 20–23
  Egyptian, [**2**], 38–40, 237
  Egyptian demotic, 39, 40
  Egyptian hieratic, 39, 40, 123, 136
  Egyptian hieroglyphic, 23, 38–40, 136
  Greek alphabetic (Ionic), 27–29, 33,
   42–43, 47, 50, 75, 90, 132–33, 262,
   298
  Greek Attic (Herodianic), 42, 82, 298
  Hebrew, 27, 75
  Hindu, 28–29, 47, 48
  Hindu-Arabic, [**7**], 6, 28, 46–49, 76–
   77, 94–95, 118, 133–34, 242, 243, 260,
   309, 365
  Mayan, [**6**], 45–56, 50, 89
  multiplicative principle of, 6, 21, 39,
   41, 43, 44
  nonpositional, 23, 27–28, 31
  positional (place-value), 4, 6, 23–25,
   28, 31, 37, 45, 47, 48, 49, 50, 90
  sexagesimal, 4, 6, 28, 36–38, 47, 49, 50,
   89, 99–100, 102, 130–32, 135, 235, 274,
   337–38, 355–56, 360, 364–65, 370, 373
  subtractive principle, 37, 41
  vigesimal, 45
Numerology. *See* Gematria

Oblong numbers, 56
Ocagne, Maurice d', *1862–1938*, 146
  Traité de nomographie, 146
Odd numbers, 16, 33, 52, 54, 60, 73
Odhner, W. T., *1878*, 112
Oldenburg, Henry, *c. 1615–1677*, 265
Omar Khayyam, *c. 1100*, 156, 177, 242,
  264, 276, 308
One-to-one correspondence, 19
On the Sphere and Cylinder (Archi-
  medes), 174
Operations, symbols for, [**37**], 139–40
Operations research, 473
Oresme (ô'râm'), Nicole, *c. 1323–1382*,
  180–81, 212, 330, 383, 384, 385, 386,
  387, 401
Ostrowski, Alexander M., *1893–*, 318,
  434
Otho, Valentin, *c. 1550–1605*, 104
Oughtred (ô'trĕd; -trĭd), William, *c.*
  1574–1660, 111, 139, 145, 153, 154,
  271, 310
  Circles of Proportion, 271
  Clavis mathematicae, 139, 271
Ozanam, Jacques, *1640–1717*, 213

Pacioli (pä·chō'lē), Luca, *c. 1445—c.*
  1514, 78, 94, 121, 133, 134, 158, 207,
  262, 291, 332
  De divina proportione, 78, 207
  Sūma, 94, 121, 133, 158, 332
Paganini, Nicolo, b. *1886,* 58
Pappus, *c. 320,* 175, 200, 203, 219, 383
  Mathematical Collection, 175, 383
Papyrus
  Moscow, [**31**], 127, 169–70, 237
  Rhind, [**30**], 93, 96, 123–27, 135–36,
   149, 169, 202, 237, 249–50, 371–72,
   377
Parabola, 276. *See also* Conic sections
  area of a segment of, 381, 403–5, 430–
   31
  etymology of, 175, 222–24
Paradox, 78, 85
Parallel axiom, 462
Parallel postulate, 10, 177, 183–84, 207–
  10, 363
  alternatives, 184
  converse of, 183–84, 208
  Playfair's, 184
Parent, Antoine, *1666–1716*, 181
Partial derivatives, [**116**], 448–50, 451
  notation for, 448–49

Pascal (pàs'kàl'), Blaise, *1623–1662*, 112, 156, 158, 159, 161, 212, 247, 264, 314, 418, 420
arithmetic triangle (*see* Pascal's triangle)
Traité du triangle arithmétique, 156
Pascal's triangle, [**46**], 156–57, 264, 314–16, 438
Peano (pä·ä'nô), Giuseppe, *1858–1932*, 468
Pell, John, *1611–1685*, 438
Pell equation, 304, 241, 438
Pellos, Francesco, *1492*, 98
Pentagon, 71
Pentagonal numbers, 55–57
Pentagram (star pentagon), 71
Percent, [**42**], 146–47
Perfect numbers, [**12**], 51, 53, 59–61, 61–62
Perigal, Henry, *1873*, 218
Peruvian quipu, [**33**], 129
Pesu calculations, 126
Peurbach (poi'ĕr·bäk), George von, *1423–1461*, 103, 371
φ (phi). *See* Golden section
π (pi), [**44**], 148–53. *See also* Circle, squaring of
approximations for, 103, 109, 110, 149–52, 169, 174, 226, 377, 403, 453–54
Archimedian approximation for, 150, 174, 403
Babylonian approximation for, 149
biblical approximation for, 149
Brouncker's expression for, 151–52, 269
Chinese value of, 109, 150
classical method of computing, 108, 150–51, 174, 226, 403
computation of, 108–10
computation of, by computer, 152–53, 174
continued fraction form of, 269–70
Egyptian approximation for, 149
Hindu approximations for, 150–51
irrationality of, 152
Lambert's expression for, 270
Leibniz' series for, 152–53, 393, 421
Ludolphian number, 109, 151
Machin's formula for, 109, 152
probability methods of computing, 153
Ptolemy's value of, 103, 109
as a ratio, 148, 150, 151, 152
symbols for, 153

transcendence of, 84, 152, 204
Viète's expression for, 151
Wallis' expression for, 152, 414
Picard (pĕ'kàr'), Émile, *1856–1941*, 442
Pike, Nicholas, *1790*, 426
Pitiscus, Bartholomäus, *1561–1613*, 104, 371
Plane Trigonometry (Todhunter), 366
Planisphaerium (Ptolemy), 340
Plato, *c. 430—c. 349* B.C., 11, 67, 71, 192, 198, 204, 378
Platonic solids. *See* Polyhedra, regular
Playfair, John, *1748–1819*
Elements of Geometry, 184
Parallel postulate, 184
Plimpton 322, 68, 215, 339
Pliny (plĭn'ĭ), *23–79*, 120
Plücker (plük'ĕr), Julius, *1801–1868*, 179, 182
Plutarch, 54
Poincaré (pwăN'kà'rä'), Henri, *1854–1912*, 2, 185, 187, 463, 468, 469
Polar coordinates, [**64**], 181, 204, 229
Polybius (pô·lĭb'ĭ·ŭs), *c. 130* B.C., 119
Polygonal numbers. *See* Numbers, figurate
Polyhedra, regular, [**60**], 206, 220–21
Polynomial equations, 83, 288–89
Poncelet (pôNs'lĕ'), Jean Victor, *1788–1867*, 179, 180, 194, 195, 230, 259
Poncelet-Steiner theorem, 195–96
circle, 196
Pons asinorum ("bridge of asses"), [**59**], 219–20
Positional numeration systems, 4, 6, 23–25, 28, 31, 37, 45, 47, 48, 49, 50, 90
Postulate of Archimedes, 174
Postulate, fifth. *See* Parallel postulate.
Pottin, Henry, *1883*, 113
Precious Mirror of the Four Elements (Chu Shih-chieh), 56
Prestet, Jean, *1648–1690*, 320
Prime numbers, [**14, 15**], 34–35, 59–60, 61–62, 62–63, 64–66. *See also* Numbers
conjectures about, 65, 66
infinitude of, [**14**], 35, 62–63, 64
large, 60, 62, 63, 66
Mersenne, 61, 62
twin, 66
unproved conjectures about, 62, 65–66
Principal value of multiplication table, 131

Principia (Newton), 395, 396, 397
Principia Mathematica (Whitehead and Russell), 15, 258, 468, 469
Principle of continuity, 407–8
Printing, invention of, 243
Probability, [**47**], 157, 158–59, 247, 470
Proclus (prō'klŭs; prŏk'lŭs), *410–485*, 66, 70, 171, 180, 184, 204, 208, 219, 362, 364
  Commentary on Euclid, Book I, 171, 178
  Eudemian Summary, 171, 172
Progressions
  arithmetic, 143–44
  geometric, 143–44
Projective geometry, 178–80, 188
Proportion, Eudoxian theory of, 71
Propositional calculus, 284–86
Ptolemy (tŏl'ĕ·mĭ), Claudius, *c. 85—c. 165*, 50, 96, 98, 175, 180, 208, 212, 242, 337, 338, 340, 342, 359, 365, 370, 374, 375
  Almagest, [**91**], 50, 90, 96, 101, 242, 337, 338, 342, 344, 359–61, 370
  Analemma, 342
  Planisphaerium, 340
  Syntaxis mathematica, 359
Ptolemy's theorem, 101–2, 360, 369–70
Pyramid, 371–72
  frustum of, 127, 169–70
  of Gizeh, 169, 206
Pythagoras (pĭ·thăg'ô·răs; pĭ-), *c. 540 B.C.,* 51, 58, 66, 119, 172, 215
  table of, 119
The Pythagorean Proposition (Loomis), 216
Pythagoreans, 220, 236, 314
  number beliefs of, [**9**], 9, 29, 3, 51–52, 58, 64, 74–75
  school of (brotherhood), 31, 32, 33, 70, 172, 206
Pythagorean school. *See* Pythagoreans, school of (brotherhood)
Pythagorean theorem, [**58**], 66, 70, 100, 169–70, 215–18, 225, 300, 338, 377, 438
Pythagorean triples, [**16**], 66–68, 79, 215, 339

Quadratic equations, 29, 145, 222, 241, 246, 261–63, 270, 291–92, 298, 301–3, 306–8, 317, 323–24
Quadratrix, 150, 193, 201, 202–3

Quadrature of the circle. *See* Circle, squaring of
Quadrilateral, complete, 343–45, 350–53
Quadrivium, 51
Quantum mechanics, 259
Quantum theory, 470
Quarter squares, 105
Quartic equations, 236, 245, 248, 253, 278, 310, 317, 318
Quaternions, [**77**], 256–59, 280, 282, 295–97
Quintic equations, 254–55, 278–79, 311
Quipu, Peruvian, [**33**], 129

Rabdologia (Napier), 98, 111, 138, 141
Radians, 366–67
Radical symbol, [**43**], 147–48
Rahn, Johann Heinrich, *1622–1676,* 65, 140
Ramanujan (rä·mä'noo·jŭn), Srinivasa (shrē'nĭ·vä'sa; srē'-), *1887–1920,* 14
Raphson, Joseph, *1648–1715,* 100
Ratio
  of circle areas, 150, 380
  mean and extreme, 204–5, 78
Rational numbers, 35, 84
Recorde (rĕk'ôrd), Robert, *c. 1510–1558,* 119, 140, 244, 263, 332
  The Ground of Artes, 119, 332
  The Whetstone of Witte, 244
Real numbers, 83
Real number system, topological structure of, 465–66
  topological structure of, 465–66
Rechentafeln (Crelle), 95
Rechnung auff der Linien vnd Federn (Riese), 119
Regiomontanus [Johann Müller], *1436–1476,* 103, 356, 371, 375, 387
Regular polyhedra, [**60**], 172, 206, 220–21
Regular sexagesimal numbers, 131
Reinhold, Erasmus, *1553*
  Canon foecundus, 103
Reitweisner, George W., 1919–, 110
Relatively prime numbers, 68, 69
Relativity, 280
  theory of, 470
Revelation, Book of, 34, 75
Reye, Karl Theodor, *1837–1919,* 179
Reyher, Samuel, *1635–1714,* 364
Reymond, Paul Du Bois-. *See* Du Bois-Reymond, Paul

Rheticus (rā'tĕ·koos) [Georg Joachim], *1514–1576,* 104, 356, 371, 373

Rhind, A. Henry, *1833–1863,* 123

The Rhind Mathematical Papyrus (Chace), 124

Rhind papyrus, [**30**], 93, 96, 123–27, 135–36, 149, 169, 202, 237, 249–50, 371–72, 377

Riemann (rē'män), G. F. Bernhard, *1826–1866,* 182, 183, 185, 186, 187, 188, 209, 259, 465, 469, 470

Ricci, M. M. G., *1853–1925,* 226

Ricci, Matteo, *1552–1610,* 280

Riemannian geometries, 183

Riemann sums, 431

Riemann surface, 186–87

Riese (rē'zĕ), Adam, *1492–1559,* 97, 119, 137

Rechnung auff der Linien vnd Federn, 119

Right angle, [**93**], 363–64

Rigor, 467–68

Robertson, John, *1775,* 111

Roberval (rô'bĕr'väl'), Gilles Persone de, *1602–1675,* 227, 387

Rolle, Michel, *1652–1719,* 399

Roman numeration system, [**3**], 23, 40–41, 90, 243

Romanus, Adrianus, *1561–1615,* 329

Rosenhead, Louis, *1906–,* 107

Rosenthal, Arthur, *1887–,* 410–11

Roth, Peter, *c. 1580—c. 1617,* 317

Rudolff, Christoff, *c. 1500—c. 1545,* 138, 148, 244, 263

Die Coss, 148, 244

Ruffini, Paolo, *1765–1822,* 253, 256, 275, 278

Rule of false position, [**90**], 237, 308, 332

Rule of signs, Descartes's, [**85**], 310, 317, 318–20

Runge-Kutta method, 115

Russell, Bertrand, *1872–,* 15, 85–86, 258, 466, 467, 468, 469

Principia Mathematica, 15, 258, 468, 469

Rusty compass, 194–96

Rybnikov, K. A., 5

Saccheri, Girolamo, *1667–1733,* 10, 184, 208, 364, 462

Euclides ab omni naevo vindicatus, 462

Saccheri quadrilateral, 208–9

Sacrobosco (săk'rô·bŏs'kō) [John of Halifax, or of Holywood], *c. 1200 —c. 1256,* 48

Algorismus vulgaris, 48

Salmon, George, *1819–1904,* 452

Modern Higher Algebra, 452

The Sand-Reckoner (Archimedes), 89–90

S.a.s., 217

Scheutz, Edward, son of George, 162

Scheutz, George, *1785–1873,* 162

Schlafli, Ludwig, *1814–1895,* 221

Schneider, Theodor, *1911–,* 84

Schoenflies, Arthur, *1853–1928,* 469

Schotten, H., *1893,* 362

Schumaker, John A., 5

Secant (function), 354

Seconds, 360, 365–66

Segner, Johann Andreas, *1704–1777,* 320

Seqt, 371–72

Series, 104, 266, 270, 358, 421, 428. *See also* Infinite series

Serret, Joseph Alfred, *1819–1885,* 256

Servois, François Joseph, *1767–1847,* 257

Set concept, 21

Sets, 284–86

Set theory, 188, 402, 463–66, 474

contradictions in, 467

Severi, Francesco, b. *1879,* 196

Severi's small arc theorem, 196

Sexagesimal numeration system, 4, 6, 28, 36–38, 47, 49, 50, 89, 99–100, 102, 130–32, 135, 235, 337–38, 355–56, 360, 364–65, 370, 373

fractions in, 131–32, 135

modern notation in, 38

origin of, 37, 365

regular numbers in, 131–32

separatrix (sexagesimal point) in, 6, 37, 38, 131–32, 135

Shakespeare, William, *1564–1616,* 119

Shanks, William, *1812–1882,* 109, 152

Side and diagonal numbers, 71

Simplicius (sĭm·plĭsh'ĭ·ŭs), *c. 520,* 212

Simson, Robert, *1687–1768,* 78, 178

Sine, [**95**], 345–50, 351, 358–59, 368–71

etymology of, 12, 350, 370

law of, 352, 375

Slaught, H. E., 471

Slide rule, [**69**], 111–13, 271–73, 310

Sluse, René de, *1622–1685,* 391

Smith, David Eugene, *1860–1944,* 106

Social implications, computing machines, 115–17

Soroban. *See* Abacus
Speidell, John, *1622*
    New Logarithms, 154
Speiser, Andreas, b. *1885,* 251
Sphaerica (Menelaus), 342
Spiral of Archimedes, 202–4
Spirals, 181
Square numbers, 33–34, 52, 53–57, 66, 67
Square root, 4, 35, 71, 76, 97, 102, 133,
    137, 239, 264–65, 268–70, 291–92, 303,
    307, 327–28, 377
    computation of, 98–101
    symbol for, 147–48
    tables of, 97
Squaring the circle. *See* Circle, squaring
    of
Star pentagram, 206
Staudt, Karl Christian von, *1798–1867,*
    179
Steiner (shtī′nēr), Jakob, *1796–1863,* 179,
    194, 195, 196
Stereographic projection, 339–42
Stereometria doliorum vinorum (Kep-
    ler), 424
Stereometrica (Heron of Alexandria),
    291
Stevin (stĕ·vīn′), Simon, *1548–1620,*
    [**99**], 6, 49, 69, 97, 137, 138, 263,
    330, 356, 405–7
    anticipation of definite integral, 406
    La Disme, 97, 138
Stieltjes, Thomas Joannes, *1856–1894,*
    270
Stifel (shtē′fĕl), Michael, *c. 1487–1567,*
    75, 140, 148, 212, 263
    Arithmetica integra, 140
Stirling, James, *1692–1770,* 428
    Methodus differentialis, 428
Stirling's formula, 428
Straightedge, marked, 200. *See also*
    Compass
Suan phan. See Abacus
Subtractive principle, 37, 41
Sullivan, J. W. N., 9
Sulvasutras, 68, 216
Sūma (Pacioli), 94, 121, 133, 158, 332
Sumerian. *See* Babylonian
Summability methods, 434
Sundials, 470. *See also* Gnomon
Surya Siddhanta, 301, 346, 349
Switching circuits, 287
Sylvester II. *See* Gerbert [Pope Syl-
    vester II]

Sylvester, James Joseph, *1814–1897,* 259,
    283, 311, 324, 451, 452
Symbol
    for addition, 139
    for angle, 362
    for "approaches a limit," 234
    for base of natural logarithms (e),
    155
    for common fractions, 136
    for congruence (number theory), 252
    for decimal fractions, 138
    for difference, 234, 271, 332
    for division, 139–40
    for equals, 244
    for exponents, 310
    for "greater than," 310
    for infinity, 414
    for integral, 396, 419, 423, 458
    for "is approximately equal to," 234
    for "less than," 310
    for multiplication, 139, 271, 310
    for partial derivatives, 448–49
    for percent, 147
    for plus, 332
    for proportion, 271, 310
    for radical, [**43**], 147–48, 244
    for right angle, 364
    for subtraction, 139
Symmetric functions, [**86**], 310, 321–22
Syntaxis mathematica (Ptolemy), 359
Systems of numeration. *See* Numera-
    tion systems

Tabit ibn-Qorra. *See* ibn-Qorra, Tabit
Table of Pythagoras, 119
Tables, 161
    astronomical, 47, 50
    of chords, 12, 98, 101, 337–38, 359–60,
    369–70
    of cube roots, 101
    of cubes, 276
    factor, 65
    of half-chords, 12, 103, 370
    of logarithms, 98, 105–6, 143–45, 154,
    248
    makers of, 108
    number of mathematical, 107
    mortality, 326
    multiplication, 93–95, 130–31
    of Pythagoras, 119
    of quarter squares, 105
    recent mathematical, 107
    of reciprocals, 93, 95, 96, 131–32

shadow, 335, 353, 373
of square roots, 97
trigonometric, 97, 101–4, 346, 353–56,
359–61, 367, 369–71, 373
for "2 over *n*," 95, 124–25
Tables of Logarithms (Gardiner), 145
Tabula beneficia (Maurolico), 103
Tait, Peter Guthrie, *1831–1901,* 258,
289
Treatise on Natural Philosophy, 366
Takebe Hikojiro Kenko, *c. 1722,* 453–
54
Tangent, [96], 335, 352, 371–74
etymology of, 373
Tangents to a curve, 390–91, 416–17
Tannery, Paul, *1843–1904,* 133
Tartaglia (tär·tä'lyä), Niccolo, *c. 1500–
1557,* 156, 157, 195, 277, 309, 326
General trattato, 326
τ (tau). *See* Golden section
Tautochrone, 228
Taylor, Brook, *1685–1731,* 114, 115, 427–
28, 429, 443
*Methodus incrementorum directa et
inversa,* 427–28, 445
Taylor series, [114], 443–46
Tchebycheff, P. L., *1821–1894,* 63
Tensor calculus, 183, 280
Thales (thā'lēz) of Miletus, *c. 600* B.C.,
171, 215, 219
Theaetetus, *c. 380* B.C., 221
Theodorus of Cyrene, *c. 390* B.C., 71
Theon of Alexandria, *c. 390,* 99, 365,
369
Theon of Smyrna, *c. 125,* 99, 101
Theorem, fundamental, of arithmetic,
35
Théorie des fonctions analytiques
(Lagrange), 400
Théorie des Nombres (Legendre), 257
Theory of ideals, 80
Theory of numbers. *See* Numbers,
theory of
Theory of perspective, 179
Theory of proportion, 71, 172
Theory of relativity, 470
Theudius, *4th cent.* B.C., 173
Thomas, Charles X., *c. 1820,* 112
Thomson, James T., 366
Thomson, William, *1824–1907,* 114
Treatise on Natural Philosophy, 366
Todhunter, Isaac, *1820–1884*
Plane Trigonometry, 366
Topology. 185–97, 188, 469, 472

Torricelli (tŏr'rê·chĕl'lê), Évangelista,
1608–1647, 227, 387, 390
Tractatus de sectionibus conicis (Wallis), 413
Traité de nomographie (Ocagne, d'),
146
Traité du triangle arithmétique (Pascal), 156
Transcendental functions, 392–93, 398
Transcendental numbers, [26], 83–84,
152, 155
Transfinite numbers, [27], 85–86
Transformations, 283
Treatise on Fluxions (Maclaurin), 445–
46
Treatise on Natural Philosophy
(Thomson and Tait), 366
Treatise on Quadrilaterals (Nasir
eddin), 375
Treviso arithmetic, 94, 97. 134
Triangle
arithmetic (*see* Pascal's triangle)
spherical, 350–53, 357, 375
Triangular numbers, 34, 51, 53–57
Trigonometry, 175, 177, 391. *See also*
Tables, trigonometric
analytic, 334, 358–59
Arabic, 334, 350–53, 370, 373
Babylonian, 336–39
of chords, 96, 335–39, 343–45,
definition in complex domain, 359
Egyptian, 336, 371–73
functions, 12–13, 169, 177, 355 (*see
also specific functions*)
functions as ratios in right triangles,
371, 373
Greek, 334, 359–61, 369–70
Hindu, 334, 336, 346–50, 370
identities, [97], 294, 355, 357, 359–61,
374–75
separated from astronomy, 333, 375
spherical, 177, 340, 342–45, 375
Le triparty en la science des nombres
(Chuquet), 329
Trisection of angle. *See* Angle, trisection of
Trisection problem. *See* Angle, trisection of
Truth tables, 284
Tsu Ch'ung-chih, *430–501,* 150
Turing, Alan Mathison, *1912–1954,* 164

Union of sets, 285–87

Use of the history of mathematics, 2–6, 16–17

Uttara Khandakhadyaka (Brahmagupta), 349

Vallée-Poussin, C. J. de La, *1866-1962*, 65

van Ceulen, Ludolph. *See* Ceulen, Ludolph van

Vander Hoecke, *1514*, 262

Van der Waerden, B. L., 242

Vandermonde (vän'dĕr'môNd'), Alexandre Théophile or Alexis, *1735-1796*, 282

Varahamihira, *505*, 374

Varignon, Pierre, *1654-1722*, 439

Veblen (vĕb'lĕn), Oswald, *1880-1960*, 187, 471

Vectors, [72], 256, 258, 279–80, 295–97

Venn, John, *1834-1923*, 286

Venn diagrams, 286

Verging, 200

Versed sine, 339, 348, 357, 370

Vieta, Francis. *See* Viète, François

Viète (vyĕt), François, or Francis Vieta, *1540-1603*, 101, 104, 138, 151, 198, 234, 243, 245, 246, 263, 275, 310, 317, 321, 330, 356, 358, 375, 387, 389, 411

De aequationum recognitione et emendatione, 245

Logistica speciosa, 245

Vigesimal numeration system, 45

Vinci (vēn'chê), Leonardo da, *1452-1519*, 78, 195, 207

Vlacq (vläk), Adriaen or Adrian, *c. 1600-1667*, 106

Volume of sphere, 381, 403, 408

Volume of square pyramid, 377

Volumes of rotation, 408

Volumes of solids, [107], 424–27

von Hohenberg, Herwart. *See* Hohenberg, Herwart von.

von Neumann, John, *1903-1957*, 86, 164, 460, 471, 473

Vorstudien zur Topologie (Listing), 186

Wallis, John, [103], 114, 145, 152, 153, 208, 213, 244, 263, 269, 292, 331, 387, 388, 392, 413–15, 447

Algebra, 213, 292

Arithmetica infinitorum, 387–88, 414

Tractatus de sectionibus conicis, 413

Wantzel, Pierre, *1814-1848*, 201

Weber (vä'bĕr), Heinrich, *1842-1913*, 260

Wefa, Abu'l. *See* Abu'l Wefa

Weierstrass (vī'ĕr·shträs'), Karl, *1815-1897*, 401, 465, 466, 468

Wessel, Caspar, *1745-1818*, 15, 246, 291, 292, 293

Weyl (vīl), Hermann, *1885-1955*, 464–65, 468, 471

The Whetstone of Witte (Recorde), 244

Whitehead, Alfred North, *1861-1947*, 13, 15, 258, 468, 470

Principia Mathematica, 15, 258, 468, 469

"Whys" in teaching mathematics

chronological, 2

logical, 2–4

pedagogical, 2, 4–5

Widmann (vĭt'män), Johann, *c. 1498*, 139, 140, 332

Wiener, Norbert, *1894-1964*

Cybernetics, 116

Wilkes, M. V., *1946*, 164

Wilson, Edwin Bidwell, *1879-1964*, 280

Witch of Agnesi, [56], 210–11

Wolfskehl, Paul, *c. 1908*, 80

Wren, Christopher, *1632-1723*, 227

Wronski (vrôn'y'·skê), Josef Maria, *1778-1853*, 282

Yang Hui, *c. 1261*, 438

Yu (yü) the Great, *c. 2200* B.C., 81

Zeno (zē'nō) of Elea, *c. 450* B.C., 85, 378

paradoxes of, 378–79, 384, 432, 467

Zermelo, Ernst, *1871-1953*, 86

Zero

as cardinal number, 25, 27, 30

concept of, 25, 27, 47, 48, 91,

name, 50

operations on, 30, 435, 437, 455–57

origin, [8], 49–50

as placeholder, 25, 27, 28–29, 30

symbol, 25, 28–29, 37, 38, 44, 45, 47, 48, 49–50, 128, 129